Organic Synthesis:
The Disconnection Approach

Organic Synthesis:
The Disconnection Approach

Stuart Warren
Department of Chemistry and Churchill College,
Cambridge University

1807 1982

175 YEARS OF PUBLISHING

JOHN WILEY & SONS
Chichester · New York · Brisbane · Toronto · Singapore

Library of Congress Cataloging in Publication Data:

Warren, Stuart G.
  Organic synthesis, the disconnection approach.

  Includes index.
  1. Chemistry, Organic — Synthesis.   I. Title
OD262.W284   547'.2   81-19694

ISBN 0 471 10160 5 (cloth)
ISBN 0 471 10161 3 (paper)

British Library Cataloguing in Publication Data

Warren, Stuart
  Organic synthesis: the disconnection
  approach.
  1. Chemistry, Organic — Synthesis
  I. Title
  547'.2'028   QD262

  ISBN 0 471 10160 5 (cloth)
  ISBN 0 471 10161 3 (paper)

Phototypeset by Dobbie Typesetting Service, Plymouth, Devon, and printed by Pitman Press Ltd., Bath, Avon

Preface

This book was written to support a lecture course on organic synthesis devised by Jim Staunton and myself. The course, my programmed book,[1] and now this more comprehensive text are all meant to help students design organic syntheses for themselves.

I thank all those who have helped to turn the original ideas into this book — my colleagues, especially Jim Staunton and Ted McDonald, our second year students from 1976 to 1981, the members of my research group, and most particularly Denis Marrian who read and corrected the many versions of the MS, made many creative suggestions, checked all the references, and generally kept the project moving with his enthusiasm.

Cambridge 1981   *Stuart Warren*

Contents

viii

Summaries of Approach to Synthesis Design

List of Tables

Introduction

Chemists synthesise compounds in just about every organic chemistry laboratory in the world. Industrial chemists synthesise pharmaceuticals, polymers (plastics), pesticides, dyestuffs, food colourings and flavourings, perfumes, detergents, and disinfectants. Research chemists synthesise natural products whose structure is uncertain, compounds for mechanistic investigations, possible intermediates in chemical and biological processes, thousands of potential drugs for every one which is used in medical practice, and even compounds which might themselves be useful for organic syntheses. Before and during these syntheses groups of chemists sitting round blackboards or piles of paper plan the work they are about to undertake. Possible routes are drawn out, criticised, modified, or abandoned until a decision is reached. The plan is tried, modified again when the behaviour of the compounds in the flask turns out to be different from what was expected, until finally success is achieved.

The aim of this book is to show you how this planning is done: to help you learn the *disconnection* or *synthon* approach to organic synthesis. This approach is analytical: we start with the molecule we want to make (the *target molecule*) and break it down by a series of disconnections into possible starting materials. In the last chapter of the book we shall discuss the synthesis of the natural product α-sinensal (1) and we shall devise a route using five different, easily available starting materials (2–7).

No-one could look at the structure of α-sinensal and immediately write down the five starting materials. We arrive at these only after a prolonged analysis with many disconnections. This book shows you the systematic approach to such analyses, starting with simple molecules and progressing to molecules like α-sinensal.

Chapters of instruction in types of disconnections alternate with *strategy* chapters which aim to put the instruction in a broader context. At four points, general strategy chapters (12, 28, 37, 40) of exceptional importance are inserted into this scheme. If you are a student, you will probably need to read all the chapters, though you may find much familiar chemistry at first. If you are a practising organic chemist, you will find the early instructional chapters elementary but, I hope, worthwhile in their relationship to the early strategy chapters.

I have assumed a basic knowledge of organic chemistry up to about first year degree level as this is not a general textbook of organic chemistry. If you cannot understand a particular reaction, a general text should help you. I have tried to give just enough explanation of the mechanism of a reaction for you to be able to use it in syntheses.

Accompanying the main text is a workbook which gives further worked examples for each chapter, problems, and solutions. Designing organic syntheses is a skill you can learn only with instruction and *practice*. It is essential that you try problems from the workbook as you go along so that you can discover whether you understand each chapter. My programmed book[1] may help you with the core of the work: the examples in it are mostly different from those in this book.

The first chapter sets the scene by looking at some completed syntheses. In Chapter 2 the serious instruction begins.

CHAPTER 1

The Disconnection Approach

This book is to help you design your own syntheses rather than tell you about those devised by others. It still contains many examples of other people's work since learning by example is as important here as elsewhere. This chapter sets the scene for what follows so that the details of the syntheses need not concern you as much as the general approach.

The ketone (1) is an important industrial compound made by the ton from cheap starting materials[2] and used to make vitamin A and some flavouring and perfumery compounds.

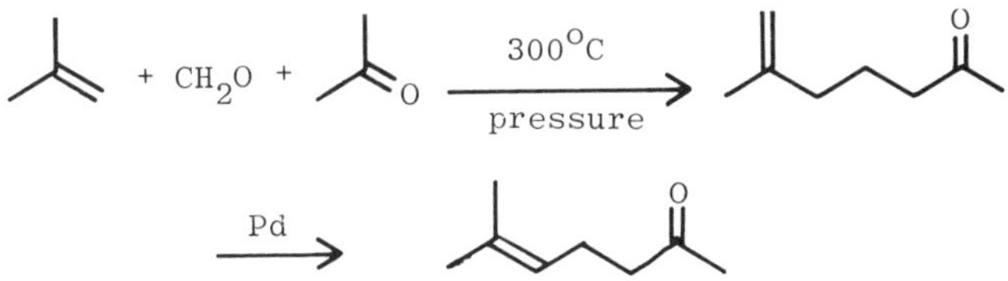

High pressure and temperatures are inconvenient in the laboratory where a simpler, though longer, route[3] uses (2) as an intermediate. This is still quite short, uses cheap starting materials, and gives high yields in each step.

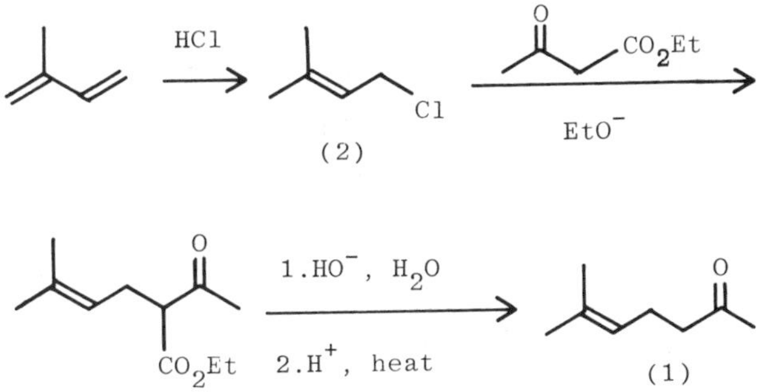

How did the workers choose these routes? The approaches to this simple molecule (1) containing only eight carbon atoms probably owed more to a comprehensive knowledge of reliable chemical reactions and reaction

1

mechanisms than to any step by step analysis. Even with the analytical approach these are still of vital importance as synthesis is largely about applying known reactions to unknown molecules.

The synthesis of the next target molecule (3) could hardly be devised in a similar way. Its greater complexity demands a more sophisticated approach.

(3)

Multistriatin (3) is one of the pheromones of the elm bark beetle, a volatile compound released by a virgin female beetle when she has found a good source of food—an elm tree. Male beetles, which carry the fungus causing Dutch elm disease, are attracted by the pheromone, the tree becomes infected and soon dies.

Multistriatin could be used to trap the beetles and so prevent the spread of the disease but there is no prospect of isolating useful amounts from the beetles. It must be synthesised. In analysing the problem we notice that C-6 has two single bonds to oxygen atoms. We therefore *recognise* an acetal *functional group*. Acetals (4) can be made by a reliable reaction from carbonyl compounds and alcohols.

(4)

Working backwards, we *disconnect* the acetal, using ⇒ to indicate the reverse of a synthetic step, and discover (5) as the intermediate from which the required acetal (3) could be made.

(3)   (5)

To make (5) we shall doubtless join two simpler fragments together by forming a C–C single bond. But which one? Bond C4–C5 is a good choice because it joins a symmetrical ketone (6) to the rest of the molecule. We can therefore disconnect this bond (5a), writing ~ across the bond and using our symbol ⇒. Before writing the fragments, we consider the synthetic step corresponding to this disconnection. The ketone group in (6) could stabilise an anion, so (7) should be a cation for an ionic reaction to take place.

(5a) (6) (7)

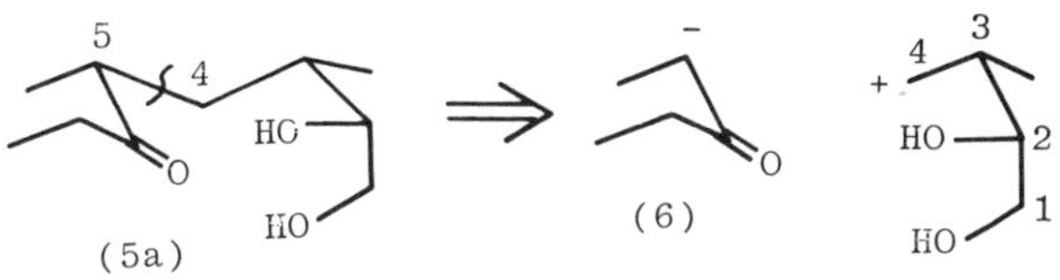

Anion **(6)** can be made from ketone **(8)** with base, but there is no simple way to make a cation at C4 of **(7)**. The solution is to attach a good leaving group to C4 giving **(9)** (X = Br, etc.) as the complete fragment.

(8) (6)

(7) =

(9) (10)

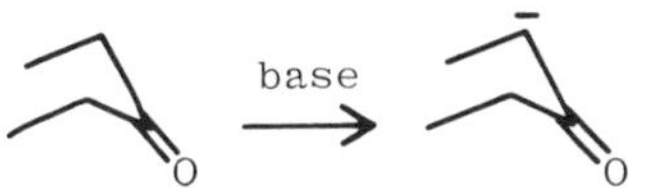

The ketone **(8)** is available, but **(9)** must be made. Once again we must recognise that **(9)** contains the 1,2-diol *functional group*, made by the hydroxylation of an alkene **(10)**, a known and reliable reaction.

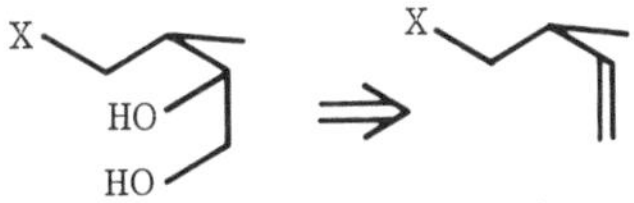

One group of workers[4] planning this synthesis decided to use the alcohol **(10)** (X = OH) as it had already been made from the acid **(11)** and to use tosylate (= toluene-*p*-sulphonate) as a leaving group. This synthesis can now be written in a forward direction. In carrying out the synthesis, they hydroxylated **(12)** with a per-acid and found that the epoxide **(13)** gave multistriatin directly on treatment with a Lewis acid.

Synthesis

(11) (10, X=OH) (10, X=OTs) (12) (13)

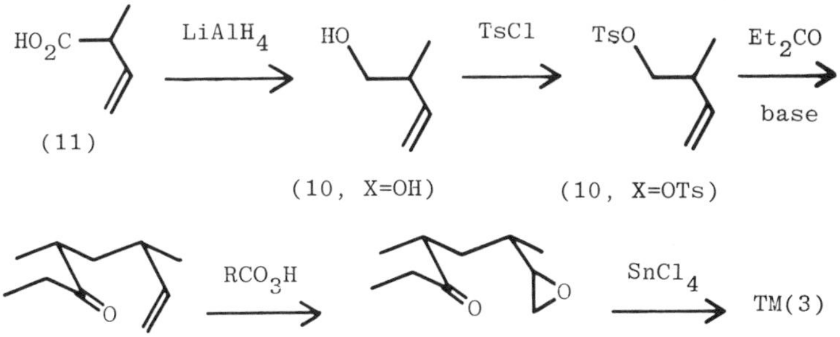

Routine for Designing a Synthesis

1. *Analysis*
    (a) recognise the functional groups in the target molecule.
    (b) disconnect by methods corresponding to known and reliable reactions.
    (c) repeat as necessary to reach available starting materials.
2. *Synthesis*
    (a) write out the plan according to the analysis, adding reagents and conditions.
    (b) modify the plan according to unexpected failures or successes in the laboratory.

We shall be using this routine throughout the book.

(14)

The synthesis of multistriatin just described has one great fault: no attempt was made to control the stereochemistry at the four chiral centres (• in **14**) and a mixture of stereoisomers was the result. Only the natural isomer **(14)** attracts the beetle and a stereoselective synthesis of multistriatin has now been devised (see Chapter 12). We must therefore add stereochemistry to the list of essential background knowledge an organic chemist must have to design syntheses effectively. The list is now:

1. an understanding of reaction mechanisms.
2. a working knowledge of reliable reactions.
3. an appreciation that some compounds are readily available.
4. an understanding of stereochemistry.

This book will show you how to apply this background knowledge to organic syntheses using the basic scheme set out above. Don't be concerned if you feel that your background knowledge is weak. In each chapter all four aspects (1–4 above) will be discussed, if appropriate, and your background knowledge should be progressively strengthened.

The elm bark beetle releases three compounds in its pheromone mixture: multistriatin **(14)**, the alcohol **(15)**, and α-cubebene **(16)**. At first we shall be looking at simple molecules such as **(15)**. We shall progress to natural multistriatin, and finally, by the end of the book, to molecules as complex as α-cubebene.

OH

(15)

H

(16)

The compounds we have met in this chapter, the ketone **(1)** and multi-striatin **(3)**, have been made many times by different methods. Synthesis is a creative science and there is no 'right' or 'best' synthesis for any molecule. I shall usually give one synthesis only for each target molecule in the book: you may be able to devise shorter, more stereochemically controlled, higher yielding, more versatile—in short better—syntheses than those already published. If so, you are using the book to advantage.

CHAPTER 2

Basic Principles: Synthesis of Aromatic Compounds

We start with aromatic compounds because the bond to be disconnected is almost always the bond joining the aromatic ring to the rest of the molecule: all we have to decide is when to make the disconnection and exactly which starting materials to use. We shall use the technical terms disconnection, functional group interconversion (FGI), and synthon in this chapter.

Disconnection and FGI

Disconnections are the reverse of synthetic steps or *reactions* and we disconnect only when we have a reliable reaction in mind. In designing a synthesis of the local anaesthetic benzocaine (1) we know that esters are made from alcohols and acids so we can write a C–O disconnection. Usually, disconnections will be labelled to show the reason for making them.

Benzocaine: *Analysis 1*

We should now like to disconnect either CO_2H or NH_2 from the aromatic ring but we know of no good reactions corresponding to these disconnections. We must therefore first do *functional group interconversion* (FGI) to change these functional groups into others which can be disconnected. Aromatic acids can be made by the oxidation of methyl groups and amino groups by the reduction of nitro groups. We can write these as follows.

Analysis 2

Now, disconnection of the nitro group is rational because we know that nitration of toluene occurs easily, and toluene is available.

6

Analysis 3

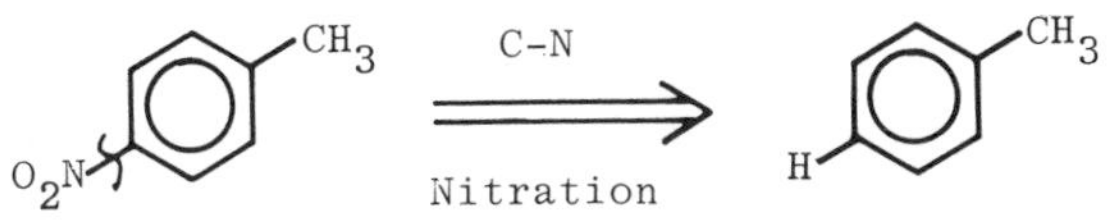

This completes the analysis and we should now write out the synthesis with suggested reagents. You should not expect to predict exact reagents and conditions and indeed no sensible organic chemist would without a thorough literature search. It is sufficient to be aware of the type of reagent needed and I shall give actual reagents and conditions to help broaden that awareness, emphasising any essential conditions.

Synthesis[5]

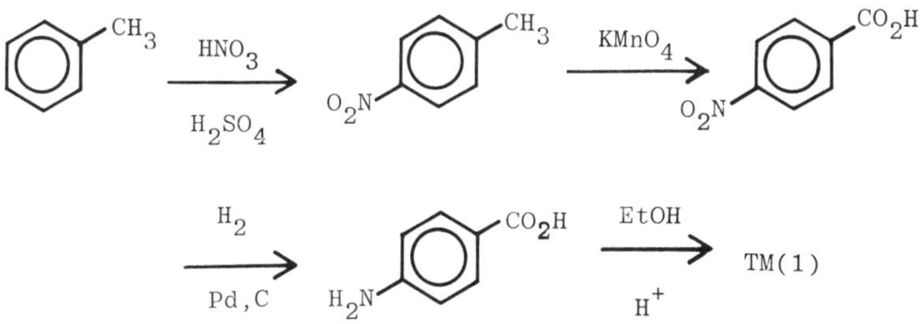

It might be possible to carry out these steps in a different order (e.g. reverse the order of the last two); decisions of this sort form part of *strategy* and are discussed in Chapter 3.

Synthons

Another useful aromatic disconnection corresponds to the Friedel–Crafts reaction which would be used in the synthesis of the hawthorn blossom perfume compound (2). The synthesis is one step from an available ether.

Analysis

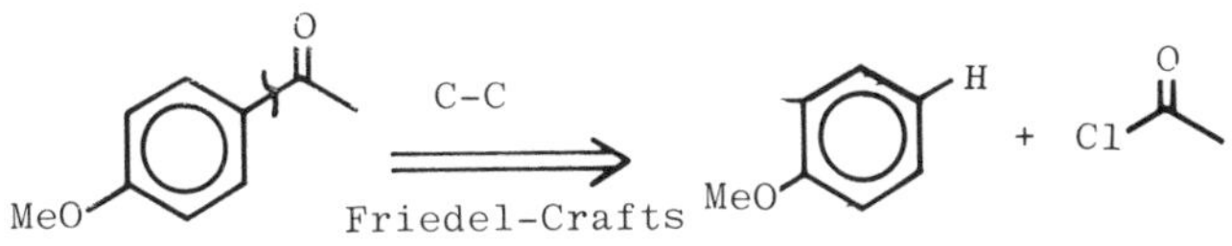

Synthesis[6]

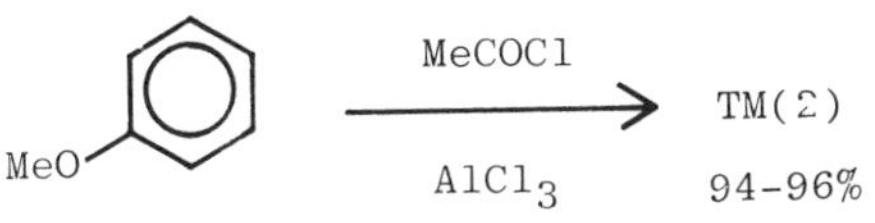

8

In both this reaction, and in the nitration we used to make benzocaine, the reagent which carries out the attack on the benzene ring is a cation, $MeCO^+$ for the Friedel–Crafts, NO_2^+ for the nitration. When we disconnect a bond to an aromatic ring we normally expect this type of reaction and so we can choose not only which bond to break but which way, electronically, to break it. Here we write (a) and not (b) because the aromatic ring behaves as the nucleophile and the acid chloride as the electrophile.

These fragments (3) and (4) are *synthons*—that is idealised fragments which may or may not be involved in the reaction but which help us to work out which reagents to use. Here, as it happens, (4), but not (3), is an intermediate in the synthesis. When the analysis is complete, the *synthons* must be replaced by *reagents* for practical use. For an anionic synthon, the reagent is often the corresponding hydrocarbon: for a cationic synthon the reagent is often the corresponding halide.

Friedel–Crafts alkylation is also a useful reaction, particularly with tertiary halides, so that the first disconnection on 'BHT' (5) (butylated hydroxytoluene —an antioxidant used in foods) can be of the tertiary butyl groups.

BHT: *Analysis*

As reagents for the *t*-butyl cation (6) we can use either *t*-BuCl and AlCl₃, or the readily available alkene (7) and protic acid.

Synthesis[7]

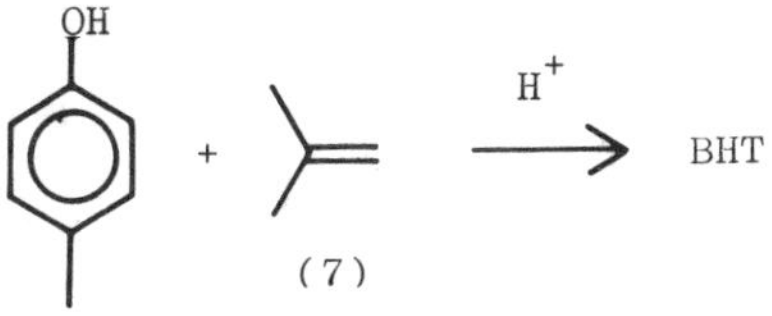

Polyalkylation, an advantage here, can be a nuisance with Friedel–Crafts alkylations as can the rearrangement of primary alkyl halides. Thus, the alkyl halide (8) gives a mixture of (9) and (10) with benzene: and if we want to make compound (11) we must use the Friedel–Crafts acylation, which suffers from neither of these disadvantages, and then reduce the carbonyl group[8] (see Chapter 24).

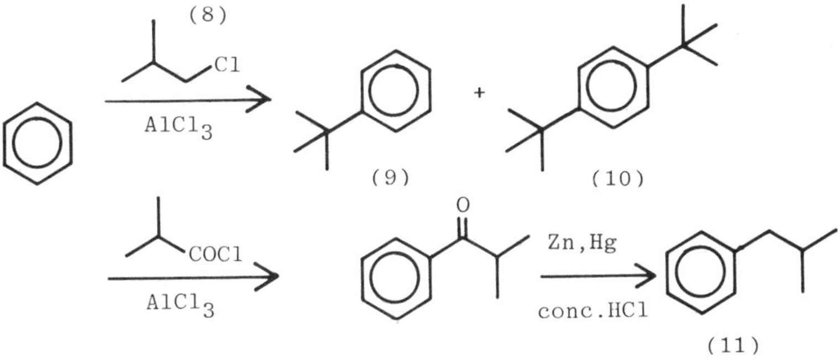

If we wish to add just one carbon atom, as in the synthesis of aromatic aldehydes, we cannot use HCOCl since it does not exist. One of the most reliable methods is chloromethylation[9] with CH₂O and HCl giving a CH₂Cl group which can easily be oxidised to CHO (FGI). The important perfumery compound piperonal (12) can be made this way. Other methods of adding one carbon atom with a functional group are given in Table 2.1.

Piperonal: *Analysis*

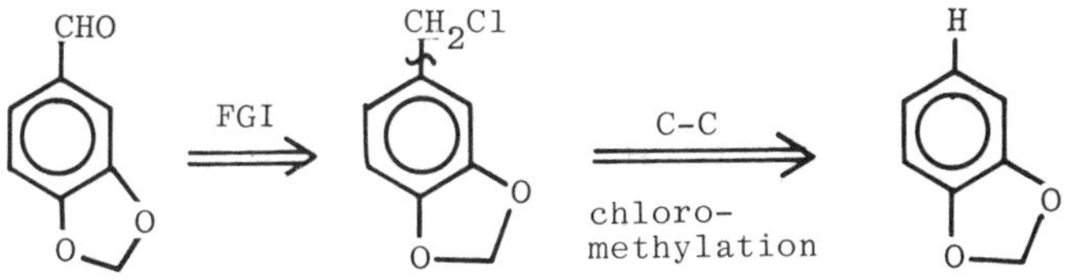

Table 2.1 One-carbon electrophiles[a] for aromatic synthesis

X	Reagent	Reaction
CH₂Cl	CH₂O + HCl + ZnCl₂	Chloromethylation
CHO	CHCl₃ + HO⁻	Reimer–Tiemann[b]
	Me₂N=CH-OPOCl₂ (Me₂NCHO + POCl₃) CO + HCl + AlCl₃ Zn(CN)₂ + HCl	Vilsmeier–Haack Formylation

[a]See also Grignard reagents in Chapter 10.
[b]Only on phenol (R = OH): the *ortho* product is favoured.

Synthesis[10]

When heteroatoms are required, nitration gives the NO_2 group and halogenation puts in Cl or Br directly (OR and I are generally added by nucleophilic substitution, see page 12). Table 2.2 gives reliable reagents for these and some other synthons for aromatic synthesis.

Table 2.2 Reagents for aromatic electrophilic substitution

Synthon	Reagent	Reaction
R⁺	RBr + AlCl₃ ROH + H⁺ Alkene + H⁺	Friedel–Crafts[11] alkylation
RCO⁺	RCOCl + AlCl₃	Friedel–Crafts[12] acylation
NO₂⁺	HNO₃ + H₂SO₄	Nitration
Cl⁺	Cl₂ + FeCl₃	Chlorination
Br⁺	Br₂ + Fe	Bromination
⁺SO₂OH	H₂SO₄	Sulphonation
⁺SO₂Cl	ClSO₂OH	Chlorosulphonation
ArN₂⁺	ArN₂⁺	Diazocoupling

Other aromatic side chains are best added by FGI on these products. Table 2.3 gives some examples.

Table 2.3 Aromatic side chains by functional group interconversion

Y	X	Reagent
Reduction		
$-NO_2$	$-NH_2$	H_2, Pd, C
		Sn, conc. HCl
$-COR$	$-CH(OH)R$	$NaBH_4$
$-COR$	$-CH_2R$	e.g. Zn/Hg, conc. HCl
		see Table 24.1
Oxidation		
$-CH_2Cl$	$-CHO$	hexamine
$-CH_2R$	$-CO_2H$	$KMnO_4$
$-CH_3$		
$-COR$	$-OCOR$	$R'CO_3H$
Substitution		
$-CH_3$	$-CCl_3$	Cl_2, PCl_5 [13]
$-CCl_3$	$-CF_3$	SbF_5 [13]
$-CN$	$-CO_2H$	HO^-, H_2O

Nucleophilic Aromatic Substitution

So far we have studied the addition of cationic synthons to the aromatic ring, but suitable reagents are not available for the synthon RO^+. If we wish to add an oxygen atom to an aromatic ring we must use the alternative approach, and add anionic reagents RO^- to an aromatic compound with a leaving group. This is nucleophilic aromatic substitution and works best when the leaving group is N_2 (diazonium salts). The synthetic sequence is nitration, diazotisation, and substitution.

The synthesis of phenol (13) can be analysed in this way, the OH reverting to NO_2. The bromine could be added at the amine or the phenol stage, but the amine stage gives better control.

Analysis

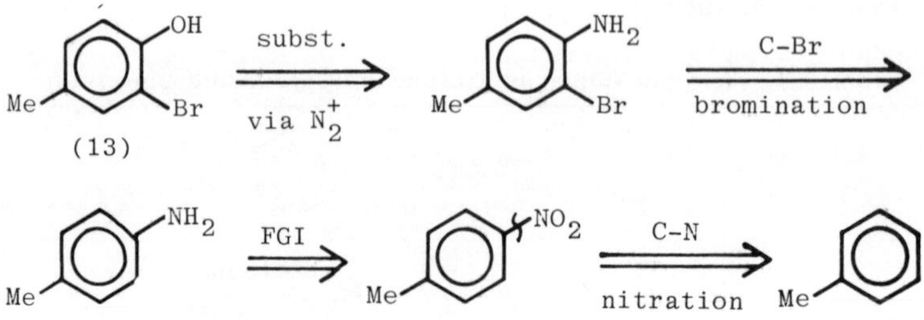

(13)

In practice, the amine was protected as an amide to prevent the bromine adding to the other ortho position as well.

Synthesis[14]

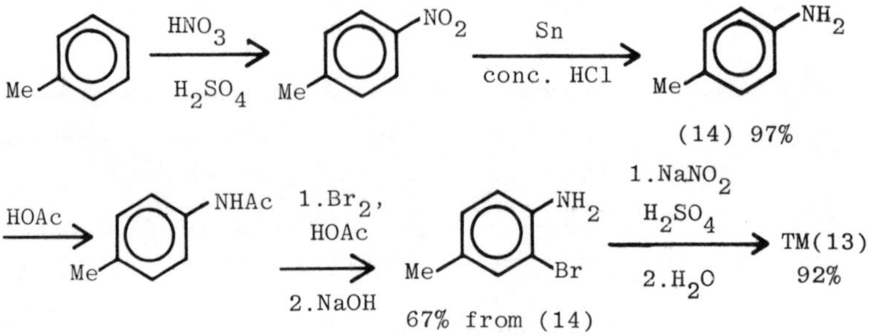

(14) 97%

67% from (14)

TM(13) 92%

Some nucleophiles (CN⁻, Cl⁻, Br⁻ for example) are best added as Cu(I) derivatives: a list of these and others appears in Table 2.4. The aromatic cyanide (15) is most easily disconnected this way.

Table 2.4 Aromatic compounds made by nucleophilic displacement of diazonium salts

$$ArNH_2 \xrightarrow{HONO} ArN_2^+ \xrightarrow{Z^-} ArZ$$

Z	Reagent
HO	H_2O
RO	ROH
CN	Cu(I)CN
Cl	Cu(I)Cl
Br	Cu(I)Br
I	KI
Ar	ArH
H	H_3PO_2 or EtOH/H⁺

Analysis

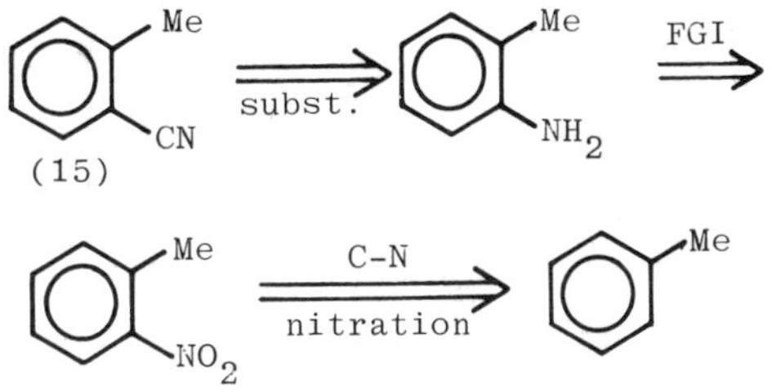

(15)

Synthesis[15]

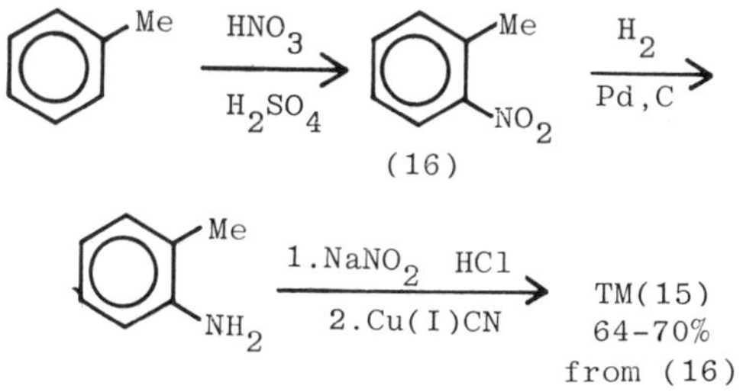

(16)

$$\text{TM(15)} \quad 64\text{-}70\% \quad \text{from (16)}$$

Nucleophilic Substitution of Halides

Direct displacement of halide from an aromatic ring is possible only if there are *ortho* and *para* nitro groups or similar electron-withdrawing groups. Fortunately these compounds are easily made by nitration:

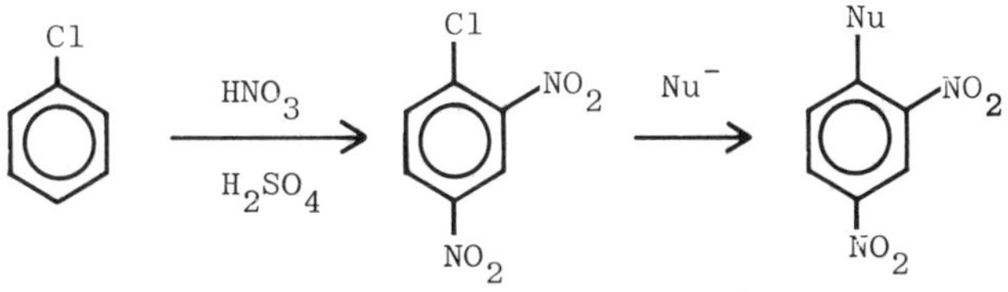

The Lilley Company's pre-emergent herbicides such as trifluralin B **(17)** are good candidates for this approach. The amino group can be added in this way and the two nitro groups put in by direct nitration. The synthesis of the starting material **(18)** is discussed in Chapter 3.

Trifluralin B: *Analysis*

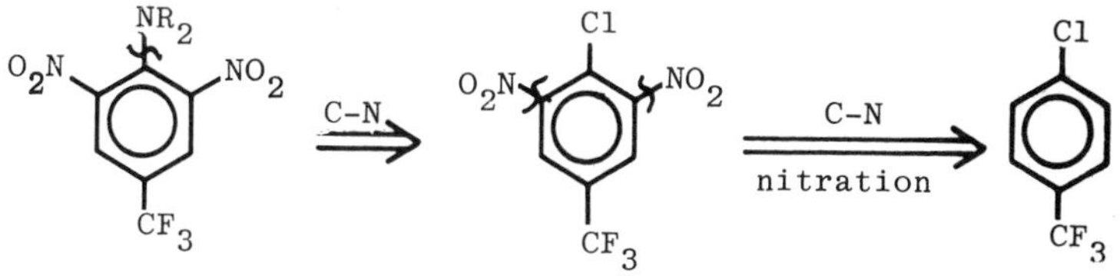

(17, R=n-propyl)

(18)

14

$$(18) \xrightarrow[\text{H}_2\text{SO}_4]{\text{HNO}_3} \quad \xrightarrow[\text{n-Pr}_2\text{NH}]{\text{base}} \text{TM(17)}$$

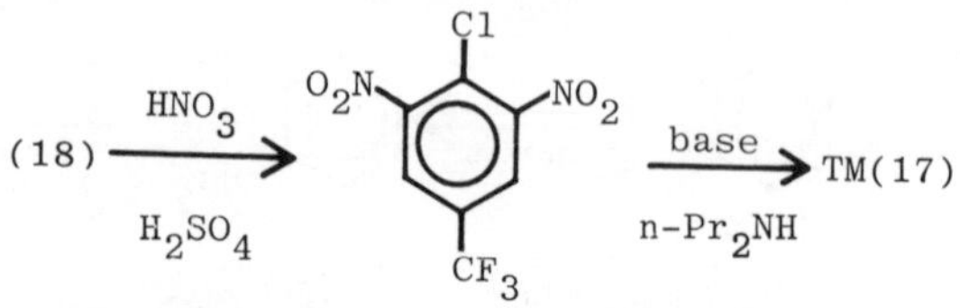

Ortho and *Para* Product Mixtures

We used the same reaction — the nitration of toluene — to make both the *ortho* (**10**) and *para* (for **14**) nitrotoluenes. In practice, a mixture is formed and must be separated to give the required isomer. In other circumstances, reactions which give mixtures of products are best avoided but aromatic substitution is so easy to carry out that separation is acceptable, particularly if it is at the first stage in a sequence. The reaction is then carried out on a large scale to get enough of the right isomer and a use sought for the other.

Saccharine (**19**) is made this way. Disconnection of the imide gives the diacid (**20**) which can be made by FGI from toluene-*ortho*-sulphonic acid.

Saccharine: *Analysis*

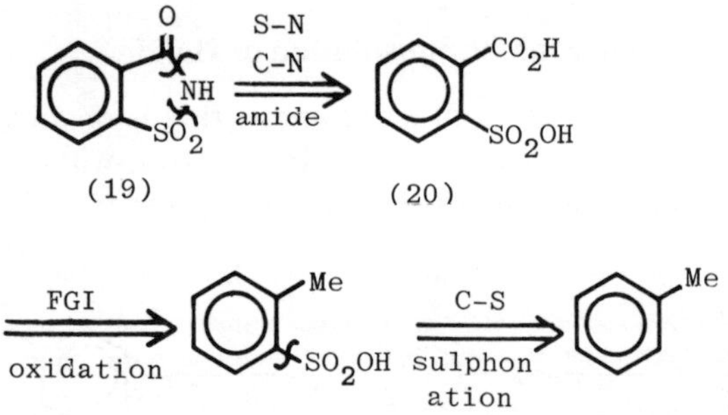

In practice it is quicker to make the sulphonyl chloride (**21**) directly and separate it from the *para* compound. The rest of the synthesis is routine.

Synthesis[17]

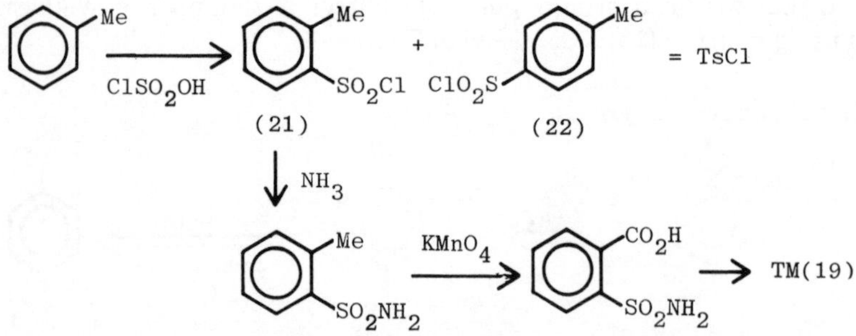

Saccharine is made on a large scale so there is plenty of toluene-*p*-sulphonyl chloride spare and it is cheap. This is one reason why the tosyl group is such a popular leaving group with organic chemists (see Chapter 4).

The question of *ortho–para* mixtures and other similar strategic questions in aromatic syntheses are the subject of the next chapter.

Technical Terms for the Disconnection Approach

Target Molecule (TM): the molecule to be synthesised.

Analysis or *Retrosynthetic Analysis*: the process of breaking down a TM into available starting materials by FGI and disconnection.

FGI (Functional Group Interconversion): the process of converting one functional group into another by substitution, addition, elimination, oxidation, or reduction, and the reverse operation used in analysis.

Disconnection: the reverse operation to a reaction. The imagined cleavage of a bond to 'break' the molecule into possible starting materials.

$\Rightarrow$: symbol for disconnection or FGI.

Synthon[a]: an idealised fragment, usually a cation or an anion, resulting from a disconnection. May or may not be an intermediate in the corresponding reaction.

Reagent[a]: compound used in practice for a synthon. Thus MeI is the reagent for the Me^+ synthon.

---

[a]Some chemists use 'synthon' to mean a useful reagent in organic synthesis.

CHAPTER 3

Strategy I: The Order of Events

Alternating with instructional chapters, like the last one, will be strategy chapters, like this one, in which some point relevant more to the overall plan than to some individual reaction is examined. In this chapter, using aromatic compounds as examples, we examine the question of the order in which reactions should be carried out.

The detergents commonly used nowadays contain sodium salts of sulphonic acids such as (1). They are made industrially[18] in two steps from benzene, a Friedel–Crafts reaction, and a sulphonation. The question is: why this order of events? Two factors influence the answer. The alkyl group is electron-donating and makes the sulphonation easier. The alternative sequence via the sulphonic acid (2) would be very difficult as the SO_2OH group is strongly electron-withdrawing and therefore deactivating. The second point is that the electron-donating alkyl group is o,p-directing (it gives only *para* product because of its size). The SO_2OH group is *meta* directing and would give a different product.

In choosing the order of events we must take both these related aspects into consideration (they are summarised in Table 3.1) and we can lay down some general guidelines based on them.

Guidelines for the Order of Events

Guideline 1

Examine the relationship between the groups, looking for groups which direct to the right position. The thorough way to do this is to disconnect all groups in turn and see if the reverse reaction would give the right orientation.

The analysis of the orris odour ketone (3) could be tackled by two possible first disconnections. One (b) gives starting materials which would react in the right orientation since the ketone group in (a) is *meta* directing. The order of events in the synthesis follows.

Analysis

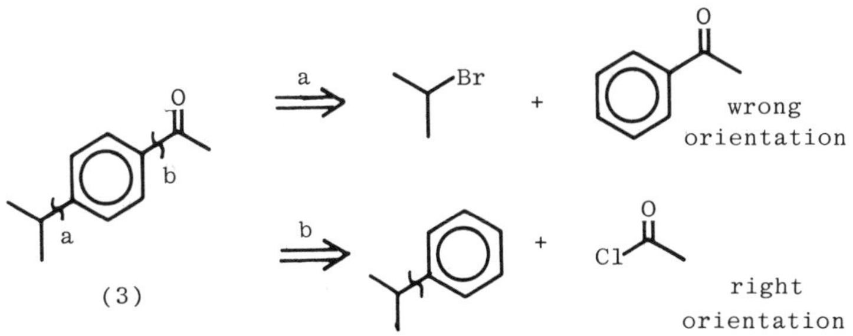

Synthesis[19]

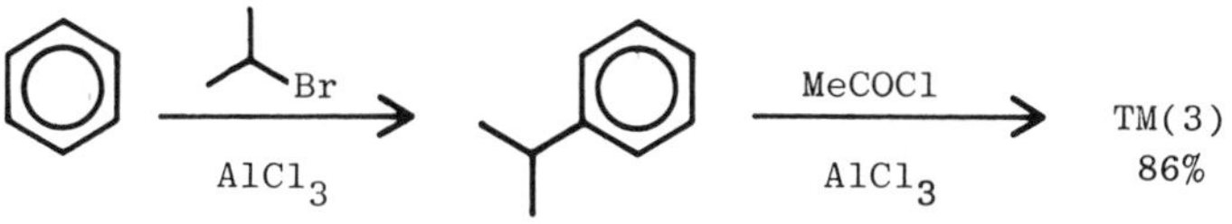

Guideline 2

If there is a choice, disconnect *first* (that is add last) the most electron-withdrawing substituent. This substituent will be deactivating so it may be difficult to add anything else once it is in.

Musk ambrette (4), a synthetic musk, essential in perfumes to enhance and retain the odour, is an aromatic compound with five substituents on the benzene ring. The nitro groups are by far the most electron-withdrawing so we can disconnect them first.

Musk ambrette: *Analysis 1*

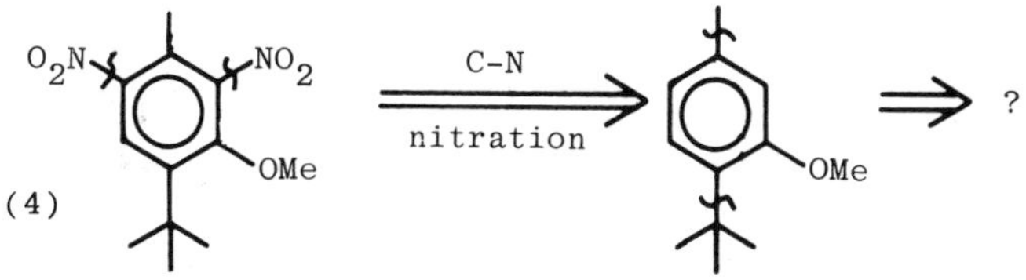

We could add either the Me or the *t*-Bu group by a Friedel–Crafts alkylation. The OMe group is strongly *o,p*-directing so only the *t*-Bu disconnection is reasonable (guideline 1).

Analysis 2

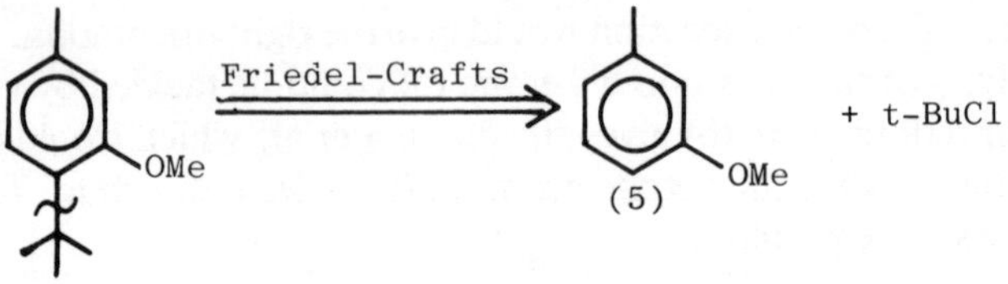

The starting material (5) is the methyl ether of readily available *meta*-cresol, and can be made with any methylating agent. Dimethyl sulphate is often used.

Synthesis[20, 21]

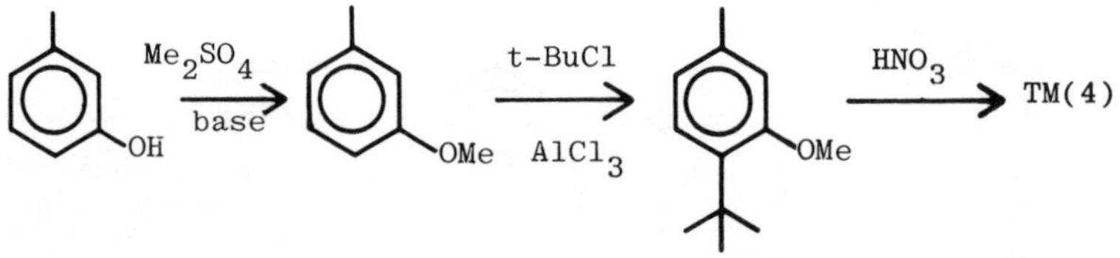

Only experience would show whether the Friedel–Crafts alkylation puts the *t*-butyl group *ortho* or *para* to the methoxy group.

Guideline 3

If FGI is needed during the synthesis, it may well alter the directing effect of the group and the other substituents may therefore be added either before or after the FGI. Some examples are:

$$o,p\text{-directing Me} \quad \rightarrow \quad CO_2H \; m\text{-directing}$$

$$Me \quad \rightarrow \quad CCl_3/CF_3$$

$$m\text{-directing NO}_2 \quad \rightarrow \quad NH_2 \; o,p\text{-directing}$$

The synthesis of (6) obviously involves chlorination of both the ring and a methyl group (FGI). CCl_3 is *m*-directing so we must reverse the FGI before we disconnect the aryl chloride.

Analysis

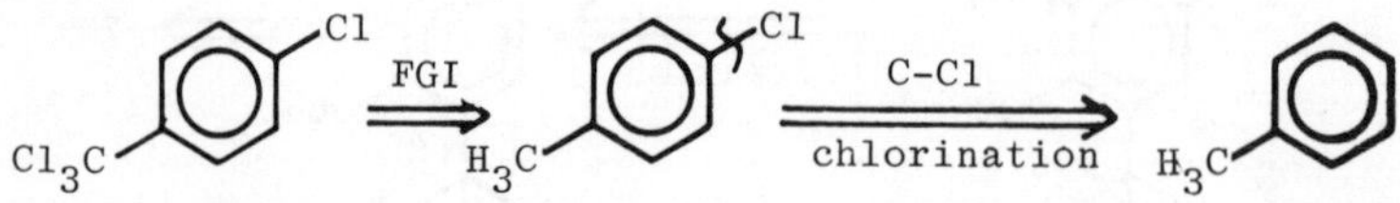

The synthesis, used to make (7), goes in excellent yield.

Synthesis[13]

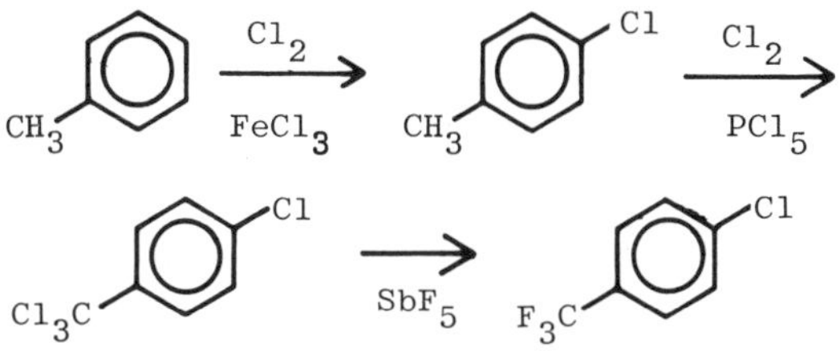

(6), 93%   (7), 95%

Guideline 4

Many groups can be added by *nucleophilic* substitution on a diazonium salt (see Chapter 2), made from an amine. Adding other groups at the amine stage may be advisable as the amino group is strongly *o,p*-directing.

Acid (8) was needed at Hull University[22] to study its liquid crystal behaviour (liquid crystals are used in digital displays). The other benzene ring is *o,p*-directing, so to get the chlorine in we must replace the CO_2H group by a *more o,p*-directing group than Ph. Amino is the obvious choice.

Analysis

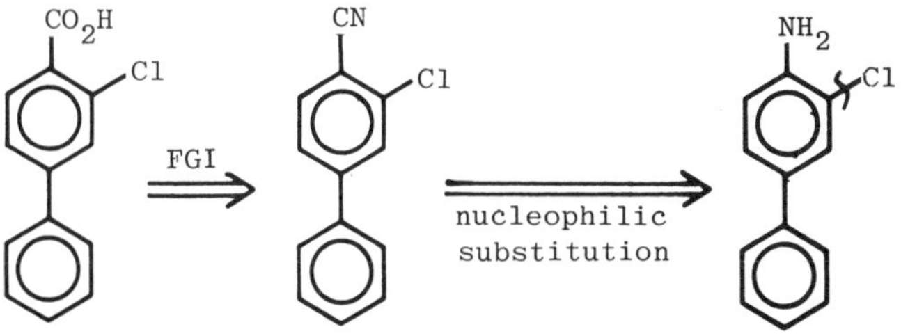

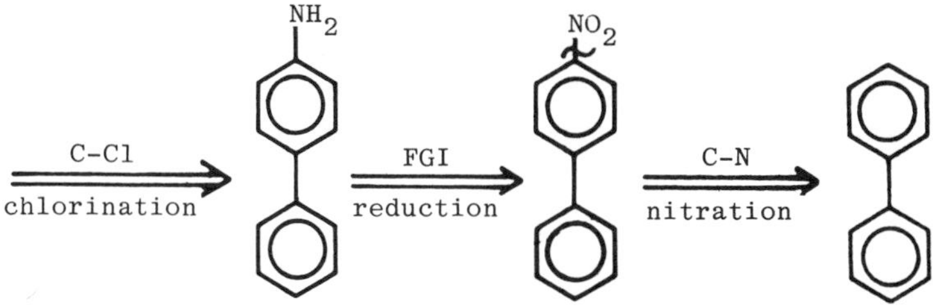

In the synthesis it will be necessary to acylate the amino group to prevent over-chlorination (cf. Chapter 2).

20

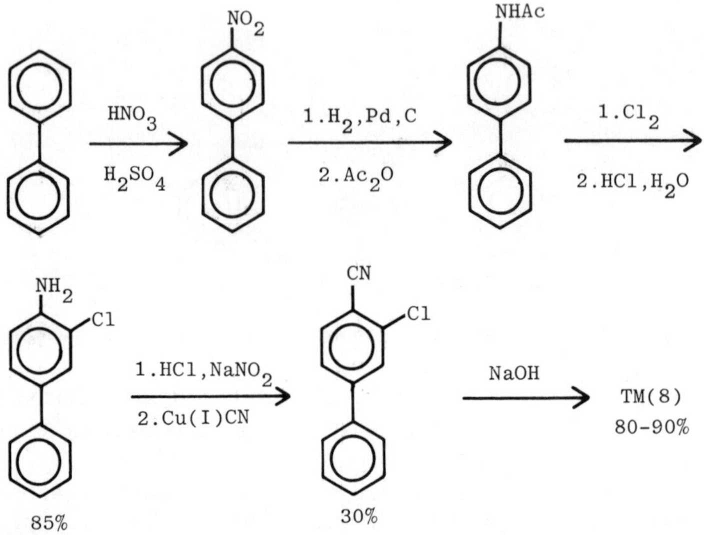

Guideline 5

As a last resort, there is a trick to solve some difficult problems, such as adding two *o,p*-directing groups *meta* to each other. A 'dummy' amino group is added, used to set up the required relationship and then removed by diazotisation and reduction:

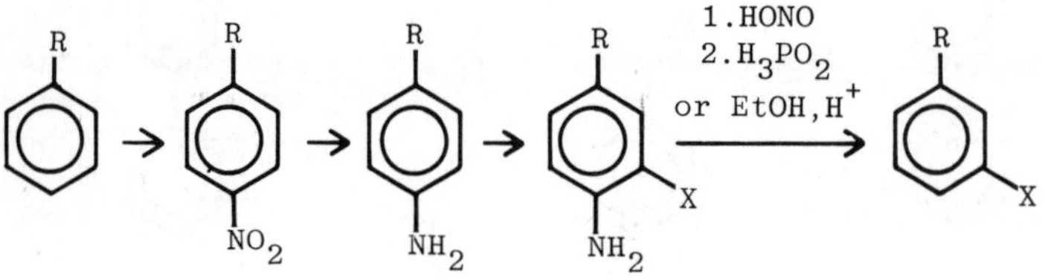

The acid **(9)** is used in the synthesis of a number of local anaesthetics[24] such as Propoxycaine **(10)**. The amino group cannot be put in by nitration of salicyclic acid **(11)** as the oxygen atom will direct *o,p* and give the wrong isomer. The problem can be solved by deliberately making the wrong isomer and nitrating that.

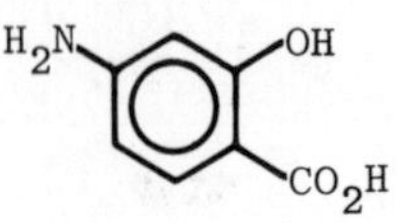

(9) 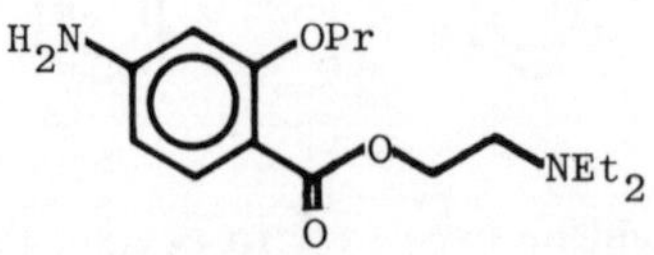 (10) 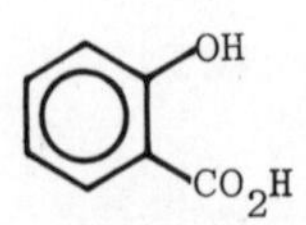 (11)

Analysis

In practice it is wise to add the alkyl group at the start to protect the hydroxyl group.

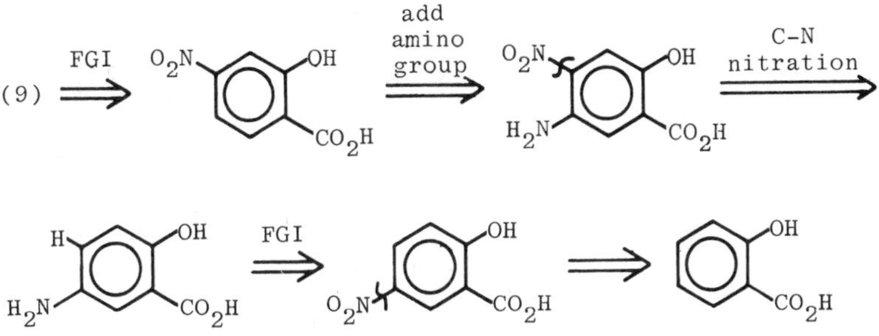

Synthesis[25]

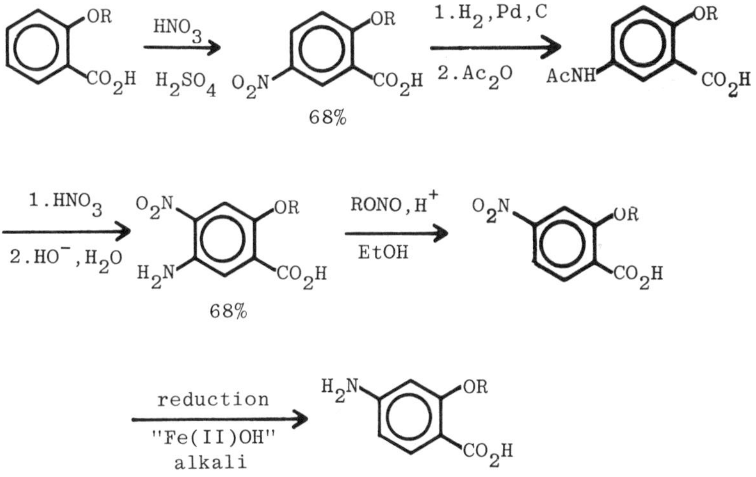

Guideline 6

Look for substituents which are difficult to add. It is often good strategy not to disconnect these at all but to use a starting material containing the substituent. OH and OR are examples. We have already used this guideline for compound (4) (substituent OMe) and for compound (8) (substituent Ph).

Guideline 7

This is an extension of guideline 6. Look for a combination of substituents present in the TM and in a readily available starting material, particularly if it would be a difficult combination to set up.

Examples are:

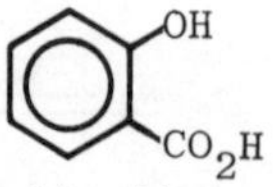

salicylic acid
and aldehyde

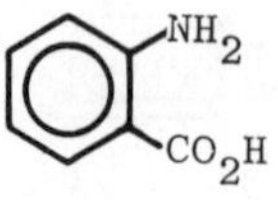

anthranilic acid

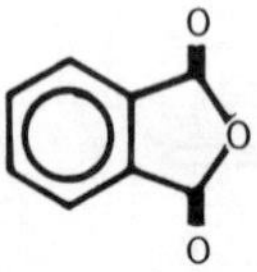

phthalic
anhydride

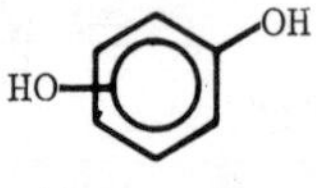

ortho, meta, and
para compounds

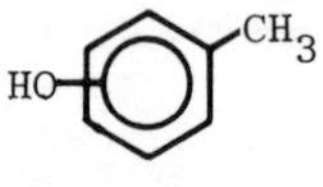

ortho, meta,
and para cresols

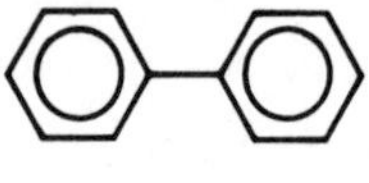

diphenyl

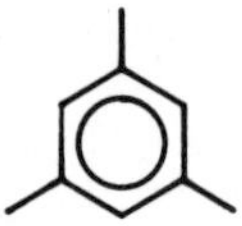

mesitylene

We have already used this guideline in syntheses of compounds (4) (from *m*-cresol), (8) (from biphenyl), and (9) (from salicyclic acid).

Another example is compound (12) needed for the synthesis of the anti-asthma drug Salbutamol (13). The acid (12) can obviously be made by a Friedel–Crafts reaction on salicyclic acid.

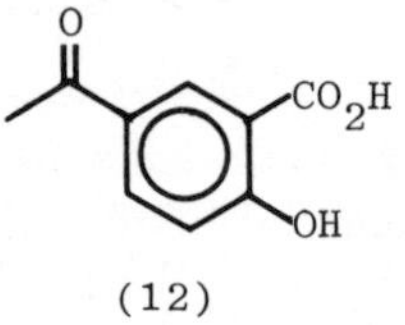

(12)

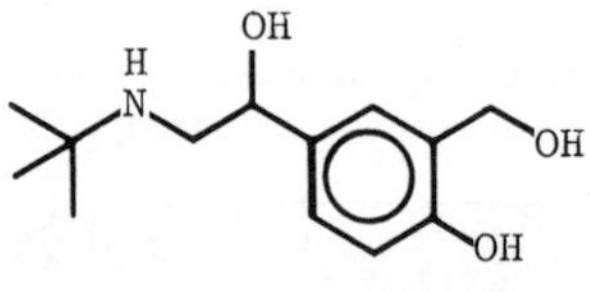

(13)

Analysis

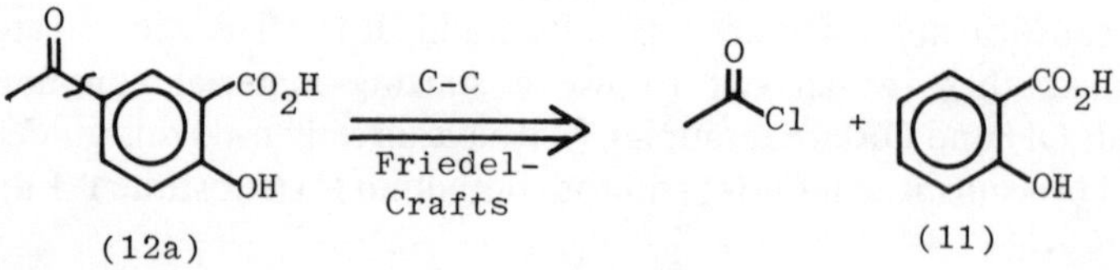

(12a)

$C-C$

Friedel–
Crafts

+

(11)

The synthesis is easier than it may seem since Friedel–Crafts acylation of phenols is best done by first making the phenolic ester and rearranging this with $AlCl_3$. In this case, the ester needed is (14) which hardly needs to be made since it is aspirin. No doubt this Salbutamol synthesis was planned with this cheap starting material in mind.

Synthesis[26]

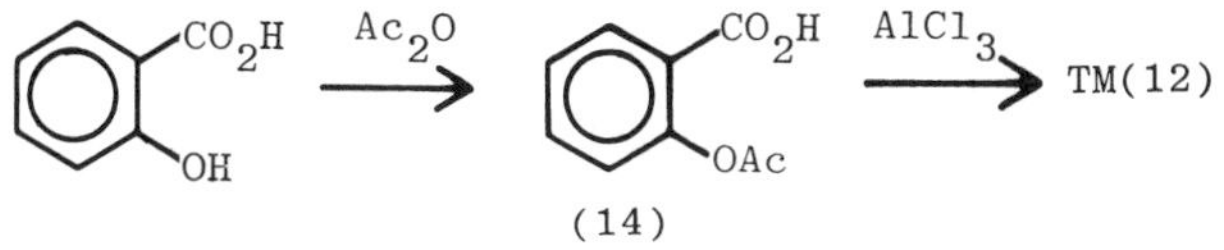

(14)

Guideline 8

Avoid sequences which may lead to unwanted reactions at other sites in the molecule. Thus nitration of benzaldehyde gives only 50% *m*-nitrobenzaldehyde since the nitric acid oxidises CHO to CO_2H. One way round this particular problem is to nitrate benzoic acid and reduce CO_2H to CHO.

A more interesting example is compound (15), needed to make amines such as (16) for trial as antimalarial drugs.[27] The OEt group is best left to appear in the starting material (guideline 6) so we have two strategies differing only in the order of events.

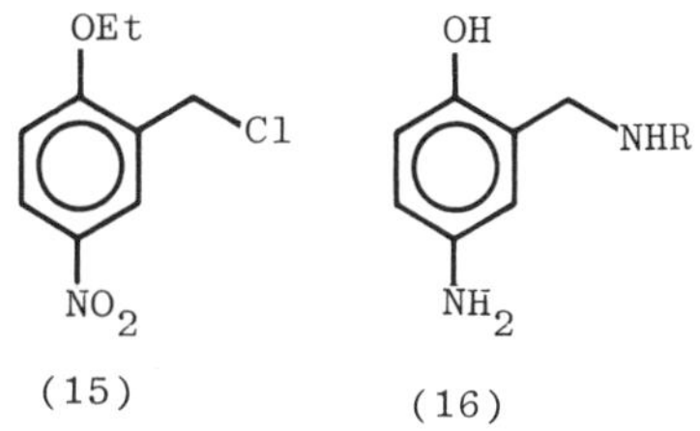

(15)   (16)

Analysis

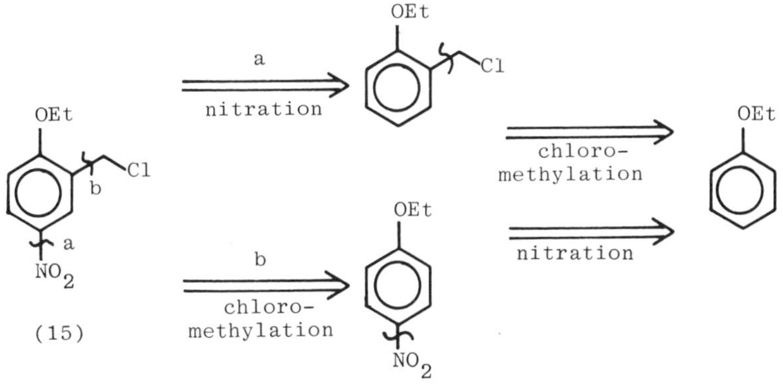

Both strategies fit the substitution pattern (OEt is more electron-donating than CH_2Cl) and strategy (a) also meets guideline (2). But CH_2Cl is oxidised easily (see Chapter 2) so nitrating conditions may destroy it. Strategy (b) gives good yields.

24

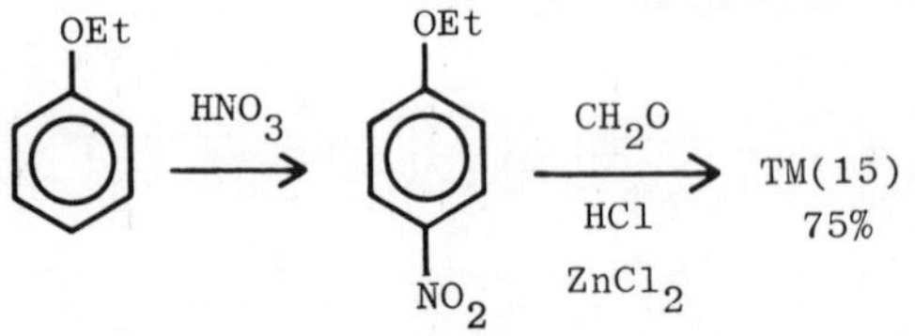

Guideline 9

If *o,p*-substitution is involved, one strategy may avoid separation of isomers in that the other position becomes blocked.

Esters of phenol **(17)** are used as garden fungicides,[28] e.g. **(18)** is Dinocap. We can disconnect the nitro group first (guideline 2) but the Friedel–Crafts reaction required would surely give mostly *para* product as the electrophile is so large.

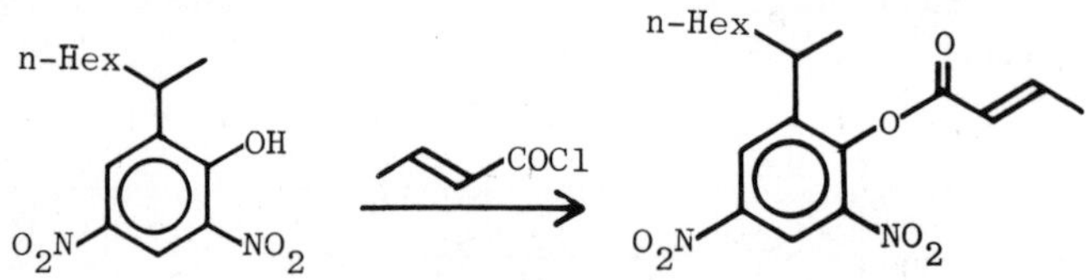

(17)   (18)

Analysis 1

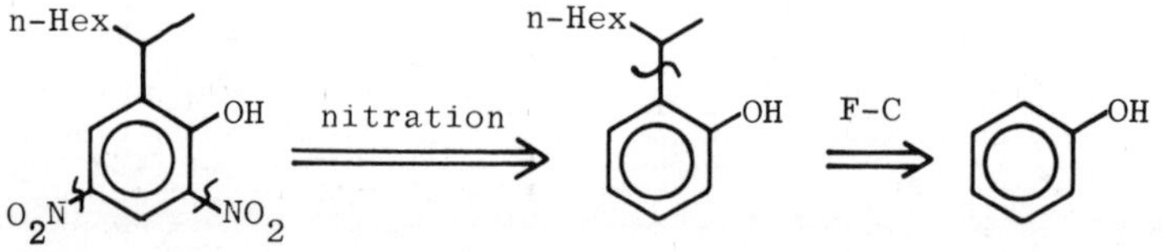

The alternative order of events, disconnecting the Friedel–Crafts first, is unusual but sensible here since the *para* position is blocked.

Analysis 2

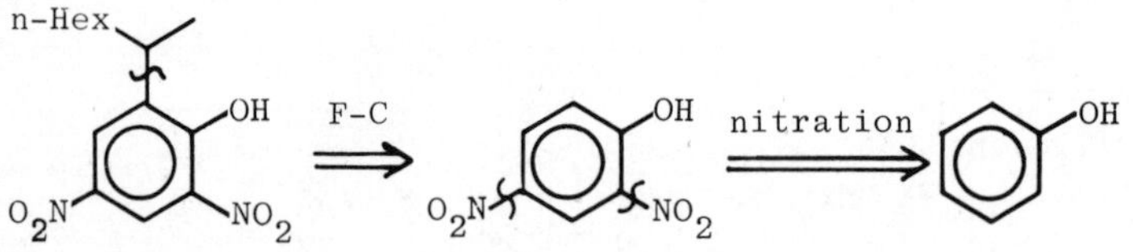

Dinocap is manufactured by the second route.

Synthesis[28]

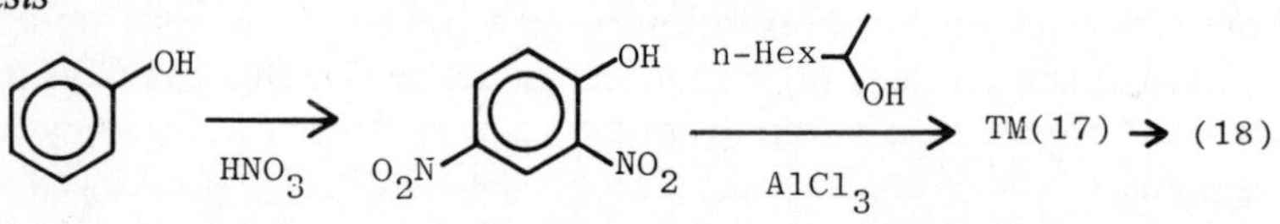

There are two reactions which can give unusually large amounts of *ortho* product: the Fries rearrangement[29] (i) (see page 22), and the Reimer–Tiemann reaction[30] (ii) These can be used to set up *ortho* substituents with other substituents present but one OH group is needed in the molecule.

Not all these nine guidelines apply to any one case—indeed some may well contradict others. It is a matter of judgement—as well as a laboratory trial and error—to select a good route. As always, several strategies may be successful.

Table 3.1 Direction and activation in aromatic electrophilic substitution. The most activating groups are at the top of the list. In general, the more activating group dominates the less activating* and the selectivity will be greater the more the difference between them

Direction	Group	Activation
o,p	R_2N, NH_2	Activating (electron-donating)
	RO, OH	
	Alkenyl	
	Aryl	
	Alkyl	
	CO_2^-, H	Electronically neutral
	Halide	
m	CX_3	
	(X=F, Cl etc)	
	CO_2H	Deactivating (electron-
	COR, CHO	withdrawing)
	SO_3H	
	NO_2	

*Ignoring steric effects.

CHAPTER 4

One-Group C–X Disconnections

We started with aromatic compounds because the position of disconnection needed no decision. We continue with ethers, amides, and sulphides because the position of disconnection is again easily decided: we disconnect a bond joining carbon to the heteroatom (X). This approach is fundamental to synthetic design and is a 'one-group disconnection' since we need to recognise only one functional group to know that we can make the disconnection. The label 'C–X' or 'C–N' etc. can be used.

The corresponding reactions are mostly ionic and involve nucleophilic heteroatoms as in alcohols (ROH), amines (RNH_2), or thiols (RSH). The disconnection will therefore give a cationic carbon synthon (1). The reagent for (1) will usually have a good leaving group attached to R (2). In other words, the reaction is a substitution of some kind and the reagents will be alkyl halides, acid chlorides, and the like and the best reagents will be those which undergo substitution most readily.

$$R \overset{C-X}{\not}X \implies X^- + R^+ = RY \qquad Y = Br, OTs, etc$$
$$(1) \quad (2)$$

Carbonyl Derivatives RCO.X

Acid derivatives are easy to disconnect since we almost always choose the bond between the carbonyl group and the heteroatom for our first disconnection (i).

$$R\overset{O}{\overset{\|}{—}}X \overset{C-X}{\implies} R\overset{O}{\overset{\|}{—}}Y + XH \quad (i)$$

The ester (3), used both as an insect repellent,[31] and as a solvent in perfumery,[32] invites this disconnection.

Analysis

(3)

26

The synthesis can be carried out in a number of ways: perhaps the acid chloride route (Y=Cl) is the easiest, with pyridine as catalyst and solvent.

Synthesis

$$Ph\diagup\diagdown OH \xrightarrow[\text{pyridine}]{PhCOCl} TM(3)$$

Acid chlorides are often used in these syntheses because they are the most reactive of all acid derivatives and because they can be made from the acids themselves and PCl$_5$ or SOCl$_2$. It is easy to move *down* the hierarchy of reactivity (see Table 4.1) and fortunately esters and amides, which are at the bottom, are the acid derivatives most usually required.

Table 4.1 Hierarchy of reactivity for acid derivatives

Most reactive

		SOCl$_2$ or PCl$_5$	
Acid chlorides	RCOCl	$\longleftarrow$	
Anhydrides	RCO.O.COR	$\xleftarrow{Ac_2O}$	
Esters	RCO.OR^1	$\xleftarrow[H^+]{R^1OH}$	RCO$_2$H Acids
Amides	RCO.NR^{1}R^2	$\xleftarrow{R^1R^2NH}$	

Most stable

not usually made directly

The weedkiller Propanil (4) used in rice fields[33] is an amide so we disconnect to an amine and an acid chloride. Further disconnection of the aromatic amine (5) follows from Chapters 2 and 3.

Propanil: *Analysis*

(4)   (5)

28

The orientation for nitration is correct: steric hindrance will prevent formation of much 1,2,3-trisubstituted compound.

Synthesis

Compound **(6)** is a more complicated example but we can recognise an ester which we can disconnect in the usual way, simplifying the problem greatly. The very cheap phthalic anhydride **(8)** is the best acid derivative here and the synthesis of the alcohol **(7)** is discussed in Chapter 10.

Analysis

Synthesis[34]

This molecule **(6)** was needed for the resolution of alcohol **(7)** into optical isomers, a derivative with an ionisable group (here CO_2H) being required.

Alcohols, Ethers, Alkyl Halides, and Sulphides

C–X disconnection in aliphatic compounds (ii) gives a nucleophile XH and an electrophilic carbon species usually represented by an alkyl halide, tosylate*, or mesylate*. These compounds can all be made from alcohols (ii) and as alcohols can be made by C–C bond formation (Chapter 10) we shall treat the alcohol as the central functional group (Table 4.2).

*Tosylate = toluene-*p*-sulphonate; mesylate = methane sulphonate. See Chapter 2 for the synthesis of TsCl.

$$RX \implies XH + R^+ = RBr \text{ or } ROTs \text{ or } ROMs$$
$$(ii)$$

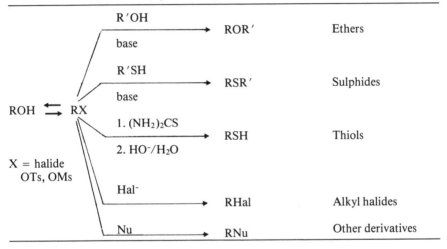

Table 4.2 Aliphatic compounds derived from alcohols

Conditions must be chosen to suit the structure of the molecules. Methyl and primary alkyl derivatives react by the S_N2 mechanism so powerful nucleophiles and non-polar solvents are effective. The nitro compound[35] **(9)** and the azide[36] **(10)** — examples of the 'other derivatives' in Table 4.2 — are easily made from the corresponding bromides by S_N2 reactions as they are both primary alkyl compounds.

Tertiary compounds react even more easily by the S_N1 mechanism *via* stable carbonium ions **(11)** generated directly from alcohols, alkyl halides, or even alkenes. Powerful nucleophiles are no help here but polar solvents and catalysis (usually acid or Lewis acid) help by making the OH a better leaving group.

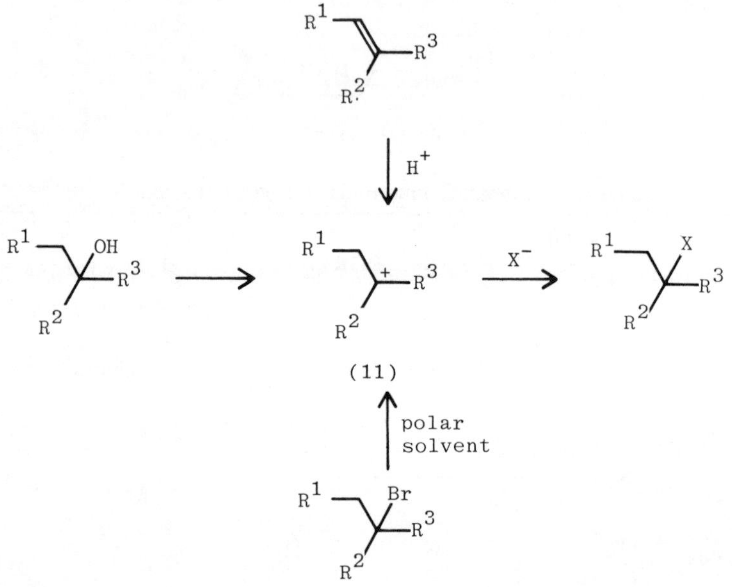

(11)

Compound (12) can obviously be made by a Friedel–Crafts reaction from benzene and the tertiary chloride (13), which comes from the alcohol (14). The only reagent needed for (14) → (13) is conc. HCl. The synthesis of compounds like (14) is discussed in Chapter 10.

Analysis

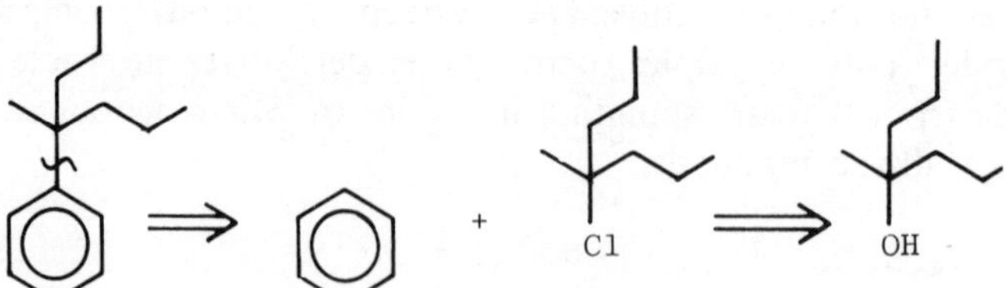

(12)   (13)   (14)

Synthesis[37]

$$ (14) \xrightarrow{\text{conc. HCl}} (13) \xrightarrow[\text{AlCl}_3]{\text{PhH}} \text{TM}(12) \quad 70\% $$

Allylic (15) and benzylic (16) derivatives react easily both by S_N1 and S_N2 mechanisms so conditions are relatively unimportant here. By contrast, secondary alkyl derivatives are the most difficult to make and conditions need to be rather harsh in these cases.

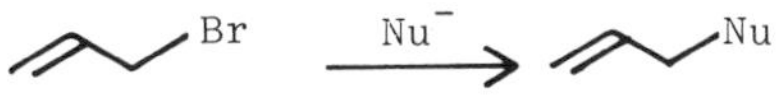

(15)

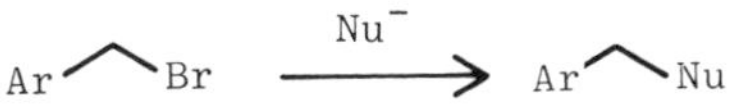

(16)

These interconversions are rather elementary in concept but are essential to synthetic planning. Compounds of the type R^1-X-R^2 offer a choice for the first disconnection and are more interesting.

Ethers and Sulphides

We can often choose our first disconnection because of the reactivity (or lack of it) of one side of the target molecule. The oxygen atom in the wallflower perfume compound (17) has a reactive side (Me, by S_N2) and an unreactive (Ar) side so disconnection is easy.

Analysis

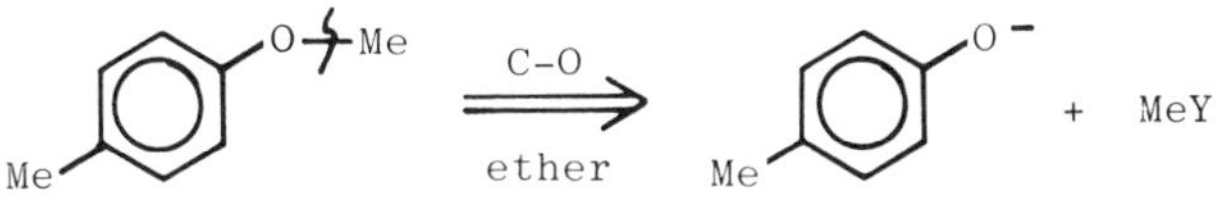

(17)

Dimethyl sulphate is used for methylation of phenols in alkaline solution where the phenol is ionised. Since the mechanism is S_N2, the more powerfully nucleophilic anion is an advantage.

Synthesis[38]

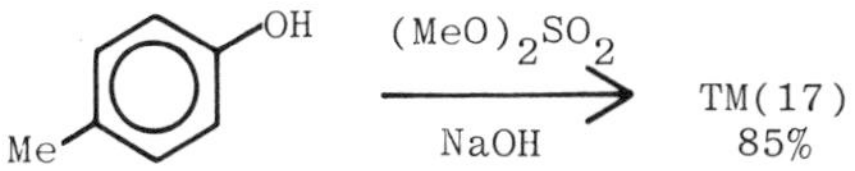

The gardenia perfume compound (18) can be disconnected on either side as both involve primary alkyl halides. The benzyl halide is more reactive but the decisive point in favour of route (b) is that route (a) might well lead to elimination.

Analysis

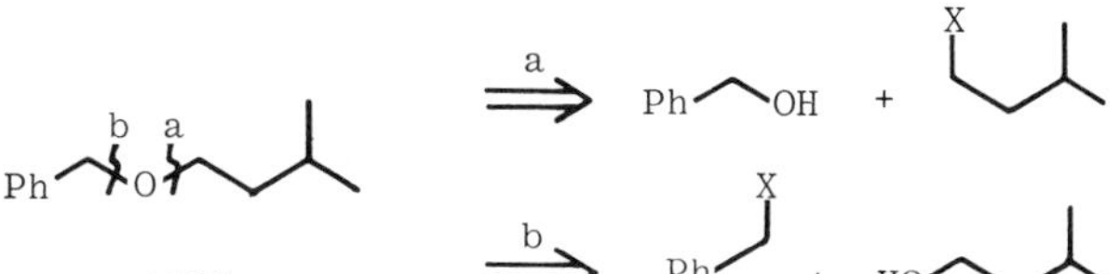

(18)

Synthesis

This is S_N2 again, so base catalysis helps.[39]

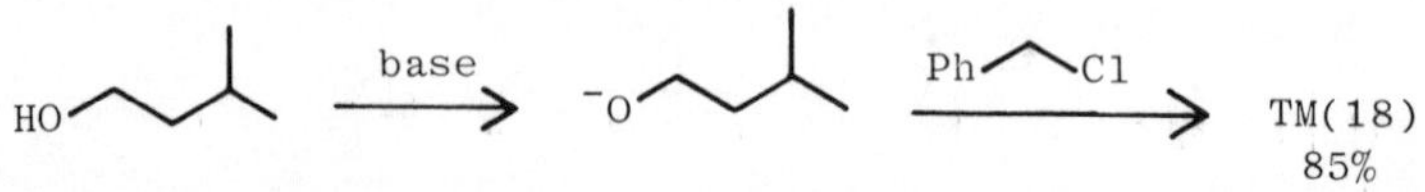

If there is no obvious preference, it is more helpful to write both fragments as alcohols and decide later which to convert into an electrophile. Baldwin[40] needed the ether **(19)** to study the rearrangement of its carbanion. Both sides are reactive so we write the two alcohols. Baldwin does not reveal[40] how he actually made the ether **(19)** — both routes look good, although the one shown is less ambigious.

Analysis

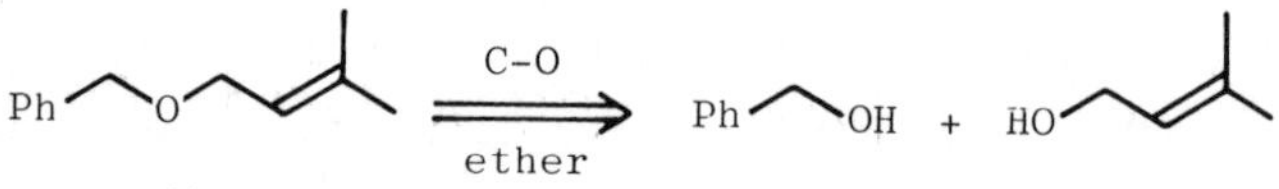

Synthesis

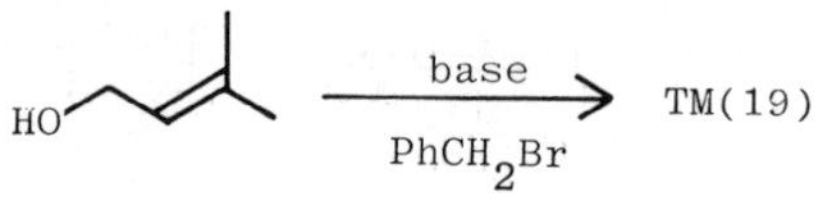

The same principles apply to sulphide (R^1SR^2) synthesis. The reaction is even easier by S_N2 as thiols ionise at a lower pK_a than alcohols, the anion **(20)** is softer than RO^- and thus more nucleophilic towards sp^3 carbon.

$$R^1-S{\not|}R^2 \implies R^1S^- + R^2Y$$
$$(20)$$

The acaricide (kills mites and ticks) Chlorbenside **(21)** is first disconnected on the alkyl rather than the aryl side. The synthesis of thiols is discussed in the next chapter.

Chlorbenside: *Analysis*

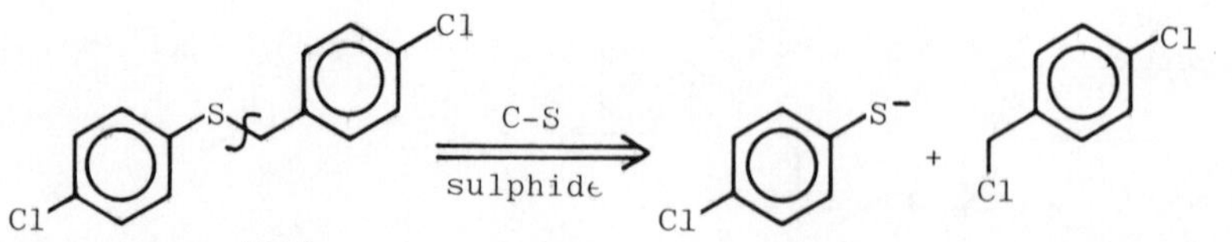

Synthesis[41]

CHAPTER 5

Strategy II: Chemoselectivity

When a molecule contains two reactive groups and we want to react one of them but not the other, the question of chemoselectivity arises. Under this heading we can consider

1. relative reactivity of two different functional groups; e.g.,

2. reaction of one of two identical functional groups; e.g.,

3. reaction of a group once when it may react again, e.g. thiol synthesis.

$$S^{2-} \xrightarrow{\text{RBr}} RS^- \xrightarrow{\text{RBr}} RSR \text{ ?}$$
$$\text{wanted}$$

We shall deal with all three cases in this chapter, and although each one needs to be taken on its merits, there are some helpful general principles.

Guideline 1

With two functional groups of *unequal* reactivity, the *more* reactive can always be made to react alone.

The acid (1), needed to synthesise the anaesthetic Cyclomethycaine (2), can be analysed as an ether (see Chapter 4) leading to simple starting materials. But will the hydroxyacid (3) react only at the OH group, or will the CO_2H group react too? In alkaline solution, when both are ionised (i.e. pH > 10), the phenolate ion is much more reactive than the carboxylate ion (pK_a difference about 5), and only the phenol is alkylated.

34

Cyclomethycaine: *Analysis*

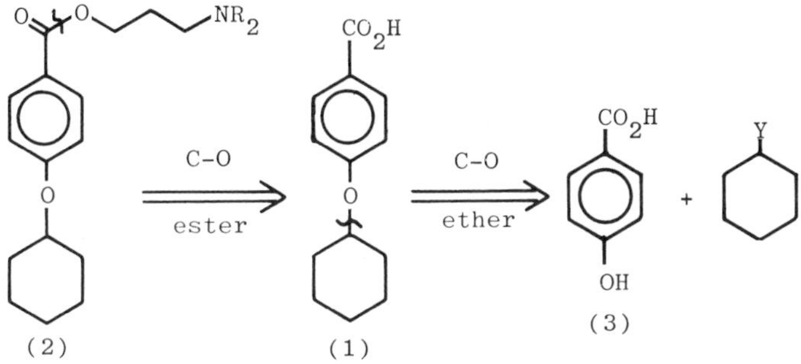

The published synthesis[42] used the alkyl iodide as it is a secondary alkyl derivative and therefore rather unreactive (Chapter 4). Iodide is a better leaving group than chloride or bromide.

Synthesis[42]

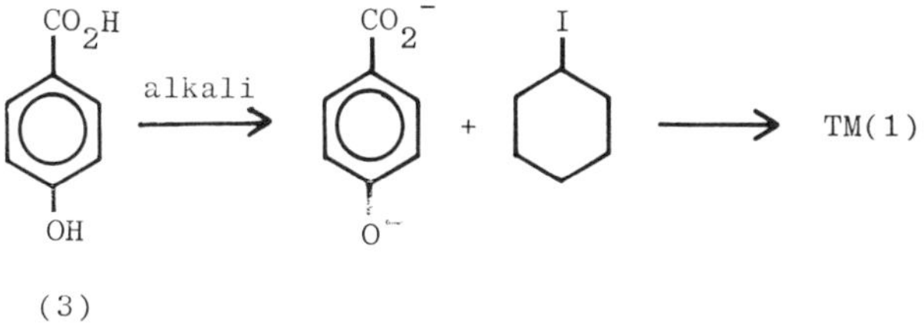

The commonly used analgesic Paracetamol (**4**) is a simple amide and should be available by acetylation of *p*-aminophenol (further analysis according to Chapters 2 and 3). Here we want to keep the phenol unionised so that NH_2 will be more reactive than OH (NH_3 is more nucleophilic than water, but less so than HO^-).

Paracetamol: *Analysis*

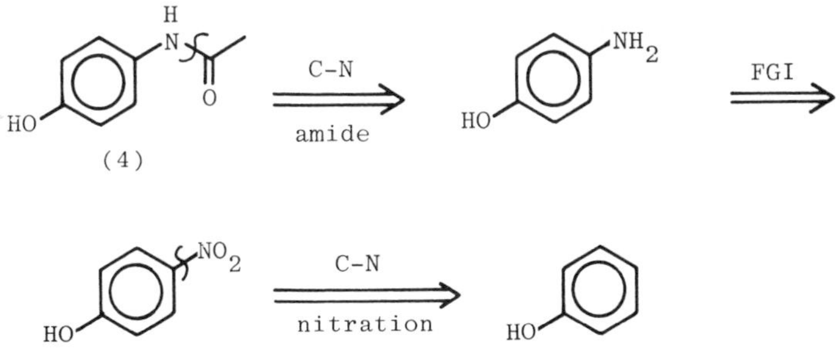

36

Synthesis[43]

separate from
ortho compound

Guideline 2

When one functional group can react twice, the starting material and first product will compete for the reagent. The reaction will be successful only if the first product is *less* reactive than the starting material.

The acid chloride (5) is used to protect amino groups in peptide synthesis. Disconnection of the ester bond gives simple starting materials, but the synthesis will require $COCl_2$ (phosgene) to react once only with $PhCH_2OH$. This succeeds since the half ester (5) is less reactive than the double acid chloride $COCl_2$, because of conjugation (6).

Analysis

Synthesis[44]

$$PhCH_2OH \xrightarrow{\;COCl_2\;} TM(5)$$

The halogenation of ketones (Chapter 7) provides another example.

Guideline 3

Unfavourable cases from guidelines 1 and 2 may be solved by the use of protecting groups.

If we wish to react the *less* reactive of two different functional groups or if the product of a reaction with one functional group is as reactive or more reactive than the starting material, then we must block the unwanted reaction with a protecting group. Amino acids (7) are the constituents of proteins and in most reactions of the CO_2H group, the more reactive NH_2 group must be protected. Compound (5) is used in this way.[44] Note that (5) could react twice with an amine, but the first product (8) is even more conjugated than (5) and so less reactive. The CO_2H group is less reactive than NH_2 and does not react. Protecting groups are systematically treated in Chapter 9.

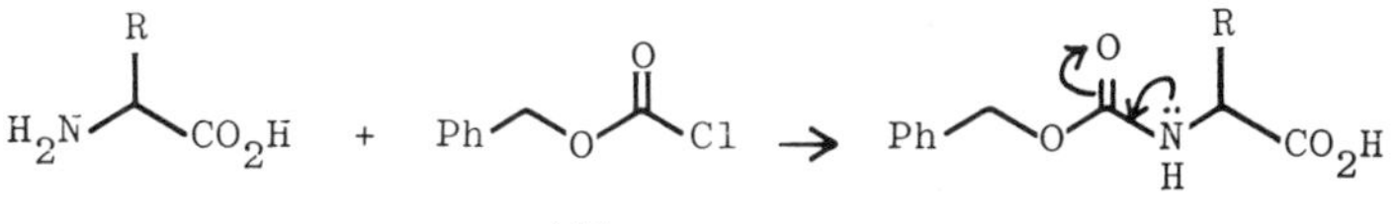

(7)   (5)   (8)

The synthesis of thiols, RSH, by direct alkylation of H_2S is not a good reaction as the product is at least as reactive as the starting material (i).

$$H_2S \xrightarrow{\text{base}} HS^- \xrightarrow{RBr} RSH \xrightarrow{\text{base}} RS^- \xrightarrow{RBr} RSR \quad (i)$$

Thiourea (9) is used as a masked H_2S equivalent, the thiouronium salt (10) being unable to react further and easily hydrolysed to the thiol.

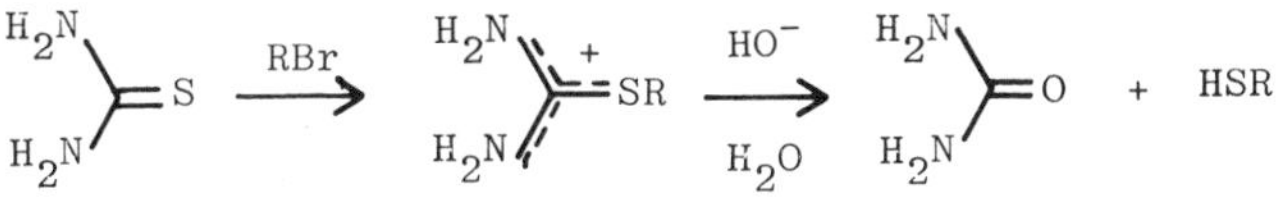

(9)   (10)

The synthesis of Captodiamine (11), a sedative and tranquilliser, illustrates this point and revises material from previous chapters. The thiol (12) is one obvious starting material, the other (13) is discussed in Chapter 6.

Captodiamine: *Analysis 1*

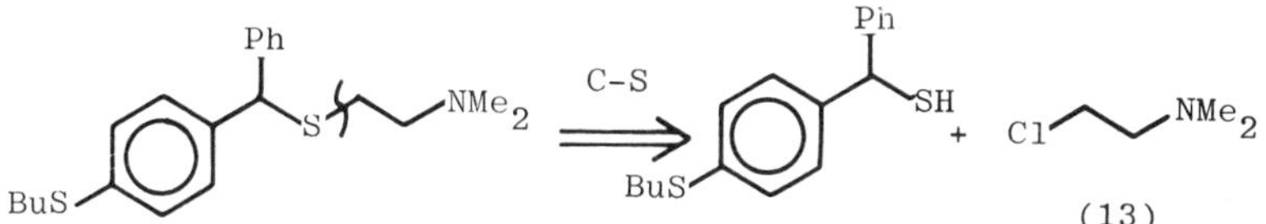

(11)   (12)   (13)

Thiol (12) is made by the thiourea method from halide (14), and this is clearly derived from the Friedel-Crafts product (15). Benzene thiol is available.

Analysis 2

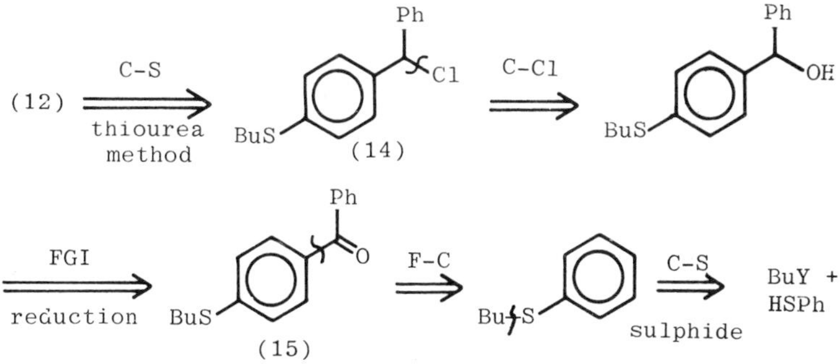

(12)   (14)   (15)

The orientation of the Friedel–Crafts reaction is correct since the lone pairs of electrons on bivalent sulphur direct *o,p.*

Synthesis[45]

$$PhSH \xrightarrow[Na_2CO_3]{n-BuCl} PhSBu \xrightarrow[AlCl_3]{PhCOCl} (15) \xrightarrow[2.SOCl_2]{1.NaBH_4}$$

$$(14) \xrightarrow[2.HO^-/H_2O]{1.thiourea} (12) \xrightarrow[base]{(13)} TM(11)$$

Guideline 4

One of two identical groups may react if the product is less reactive than the starting material.

Partial reduction of *m*-dinitrobenzene is an example. Reduction involves acceptance of electrons from the reducing agent. The product has only one electron-withdrawing nitro group and so is reduced more slowly than the starting material. The best reducing agent for this purpose is sodium hydrogen sulphide.[46]

This product is useful as the amino group can be used to direct electrophilic substitution and can itself be replaced by nucleophiles after diazotisation. Its availability adds extra versatility to Chapter 3.

The soluble dye (17) is clearly a diazo-coupling product from (18) and (19) (see Table 2.2). Further analysis by standard aromatic disconnections leads to *m*-nitroaniline (16) and available β-naphthol (20).

Analysis

Synthesis[47]

$$(20) \xrightarrow{H_2SO_4} (18)$$
$$(16) \xrightarrow{HNO_2} (19)$$
$$\left. \right\} \longrightarrow TM(17)$$

Guideline 5

One of two identical functional groups may react with one equivalent of reagent using the statistical effect.

This is an unreliable method, but if successful it avoids protecting groups or roundabout strategy. The two groups must be identical and must be separated from one another. The diol **(21)** can be monoalkylated in reasonable yield[48] by using one equivalent of sodium in xylene to generate mostly the monoanion **(22)**. Although this will be in equilibrium with dianion and **(21)**, adding the alkyl halide gave an acceptable yield of hydroxy ether **(23)**, used in the synthesis[48] of vitamin E.

HO‿‿‿OH $\xrightarrow[\text{xylene}]{Na}$ HO‿‿‿O⁻ $\xrightarrow{EtBr}$ HO‿‿‿OEt

(21)　　　　　　(22)　　　　　　(23) 62%

Guideline 6

A more reliable method with two identical functional groups is to use a derivative which can react once only. A cyclic anhydride is the most important example. When the anhydride has combined once with a nucleophile (e.g. to give **24**) the product is no longer reactive. Further reactions can maintain the distinction (e.g. to give the half ester, half acid chloride, **25**).[49]

(CO₂H / CO₂H) $\xrightarrow{Ac_2O}$ (anhydride) $\xrightarrow{MeOH}$ (CO₂Me / CO₂H) $\xrightarrow{SOCl_2}$ (CO₂Me / COCl)

(24)　　　　　　(25)

The Friedel–Crafts reaction is also effective on anhydrides and goes once only with cyclic anhydrides. Compound **(26)** was used in the synthesis of fungicidal compounds.[50]

(26)

Guideline 7

When two groups are *nearly* but not quite identical, as in **(27)** and **(28)**, avoid attempts to make only one of them react.

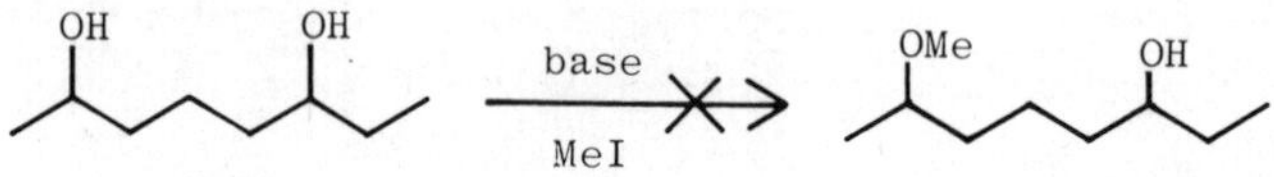

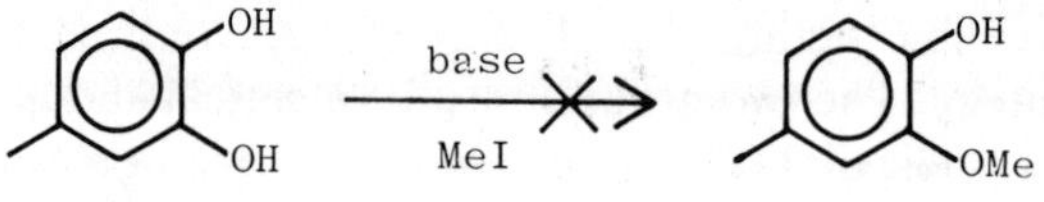

CHAPTER 6

Two-Group C–X Disconnections

1,1-Difunctionalised Compounds

All the disconnections we have used so far have been 'one-group' disconnections, that is we have recognised a single functional group and the disconnection corresponded to a reliable reaction to make that functional group. An important extension of this method is to use one functional group to help disconnect another elsewhere in the molecule. One example we have already met is the synthesis of acetals (1). These compounds have four C–O bonds, all candidates for disconnections if we regard the compound as an ether. If we recognise that one carbon atom (marked • in 2) has two C–O bonds, we can use one oxygen atom to help disconnect the other (2) and discover that we have an acetal. Both C–O bonds should therefore be

```
  {   X OMe        X .. OMe                    +
  {     OMe          G OMe    ⟹  ⟩=OMe  ⟹  ⟩=O    +    MeOH

     (1)              (2)
```

disconnected and we can label the operation '1,1-diX' to show what we mean.

We have already met one important acetal in multistriatin, the insect pheromone discussed in Chapter 1. Another is 'green leaf lilac' perfume (3). The acetal group is easily recognised and the synthesis straightforward.

Vert de lilas: *Analysis*

```
Ph   OMe      1,1-diX      Ph   CHO + 2MeOH
       OMe     ⟹
       OMe          acetal
    (3)
```

Synthesis[51]

```
                    MeOH
Ph   CHO   ──────→  TM(3)
                    H+
```

41

'Vert de lilas' is useful as an additive in soaps since acetals, unlike aldehydes and ketones, are stable to the alkali in soaps. The main use for acetals in synthesis is as protecting groups for aldehydes and ketones (see Chapter 9). Cyclic acetals (e.g. **4**) are usually used for ketones (Chapter 7): the disconnection is the same once the carbonyl carbon has been discovered.

Analysis

(4)

Synthesis

(5)

Compound (4) was to be converted into a Grignard reagent[52] and so the ketone had to be protected or it would have reacted with itself. The synthesis of chloro ketone (5) is discussed in Chapter 25, and more details of protecting groups appear in Chapter 9.

Acetals are examples of a general type of molecule (6) in which two heteroatoms are both joined to the same carbon atom. This carbon atom (• in **6**) is then at the oxidation level of a carbonyl group, and the molecule is made from a carbonyl compound and two nucleophiles.

(6)

(7)

If one of the heteroatoms is present as an OH group then only one nucleophile is involved and molecules such as cyanohydrins (7) are obviously made from carbonyl compounds and HCN. Hence hydroxy amine (8), needed[53] for a ring expansion (see Chapter 30), can be made by reduction of (9) (see Chapter 8) and hence from cyclohexanone.

(8)

Analysis

(8) $\xrightarrow{\text{FGI}}$ (9) $\xrightarrow{\text{1,1-diX}}$ + HCN

(9)

Synthesis[54]

$\xrightarrow[\text{H}^+]{\text{KCN}}$ (9) $\xrightarrow[\text{PtO}_2]{\text{H}_2}$ TM(8)
60%

When both oxygen atoms have been replaced by other groups, 1,1-disconnections may be more difficult to spot. The synthesis of α-amino acids **(10)** is important enough for a special method based on this disconnection to be memorised. With cyanide as one 'heteroatom' and nitrogen as the other, disconnection gives an aldehyde, ammonia, and cyanide. In the synthesis (known as the Strecker synthesis) the amino cyanide **(11)** is made in one step from the aldehyde and hydrolysed in acid or base.

Analysis

(10) $\xrightarrow{\text{FGI}}$ (11) $\xrightarrow{\text{1,1-diX}}$ RCHO + NH$_3$ + HCN

(10)   (11)

Synthesis

RCHO $\xrightarrow[\text{HCN}]{\text{(NH}_4)_2\text{CO}_3}$ (11) $\xrightarrow[\text{or H}^+,\ \text{H}_2\text{O}]{\text{NaOH, H}_2\text{O}}$ TM(10)

The amino acid **(12)** an analogue of the metabolite 'dopa' was needed for trial as a treatment for Parkinson's disease.[55] Strecker disconnection takes us back to ketone **(13)**, and the compound could indeed be made this way.

Analysis

(12)   1,1-diX / Strecker   (13)

Synthesis[55]

$$(13) \xrightarrow[\text{2.conc.HCl}]{\text{1.NH}_4\text{Cl, KCN}} \text{TM}(12) \quad 94\%$$

α-Halo ethers **(14)** are more obviously derived from aldehydes or ketones, the other reagents being HCl and an alcohol ROH. This is of course the same sort of reaction as would produce an acetal, but the acid has a nucleophilic counter ion (Cl⁻) and the proportions of the reagents are changed. For acetal formation a large excess of the alcohol is used and an acid such as TsOH with a non-nucleophilic counter ion.

(14)

Important examples include MeOCH₂Cl, available commercially (but a dangerous carcinogen) and compound **(15)** used by Corey[56] to introduce his 'MEM' protecting group for alcohols (see Chapter 9). The synthesis of alcohol **(16)** is discussed in the next section of this chapter.

MEM Reagent: *Analysis*

(15)   1,1-diX   (16)   + CH₂O + HCl

Synthesis[56]

$$(16) \xrightarrow{\text{CH}_2\text{O, HCl}} \text{TM}(15) \quad 88\%$$

1,2-Difunctionalised Compound

Alcohols

Compounds with heteroatoms on adjacent carbon atoms, e.g. **(17)** and **(18)**, are most helpfully considered as derivatives of alcohols. Disconnection gives the synthon **(19)**, the reagent for which is the epoxide **(20)**, the compound you would get if you tried to make **(19)** itself.

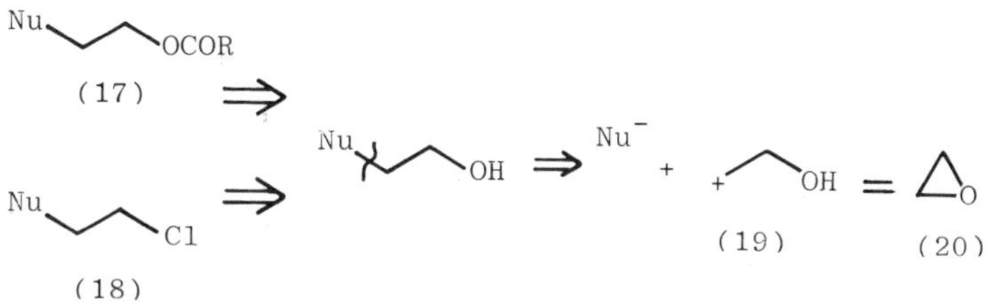

The amines **(21)** are very important members of this group as they are used as essential parts of many drug molecules, having the right balance of hydrophilic and hydrophobic properties to carry the drug molecule into the living cell. A series of anaesthetics contains various positional isomers of Proparacaine[57] **(22)** (see also Chapter 3).

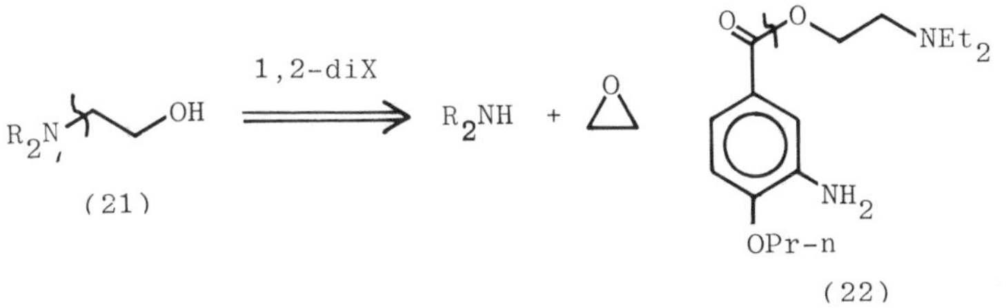

We recognise ester, ether, and amine in **(22)** but the ester is the obvious place to start. The acid **(23)** can be made by standard aromatic disconnections and the alcohol **(24)** is a 1,2-diX compound.

Proparacaine: *Analysis*

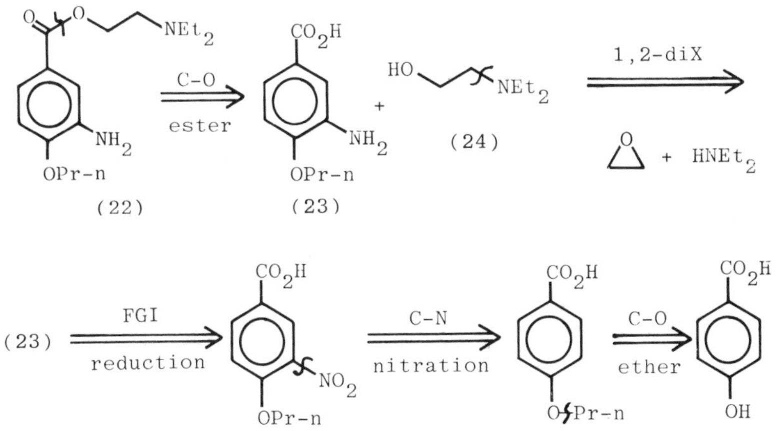

The order of events in the synthesis is chosen to minimise unwanted reactions: the free amino group in **(23)** might interfere with the esterification so reduction is kept to the end.

46

Synthesis[57]

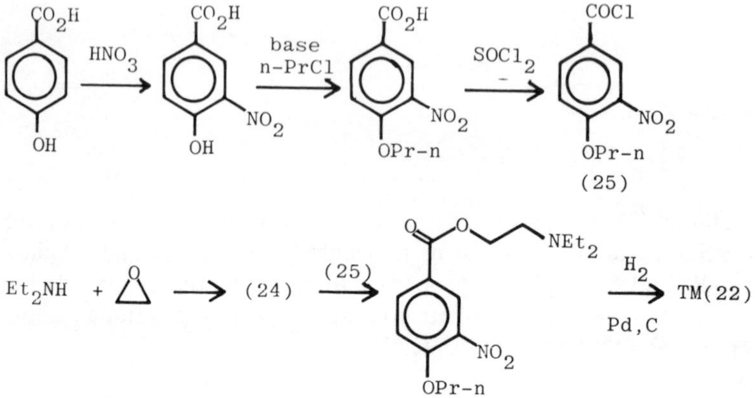

(25)

Et_2NH + [epoxide] $\longrightarrow$ (24) $\xrightarrow{(25)}$ [structure] $\xrightarrow[Pd,C]{H_2}$ TM(22)

A number of β-chloro amines have physiological activity, often anti-tumour activity, and compound **(26)** is one of these.[58] The chlorine atom must come from the alcohol **(27)** which is clearly an epoxide adduct of amine **(28)**.

Analysis 1

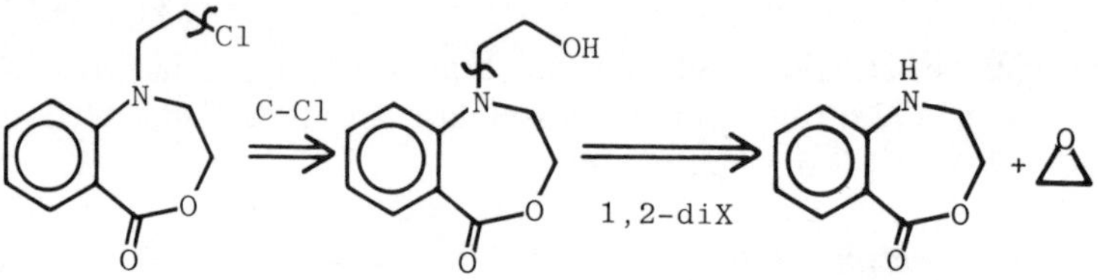

(26)　　　　　(27)　　　　　(28)

Cyclic compounds are no harder to make than open chain compounds: often they are easier because cyclisation is such a good reaction (see Chapter 7). Amine **(28)** is also an ester so we have an obvious disconnection to **(29)**, another epoxide adduct, this time of readily available anthranilic acid.

Analysis 2

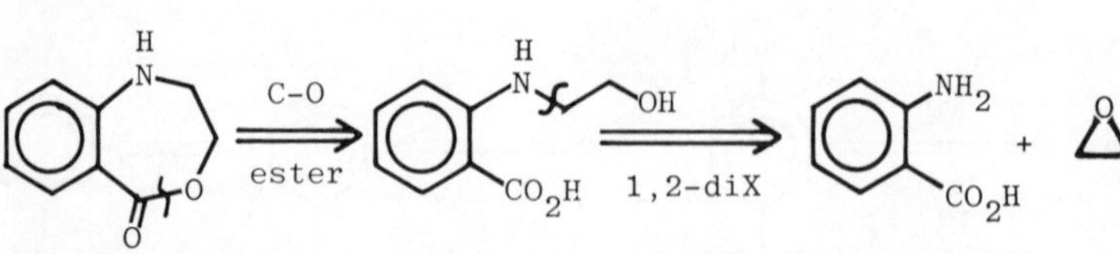

(28)

The synthesis turns out to be easier than expected as treatment of anthranilic acid with an excess of ethylene oxide gives **(27)** directly. $POCl_3$ was used instead of the more usual $SOCl_2$ to make the chloride.

Synthesis[58]

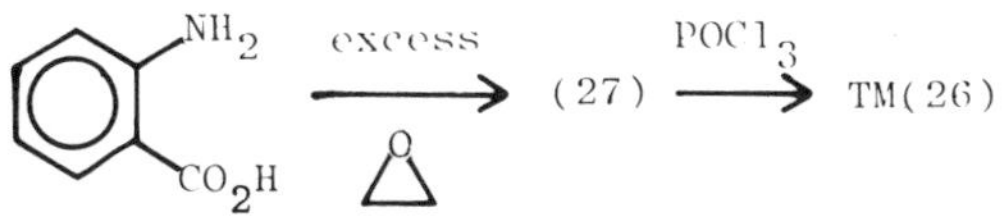

Unsymmetrical epoxides are attacked by nucleophiles at the less substituted carbon atom. Hence the compound **(29)**, needed[59] to study the Claisen rearrangement (see Chapter 34), can be disconnected as a 1,2-diX compound since the epoxide **(30)** will be attacked in the right position.

Analysis

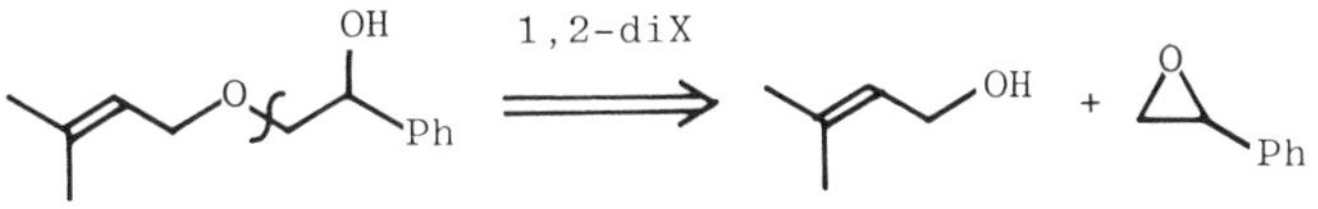

Synthesis[59]

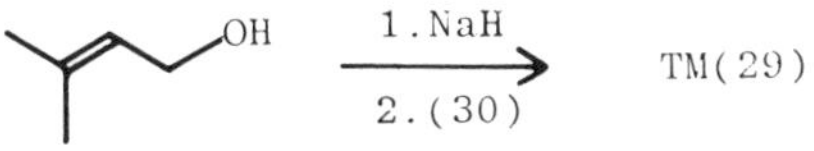

Carbonyl compounds

At a higher oxidation level **(31)**, the electrophilic synthon would be the α-carbonyl cation **(32)**, a very unstable species. The best reagents for this synthon are the α-halo carbonyl compounds, easily made (see Chapter 7) and readily attacked by nucleophiles.

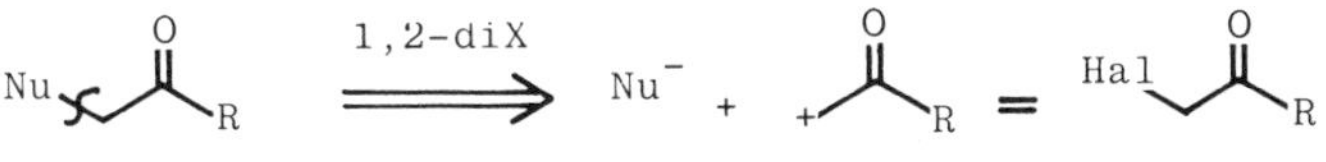

The herbicide '2,4-D' **(33)**, one of the most widely used pesticides, has an obvious disconnection by this approach leading eventually back to phenol.

2,4-D: *Analysis*

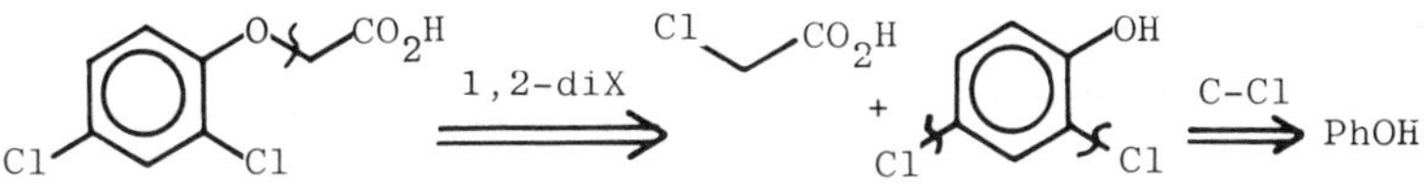

48

Synthesis[60]

Chlorination of phenol can be controlled to give mostly the 2,4-dichloro compound as each chlorine atom decreases the reactivity of the molecule towards further chlorination. The anion of the phenol will be needed for the substitution.

$$PhOH \xrightarrow[Fe]{Cl_2} \text{[2,4-dichlorophenol]} \xrightarrow[2.ClCH_2CO_2H]{1.NaOH} TM(33)$$

These α-halo carbonyl compounds are reactive enough for us to consider an alternative disconnection of certain esters (34).

Analysis

$$\text{(34)} \xRightarrow{1,2-diX} \text{(35)} + \text{[}{}^-O{-}CO{-}R\text{]}$$

(34)   (35)

The reaction is simple to carry out and is a way of making crystalline derivatives of liquid carboxylic acids for purification, identification, and protection.[61]

Synthesis

$$\text{(35)} \xrightarrow[NaHCO_3]{RCO_2H} TM(34)$$

The synthesis of the reagent (35) is described in Chapter 7.

1,3-Difunctionalised Compounds

These compounds can be disconnected only at the carbonyl oxidation level (e.g. 36) where unsaturated compounds (38) are reagents for the synthon (37). This is the Michael reaction, effective for all carbonyl compounds, cyanides, nitro compounds etc., and most nucleophiles.

$$\text{(36)} \Rightarrow Nu^- + \text{(37)} = \text{(38)}$$

(36)   (37)   (38)

Amines of type (39) can be made by reduction of cyanides (40) and these in turn by the Michael reaction. Base catalysis is required in this synthesis as RO^- is a better nucleophile than ROH.

Analysis

$$RO\diagdown\diagup\diagdown^{NH_2} \xrightarrow[\text{reduction}]{\text{FGI}} RO\diagdown\diagup^{CN} \Longrightarrow RO^- + \diagup\diagdown^{CN}$$

(39) (40)

Synthesis[62]

One example is:

$$\text{(cycloheptyl)}{-}OH \xrightarrow[\diagup\diagdown_{CN}]{\text{NaOMe}} \text{(cycloheptyl)}{-}O\diagdown\diagup^{CN} \xrightarrow[\text{Rh, Al}_2O_3]{H_2}$$

76%

$$\text{(cycloheptyl)}{-}O\diagdown\diagup\diagdown^{NH_2}$$

76%

The Grignard reagent from **(41)** has been widely used in synthesis. We recognise a bromide and an acetal: disconnecting the acetal reveals a β-bromo aldehyde available by Michael addition of Br⁻ to acrolein **(42)**

Analysis

$$Br\diagdown\diagup\diagdown\underset{(41)}{\overset{\diagdown O\diagup O\diagdown}{}}H \underset{\text{acetal}}{\overset{1,1-\text{diX}}{\Longrightarrow}} Br\diagdown\diagup\diagdown\overset{O}{\diagdown}H \overset{1,3-\text{diX}}{\Longrightarrow} Br^- \quad \diagdown\diagup\diagdown\overset{O}{\diagdown}H$$

(42)

All the simple acrylic derivatives CH₂=CH.COR (R = OH,OR,H,Me) are available commercially as they are the monomers from which acrylic polymers are made. In this synthesis we shall need acid catalysis as Br⁻ is a very weak nucleophile.

Synthesis[63]

$$(42) \xrightarrow{\text{HBr}} Br\diagdown\diagup\diagdown\overset{O}{\diagdown}H \xrightarrow[H^+]{\text{HO} \quad \text{OH}} TM(41)$$

To disconnect a 1,3-diX compound not at the carbonyl oxidation level we must first alter the oxidation level by FGI. If the TM contains no oxygen substituent, we must produce one by substitution. Japanese chemists[64] wished to study the stereochemistry of the Friedel–Crafts alkylation and chose to make optically active **(43)** for this purpose. Disconnection of the ester reveals a 1,3-diX relationship, and adjustment of the oxidation level to the acid **(44)** takes us back to simple starting materials.

50

Analysis

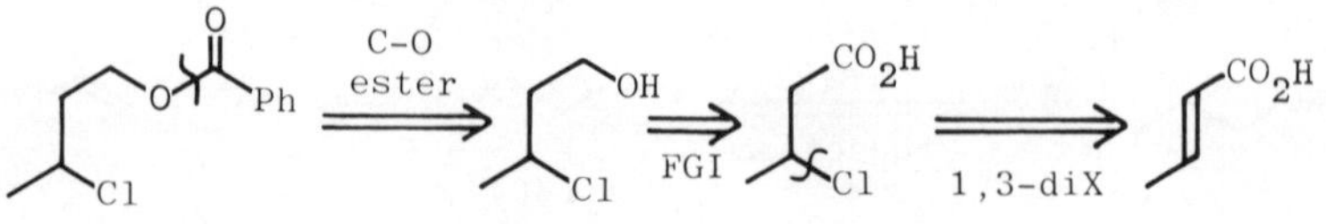

(43)   (44)

Another advantage of using the acid **(44)** is that it can be resolved into optical isomers before making the target molecule (see Chapter 12 for a more detailed discussion of resolutions).

Synthesis[64]

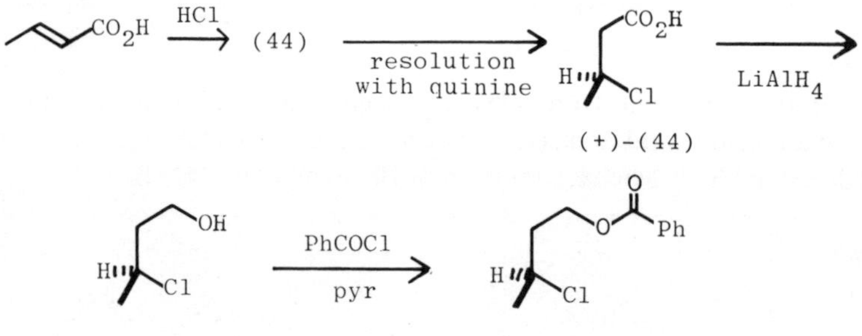

(+)-(44)

(+)   (+)-TM(43)

We have assumed in this chapter that nucleophiles add to the C–C double bond and not directly to the carbonyl group. This question is explored more fully in Chapter 14.

CHAPTER 7

Strategy III: Reversal of Polarity, Cyclisation Reactions, Summary of Strategy

This chapter collects two general strategic points which emerge from the discussion of C–X disconnections and adds a summary of the stage we have reached in our overall planning of syntheses.

Reversal of Polarity:
The Synthesis of Epoxides and α-Halo Carbonyl Compounds

In Chapter 6 we needed four types of synthons ((1)–(4)), see Table 7.1). The synthons for the 1,1- and 1,3-diX relationships could be turned into reagents simply by using the natural electrophilic properties of ketones and enones — the atoms marked + in (1) and (4) are naturally electrophilic. Synthons (2) and (3) could not be expressed as reagents so easily: indeed (3) is so unlikely an intermediate that it cannot be made.

Table 7.1 Synthons for 1,n-diX synthesis

2-Group relationship	Synthon	Reagent
1,1	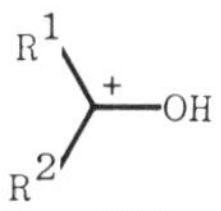 $R^1R^2C^+$–OH (1)	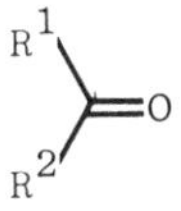 $R^1R^2C{=}O$
1,2	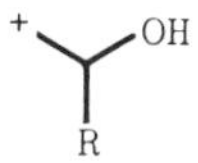 ^+CR–CH$_2$OH (2)	 epoxide (R)
1,3	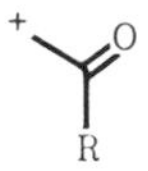 ^+CR–CH$_2$–C(=O) (3)	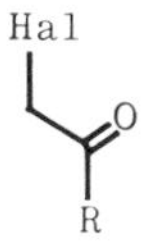 Hal–CH$_2$–C(=O)R
1,3	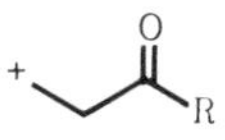 $^+CH_2$–CH$_2$–C(=O)R (4)	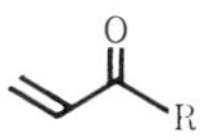 CH$_2$=CH–C(=O)R

51

We solved the problem by using a three-membered ring (epoxide) for **(2)** and an α-halo carbonyl compound for **(3)**: two apparently different devices which in fact rely on the same principle. The atoms marked + in **(2)** and **(3)** are easily made nucleophilic—by enolisation (i) for example—and the common principle is to use a preliminary nucleophilic attack on a heteroatom to reverse the natural polarity of the atom from nucleophilic to electrophilic. Halogenation (i, E = Br) of a ketone provides a reagent for **(3)** and epoxidation (ii) of an alkene provides the reagent for **(2)**.

(i)

(ii)

Epoxide **(5)**, used in Chapter 6, is made from readily available styrene **(6)**—the monomer of polystyrene. Commercially available *meta*-chloroperbenzoic acid (MCPBA) **(7)** is often used for such epoxidations.

Halogenation of ketones

The α-halogenation of ketones (Chapter 5) must be carried out in acid solution to avoid polyhalogenation. Hence the full analysis of the reagent **(8)**, used in Chapter 6 to make derivatives of carboxylic acids, is simple providing that we notice the directing effects of the two groups on the benzene ring and disconnect by Friedel–Crafts first.

Analysis

Synthesis[65]

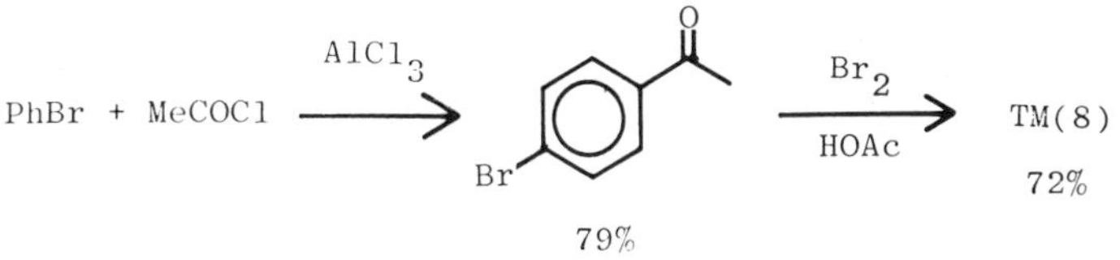

This bromination was unambiguous as the ketone could enolise on one side only. In general the reaction is suitable only for ketones which are symmetrical (e.g. **9**),[66] blocked on one side (e.g. **10**),[67] or which enolise specifically (e.g. **11**)[68] (see Chapters 13 and 19).

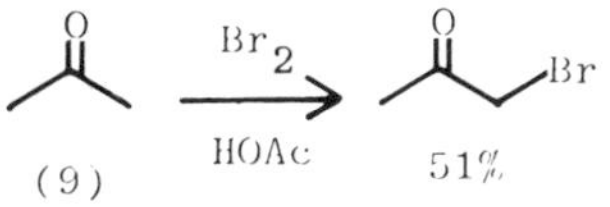

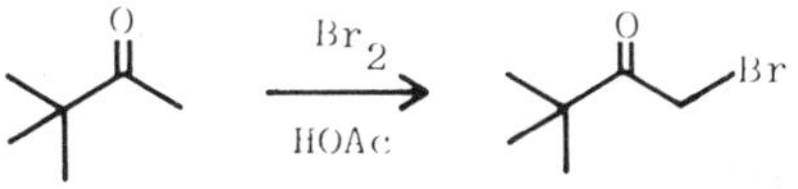

Halogenation of acids

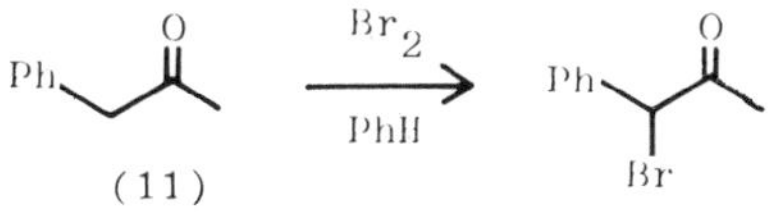

There is no ambiguity in the halogenation of acids which can of course enolise in one direction only. A reliable method is bromination with red phosphorus catalyst via the acid bromide*. This leads directly to α-bromo esters.

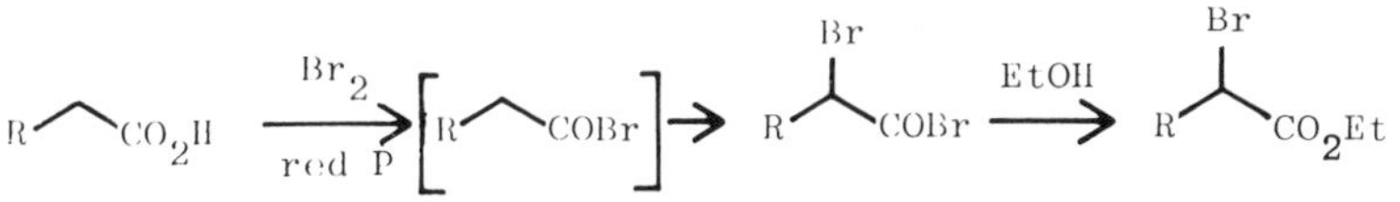

α-Chloro acids are also available, often commercially, from chlorination of acids, and chloroacetyl chloride (**12**) is made industrially[70] from acetic acid.

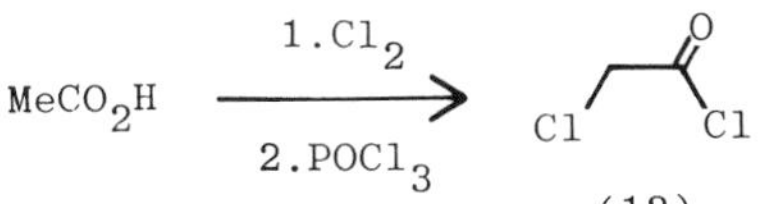

*The α-bromo acid may be made from the acid by bromination with bromine and PCl₃.[69]

54

The α-chloro amide (13) needed to synthesise some analeptic tetrazoles[71] is best treated as an amide since the acid chloride is available chloracetyl chloride.

Analysis

It is best to acetylate the amine before nitration: this ensures mononitration and increases the size of the amino group so that more *para* product is formed (see Chapter 2). Amines are hard[72] basic nucleophiles and so (14) will attack the carbonyl group of (12) and not the α-carbon.

Synthesis[71,73]

Cyclisation Reactions

The ease of ring formation helped us to make five-membered cyclic acetals from ketones and a seven-membered heterocyclic compound (see Chapter 6). In this chapter we have seen how three-membered epoxides are formed. Ring formation is generally preferred to bimolecular reactions forming open chain compounds providing that the ring is three-, five-, six-, or seven-membered. Four-membered ring are a special case discussed in Chapter 29 where all these points are developed more fully.

The synthesis of morpholines (15) is an important example of easy cyclisation. Substituted morpholines are often found as parts of drug molecules — the analgesic Phenadoxone[74] **(16)** is an example. Compound **(15)** is both an amine and an ether. Disconnection of the ether is easier, and we preserve the symmetry by writing the diol as intermediate (see Chapter 4). This is clearly an adduct of an amine and two molecules of ethylene oxide.

Analysis

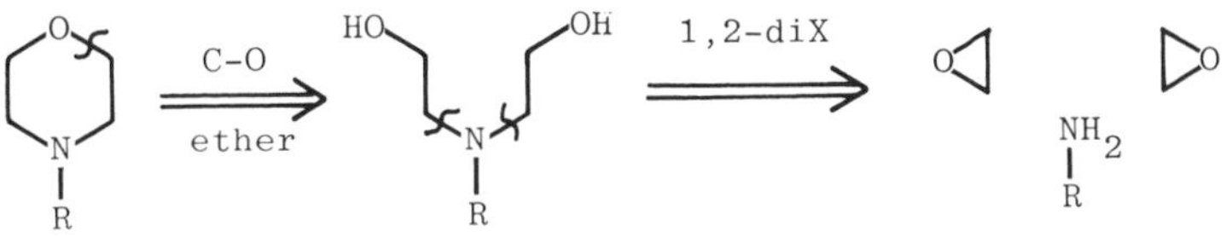

The synthesis involves a cyclisation, so it will not be necessary to turn one alcohol into a tosylate or halide: simple acid treatment will catalyse the formation of the stable six-membered ring.

Synthesis[75]

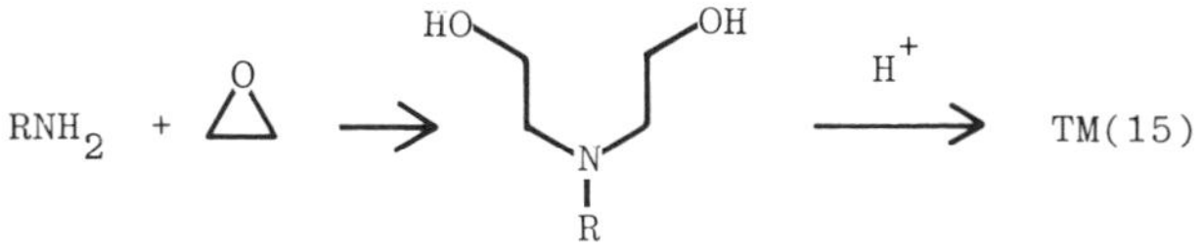

Because cyclisation is so easy, it can help us find better routes to compounds by alternative strategies. The obvious disconnection on the cyclic ether **(17)** requires an *ortho*-disubstituted benzene which could no doubt be made, but so too would much more of the *para* compound.

Analysis 1

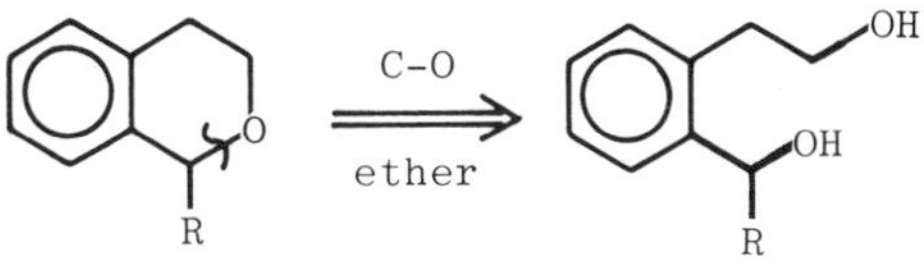

The alternative Friedel–Crafts disconnection requires **(18)** — a rather unstable compound. This is hardly a problem as **(18)** is unstable because it cyclises readily to **(17)**. The reaction occurs at 35°C and **(18)** need not be isolated.

Analysis 2

+ RCHO + HCl

(17)   (18)

Synthesis[76]

Summary of Strategy

In Chapter 1 we set out the outlines of a routine to design a synthesis. We can now fill in the picture by adding the main points from Chapters 2-7.

Analysis

1. Recognise the functional groups in the target molecule.
2. Disconnect by known reliable methods, using FGI if necessary to produce the right FG for disconnection. Disconnect:
   (a) bonds joining an aromatic ring to the rest of the molecule, whether Ar–C or Ar–X (Chapters 2 and 3).
   (b) any C–X bond (Chapter 4) especially
      (i) bonds next to carbonyl groups RCO–X (Chapter 4);
      (ii) two-group disconnections of 1,1-, 1,2-, or 1,3-diX (Chapter 6);
      (iii) bonds within rings as cyclisation reactions are favourable (Chapter 7).
3. Repeat as necessary to reach available starting materials.

Synthesis

1. Write out the plan according to the analysis, adding reagents and conditions.
2. Check that a rational order of events has been chosen (Chapter 3).
3. Check that aspects of chemoselectivity (Chapter 5) have been taken into account, in particular that unwanted reactions will not occur elsewhere in the molecule. If necessary use protecting groups (see Chapter 9).
4. Modify the plan according to 2 and 3 and to unexpected failures or successes in the laboratory.

We shall develop this routine as the book progresses.

Example: Salbutamol

The asthma drug salbutamol **(19)** is closely related to adrenaline **(20)**. The extra carbon atom (• in **19**) prevents dangerous side effects on the heart and the *t*-butyl group makes the drug longer-lasting. Salbutamol has three hydroxyls and an amine functional group but the amine is part of the only

(19.)   (20)

1,2-diX relationship in the molecule. This disconnects to epoxide **(21)**, and this approach is successful (Chapter 30) but involves methods we have not yet discussed. An alternative is to use FGI first, and operate at the carbonyl oxidation level, going back to the α-bromo ketone **(22)** which can be made from the ketone itself **(23)**.

Salbutamol: *Analysis 1*

Analysis 2

Ketone **(23)** is clearly made by a Friedel–Crafts reaction, but how are we to make the starting diol **(24)**? Looking back at our guidelines for aromatic synthesis (Chapter 3) we find that a good strategy is to use available starting materials with such *ortho* substitution patterns already established and an obvious candidate here is salicylic acid **(25)**.

Analysis 3

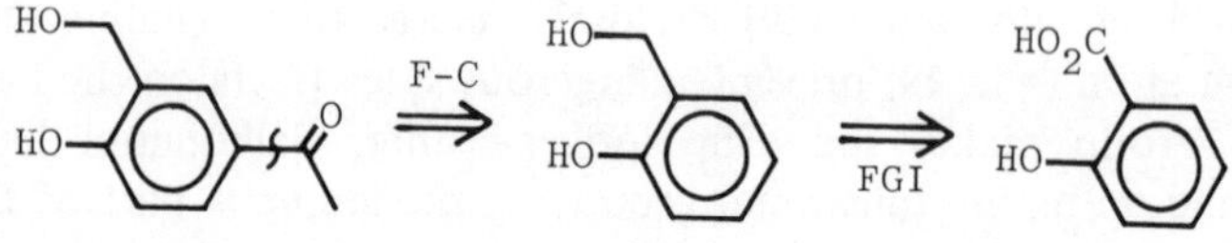

(23)　　　　　　(24)　　　　　　(25)

The conversion of salicylic acid to ketone **(26)** was discussed in Chapter 3.

Synthesis 1

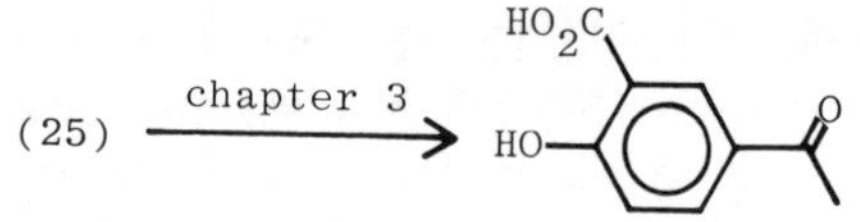

(26)

It will save a step if we reduce both the acid and the ketone together at the end. The complete plan is now:

Synthesis 2

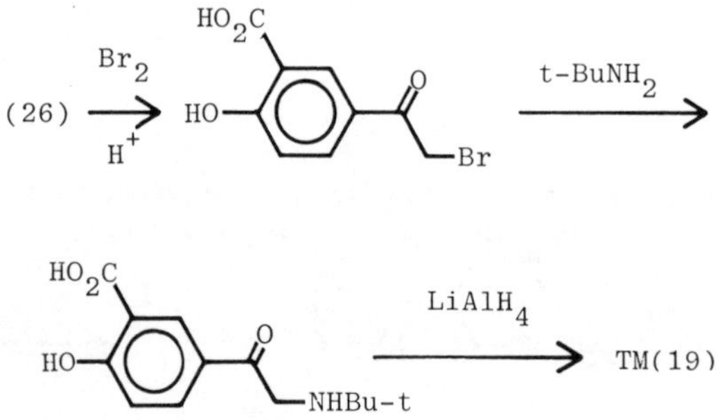

(27)

Checking this for chemoselectivity problems it is possible that the NH group in **(27)** might react again with the bromoketone or elsewhere and it is better to block it with a protecting group. These will be discussed in Chapter 9, where the benzyl group, removable by hydrogenation, will be introduced. In the laboratory, it proved better to brominate **(26)** in neutral solution and the final scheme was as follows.

Synthesis 3[77]

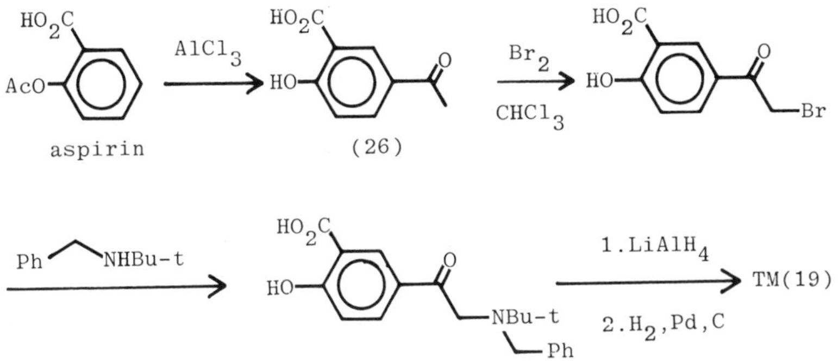

This synthesis is short and high yielding, makes good use of the strategic points noted so far in this book, and it introduces the subjects of the next two chapters: amine synthesis and the use of protecting groups.

CHAPTER 8

Amine Synthesis

The synthesis of amines warrants a separate chapter because the C-X disconnection (1a) used for ethers, sulphides, and the like is not satisfactory. The problem is that the product (1) of the reaction is at least as reactive as the starting material (2) (if not more so because of the inductive effect of the methyl group) and reacts further to give (3) and even (4).

Analysis

$$R-NH \!\!\!\not\,\,Me \overset{C-N}{\Longrightarrow} RNH_2 \ + \ MeI \qquad NOT \ useful$$

$$(1a)$$

Synthesis

$$RNH_2 \overset{MeI}{\longrightarrow} RNHMe \overset{MeI}{\longrightarrow} RNMe_2 \overset{MeI}{\longrightarrow} R\overset{+}{N}Me_3 \ I^-$$

$$(2) \qquad (1) \qquad (3) \qquad (4)$$

It is no use adding only one equivalent of MeI since the first molecule of (1) formed in the reaction will compete with (2) for MeI.

This reaction (alkylation of an amine with an alkyl halide) can occasionally be used if the product is *less* reactive than the starting material for electronic (e.g. 5) or steric (e.g. 6) reasons, or if it is intramolecular. This second example (6) is from the Salbutamol synthesis discussed in Chapter 7. Unless you can see a specific reason for success, avoid the reaction.

$$NH_3 \ + \ Cl\overset{\frown}{}CO_2h \longrightarrow H_2N\overset{\frown}{}CO_2H \rightleftharpoons H_3\overset{+}{N}\overset{\frown}{}CO_2^-$$

$$(5)$$

$$(6)$$

60

The general answer to this problem is to avoid alkyl halides and to use instead electrophiles which give relatively unreactive products with amines. The best examples are acyl halides, aldehydes, and ketones. The products, amides (7) and imines (9), can be reduced to amines.[78] The amide method inevitably produces a CH_2 group (8) next to the nitrogen atom, but the imine route is suitable for amines with branched chains (10).

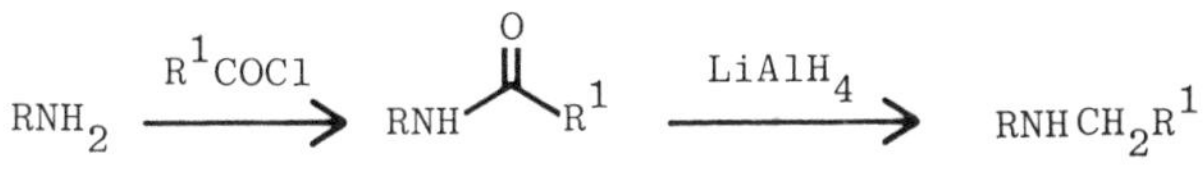

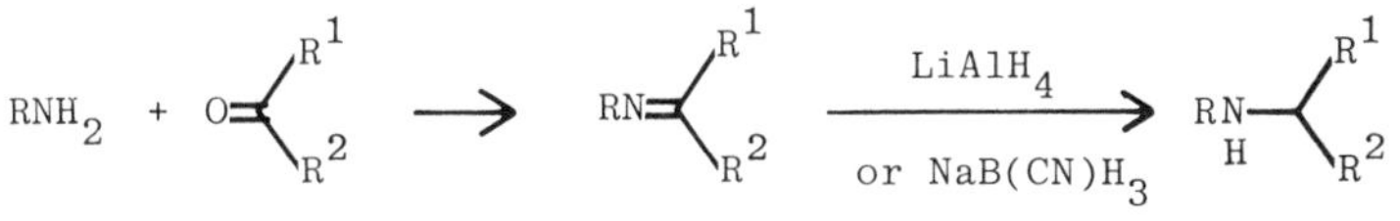

A preliminary FGI is therefore needed before we apply the C–N disconnection. Amine (11) could be disconnected by either method. It has been synthesised[79] by route (b), the reduction being carried out without isolating the imine. No doubt the amide route would be equally successful.

Analysis

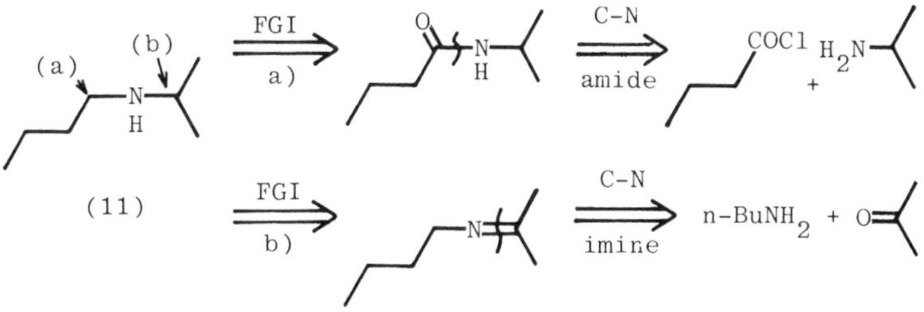

Synthesis[79]

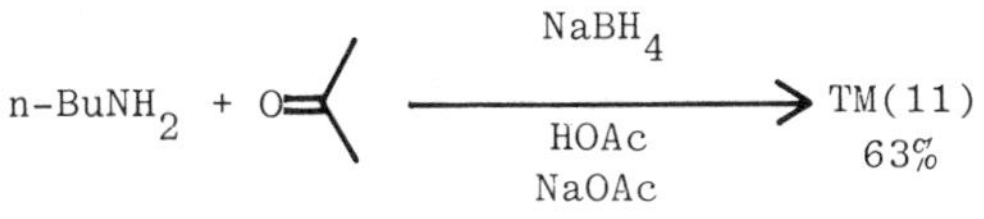

An example more suited to the amide approach is the cyclic amine (12). We choose the exocyclic CH_2 group as a site for FGI since the cyclic amine piperidine (13) is readily available.

Analysis

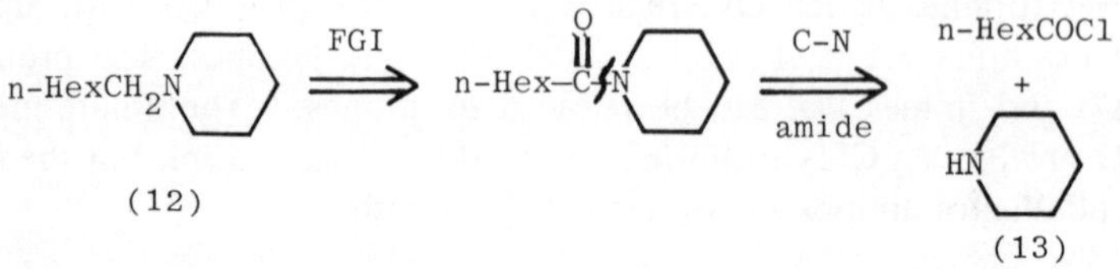

(12)

(13)

Synthesis[80]

Catalytic reduction was used in the published synthesis: LiAlH$_4$ would be a commoner choice nowadays.

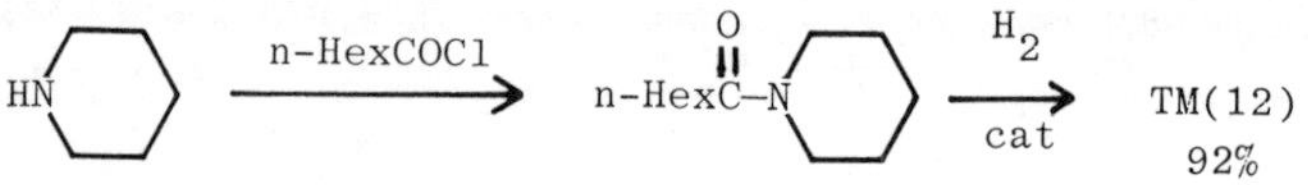

TM(12)
92%

Primary Amines RNH$_2$

Unsubstituted imines (14) are unstable and cannot usually be made in good yield, but primary amines can still be made in a one step reductive amination in which the imine is not isolated.

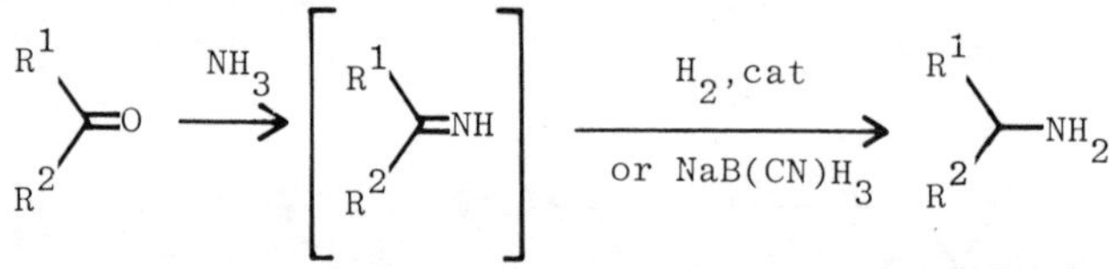

(14)

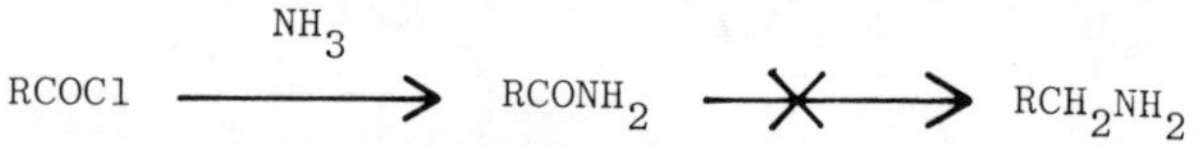

(15)

Primary amines are not usually made by reduction of amides (15) but by other reductive processes which are minor variations on this scheme. For unbranched amines (16) we can reduce cyanides.[81] This method is especially suitable for benzylic amines since aryl cyanides (17) can be made from diazonium salts (see Chapter 2), and for the homologous amines[82] (18) since cyanide ion reacts easily with benzyl halides.

$$RBr \xrightarrow{KCN} RCN \xrightarrow[\text{or } LiAlH_4]{H_2, PtO_2, H^+} RCH_2NH_2$$

(16)

$$ArNH_2 \xrightarrow[\text{2.Cu(I)CN}]{\text{1.HONO}} ArCN \xrightarrow[H^+]{H_2, Pd\text{-}C,} ArCH_2NH_2$$

(17)

Ph∼Cl $\xrightarrow{CN^-}$ Ph∼CN $\xrightarrow[AlCl_3 \ Et_2O]{LiAlH_4}$ Ph∼∼NH$_2$

(18) 83%

Disconnection again requires a preliminary FGI. We have already met examples in Chapter 6.

For branched chain primary amines (20), oximes (19) are good intermediates since they can be made easily from ketones and reduction cleaves the weak N-O bond as well as reducing the C-N bond. FGI is again required before disconnection.

$$\underset{R^2}{\overset{R^1}{>}}{=}O \xrightarrow[NaOAc]{NH_2OH.HCl} \underset{R^2}{\overset{R^1}{>}}{=}N\text{-}OH \xrightarrow[\text{or } H_2, cat]{LiAlH_4} \underset{R^2}{\overset{R^1}{>}}{-}NH_2$$

(19)　　　　　　　　(20)

The synthesis of Fenfluramine (21), a drug acting on the central nervous system, illustrates two amine disconnections. The ethyl group can be removed by the amide method leaving the branched chain primary amine (22) available from the ketone (23) by the oxime method.

Fenfluramine: *Analysis*

(21) $\xrightarrow{FGI}$ ... $\xrightarrow[\text{amide}]{C\text{-}N}$

(22) $\xrightarrow{FGI}$ ... $\xrightarrow[\text{oxime}]{C\text{-}N}$ (23)

Synthesis[83]

Neither oxime nor amide need be isolated—the published synthesis uses different methods of reduction in the two cases, no doubt developed by experiment.

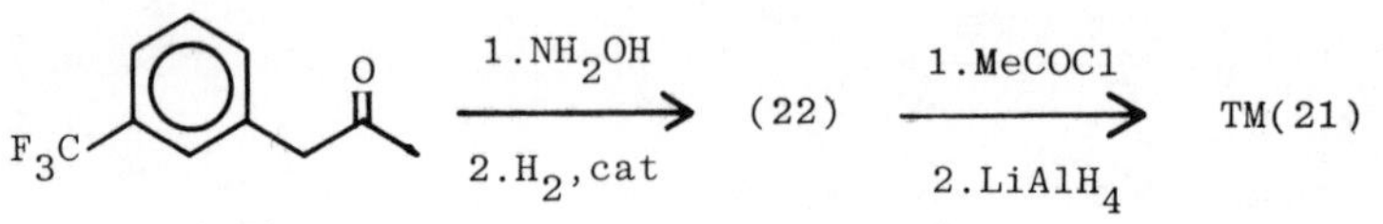

The alkylation and reduction of aliphatic nitro compounds is one route to *t*-AlkNH$_2$ and is discussed in Chapter 22. Another route uses the Ritter reaction followed by hydrolysis of the amide.

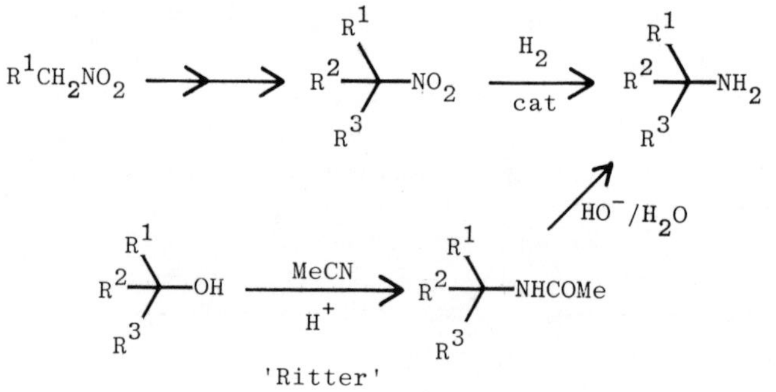

Other Routes to Amines using Reduction

We have already seen that aromatic amines are made by the reduction of nitro compounds (Chapter 2) and aliphatic nitro compounds can be used in the same way (Chapter 22).

Azides **(24)** can also be reduced to amines:[78, 84] the importance of this method is that the azide ion, N$_3^-$, acts as a reagent for NH$_2^-$, so that the disconnection is the normal one for C–X bonds. Other reagents for this synthon are discussed in the next section.

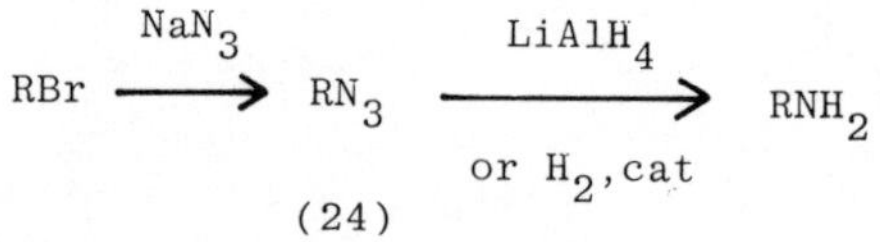

Amines of type **(25)** can therefore be made by reduction of azides **(26)** which can be derived from epoxides and azide ions.

Analysis

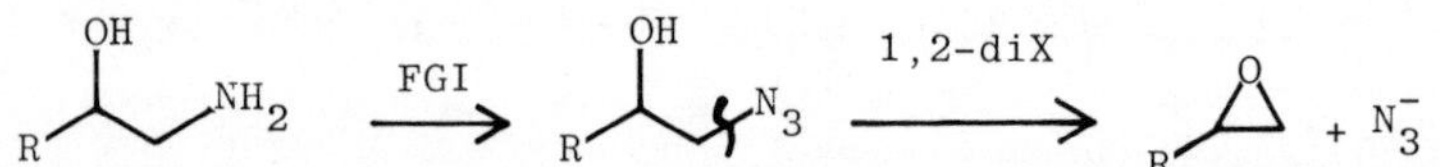

Synthesis[85]

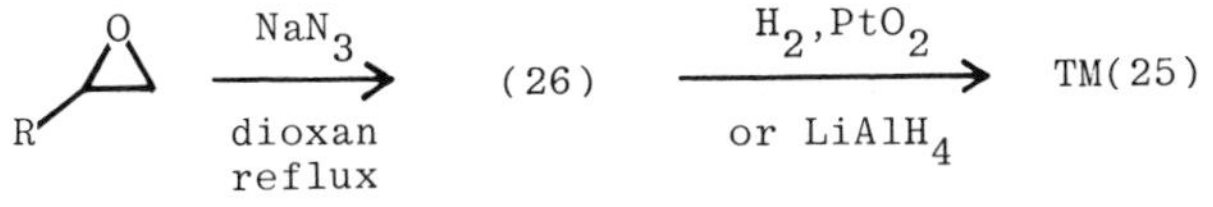

Reagents for the Synthon NH₂⁻

Though NH_2^- can be made ($NaNH_2$ is commercially available) it is very basic and normally attacks a proton (causing elimination) rather than displacing a halide ion. There are, however, several reagents available to take the role of this synthon, of which phthalimide ion **(27)** is probably the best known.

Phthalimide is available from phthalic anhydride and the two carbonyl groups both help to stabilise the anion **(27)** so that the potassium salt is a stable compound.[86] This anion **(27)** is blocked so that it can react only once with an alkyl halide forming the substituted phthalimide **(28)** which can be cleaved by hydrazine (NH_2NH_2) to release the primary amine **(29)**.

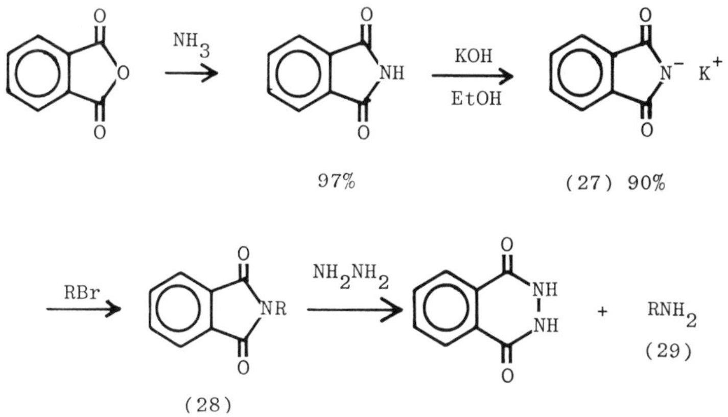

The phthalimide method can be viewed as the use of a protecting group — these are more fully discussed in the next chapter.

CHAPTER 9

Strategy IV: Protecting Groups

Protecting groups have been mentioned occasionally in previous chapters: in this chapter the ideas behind their use are systematically presented. Protecting groups allow us to overcome simple problems of chemoselectivity (Chapter 5). It is easy to make alcohol (2) from keto ester (1) by reducing the more reactive carbonyl group. Making alcohol (3) by reducing the *less* reactive carbonyl group is not so easy but can be accomplished by using the greater reactivity of the ketone to introduce a *protecting group* which does not react with LiAlH$_4$, the chosen reducing agent. The obvious protecting group to use is the acetal.

Synthesis

A protecting group must be:

1. easy to put in and remove;
2. resistant to reagents which would attack the unprotected functional group;
3. resistant to as wide a variety of other reagents as possible.

In the synthesis of (3), the acetal is easily made and easily hydrolysed, both in good yield, resists reagents such as bases, nucleophiles, and reducing agents which would attack the unprotected ketone, and resists LiAlH$_4$ when it attacks the ester. Protecting groups[87] are available nowadays for all functional groups. Table 9.1 gives some of the more important. We have already seen examples of some of these in action:

66

Chapter 5: benzyl chloroformate to protect amines;
Chapter 5: thiouronium salts to protect thiols;
Chapter 6: further discussion of the acetal protecting group;
Chapter 6: devising a new protecting group;
Chapter 7: benzyl as a protecting group in amine synthesis;
Chapter 8: many examples of amides etc. as protecting groups in amine synthesis.

In this chapter I shall select examples to illustrate other types of protecting groups and more examples will appear throughout the rest of the book.

Acetals can also be used to protect diols using a readily available carbonyl compound such as acetone or benzaldehyde—as in the synthesis of the Salbutamol intermediate (4) (see Chapters 3 and 7). The bromine atom is in the right position for immediate disconnection, and the starting material (5) comes in one step from salicylic acid (strategy of Chapter 3).

Analysis

(4)   (5)   salicylic
                                         acid

Bromination might well oxidise the reactive benzylic alcohol in (5) so an acetal protecting group is added. This could be removed to give (4), but as a protecting group was needed later in the synthesis, it was left in place.

Synthesis[88]

It is more surprising that acetals can be used to protect simple alcohols as well as diols. Two of the best are the 'THP' (6) and 'MEM' (7) derivatives. These are often preferred to ethers (Table 9.1) because they can be removed under such mild conditions. The reagent for the MEM protecting group is discussed in Chapter 6.

Table 9.1 Protecting groups

Group	Protecting group (PG)	To add	To remove	PG resists	PG reacts with
Aldehyde RCHO	Acetal RCH(OR')$_2$	R'OH, H$^+$	H$^+$/H$_2$O	Nucleophiles, bases, reducing agents	Electrophiles, oxidising agents
Ketone	Acetal (ketal)		H$^+$/H$_2$O	Nucleophiles, bases, reducing agents	Electrophiles, oxidising agents
Acid RCO$_2$H	Ester RCO$_2$Me RCO$_2$Et RCO$_2$CH$_2$Ph RCO$_2$Bu-t RCO$_2$ CH$_2$CCl$_3$	CH$_2$N$_2$ EtOH/H$^+$ PhCH$_2$OH/H$^+$ H$^+$, t-BuOH Cl$_3$CCH$_2$OH	HO$^-$/H$_2$O H$_2$, cat. or HBr H$^+$ Zn, MeOH	Weak bases, electrophiles	Strong bases, nucleophiles, reducing agents
	Anion RCO$_2^-$	Base	Acid	Nucleophiles	Electrophiles
Alcohol ROH	Ethers ROCH$_2$Ph	PhCH$_2$Br base	H$_2$, cat. or HBr	Electrophiles, bases, oxidation	HX (X is a nucleophile)
	Acetals THP		H$^+$/H$_2$O	Bases	Acids
	MEM	See Chapter 6	ZnBr$_2$	Bases	Acids
	Esters RCO$_2$R'	R'COCl pyridine	NH$_3$, MeOH	Electrophiles, bases, oxidation	Nucleophiles
Phenol ArOH	Ether ArOMe	Me$_2$SO$_4$ K$_2$CO$_3$	HI, HBr, or BBr$_3$	Bases, weak electrophiles	Attack by electrophiles on ring

			Bases, weak electrophiles	Attack by electrophiles on ring
Amine RNH$_2$	Acetal ArOCH$_2$OMe	MeOCH$_2$Cl Base	HOAc, H$_2$O	Bases, weak electrophiles	Attack by electrophiles on ring
	Amides RNHCOR′	R′COCl	HO⁻/H$_2$O H⁺/H$_2$O	Electrophiles	
	Urethanes RNHCO.OR′	Chloroformates; R′OCOCl See Chapter 5	R′ = CH$_2$Ph H$_2$, cat. or HBr; R′ = Bu-t H⁺	Electrophiles	Bases, nucleophiles
	Phthalimides	Phthalic anhydride see Chapter 8	NH$_2$NH$_2$	Electrophiles	Bases, nucleophiles
Thiols RSH		See Chapter 5	HO⁻/H$_2$O	Electrophiles	Oxidation
	AcSR	RSH + AcCl + base, see Chapter 9	HO⁻/H$_2$O	Electrophiles	Oxidation

$$ROH + \quad \underset{O}{\bigcirc} \quad \xrightarrow{H^+} \quad \underset{RO \quad O}{\bigcirc} \quad = ROTHP$$

(6)

$$ROH + Cl\diagup\diagdown O\diagup\diagdown\diagup OMe \xrightarrow{Et_3N} RO\diagup\diagdown O\diagup\diagdown\diagup OMe = ROMEM$$

(7)

$$\underset{\overset{|}{OH}}{\diagup}(CH_2)_n CH_2^+ \qquad \underset{\overset{|}{OTHP}}{\diagup}(CH_2)_n CH_2 Br$$

(8)   (9)

The THP protecting group has been used in reagents for the synthons **(8)**. The bare OH group would react to give a cyclic ether so the protected bromo alcohols **(9)** are the chosen reagents. They are usually made from the keto esters **(10)**.

Synthesis (for $n = 3$)[89]

$$\underset{(10)}{\overset{O}{\diagup}\diagdown\diagup CO_2Et} \xrightarrow{NaBH_4} \underset{OH}{\diagup}\diagdown\diagup CO_2Et \xrightarrow[H^+]{\underset{O}{\bigcirc}}$$

$$\underset{OTHP}{\diagup}\diagdown\diagup CO_2Et \xrightarrow{LiAlH_4} \underset{OTHP}{\diagup}\diagdown\diagup OH \xrightarrow[CBr_4]{PPh_3} \underset{OTHP}{\diagup}\diagdown\diagup Br$$

(11)

The reagent (PPh$_3$ + CBr$_4$) used to convert OH to Br is one which takes advantage of the affinity of phosphorus for oxygen. The other product is Ph$_3$PO. Reagent **(11)** has been used in a synthesis[90] of the enzyme inhibitor diplodialide-A.

When a molecule contains several similar functional groups, it may be necessary to protect them each in a different way so that each may be removed under different conditions. In Corey's gibberellic acid synthesis[91] he needed the triol **(15)** with only one free hydroxyl group. Compound **(12)** was available (Chapter 35) so he protected the free OH group with the easily removed MEM group[92] to give **(13)**, ozonised the double bond, introduced and protected a new OH group, and removed the MEM group. Notice that the three OH groups in **(14)** are each protected in a different way.

Synthesis

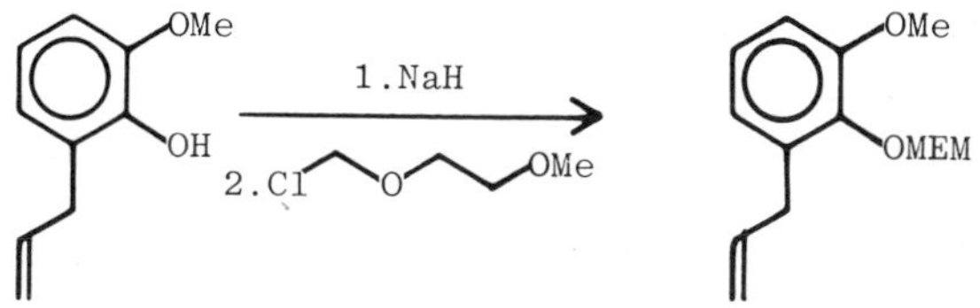

(12) → (13) 75%

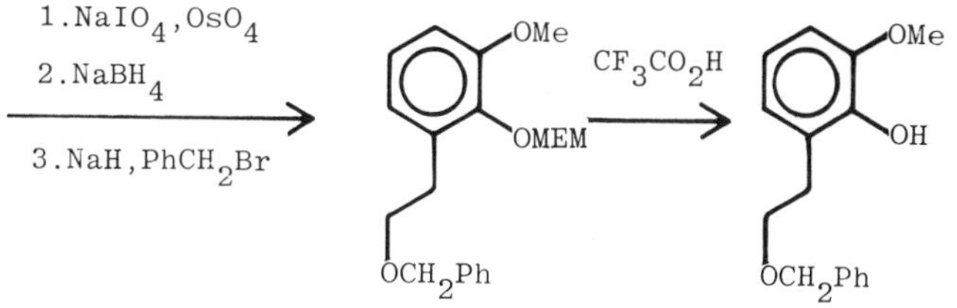

1.NaH

2.Cl~~~O~~~OMe

1.NaIO$_4$,OsO$_4$

2.NaBH$_4$

3.NaH,PhCH$_2$Br

CF$_3$CO$_2$H

(14)

(15)
74% from (13)

Alcohols can also be protected as esters but these are more widely used as protecting groups for acids. Table 9.1 gives four different types of ester, each removed under quite different conditions. Normal esters (e.g. RCO$_2$Et) can be hydrolysed in aqueous acid or base under fairly vigorous conditions. *t*-Bu esters are very easily hydrolysed in acid by a special mechanism (a$_{AL}$1) via the *t*-butyl cation (i). If hydrolysis even under mild conditions is too vigorous, a benzyl ester (like a benzyl ether or amine) can be cleaved by hydrogenolysis (ii). Even hydrogenolysis might attack an alkene elsewhere in the molecule so the trichloroethyl ester could be used: this is cleaved by zinc metal (iii). One of these sets of conditions ought to suit all cases.

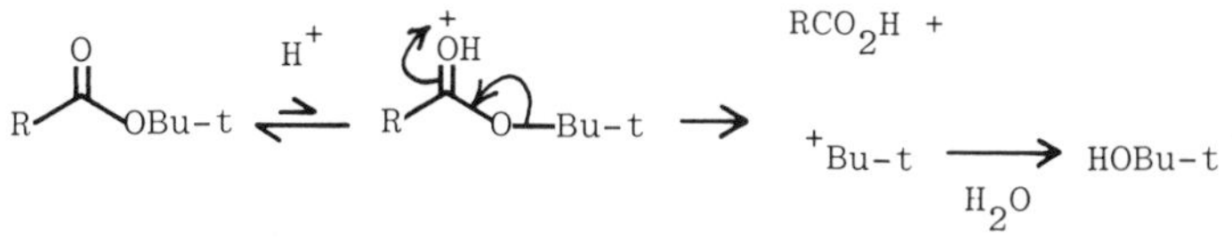

RCO$_2$H +

$^+$Bu-t $\longrightarrow$ HOBu-t (i)

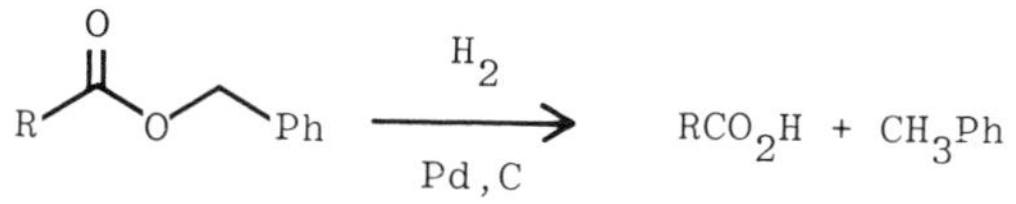

H$_2$O

H$_2$

Pd,C

RCO$_2$H + CH$_3$Ph (ii)

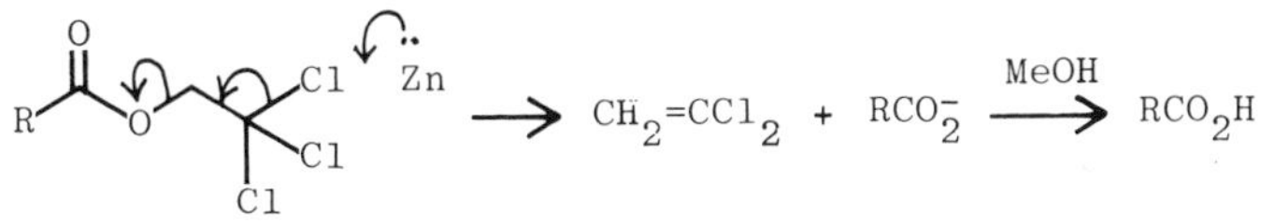

Zn $\longrightarrow$ CH$_2$=CCl$_2$ + RCO$_2^-$ $\xrightarrow{\text{MeOH}}$ RCO$_2$H (iii)

Protecting groups for carboxylic acids are vital in peptide synthesis. The problem is obvious if you consider even dipeptide synthesis. The dipeptide ester Asp–Phe–OMe **(16)** is a sweetening agent 150 times sweeter than cane sugar.[93] Only one disconnection is reasonable—if only because the individual amino acids Asp and Phe are readily available. But how can we make this particular amino group combine with that particular CO_2H group? The dimers Asp-Asp and Phe-Phe can be formed as well as the 'wrong' product Phe-Asp, and Asp has a second CO_2H group too. Protection is the answer.

Analysis

(16)

Asp

Phe

The Asp can be fully protected, the CO_2H groups as esters and the NH_2 group as a urethane, and the CO_2H of Phe similarly protected, leaving only the NH_2 group of Phe free. The problem is then to release the α-CO_2H of Asp while retaining the protection on the other CO_2H group. Peptide chemists know that this can be done by gentle alkaline hydrolysis. The non-expert cannot be expected to know this, though he might spot the possibility of neighbouring group participation in the α-ester hydrolysis. He would look up the conditions in the literature.

Protection of Asp[94]

L–Asp

96%

(17) 65%

Protection of Phe[95]

(18)

We still need to activate the free CO_2H group in (17) and we must make it more reactive than a simple ester since the NH_2 group will react with an ester given the right conditions (Chapter 4). The peptide chemists[93] here chose to use the reactive trichlorophenyl ester (19) for this purpose, though there are many alternatives. After (18) and (19) are coupled, all that remains to be done is to remove the protecting groups: these have been chosen as benzyl groups so that all can be removed in one step by hydrogenolysis.

Synthesis[93]

(19) 88%

96%

Note that the strategically poor, but essential, partial hydrolysis to give (17) is the only low yielding step. Other yields are very high so there is little loss in adding or removing protecting groups.

Although protecting groups are essential in peptide synthesis as in many other fields, the fact remains that their use is a confession of failure by the chemist as extra unproductive steps are needed to add and remove the protecting groups. Sometimes a step can be saved by combining the removal of a protecting group with some other more productive step.

In the synthesis of the heterocycle (20), compound (21) was needed[96] as an intermediate. Analysis is straightforward as it is a 1,3-diX compound (Chapter 6). We have already seen that it is necessary (Chapter 5) to protect HS⁻ against over-reaction and we used thiourea as a protected version of HS⁻. In this synthesis the acetyl protecting group is better as reduction with LiAlH₄ reduces the ketone and removes the protecting group in the same step.

Analysis

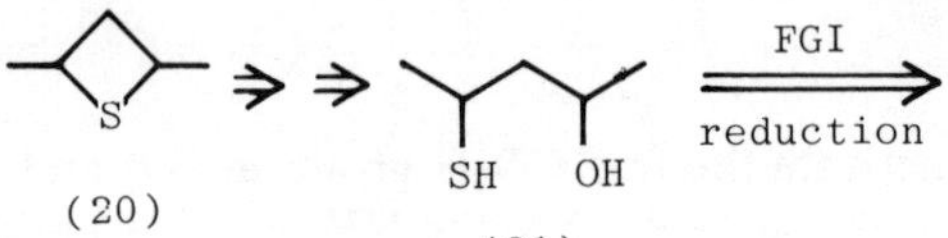

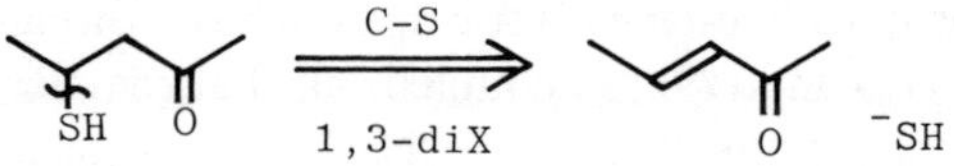

Synthesis[96]

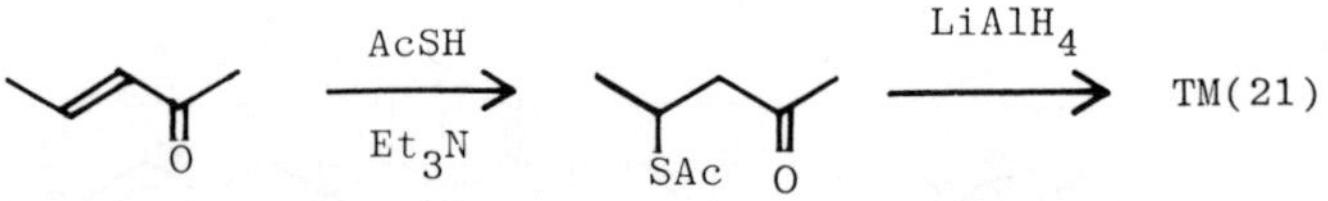

CHAPTER 10

One-Group C–C Disconnections I: Alcohols

We now leave disconnections of bonds between carbon and other atoms and turn to the more challenging C–C disconnections. These are more challenging because organic molecules contain many C–C bonds and we must learn which to disconnect. In one way they are easier than C–X disconnections. Reagents are available for both electrophilic (e.g. RBr) and nucleophilic (e.g. RMgBr) carbon whereas heteroatoms are almost always added as nucleophiles. The disconnections we met in Chapter 6, summarised in Table 10.1, also apply to C–C disconnections of types (a), (b), and (d). For type (c) it makes sense to avoid the reversal of polarity (Chapter 7) and use the natural polarity of the enolate ion (1). Though we are using logic developed from our treatment of two-group C–X disconnections, there is now only one functional group and so we must call these one-group disconnections. Disconnections (a) and (b) are discussed in this chapter; (c) and (d), in Chapter 13.

Table 10.1 One-group C–C disconnections

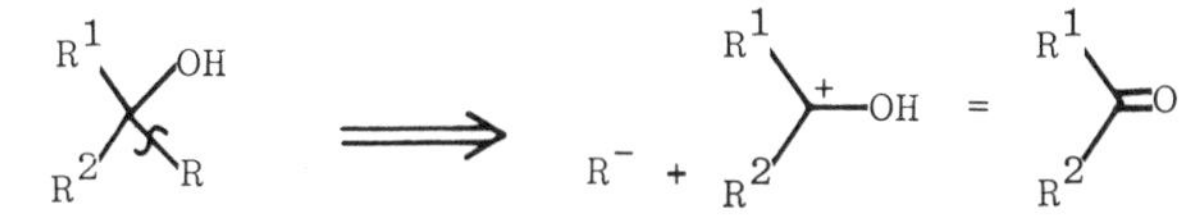

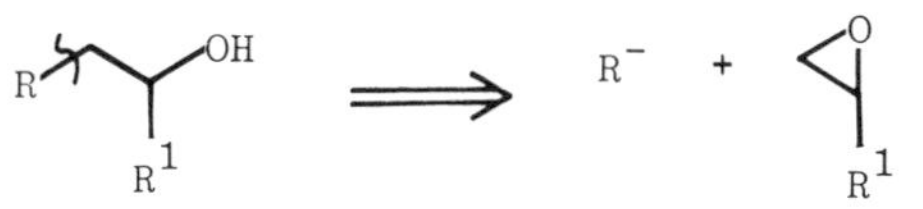

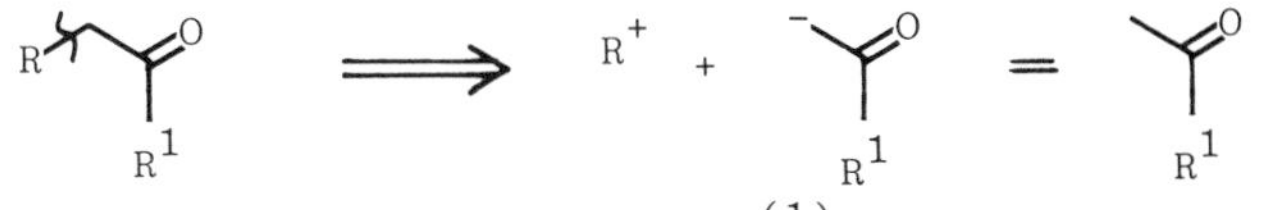

Reagents for the Carbanion Synthon

Disconnections (a), (b), and (d) in Table 10.1 all require reagents for the carbanion synthon R⁻. Simple carbanions are almost never formed in reactions so we shall need reagents in which carbon is joined to a more electro*positive* atom such as a metal. The most popular are Li and Mg. Butyl lithium (BuLi) is commercially available and other alkyl lithiums can be made from it by exchange **(i)**. Grignard reagents **(2)** are usually made directly from alkyl halides and magnesium metal (iii)—a method also available for RLi (ii). These methods are available for aryl compounds too. Transformation of RHal into RLi or RMgBr involves a formal inversion of polarity.

$$RCl + BuLi \longrightarrow BuCl\uparrow + RLi \qquad (i)$$

$$RHal + Li \xrightarrow{\quad} RLi \qquad (ii)$$
$$THF*$$

$$RHal + Mg \xrightarrow{\quad Et_2O \quad} RMgHal \qquad (iii)$$
$$or\ THF* \qquad (2)$$

1,1 C–C Disconnections

Synthesis of alcohols

Any alcohol can be disconnected at the C–C bond next to oxygen ((a) in Table 10.1) to give an aldehyde or ketone and a Grignard reagent as starting materials.

In 1963 some chemists[97] wanted to study the possibility of controlling the oxidation of a hydrocarbon chain by a hydroxyl group in the same molecule **(3)**. They decided to make alcohols with branched **(4)** and unbranched **(5)** side chains for this purpose.

(3) (4) (5)

Disconnections of the C–C bond next to the hydroxyl group reveals that both these compounds can be made from acetone and a Grignard reagent. Both alkyl halides are commercially available so the syntheses are trivial. The full method, drawn out for TM**(4)**, is usually summarised as drawn for TM**(5)**.

*THF, tetrahydrofuran

Analysis

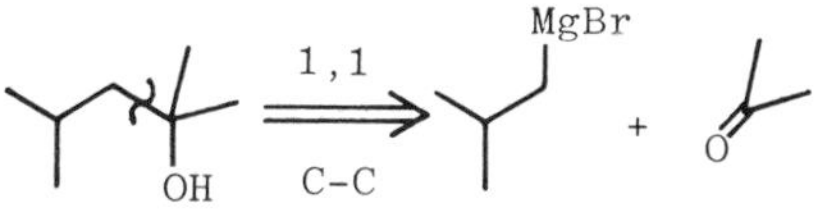

(4)

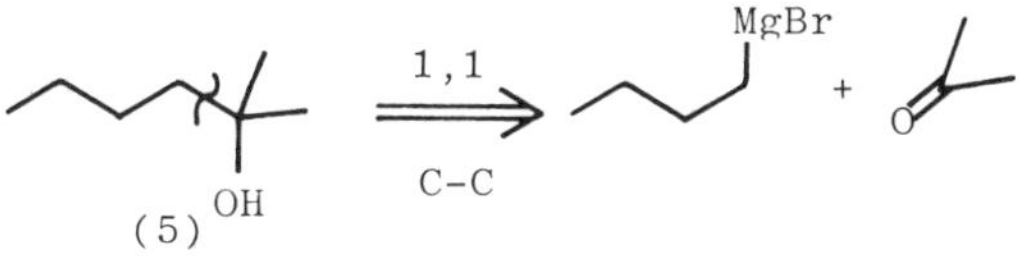

(5)

Synthesis

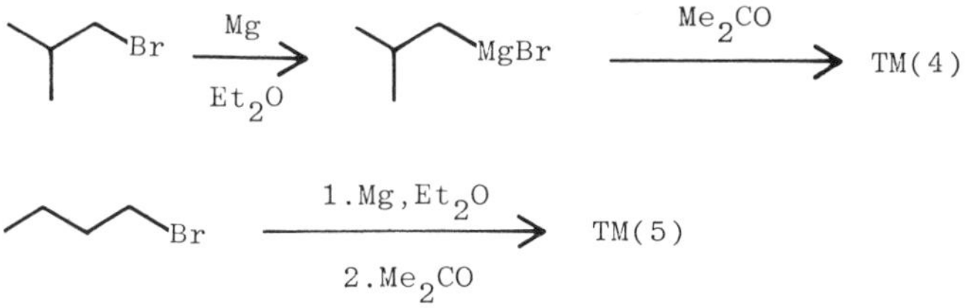

In more complicated examples, further disconnections or FGIs may be needed before and after the Grignard step. Compound **(6)** was needed[98] to study its potential as an analgesic. Disconnecting the ester reveals an alcohol and a further C–C disconnection of the phenyl group gives a simple ring compound **(7)**, a clear case for 1,3-diX disconnection (see Chapter 6).

Analysis

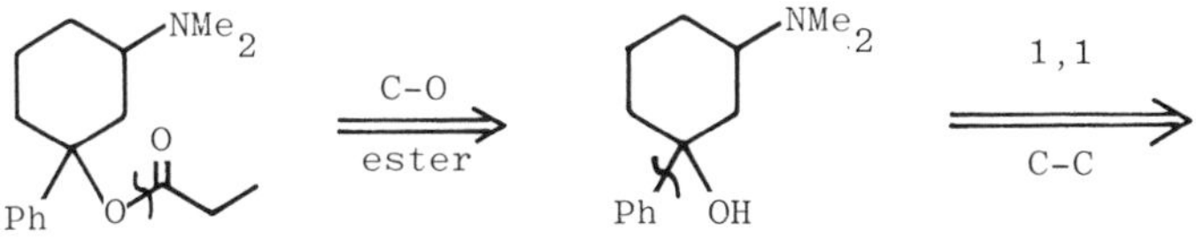

(6)

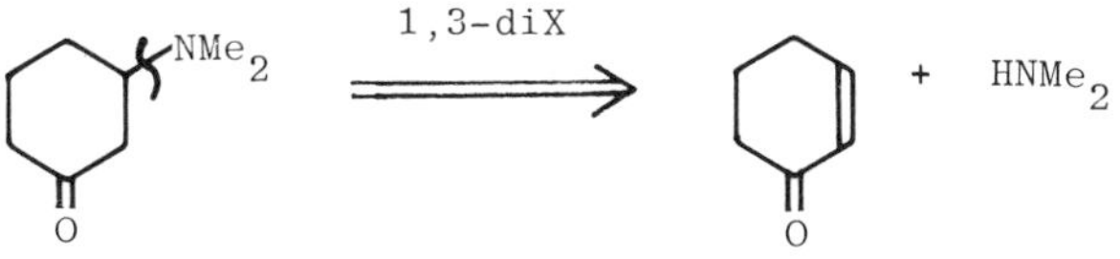

(7)   (8)

This synthesis has been carried out using PhLi for Ph⁻ and the anhydride for the ester formation. Enone **(8)** is available.

Synthesis[98]

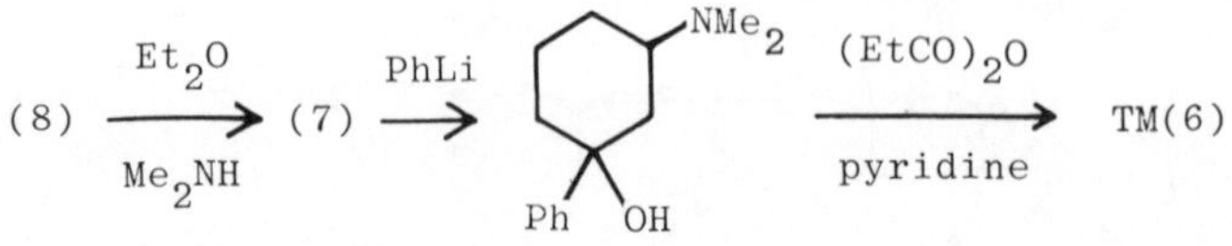

60% from (8)

An alternative approach to alcohols having two identical R groups **(9)** is to disconnect both at once: the ester **(10)** is the starting material. In the reaction, one molecule of Grignard displaces EtO⁻ from the ester **(10)** giving ketone **(11)**. This is more reactive than the ester and immediately captures a second Grignard molecule.

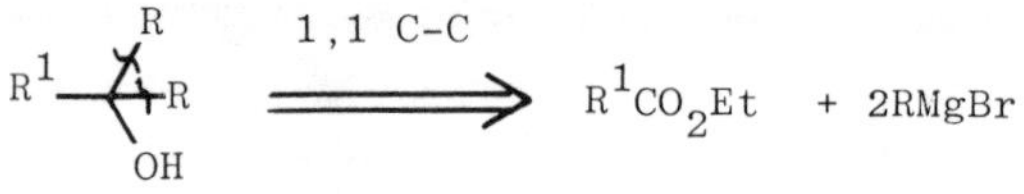

(9) (10)

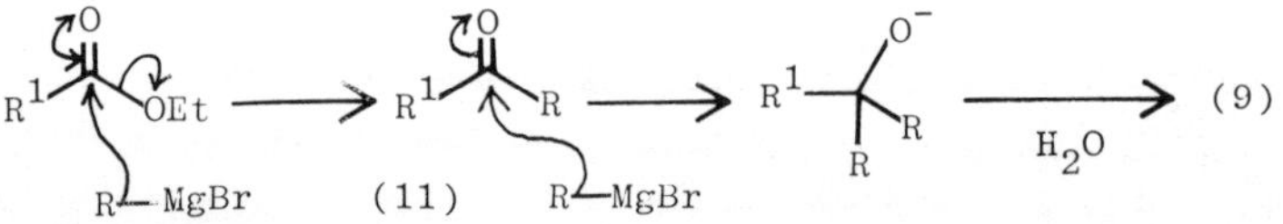

(11)

The perfumery alcohol **(12)**, whose acetate has[99] a 'unique cloudy hyacinth-lily of the valley' smell, can be disconnected in two ways. Syntheses by both routes are successful.

Analysis

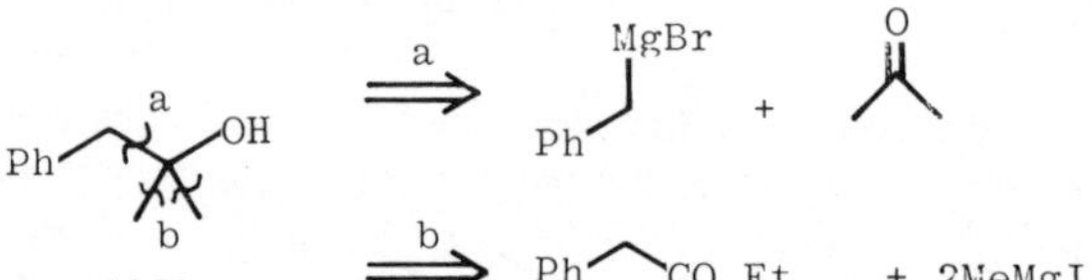

(12)

The muscle relaxant Pirindol[100] **(13)** invites disconnection of both phenyl groups to give ester **(14)**, easily made from piperidine **(15)** and an acrylate ester.

Pirindol: *Analysis*

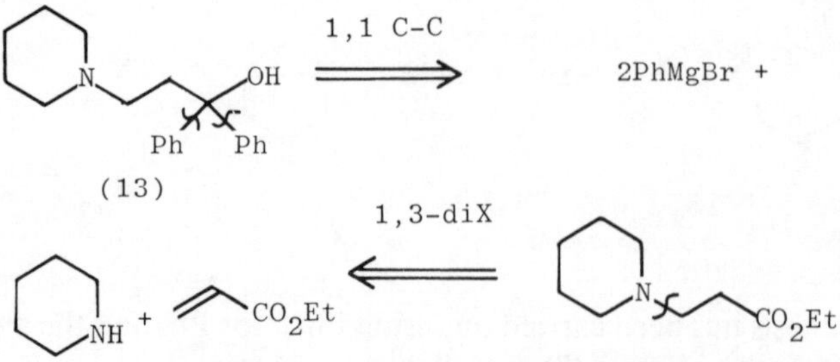

(13) (15) (14)

Synthesis[101]

$$\text{CH}_2\text{=CH-CH}_2\text{-CO}_2\text{Et} \xrightarrow{(15)} (14) \xrightarrow{\text{PhMgBr}} \text{TM}(13)$$

Synthesis of aldehydes and ketones

The easiest approach to aldehydes and ketones *via* this disconnection is by oxidation of the corresponding alcohols. Lythgoe[102] chose this route when he needed ketone (16) to demonstrate a new alkyne synthesis. Returning to the alcohol (17) by FGI followed by disconnection of the side chain gives the aldehyde (18) which can be made in the same way.

(16)

Analysis

The most popular oxidising agents are based on chromium(VI) and modern variants include PCC[103] (pyridinium chlorochromate) and PDC[104] (pyridinium dichromate). Table 10.2 gives others you are likely to meet.

Table 10.2   Oxidising agents for conversion of alcohols into aldehydes and ketones*

Name	Method	For RCH_2OH to $RCHO$
—	$Na_2Cr_2O_7$, H^+	Distil out $RCHO$ as formed
Jones	CrO_3, H_2SO_4, acetone	Distil out $RCHO$ as formed
Collins	CrO_3, pyridine	Use in CH_2Cl_2 solution
PCC[103]	CrO_3.pyr.HCl	No modification needed
PDC[104]	$2pyr.H^+\ Cr_2O_7^{2-}$	Use in CH_2Cl_2 solution
Moffatt	Me_2SO + RN=C=NR (DCC)	No modification needed

*See House, Chapters 5, 6, and 7 for a full discussion.

Synthesis[102]

$$\text{(Br-cyclohexane)} \xrightarrow[\substack{2.CH_2O \\ 3.PCC}]{1.Mg,Et_2O} \quad (18) \quad \xrightarrow[2.PCC]{1.n-HexMgBr} \quad TM(16) \quad 68\%$$

Carboxylic acids

Direct disconnection is possible at this oxidation level since CO_2, especially convenient as solid 'dry ice', reacts once only with Grignard reagents or RLi (iv). This method complements hydrolysis of cyanides (v) since the disconnection is the same but the polarity is different. Hence the *t*-alkyl acid (19) could not be made by the cyanide method as displacement at the tertiary centre would be difficult. The Grignard method works well.[105]

$$\left.\begin{array}{l} RMgBr \xrightarrow{CO_2} RCO_2MgBr \\ RLi \xrightarrow{CO_2} RCO_2Li \end{array}\right\} \xrightarrow[H_2O]{H^+} RCO_2H \qquad (iv)$$

$$RBr \xrightarrow{CN^-} RCN \xrightarrow{HO^-/H_2O} RCO_2H \qquad (v)$$

$$\underset{}{\text{>}}{-}Cl \xrightarrow[2.CO_2]{1.Mg,Et_2O} \underset{}{\text{>}}{-}CO_2H$$

$$(19) \quad 70\%$$

The cyanide method[106] is better for reactive allylic[107] (20) and benzylic[107] (21) halides and has the advantage that esters (e.g. 22) can be formed directly from the cyanide if required.[107]

$$\underset{(20)}{\text{Br}} \xrightarrow{CuCN} \underset{84\%}{\text{CN}} \xrightarrow[HCl]{conc} \underset{82\%}{\text{CO}_2H}$$

$$Ph\underset{(21)}{\diagdown}Cl \xrightarrow[H_2O,EtOH]{NaCN} Ph\underset{90\%}{\diagdown}CN \xrightarrow[H^+]{EtOH} Ph\underset{(22)\ 87\%}{\diagdown}CO_2Et$$

Acids can be converted into a range of derivatives (Chapter 4) often *via* the acid chloride so that FGI or C–X disconnections may be needed before and after the C–C disconnection. The acid bromide (23) requires the Grignard disconnection as nucleophilic displacement of aryl halides is not a good reaction.

Analysis

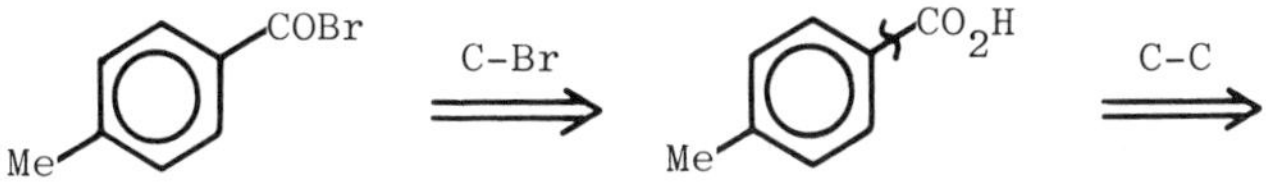

(23)

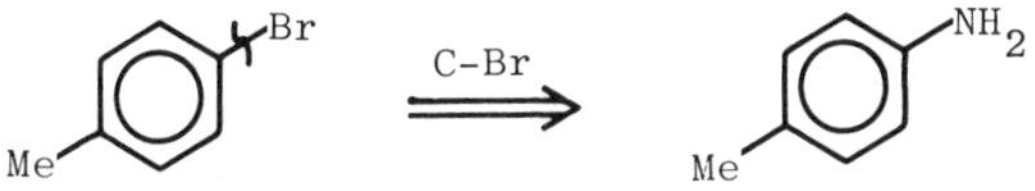

Synthesis[105, 108]

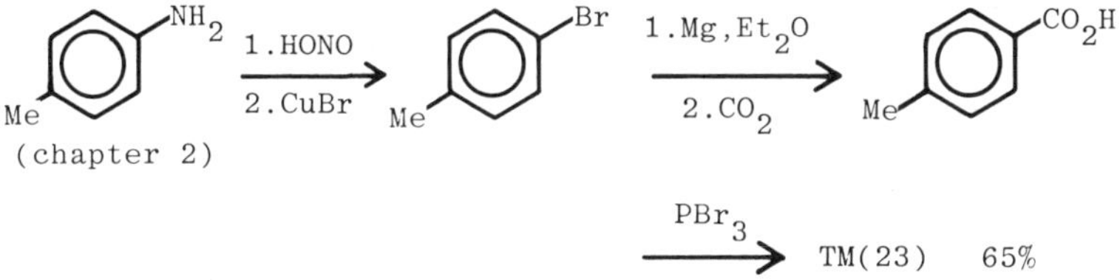

The anhydride **(24)** needs the acid chloride **(26)**, made from the acid **(25)** with $SOCl_2$. This aliphatic acid can be made by the cyanide method.

Analysis

$$(n\text{-HexCO})_2O \implies n\text{-HexCO}_2H \implies n\text{-HexBr} + CN^-$$

$$(24) \qquad\qquad (25)$$

Synthesis[109]

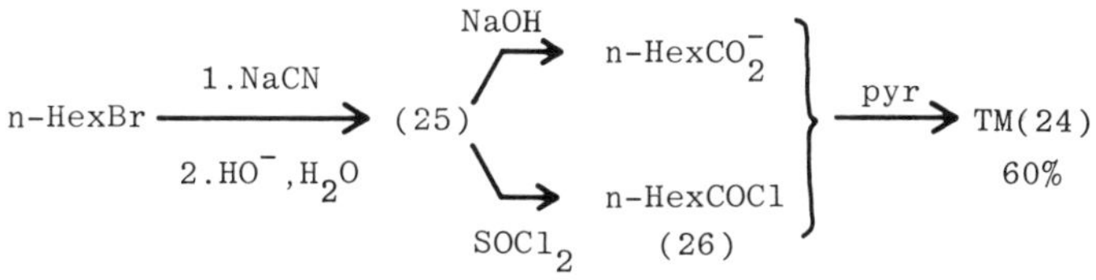

1,2 C–C Disconnections

Synthesis of alcohols

The epoxide route (vi) works very well here providing that the epoxide does not have too many substituents. The alcohol **(27)**, used in perfumery,[110] can be made this way.

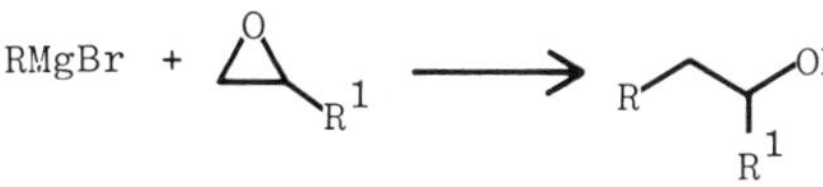

Analysis

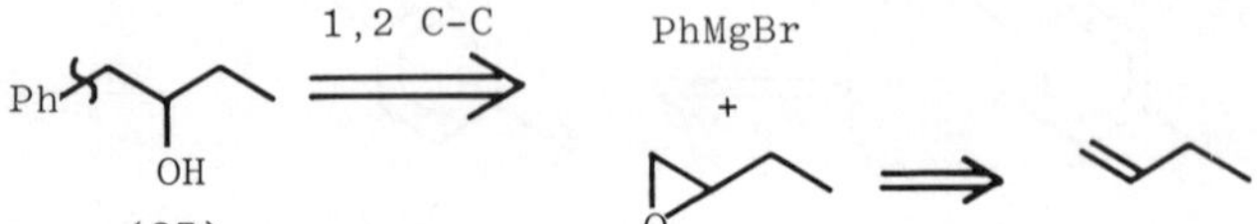

(27)

Synthesis[110]

This alcohol synthesis has the advantage that it is stereospecific and is used in Chapter 12 in this way.

Synthesis of carbonyl compounds

Carbonyl compounds can again be derived from alcohols by oxidation so that the same disconnection can be used (vii). The acid **(28)** is an example.

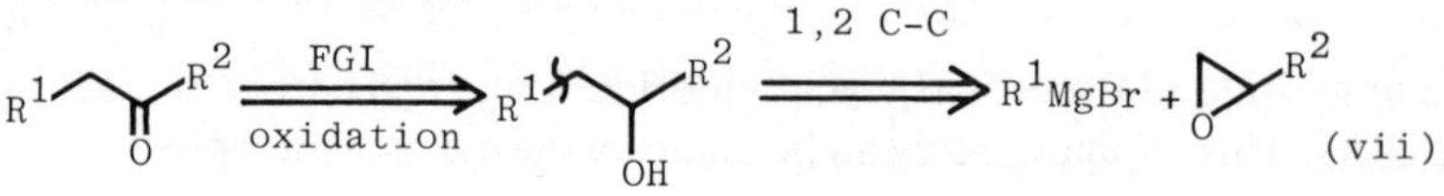

Analysis

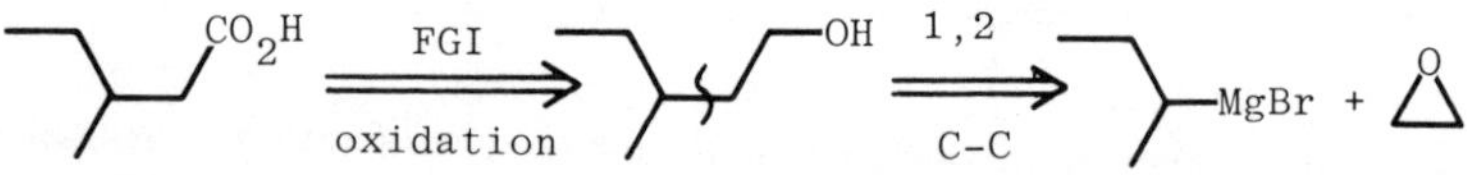

(28)

Synthesis[111]

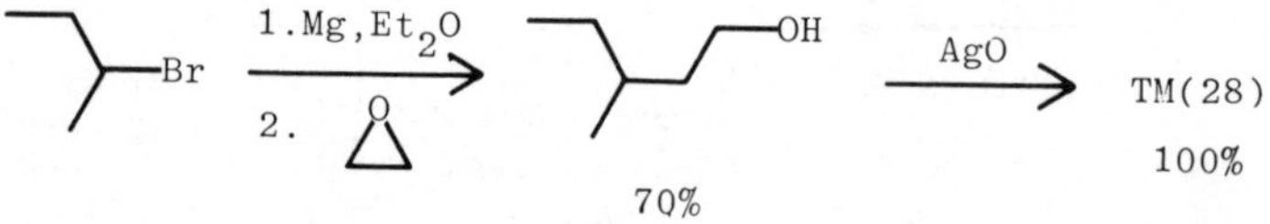

70%

A more direct approach is to reverse the polarity (see (c) in Table 10.1) as discussed in Chapter 13.

$$R \diagup CO_2H \implies R^+ + {}^-CH_2CO_2H$$

Other Compounds Made from Alcohols

Most of the C–X disconnections from Chapters 2–9 eventually led back to alcohols as starting materials. This chapter has introduced the building up of the carbon skeleton of these alcohols and so complete syntheses of a large number of compounds are now possible. Table 10.3 lists compounds derived from alcohols.

Table 10.3 Compounds derived from alcohols

Reaction type	Product	Chapter	Further products	Chapter
Oxidation	Aldehydes	10	Amines by reduction	8
	Ketones	10	of imides	
	Acids	10	Amines by reduction	8
			of amides	
Addition of acid derivatives	Esters	4	Amines by reduction of amides	8
Tosylation (TsCl, pyr)	Tosylates	4	Other substitutions (see below)	4
Substitution				
PBr₃ or HBr	Bromides	4	Ethers	4
SOCl₂	Chlorides	4	Thiols	5
			Sulphides	4
			Cyanides	10

In 1979 chemists wished to study[112] the effects of electron-withdrawing groups on S_N1 reactions and chose to make (29) for this purpose. The tertiary alkyl chloride must come from the alcohol (30) and we can continue by disconnecting the molecule of acetone (this and an alternative are mentioned on page 78).

Analysis

The nitro group can be put in at any stage as all the other groups are *o,p*-directing. The CH₂Cl group can be introduced by chloromethylation (Chapter 2) *before* nitration as the nitro group is *m*-directing. As the chemists wanted other substituted derivatives too, they chose to build the basic skeleton (31) first and nitrate last.

Synthesis[112]

A second example is the antihistamine drug[113] **(32)**. Disconnection of the ether, writing OH at both ends (Chapter 4), gives two reasonable fragments. One **(34)** is an amine–epoxide adduct (Chapter 6); the other **(33)** is clearly the product of a Grignard reaction. It is best to disconnect the phenyl group as we shall not then have the problem of making a mono-Grignard from *p*-dichlorobenzene. We can use the Friedel–Crafts disconnection on the ketone **(35)**.

Analysis

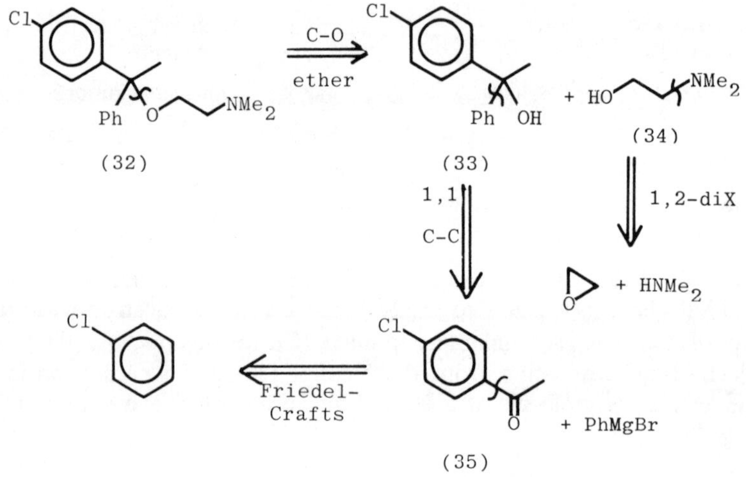

Synthesis[113]

The halide needed for the ether synthesis is most easily made from the amino alcohol.

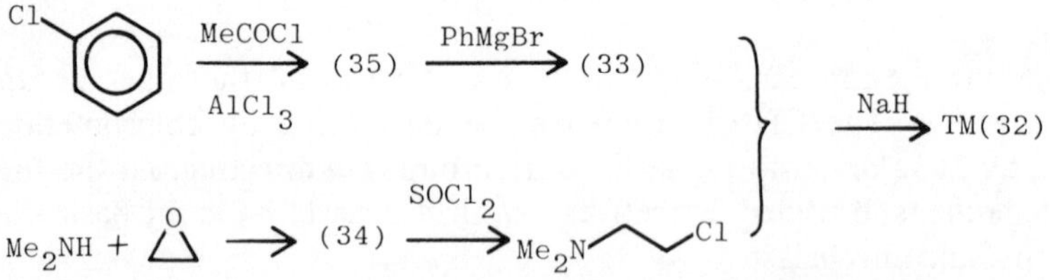

Iteration

The starting materials for the Grignard alcohol synthesis are alkyl halides and aldehydes or ketones. These starting materials are themselves made from alcohols by substitution or oxidation. It is possible therefore to build up large molecules by iteration (repetition). A simple example is alcohol **(36)** needed as part of a project[114] to synthesise authentic samples of all branched octanols for comparison with unknown samples. Grignard disconnection gives a halide **(37)** made from another alcohol **(38)** which can be made by another Grignard reaction.

Analysis

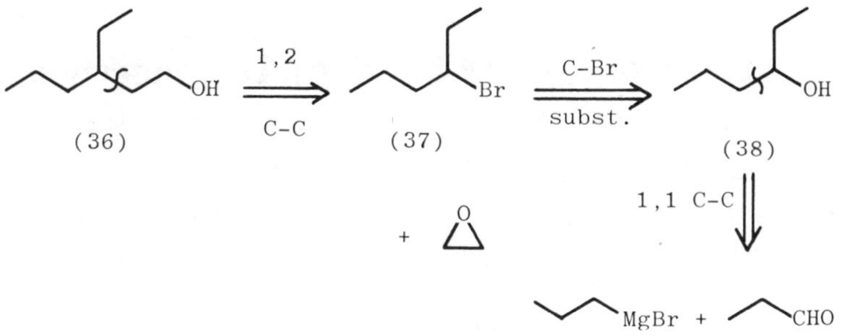

Synthesis[114]

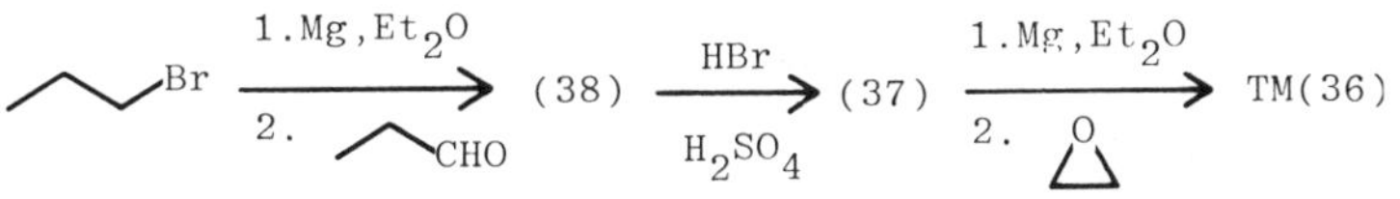

CHAPTER 11

General Strategy A: Choosing a Disconnection

This is the first of four general strategy chapters in which important points are introduced which apply to the whole of synthetic design rather than one particular area. This chapter concerns general principles to help in choosing one disconnection rather than another.

The main choice is between the various C–C disconnections. Even such a simple molecule as alcohol (1), introduced in Chapter 1 as a component of the elm bark beetle pheromone, can be disconnected at any of the five marked bonds.

$$\text{(structure of 1)} \implies \text{?}$$

(1)

Greatest Simplification

Only one of the five is a good choice (i) for two reasons. We want to get back to simple starting materials and we shall do so most quickly if we disconnect bonds which are:

1. towards the *middle* of the molecule thereby breaking it into two reasonably equal halves rather than chopping off one or two carbon atoms from the end; and
2. at a *branch point* as this is more likely to give straight chain fragments, and these are more likely to be available.

Analysis

$$\text{(structure 1a)} \underset{\text{C–C}}{\overset{1,1}{\implies}} \text{CHO} + \text{BrMg} \text{(structure)} \quad \text{(i)}$$

(1 a)

The disconnection we have chosen (i) fits both these criteria and the synthesis is a simple one step procedure.

86

Synthesis[115]

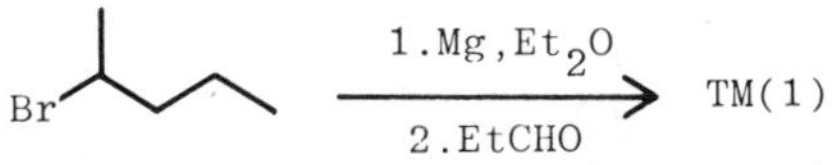

We can extend guideline 2 when we realise that the junction between a ring and a chain must always be a branch point. There is, for example, a series of drugs based on structure **(2)** and here we can disconnect one ring from the other.

Analysis

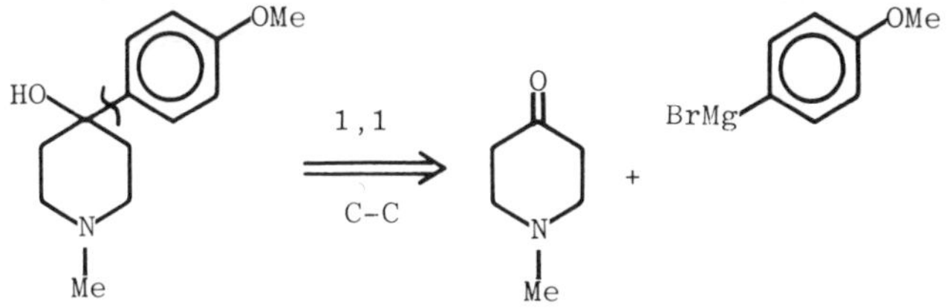

We shall analyse compounds like **(3)** in Chapter 19. The Grignard synthesis is trivial.

Synthesis[116]

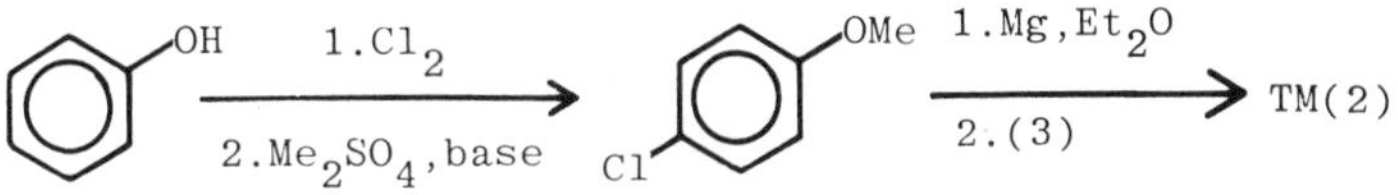

Symmetry

A second guideline is to use the symmetry of the target molecule. Thus the tertiary alcohol **(4)** is best disconnected twice at the branch point to preserve its symmetry. This synthesis was carried out by Grignard himself.[117]

Analysis

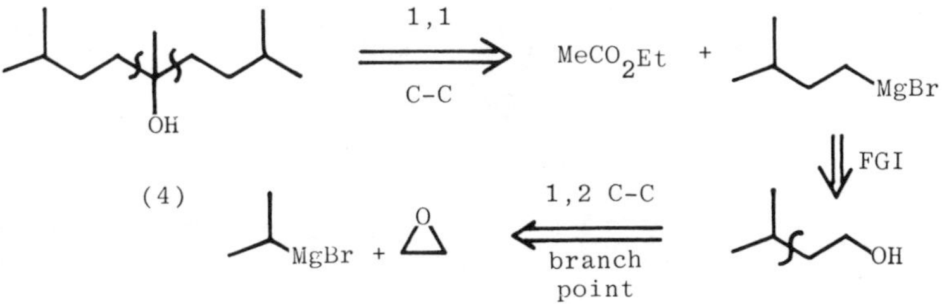

Synthesis[117]

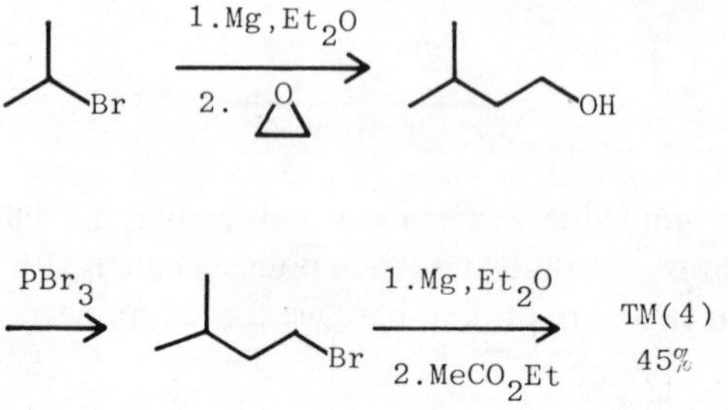

The symmetrical ether (5) can be made from the symmetrical diol (6) which can in turn come from the symmetrical diester (7) which is readily available (succinate ester). The symmetry avoids problems of selectivity.

Analysis

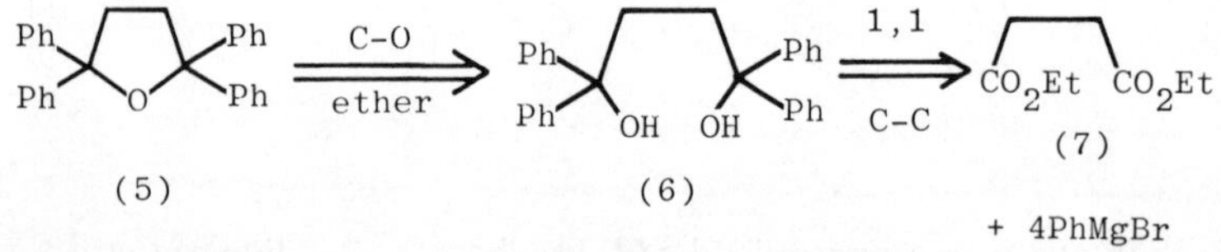

Synthesis[118]

$$(7) \xrightarrow[\text{2.H}^+]{\text{1.PhMgBr}} \text{TM(5)}$$

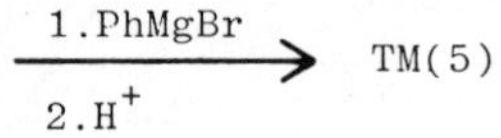

High Yielding Steps

A difficult guideline to follow when designing a synthesis, but very important when carrying it out, is to choose the disconnection corresponding to the reaction which works best in practice. This may well mean writing two or more possible routes, and trying them both in the laboratory, or at least looking for close analogies in the literature. Both disconnections (a) and (b) look all right on alcohol (8): in practice one gives a good yield of (8) whilst the other gives a quite different product.

Analysis

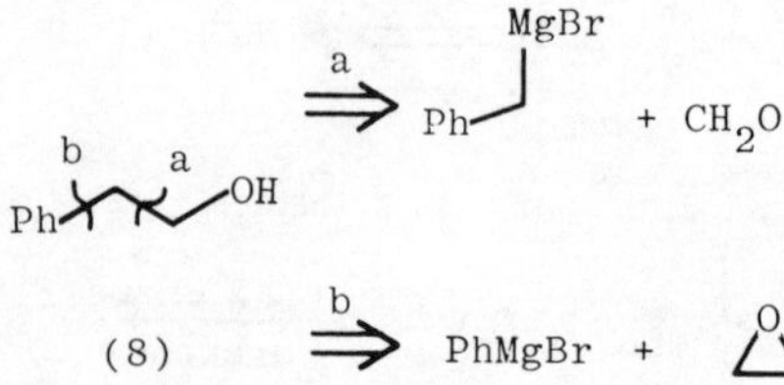

Synthesis (a)[119]

$$Ph\diagdown Br \xrightarrow[\text{2.CH}_2\text{O}]{\text{1.Mg,Et}_2\text{O}}$$

(structure: benzene ring with two substituents, one CH_2OH group)

Synthesis (b)[120]

$$PhBr \xrightarrow[\text{2. (epoxide)}]{\text{1.Mg,Et}_2\text{O}} \text{TM(8)} \quad 75\%$$

Recognisable starting materials

Another practical guideline is to look for disconnections which lead back to readily available starting materials. It is obviously impossible to give a complete list of cheap aliphatic starting materials, but as this is one point on which the newcomer to synthesis feels most ignorant, Table 11.1 gives some ideas. Readily available aromatic compounds were discussed in Chapter 3. A supplier's catalogue is the guide used by chemists themselves. This aspect of strategy is developed further in Chapter 40.

Analysis of alcohol (9) could be guided by symmetry or by recognising a molecule of methacrylate (10) embedded in its framework: the result is the same and the synthesis has been carried out[121] using BuLi.

Analysis

$$\text{(9)} \xrightarrow[\text{C–C}]{1,1} \quad + \quad \text{(10)}$$

(9) (structure with OH)

(10) (structure with CO_2R)

Synthesis[121]

$$\text{(structure with } CO_2Me) \xrightarrow{\text{BuLi}} \text{TM(9)}$$

If disconnection back to available starting materials is impossible in one step, disconnection to give compounds whose synthesis will be easy is often possible. Fragments like (11) or (12) can easily be made by two-group C–X disconnections so a disconnection leading to (11) or (12) is good strategy.

$$R_2N\diagdown\diagdown OH \qquad Br\diagdown\diagdown\overset{O}{\underset{}{C}}R$$

(11)   (12)

90

Table 11.1 Some readily available aliphatic starting materials

Straight chain compounds: C1 to about C8
  alcohols, alkyl halides, acids, aldehydes, amines

Branched chain compounds: as above, based on the following skeletons:

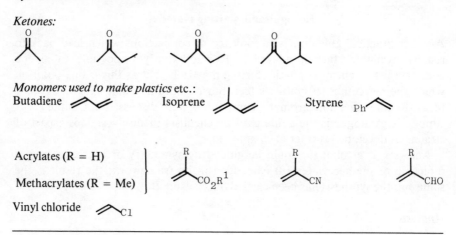

Cyclic alcohols and ketones: C4 to C8

Ketones:

Monomers used to make plastics etc.:
Butadiene   Isoprene   Styrene Ph

Acrylates (R = H)

Methacrylates (R = Me)

Vinyl chloride Cl

Readily available aromatic compounds are discussed in Chapter 3.

The hydroxyaldehyde **(13)** was used by Büchi[122] as an intermediate in his synthesis of the natural product nuciferal **(14)**. The tertiary alcohol is the obvious place to disconnect and disconnection (a) severs the ring from the chain (guideline 2). Fragment **(15)** will be easy to make but fragment **(16)** is outside our experience. The alternative disconnection (b) gives two recognisable fragments—a Friedel–Crafts product **(17)** and a compound **(18)** of type **(12)**.

(13)

(14)

Nuciferal intermediate: *Analysis*

In the synthesis we shall have to protect the aldehyde in **(18)** as an acetal before making the Grignard reagent as otherwise it will react with itself. All groups which react with the strongly basic and nucleophilic Grignard reagents must be protected if they are in the same molecule: examples include all carbonyl groups, alcohols, epoxides. The acetal could be removed at the end to liberate **(13)** but Büchi retained it for use later in the nuciferal synthesis.

Synthesis[122, 123]

Table 11.2 Guidelines to good disconnections

---

1. Make the synthesis as short as possible.
2. Use only disconnections corresponding to known reliable reactions.
3. Disconnect C–X bonds, especially two-group disconnections. This includes RCO–X.
4. Disconnect C–C bonds according to the FGs in the molecule.
   If possible:
   (a) aim for the greatest simplification
      — disconnect in the middle of the molecule;
      — disconnect at a branch point;
      — disconnect rings from chains.
   (b) Use the symmetry, if any.
5. Choose the disconnection corresponding to the highest yielding reaction, if known.
6. Disconnect back to recognisable starting materials, or to compounds which can easily be made.

All these guidelines help to make the synthesis as short as possible—this is vitally important as each step means some loss of yield, so the fewer steps the better. Table 11.2 summarises the guidelines from previous chapters and those deduced in this chapter. Remember that only some of these guidelines will apply to any one synthesis and that they may well contradict each other. Developing judgement in choosing the right disconnection can only come with practice.

CHAPTER 12

Strategy V: Stereoselectivity A

The biological properties of organic molecules depend on their stereo-chemistry. This is true for drugs, insecticides, plant growth regulators, perfumery and flavouring compounds, and all compounds with biological activity. The *cis** hydroxy aldehyde (1) has a strong and pleasant smell and is used in lily of the valley perfumes, whereas the *trans** isomer (2) is virtually odourless.[124] Any synthesis must give pure (1), not a mixture of (1) with the more stable (2)—at equilibrium there is 92% of (2). The synthesis must be stereoselective.

(1)

(2)

(3)

The elm bark beetle pheromone multistriatin (3) is a more complicated example. You may remember from Chapter 1 that a single stereoisomer alone attracts the beetle. Making one diastereoisomer by a stereoselective synthesis is not enough. The compound must be a single enantiomer, that is it must be optically active, too. In this chapter we shall consider the question of achieving the correct relative stereochemistry at several chiral centres (such as the four in multistriatin, marked •) and first, the question of making optically active compounds.

*These are diastereoisomers and the question of optical activity does not arise as neither (1) nor (2) is chiral.

Optically Active Compounds

If a single enantiomer is needed, regardless of the number of chiral centres, we can either use a naturally occurring optically active starting material, or we can resolve during the synthesis.

Resolution

In principle, resolution should be at an early stage in the synthesis as this avoids carrying through unwanted material (the wrong enantiomer). However, we should also consider whether at any stage we can recycle the unwanted enantiomer or resolve easily because the functional groups are suitable.

When Cram[125] wished to study the stereochemistry of elimination reactions, he wanted a strong optically active base which would not substitute. Bases derived from hindered secondary amines (4) are often used for elimination reactions and Cram selected amine (5) for his purpose.

$$R^1R^2NH \xrightarrow{\text{BuLi}} \begin{matrix} R^1 \\ \diagdown \\ N-Li \\ \diagup \\ R^2 \end{matrix}$$

$$(4)$$

Analysis by the amide method (see Chapter 8) leads back to amine (6) which can be made from ketone (7), a simple Friedel–Crafts product.

Analysis

Amine (6) is the first chiral intermediate so resolution was carried out at this stage using a naturally occurring optically active compound.

Synthesis[125]

$$(7) \xrightarrow[\text{2.reduce}]{1.\text{NH}_2\text{OH}} (6) \xrightarrow{\text{resolve}} \begin{array}{c} \text{Ph} \\ \text{H}_2\text{N} \overset{\text{H}}{\underset{\text{Me}}{\wedge}} \\ (+) \end{array} \xrightarrow{\text{COCl}}$$

(+) 90%

$$\xrightarrow{\text{LiAlH}_4}$$

(+)-(5) 97%

Optically active starting materials

There is a range of naturally occurring optically active compounds available in large quantities, the amino acids and the sugars being notable examples. One strategy in synthesising an optically active target molecule is to disconnect to the skeleton of one of these available compounds and to use stereospecific reactions (more details in the next section) in the synthesis.

In connection with his work on insect pheromones,* Silverstein[126] wanted a general synthesis of optically active lactones (8). For a general synthesis, disconnection of R is best and as this also leaves a fragment (9) having the skeleton of available glutamic acid (10), this route was chosen. A series of trivial disconnections is needed, but this is a price worth paying for a readily available, optically active, starting material.

Analysis

(8)　　　(9)

$$\text{EtO}_2\text{C} \xrightarrow[\text{ester}]{\text{C-O}} \text{HO}_2\text{C} \xrightarrow[\text{subst.}]{\text{C-O}} \text{HO}_2\text{C} \overset{\text{H}}{\underset{}{\wedge}} \text{NH}_2$$

S-(+)-Glu (10)

Displacements at the α-carbon of amino acids after diazotisation of the amino group are known to go with retention of configuration, and thereafter

*Described in more detail in Chapter 25.

the chiral centre is unaffected. Available S-(+)-Glu is known to have the absolute configuration shown (10) so the configuration of (8) made by this route is also known. Silverstein chose to make X = OTs in (9) and to use an organocopper reagent for the synthon R⁻ as Grignard reagents often behave poorly in alkylations, and would anyway react with the lactone.

Synthesis[126]

(+) 55% (+) 77%

1.TsCl,pyr

2.R$_2$CuLi

(+) 70% (+)-(8)
e.g. R=n-Bu, 41%

Stereospecific Reactions

We shall use stereospecific to describe a reaction whose mechanism demands a specific stereochemical outcome. This must result whether it leads to the more stable product or not and carries with it the idea that each stereoisomer of the starting material gives a different stereoisomer of the product. These may be enantiomers or diastereoisomers. In the S$_N$2 reaction the mechanism demands 'attack from the back' and hence inversion, so that R-(11) gives S-(12) whilst S-(11) gives R-(12).

R-(11) S-(12)

S-(11) R-(12)

In a molecule with two 'chiral' centres, such as (13), *cis* tosylate gives *trans* acetate and vice-versa. These are diastereoisomers and are not chiral.

cis-(13)   trans-(14)

trans-(13)   cis-(14)

Some familiar stereospecific reactions appear in Table 12.1. There are, of course, many more sophisticated examples and we shall meet some of these later in the book. The last entry may seem frivolous, but it is the safest way of transmitting a chiral centre. It is also vital to check that a chiral centre laboriously set up at an early stage is not to be affected later on by some other reaction.

Hydroxylation is particularly useful as methods are available for both *cis* and *trans* hydroxylation. Alternatively both *E*- and *Z*-alkenes* may be used as starting materials. When Crout[127] needed both (16) and (17) for some enzyme experiments, he made them by stereospecific hydroxylation of the two isomers of (15).

E-(15)   Z-(15)

(16)   (17)

Use of epoxides

Epoxides form a bridge between alkene geometry and sp^3 stereochemistry. An alkene of known geometry (e.g. 18) forms an epoxide by stereospecific *cis* addition. Nucleophilic attack (s_N2) on this epoxide (19) now sets up *two* chiral centres of known stereochemistry.

*In an *E*-alkene, the two highest priority groups are *trans*, in a Z-alkene they are *cis*.

Table 12.1 Stereospecific reactions

Substitution: S_N2 Nu: ⤻ R⟍X Nu: ↷ (R, O chiral centre diagram)	Inversion at chiral centre under attack
S_N2 with neighbouring group participation	Retention at chiral centre under attack
Elimination: E2 B: ⤻ (H, X anti-periplanar diagram)	Anti-periplanar arrangement of H and X

Electrophilic addition to alkenes

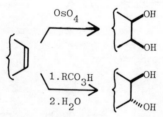

Bromination Br_2 → Br / Br	*Trans* addition
Epoxidation RCO_3H → H / O / H	*Cis* addition
Hydroxylation OsO_4 → OH / OH	*Cis* addition
1. RCO_3H 2. H_2O → OH / OH	*Trans* addition

Hydrogenation

A / B H_2, Pd catalyst → H / A / B / H	*Cis* addition

Rearrangements

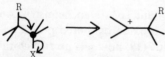

R (X diagram) → R (+ diagram)	Retention at migrating group R
	Inversion at migration terminus •

Reactions not involving the chiral centre
Any reaction

Retention

(18) MCPBA → (19)

In Chapter 1 we discussed the synthesis of multistriatin (20), a pheromone of the elm bark beetle. The time has come for a stereochemical analysis of this problem. The molecule has four chiral centres (• in 20). One of them (a) turns out to be unimportant as disconnection of the acetal reveals (21) as the true-target. If (21) cyclises to form an acetal it must give (20)—no other stereo-chemistry is possible.

(20)

Multistriatin: *Analysis 1*

(20) $\xrightarrow[\text{acetal}]{\text{1,1-diX}}$ (21) ⇒ (22) + (23)

Two of the remaining chiral centres (c and d in 21) are adjacent; the third (b) is separated by a CH_2 group. It will be easy therefore to control (c) and (d) by stereospecific reactions but (b) will be difficult. In addition (b) is next to a carbonyl group and has a proton (H in 21) so that enolisation would destroy any stereochemistry set up here. A good strategy will therefore be to disconnect (21) to the symmetrical ketone (22) and an intermediate (23) containing both the chiral centres we aim to control. So far the analysis is the same as that in Chapter 1.

The leaving group X in (23) will be derived from a hydroxyl group as usual. Hence (24) could be made from the symmetrical epoxide (25) choosing the correct stereochemistry to give (24) by S_N2 inversion. The required stereochemistry is *cis* so that we can make (24) from the readily available acetylenic diol (26) by simple steps.

Analysis 2

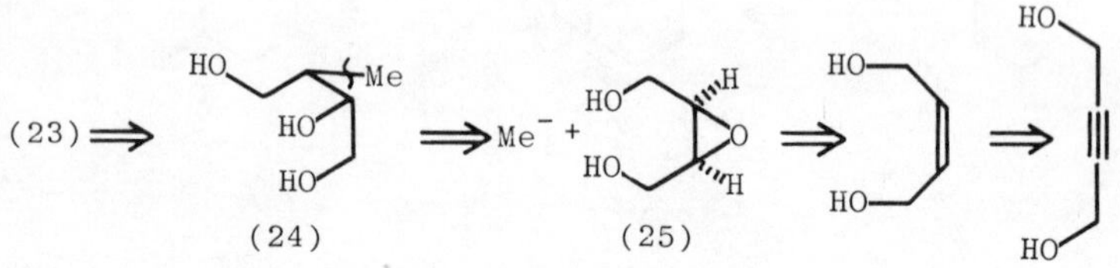

Turning this analysis into a synthesis requires two decisions — how do we protect, and what reagent do we use for Me⁻? A Grignard reagent might do, but an organocopper reagent is in fact better.[128] Protection of the hydroxyls in **(25)** is most simply done by an acetal, and this is best put in before the epoxidation.

Synthesis 1[128]

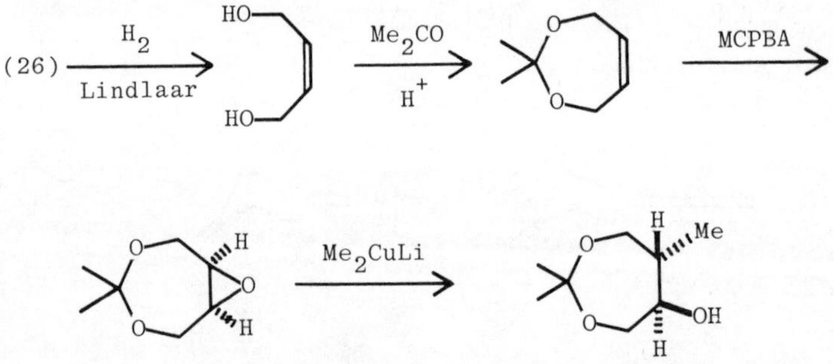

We now need to remove the acetal from **(27)** and replace it by a group protecting the two OH groups not needed for the next reaction. This can be done by thermodynamic control: treatment of **(27)** with acid rearranges it to **(28)** since a five-membered ring is more stable than a seven-membered ring. The free hydroxyl can now be converted into the iodide **(29)** with two known chiral centres. These result directly from the stereospecific opening of the epoxide with Me₂CuLi to give **(27)**.

Synthesis 2[128]

(27) $\xrightarrow{\text{HCl}}$ (28) $\xrightarrow[\text{2.NaI}]{\text{1.TsCl,pyr}}$ (29)

(22) $\xrightarrow[\text{2.(29)}]{\text{1.R}_2\text{NLi}}$ (30) $\xrightarrow{\text{H}^+}$ TM(20)

The final stages require alkylation of the ketone **(22)** with **(29)** to give **(30)** and reorganisation of the acetal. This can again be done in acid as the bicyclic multistriatin is more stable than **(30)**. This synthesis gives 85% of the natural isomer with only 15% of the wrong stereochemistry at centre (b). The natural isomer predominates because it has an equatorial methyl group.

Stereoselective Reactions

We shall use stereoselective to describe reactions whose mechanism offers alternative, chemically equivalent pathways so that the reaction may *select* the most favourable pathway (kinetic control) or the most stable product (thermodynamic control). These reactions commonly involve setting up one chiral centre in the presence of others.

Ketone **(31)** can be reduced to two alcohols, *cis* **(32)** or *trans* **(33)**. The diequatorial *trans* alcohol **(33)** is more stable and is preferred under equilibrating conditions—e.g. by reduction[129] with Al(OPr-*i*)$_3$. However, the line of approach **(34)** to form the *cis* alcohol **(32)** is better and this is preferred under conditions of kinetic control—e.g. by reduction with the reactive bulky reagent[130] LiAlH(OBu-*t*)$_3$.

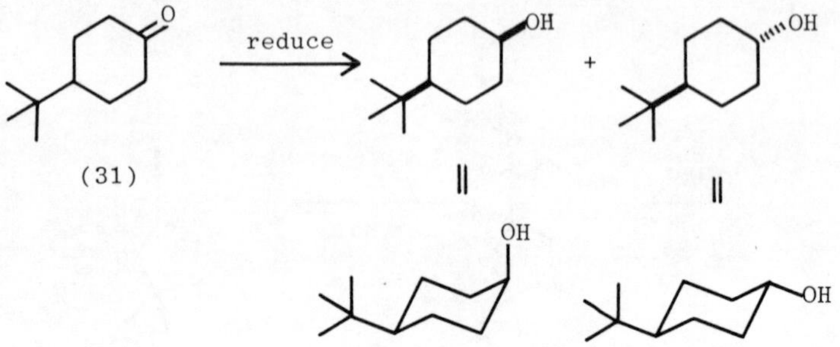

(31)

(32)   (33)

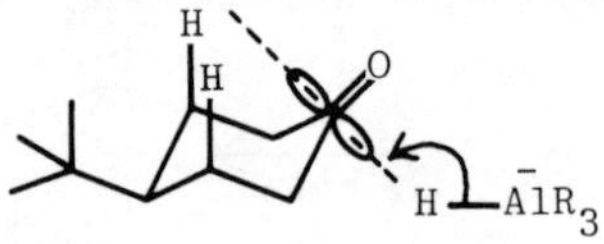

(34)

Occasionally both isomers of a compound are needed and then stereoselective reactions with a poor degree of selectivity are an advantage! Both *cis* **(35)** and *trans* **(36)** tosylates were needed[131] to study their reactions with base. The tosylates can obviously be made from the alcohols **(37)** by chemoselective (Chapter 5) tosylation of the less hindered primary OH groups, and reduction of the readily available (Chapter 19) keto ester **(38)** is a suitable weakly stereoselective route to **(37)**.

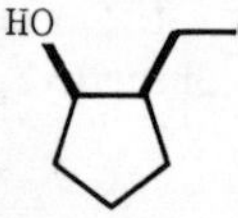

(35) 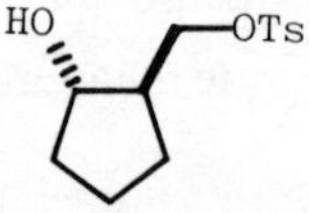   (36)

Analysis

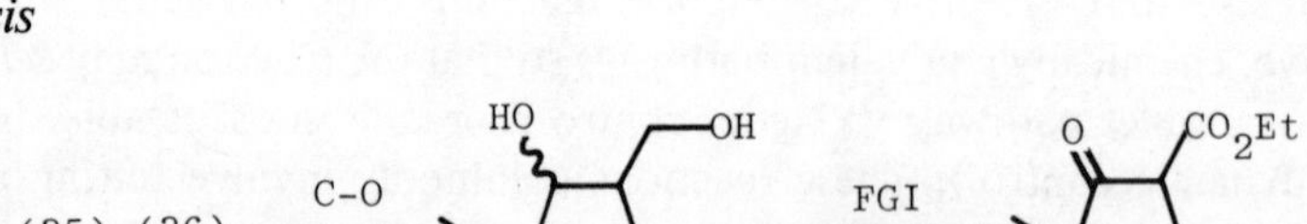

(35),(36) $\xrightarrow[\text{subst.}]{\text{C-O}}$ (37) $\xrightarrow[\text{reduction}]{\text{FGI}}$ (38)

Experiments showed[131] that reduction of **(38)** with LiAlH$_4$ gave substantial amounts of both diols which could be separated by chromatography. Tosylation gave the required **(35)** and **(36)**. The effort was worthwhile as dramatically different reactions occurred on treatment with base to give **(39)** and **(40)**.

Synthesis[131]

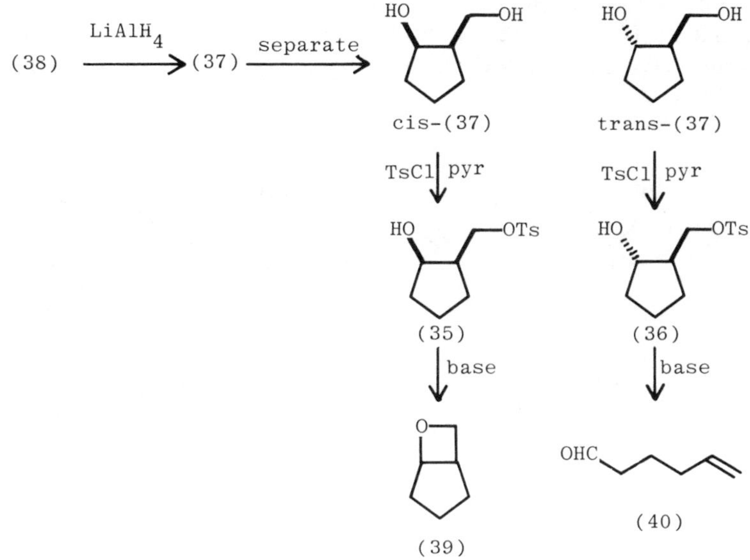

$$(38) \xrightarrow{\text{LiAlH}_4} (37) \xrightarrow{\text{separate}}$$

cis-(37)   trans-(37)

TsCl│pyr   TsCl│pyr

(35)   (36)

│base   │base

(39)   (40)

Control in Stereoselective Reactions

Conformational control in six-membered rings

Reduction of ketones and nucleophilic attack on ketones are classic cases where prediction is difficult as two factors, kinetic and thermodynamic, work in opposition. The alcohol **(41)**, needed to make an analgesic[98] (see Chapter 10), can obviously be made by addition of Ph⁻ to the amino ketone **(42)**, a simple Michael addition product (see Chapter 6). The stereochemistry required is axial addition on the opposite side from the Me₂N group, and kinetic control using PhLi as the reagent[98] gives mostly this result.

Analysis

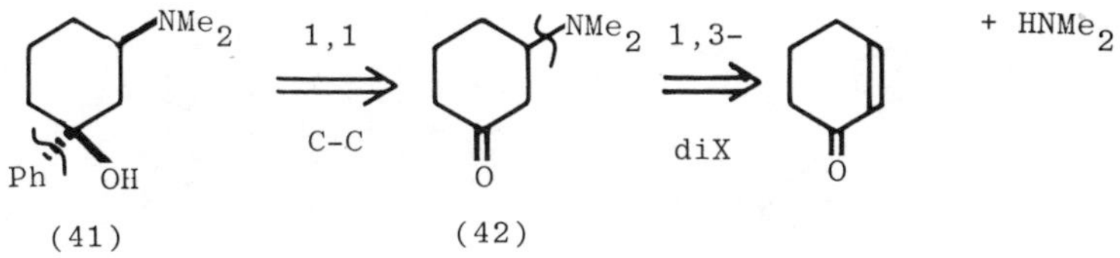

(41)   (42)   + HNMe₂

Synthesis[98]

$$\xrightarrow{\text{HNMe}_2} (42) \xrightarrow{\text{PhLi}} \text{TM(41)}$$

60% from cyclohexenone

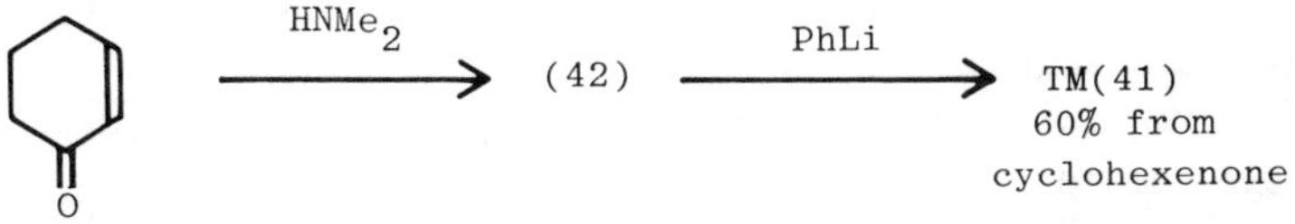

Nucleophilic attack on cyclohexenones (43) or cyclohexene oxides (44), or electrophilic attack on cyclohexenes (45) is strongly selective in favour of the axial products, though these may 'flip' to the equatorial isomer (e.g. 46) after the reaction. Examples follow throughout the book, the first in Chapter 13, page 112.

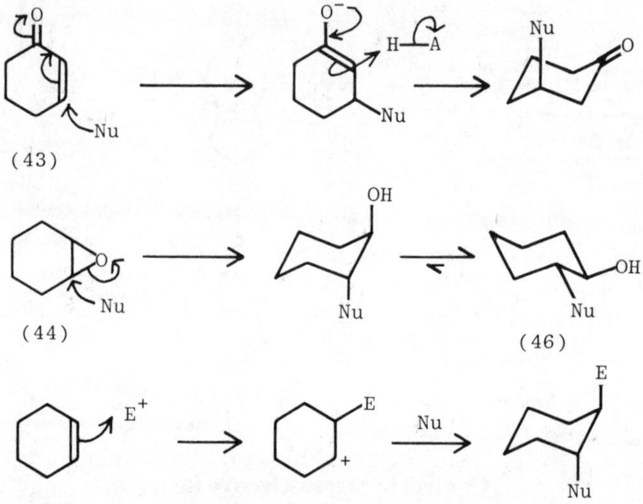

(43)

(44)   (46)

(45)

Control in other compounds

In other cases, the decisive factor is usually attack on the less crowded side of the molecule. Hydrogenation of an alkene is a good example because one side of the flat alkene must be in contact with the catalyst and selectivity in favour of the less crowded side is usually high. Hydrogenation of (47) gives[132] a large proportion of the less stable *cis* decalone (48).

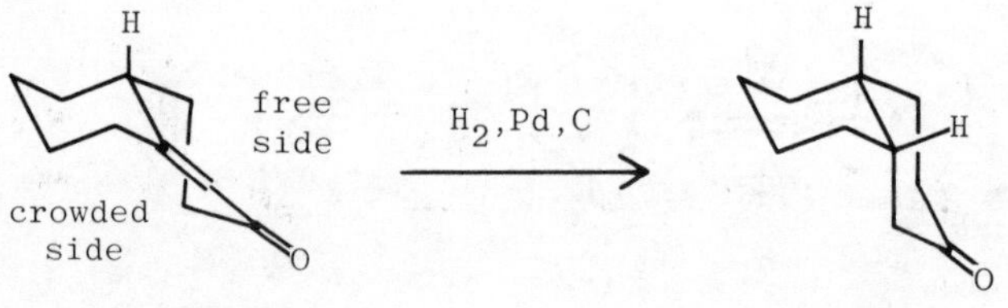

(47)   (48)

The synthesis of the *cis* amines (49) as potential analgesics by Allen and Hanbury's[133] is a case in point. The amine must be put in by reduction (Chapter 8) and hydrogenation of the oxime (50) adds hydrogen to the opposite side from the Ar group.

Analysis

(49)   (50)

Synthesis[133]

CHAPTER 13

One-Group C–C Disconnections II:
Carbonyl Compounds

In Chapter 10 the routes to carbonyl compounds were based on oxidation of alcohols. This chapter groups together a number of direct approaches to carbonyl compounds by C–C disconnections, and examines the thinking behind them. A new type of selectivity, regioselectivity, will be introduced here and treated more fully in the next chapter.

Carbonyl Compounds by 1,1 C–C Disconnections

1,1-Disconnection of ketones gives the synthon (1), for which we have previously used an ester, and a Grignard reagent. However, this approach is doomed from the start (see Chapter 10) as the ketone is formed in the presence of the Grignard reagent and is more reactive than the ester. Alcohols are formed.

$$R^1 \underset{}{\overset{O}{\diagup\hspace{-0.5em}\diagdown}} R^2 \implies R^1 \underset{}{\overset{O}{\diagup\hspace{-0.5em}\diagdown}}_+ \quad + \quad R^2MgBr$$

(1)

There are two ways round this problem. One is to use a less reactive organometallic reagent which reacts with acid chlorides but not with the less reactive ketones. Cadmium is a good metal for this purpose and the reagents (2) can be made from Grignard reagents or organolithiums.

$$R^2MgBr \searrow$$
$$\xrightarrow{\text{Cd(II)}} R^2_2Cd \xrightarrow{R^1COCl} R^1 \underset{}{\overset{O}{\diagup\hspace{-0.5em}\diagdown}} R^2$$
$$R^2Li \nearrow \quad (2)$$

The synthesis[134] of the ant alarm pheromone (3) used this strategy because the optically active acid (4) was available and only the right enantiomer alarms the ants.

Analysis

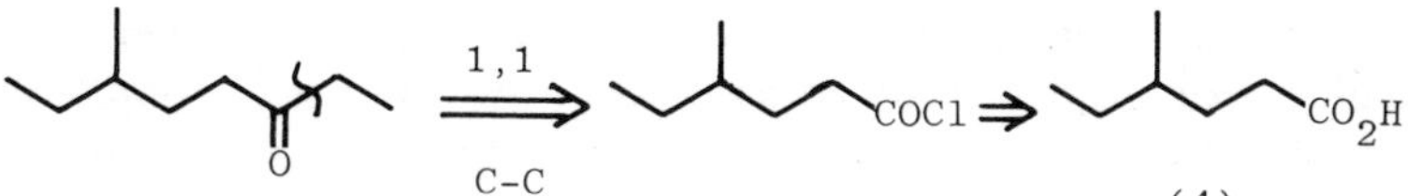

(3)   1,1   C–C   (4)

Synthesis[134]

$$(+)-(4) \xrightarrow[\text{2.Et}_2\text{Cd}]{\text{1.SOCl}_2} \text{TM(3)}$$
$$(+)$$

The other way is to use a reagent for **(1)** which does not form the ketone in the presence of the Grignard reagent. Nitriles[135] (cyanides) are ideal as the ketone is formed only when the imine **(5)** is hydrolysed during acidic work-up. The reaction between a carboxylate salt **(6)** and an alkyl-lithium achieves the same result.[136]

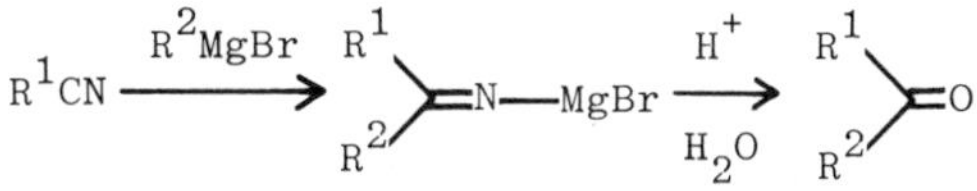

(5)

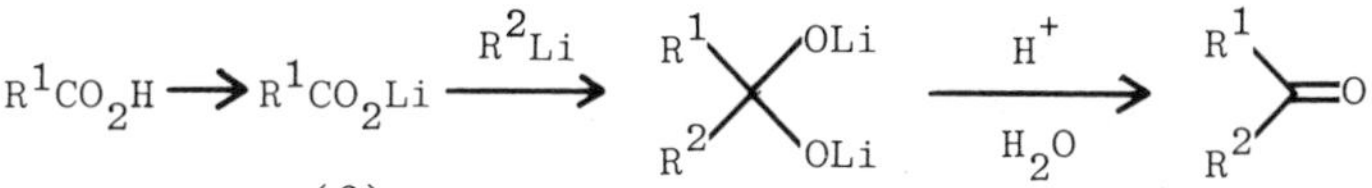

(6)

Carbonyl Compounds by Alkylation of Enols[137]

The logical approach by the 1,2 C–C disconnection requires the alkylation of an enol or enolate anion **(7)** by an alkyl halide (strategy (c) in Table 10.1). This is the natural polarity of the α-carbon in a carbonyl compound.

Analysis

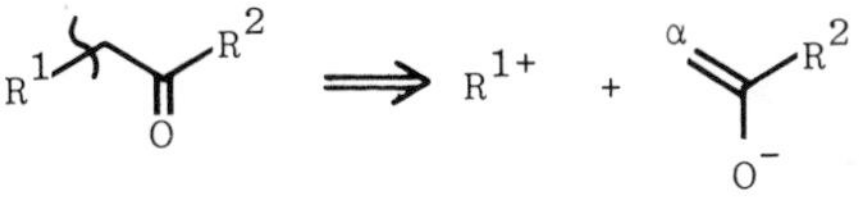

(7)

Synthesis (no good)

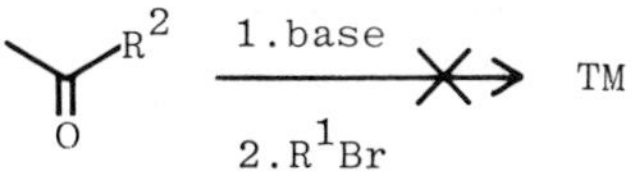

Natural and logical this idea may be, but there are serious difficulties in its execution. The product may alkylate too, and the enolate ion **(7)** may prefer to react with either ketone (starting material or product) rather than with the alkyl halide. For alkylation to be successful, an activating group, usually CO_2Et, must be present on the α-carbon atom. The reaction then becomes:

Synthesis (successful)

The activating group stabilises the enolate anion **(9)** by conjugation so that the keto ester **(8)** can be converted entirely into **(9)** with EtO^-. No reaction of **(9)** with **(8)** occurs, partly because **(9)** is so much more stable than **(7)**, and partly because no **(8)** is left. The alkyl halide is added in a separate step so that no base remains to form an anion of the product.

The activating group, in contrast to a protecting group, enhances reaction in one direction (enolisation here) but, like a protecting group, it must be easy to remove once the reaction is over. Hydrolysis of **(10)** gives the acid which loses CO_2 on heating **(11)** to give the ketone.

The equally easy addition of the activating group is treated in Chapter 20. In this chapter we shall use readily available malonate **(12)** and acetoacetate **(14)** esters, reagents for the synthons **(13)** and **(15)** respectively, that is activated versions of acetic acid and acetone.

$$\text{EtO}_2\text{C}\diagdown\diagup\text{CO}_2\text{Et} \quad \text{HO}_2\text{C}—\text{CH}_2^-$$

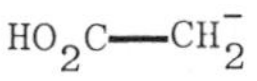

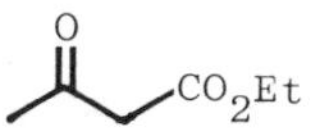

$$(12) \hspace{3cm} (13)$$

$$(14) \hspace{3cm} (15)$$

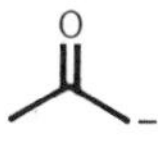

Previously (see Chapter 10) we synthesised acid **(16)** by oxidation of an alcohol. An alternative is to disconnect at the branch point and use malonic ester.[138]

Analysis

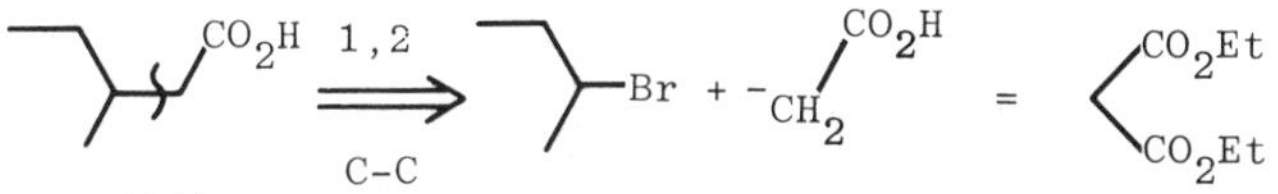

$$(16)$$

Synthesis[138]

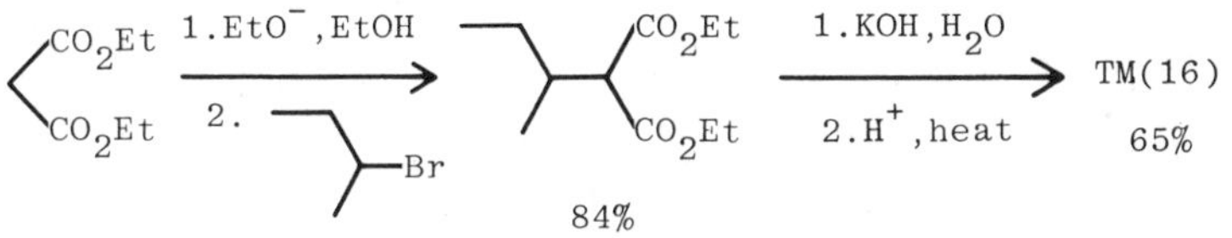

The process can be repeated to make acids with branched chains at the α-atom: the order of the disconnections is not important. The long chain fatty acid **(17)** has been made[139] by this method.

Analysis

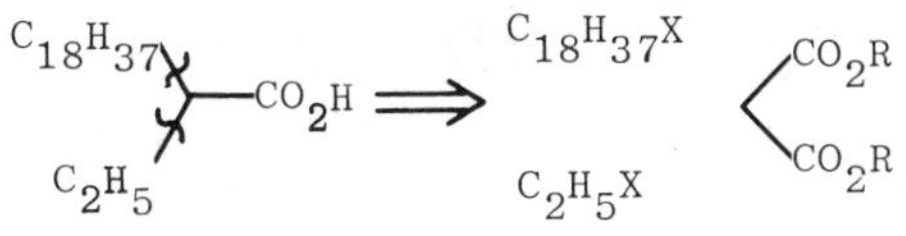

$$(17)$$

As it happens, the methyl ester was used here, with *t*-butoxide as the base in the second step. There is no great significance in these changes.

Synthesis[139]

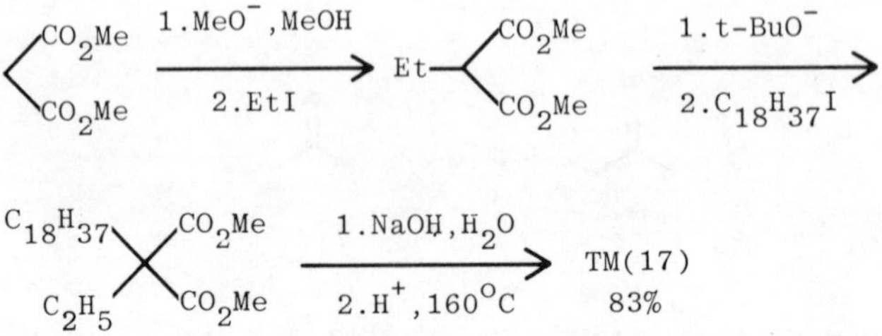

A good example of a ketone made by this approach is the industrial compound **(18)** discussed in Chapter 1. The alkylating agent is an allylic halide so the reaction should be rapid.

Analysis

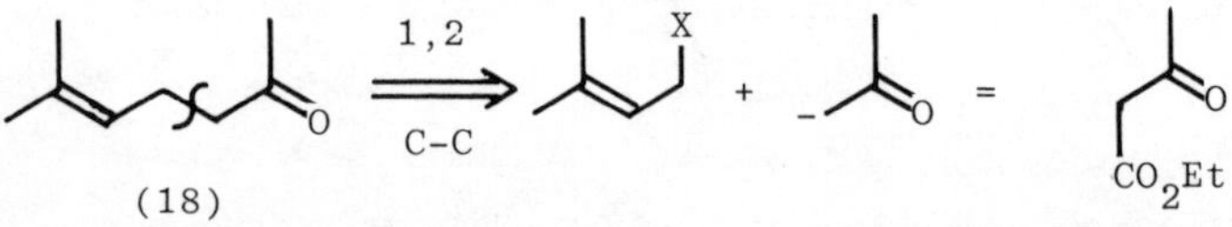

Synthesis[140]

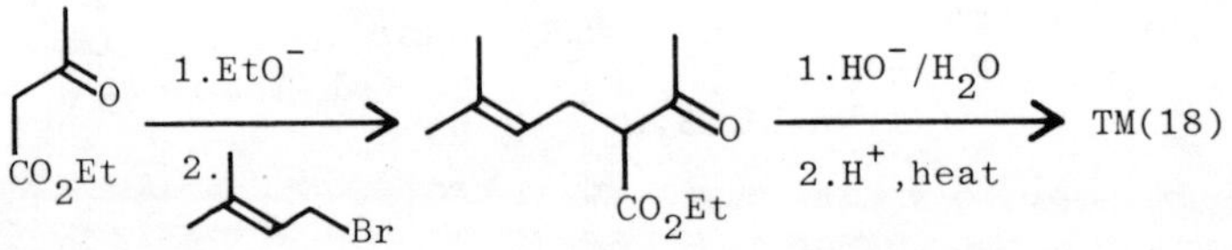

All the usual FGIs can be carried out on these alkylated products. One which deserves special mention is reduction instead of removal of the activating group. This inevitably gives a 1,3-diol, such as **(19)**, needed for the synthesis of tetracycline antibiotics.[141] Reduction of malonate **(20)** gives **(19)** and a normal malonate disconnection completes the analysis.

Analysis

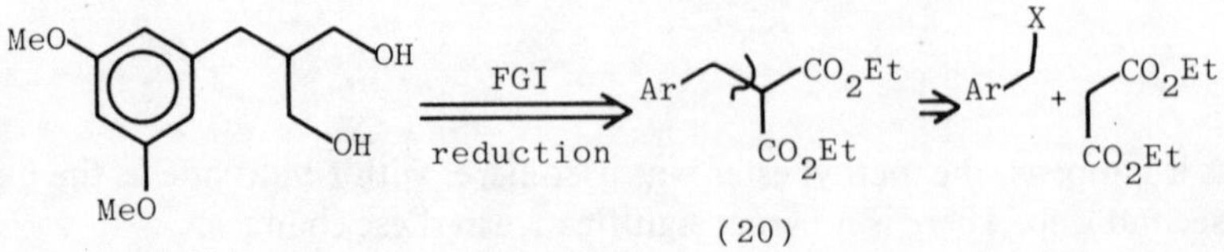

Synthesis[141]

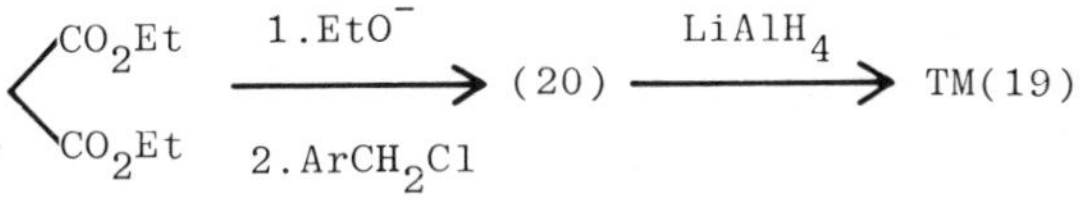

CO2Et
CO2Et
1.EtO⁻
2.ArCH2Cl
(20)
LiAlH4
TM(19)

Carbonyl Compound Synthesis by Michael Addition

R1
R2
O
1,3
C-C
R1-
R2
O

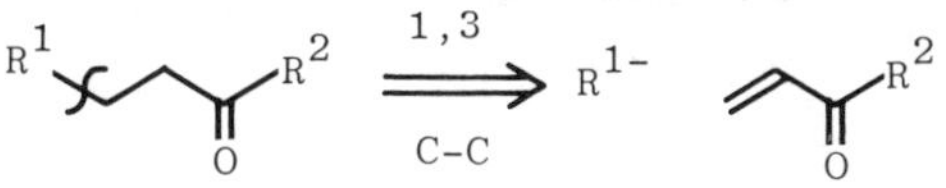

(21)

The 1,3-disconnection we met in Chapter 6 is also effective for carbon nucleophiles **(21)**. The reaction is the Michael addition of carbanions to α,β-unsaturated carbonyl compounds and we may expect Grignard reagents or RLi to do this reaction. We shall look for this disconnection when there is a branch point at the β or γ carbon atoms, and particularly when the bond to be disconnected joins a ring to a chain. Ketone **(22)** can clearly be made this way.

Analysis

O
R
1,3 C-C
O
+ R⁻

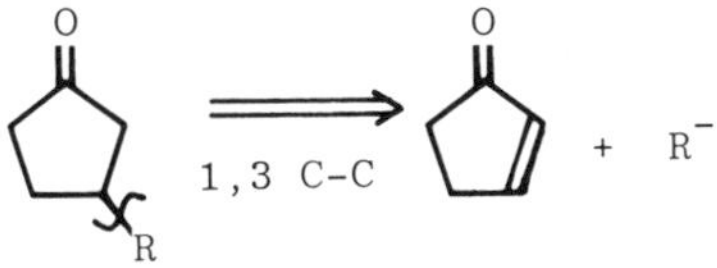

(22)

The published synthesis[142] uses a curious catalyst but follows the same strategy. We used compounds like **(22)** in Chapter 12.

Synthesis[142]

O
RLi
n-Bu3P,Cu(1)I
TM(22)

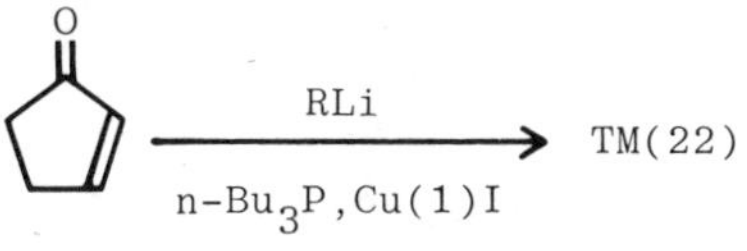

Where we have a choice, it is better to disconnect the longer chain to get back to simple starting materials more quickly, as with ester **(23)**. Here the Grignard reagent was used[143] without catalyst.

Analysis

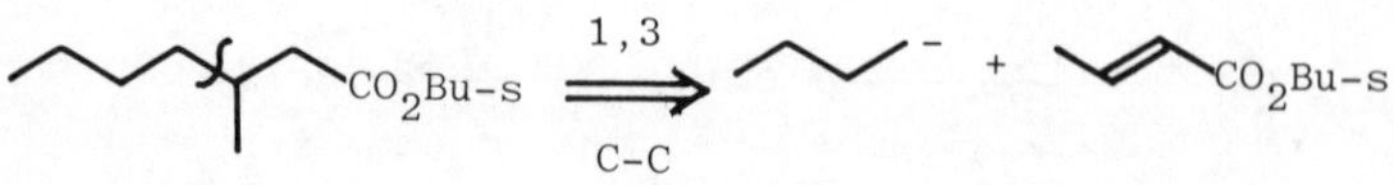

CO_2Bu-s
1,3
C-C
CO_2Bu-s

(23)

Synthesis[143]

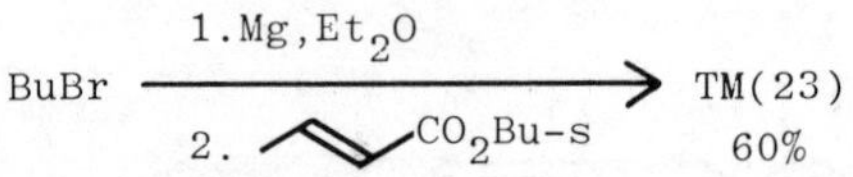

BuBr
1. Mg, Et_2O
2. CO_2Bu-s
TM(23)
60%

Aromatic compounds are good enough nucleophiles to add in this way under Friedel–Crafts conditions so that it is not necessary to make an organometallic reagent in the synthesis[144] of acid **(24)**.

Analysis

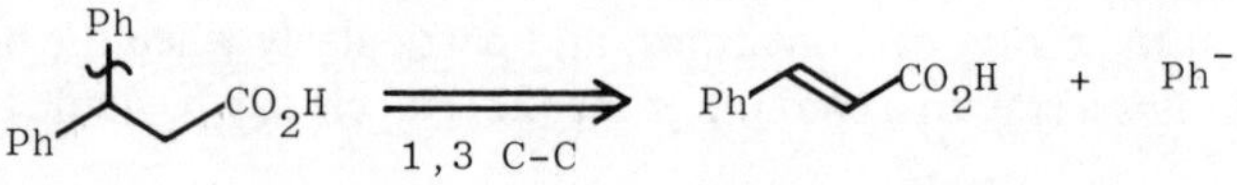

Ph
Ph
CO_2H
1,3 C-C
Ph
CO_2H
+
Ph^-

(24)

Synthesis[144]

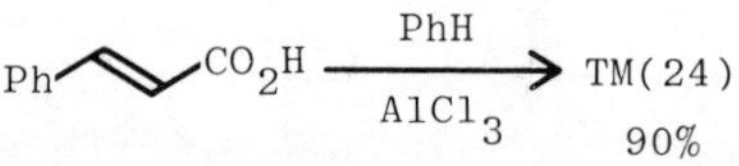

Ph
CO_2H
PhH
$AlCl_3$
TM(24)
90%

An example where stereochemistry is important is the diol **(25)**. This could be made by reduction of the keto ester **(26)** in the manner described in Chapter 12, and the stereochemistry of the cyclic alcohol is right for kinetically controlled reduction (see Chapter 12).

Analysis 1

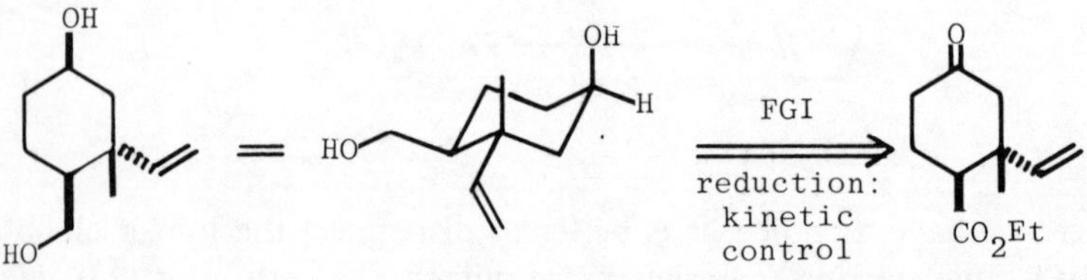

OH
HO
HO
OH
H
FGI
reduction:
kinetic
control
O
CO_2Et

(25) (26)

The two remaining chiral centres are adjacent so we should get high stereo-selectivity if we disconnect the group (vinyl) on the opposite side from the CO_2Et group. The starting material is the readily available Hagemann's ester **(27)**.

Analysis 2

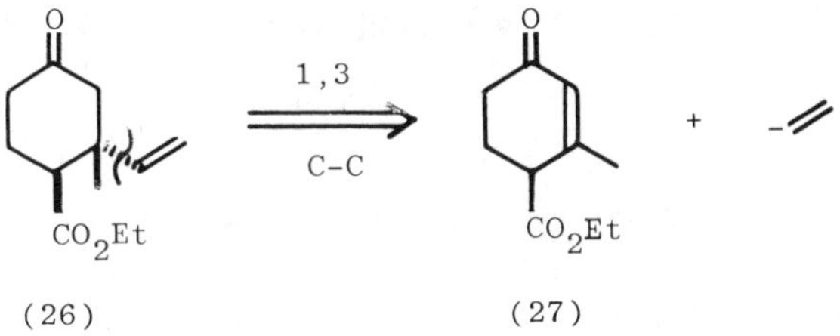

A copper catalyst is again used[145] in the Michael addition step. This catalyses Michael rather than direct addition to the ketone and is discussed in the next chapter. Under these conditions the additions is indeed highly stereoselective in favour of the required isomer. Reduction with $LiAlH_4$ favoured the axial alcohol as expected. Diol **(25)** was used in a synthesis of steroid analogues.

Synthesis[145]

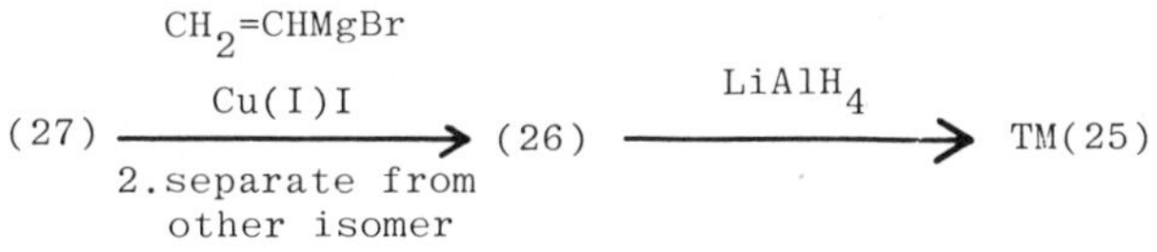

CHAPTER 14

Strategy VI: Regioselectivity

Chapter 5 dealt with *chemo*selectivity: how to react one FG and not another. Now we must face a more subtle and demanding problem—*regio*selectivity —how to react one specific part of a single FG and no other. We have already met problems of this kind. We have seen that phenolate ions (1) react with alkylating agents at oxygen but that enolate ions (2) usually react at carbon.

O-alkylation:

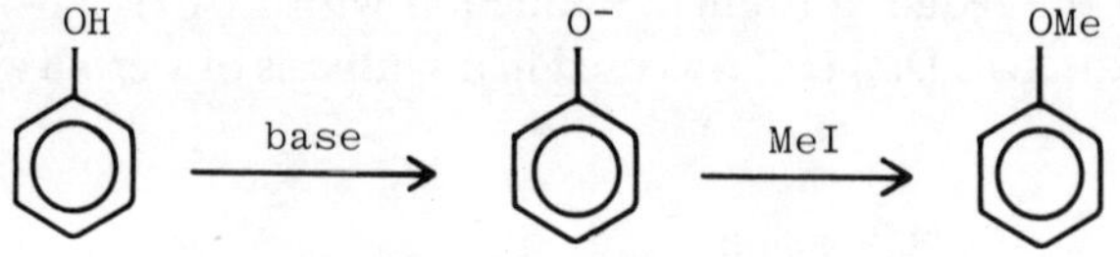

(1)

C-alkylation:

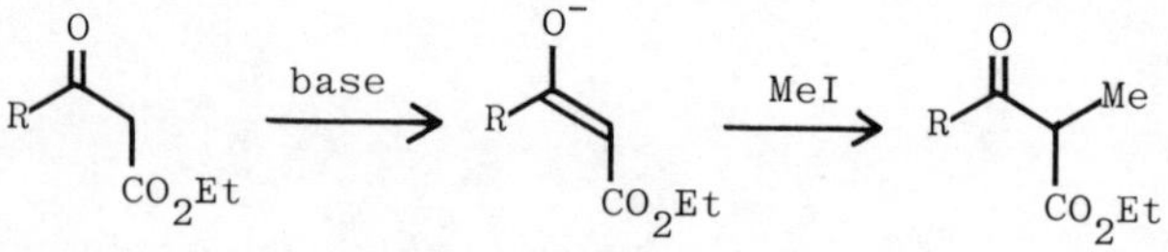

(2)

These regioselectivities can be changed but the natural results are the most helpful. In this chapter we shall look at two problems where both possibilities are useful. How can we alkylate (or brominate etc.) one side of an unsymmetrical ketone (3) and how can we control the addition of nucleophiles to unsaturated carbonyl compounds (4) to get either Michael (5) or direct (6) addition?

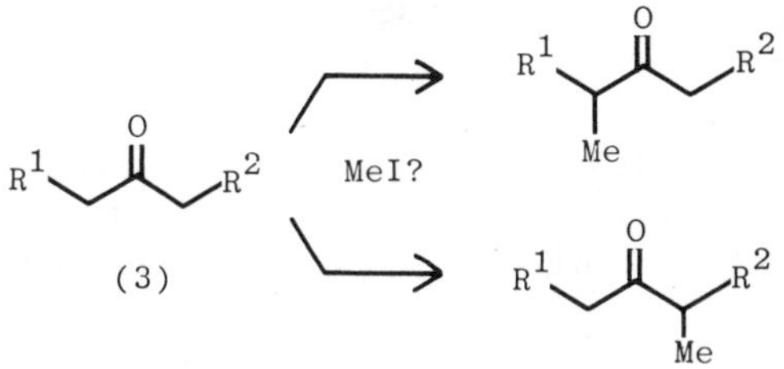

Regioselective Alkylation of Ketones

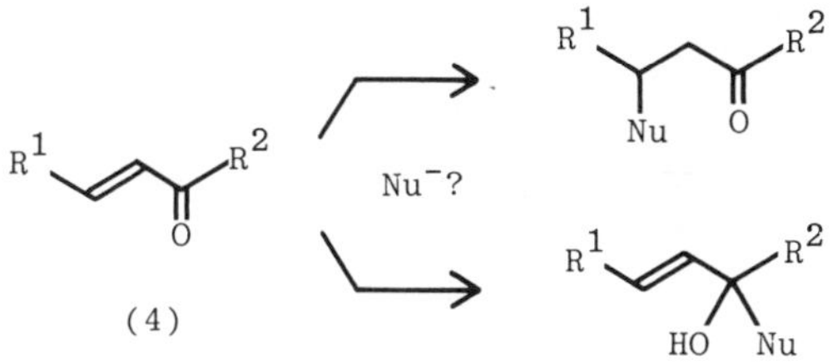

This problem as outlined above **(3)** cannot be solved within the scope of the book. However, it is possible by strategy to synthesise either target molecule by alkylation. Suppose ketone **(7)** is the target: this might be made by benzylation of the ketone **(8)**. We cannot selectively benzylate **(8)**, but if we make **(8)** by alkylation too, then the synthesis is achieved.

Analysis

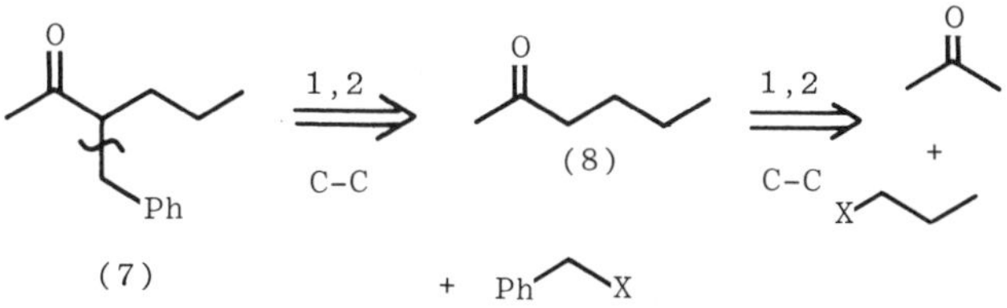

We shall need an activating group for alkylation (Chapter 13) and the same activating group can be used for both alkylations. The starting material must then be our synthetic equivalent for acetone—ethyl acetoacetate.

Synthesis[146]

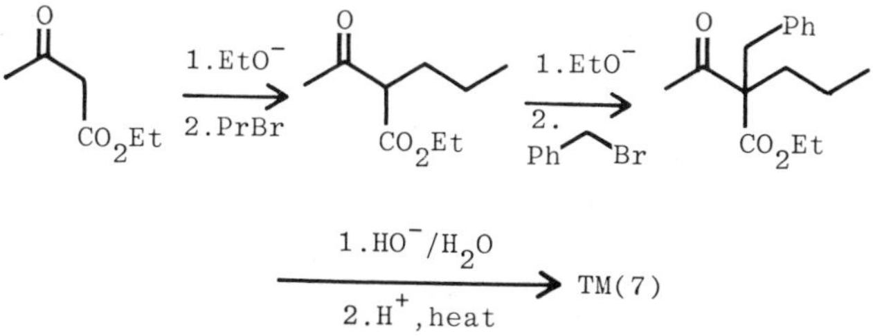

The order of events may be changed. The regioselectivity problem disappears in the alkylation of an ester since esters can enolise on one side only. Therefore, if we complete all our alkylations before making the ketone, we can start from malonate and make unsymmetrical ketones. This means disconnecting the ketone (9) first by a 1,1 C–C process (see Chapter 13). In later chapters devices like this will become more important as target molecules get larger.

Analysis

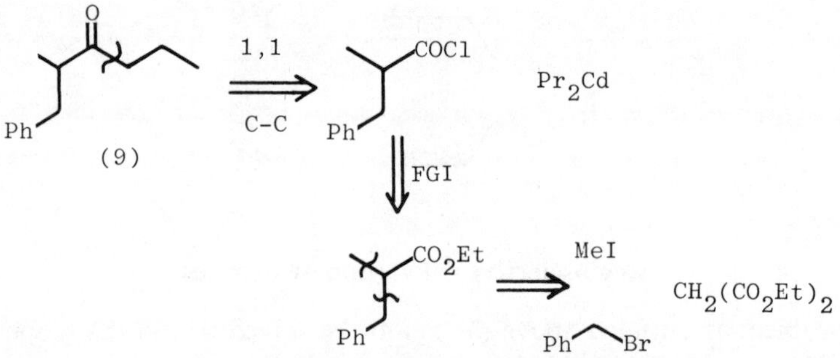

Synthesis[147]

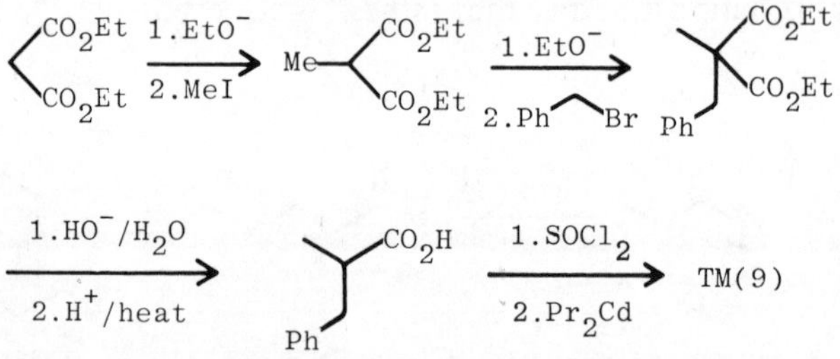

Regioselectivity in Michael Reactions

The problem of direct (1,2) versus Michael (1,4 conjugate) addition to α,β-unsaturated carbonyl compounds can be solved without finding another strategy. The general mechanistic principles follow.

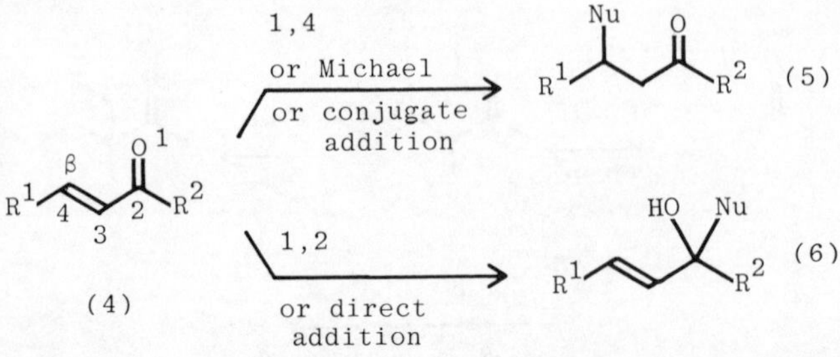

1. The Michael product (6) is the thermodynamic product since the more stable C=O bond is preserved and the weaker C=C bond destroyed.
2. Direct addition is more easily reversed than Michael addition. Hence the more stable the nucleophile, or the more reversible the addition, the more Michael addition is favoured.
3. Kinetically, the C=O is the harder site and the β carbon atom the softer. Therefore strongly basic nucleophiles tend to attack directly whereas weakly basic nucleophiles tend to attack in Michael fashion.

Among Michael acceptors, direct addition is most likely with unsaturated aldehydes and acid chlorides, Michael addition most likely with ketones or esters. Hence Grignard reagents add directly[148] to aldehyde (10) and Michael fashion[143] to ester (11).

CHO   RMgBr   HO R

(10)   85–90%

BuMgBr

CO_2Bu-s Bu CO_2Bu-s

(11)   60%

Among nucleophiles,[149] RLi, NH_2^-, RO^-, and hydride reducing agents are more likely to add direct: RMgBr, neutral amines, RS^-, and stable carbanions are more likely to add Michael fashion, and we have seen many examples of these in Chapters 6, 10, and 13. One useful example is the control over reduction of aldehydes, ketones, and esters (12) by choice of reducing agent. Hydrogenation[150] is not an ionic reaction—it selects the weaker bond and this is invariably the C=C bond. The alternative result is easily achieved[151] with $LiAlH_4$ or, for aldehydes and ketones, $NaBH_4$. Thus the ester (12) can give either saturated ester (13) or allylic alcohol (14) depending on the reagent used.

H_2 / Pd,C   R CO_2Et (13)

(12)

$LiAlH_4$ R OH   (14)

A dramatic example is the alcohol (15) whose benzyl ether was needed for a Diels–Alder reaction (Chapter 17). Reduction of the ester* (16) should be

*See workbook for Chapter 15.

regioselective in favour of carbonyl reduction if $LiAlH_4$ is used: in practice 87% of (15) can be isolated.[152]

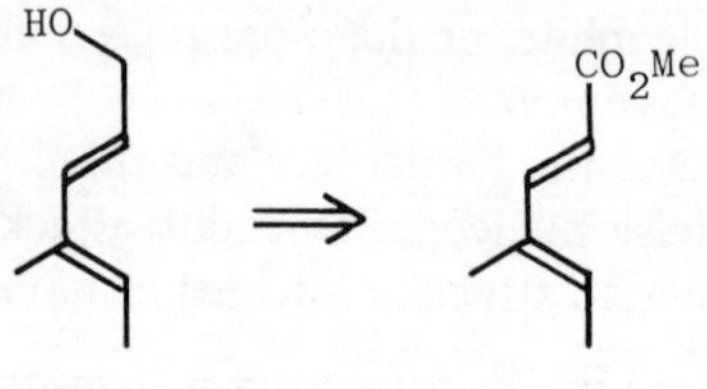

(15)   (16)

There is one reliable method for making Grignard or organolithium reagents add 1,4. That is to exchange the metal (Mg or Li) for Cu(I).[153] The exact mechanism is not known, but it almost certainly involves one-electron transfer. This is the role played by the 'curious' catalysts in Chapter 13. One can either use a Grignard reagent in the presence of Cu(I) salts, or an organocuprate (R_2CuLi) made from RLi and the Cu(I) salt.

Hence γ,δ-unsaturated ketone (17) can be disconnected at the branch point where ring and chain join to give cyclohexenone and a vinyl carbanion as starting materials. The synthesis has been carried out with the vinyl cuprate (18).[154]

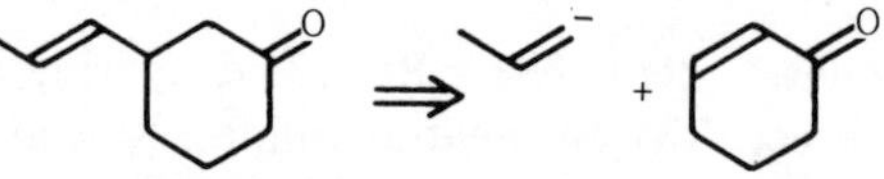

(17)

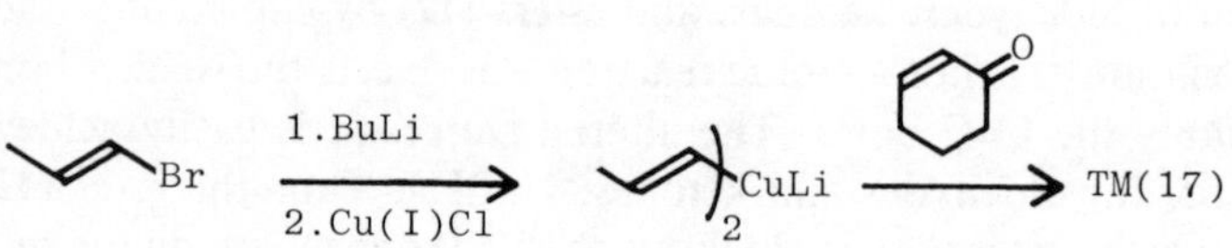

(18)

CHAPTER 15

Alkene Synthesis

By Elimination from Alcohols and Derivatives

Alkenes (olefins) can be made by the dehydration of alcohols (e.g. **1**), usually under acidic conditions, the alcohol being assembled by the usual methods. This route is particularly good for cyclic or branched olefins (e.g. **2**).

(1)   (2)

Acids must be fairly strong for this job and must have a non-nucleophilic counter ion or substitution may occur. Popular ones are $KHSO_4$ (crystalline and easier to handle than H_2SO_4) and phosphoric acid, or the less acidic $POCl_3$ in pyridine. Very little control is available over the position or geometry of the double bond but in many simple cases this is unimportant. In **(1)** for example, a double bond inside a six-membered ring is greatly preferred to one outside, and only *cis*-**(2)** is possible.

When Zimmermann[155] wished to study the photochemistry of a series of alkenes of general structure **(3)**, he could have put the OH group at either end of the double bond. Putting OH at the branch point is better strategy as disconnection of the alcohol **(4)** then gives simpler starting materials.

Analysis

(3)   (4)

The dehydration of this tertiary alcohol **(4)** will be very rapid by an E1 mechanism and there is no doubt about the position or geometry of the double bond.

119

Synthesis[155]

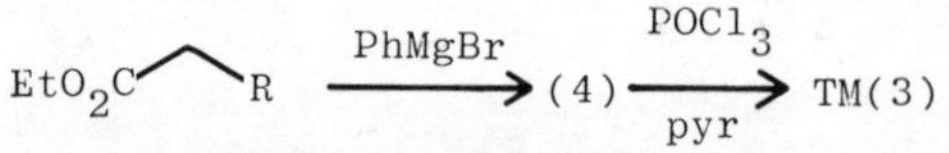

$$\text{EtO}_2\text{C} \diagdown \text{R} \xrightarrow{\text{PhMgBr}} (4) \xrightarrow[\text{pyr}]{\text{POCl}_3} \text{TM(3)}$$

Elimination reactions on alkyl halides follow essentially the same strategy as the halide is usually made from an alcohol. Eliminations of primary groups are better done this way with base instead of acid catalysis.

Analysis

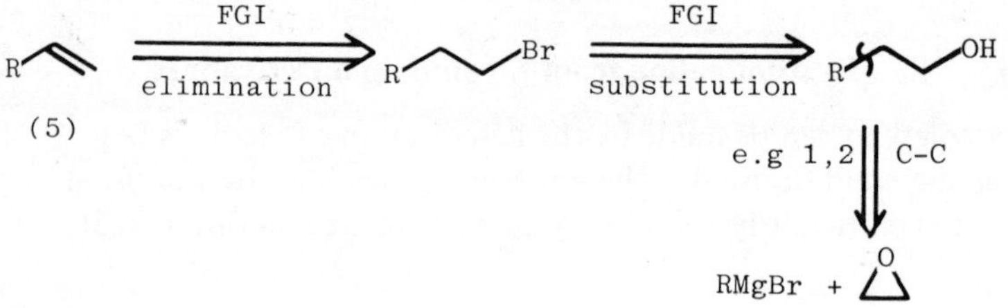

Synthesis

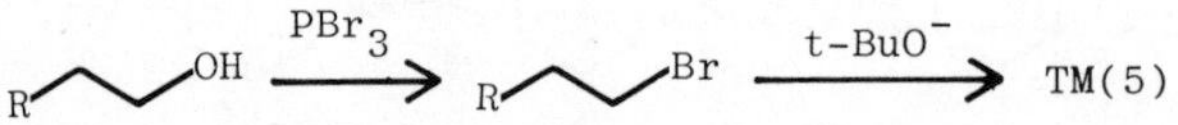

Dienes can be made by this approach if vinyl Grignards are used because the vinyl group blocks dehydration in one direction and makes the reaction faster by E1 as the intermediate is an allylic cation. An interesting example is the four-membered ring compound[156] **(6)**: note that the OH group is again added at the branch point.

Analysis

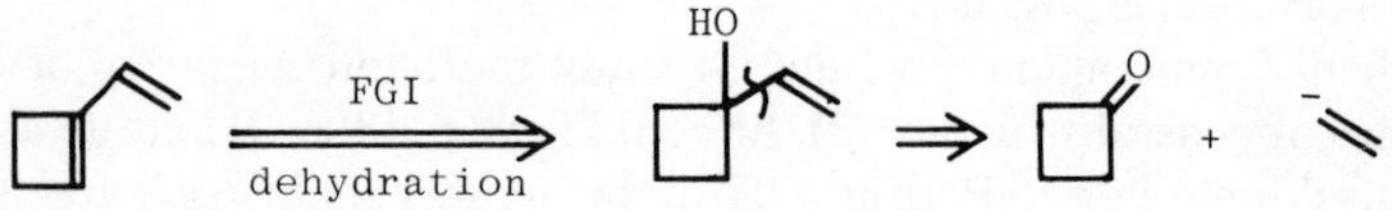

Synthesis[156]

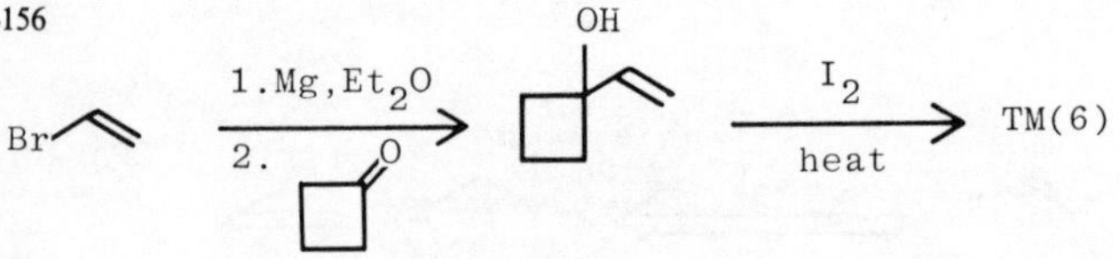

The Wittig Reaction

These methods of olefin synthesis have now largely been superseded by the Wittig[157] method which gives total control over the position of the double bond and partial control over its geometry. This reaction may be new to you: its mechanism follows.

The Wittig reagent

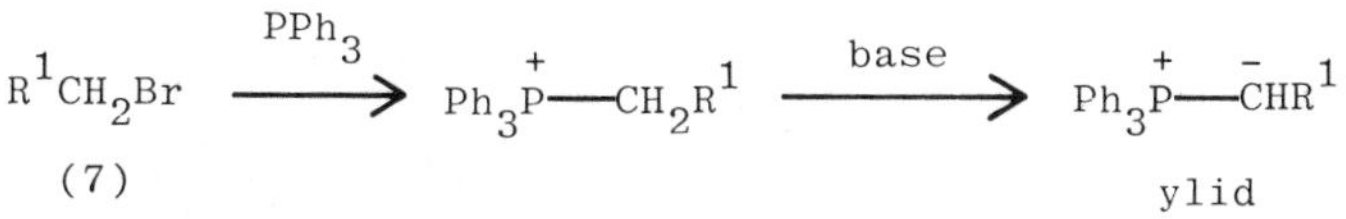

$$R^1CH_2Br \xrightarrow{PPh_3} Ph_3\overset{+}{P}\text{---}CH_2R^1 \xrightarrow{base} Ph_3\overset{+}{P}\text{---}\overset{-}{C}HR^1$$

(7)   ylid

The Wittig reaction

(8)

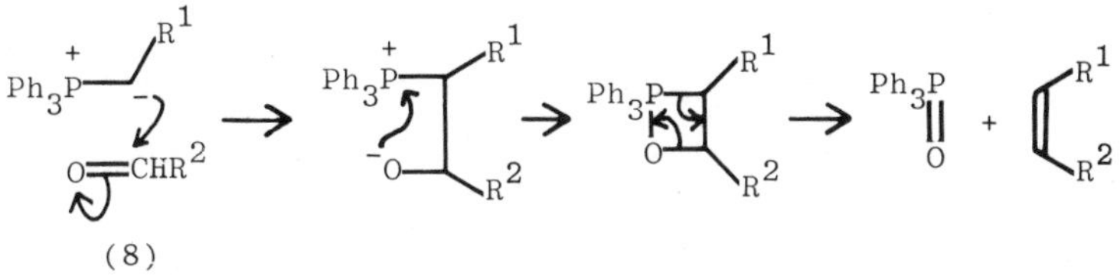

The Wittig reaction forms both σ and π bonds in one reaction so the disconnection is at the double bond with a nearly free choice of which end comes from the alkyl halide (7) and which from the carbonyl compound (8). Hence the exo-olefin (9), all but impossible to make by elimination, is easily made by either Wittig route. Route (a) is perhaps easier as cyclohexanone is easier to handle than formaldehyde.

Analysis

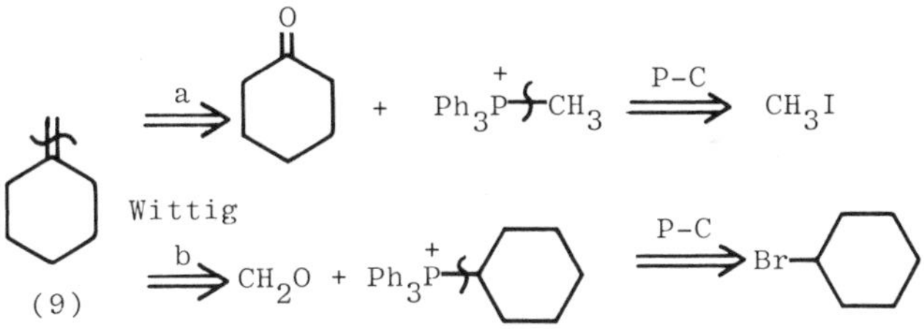

Synthesis[158]

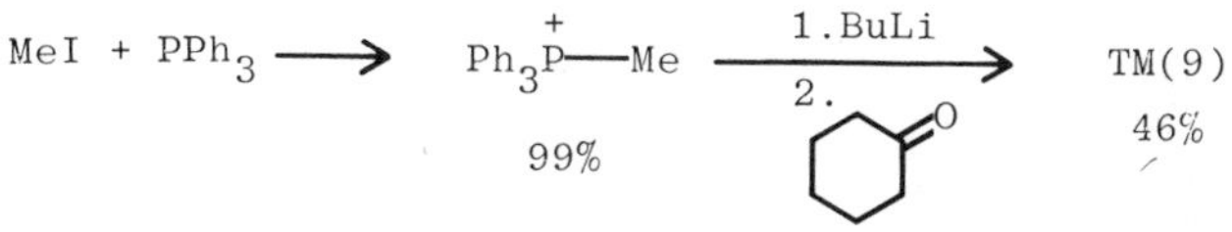

$$MeI + PPh_3 \longrightarrow Ph_3\overset{+}{P}\text{---}Me \xrightarrow[2.]{1.BuLi} TM(9)$$

99%   46%

Branched chains, e.g. (10), are no trouble as either a secondary halide (11) or a ketone can be used. This is another case where dehydration of even the branch point alcohol (12) would give a mixture of positional isomers.

Analysis

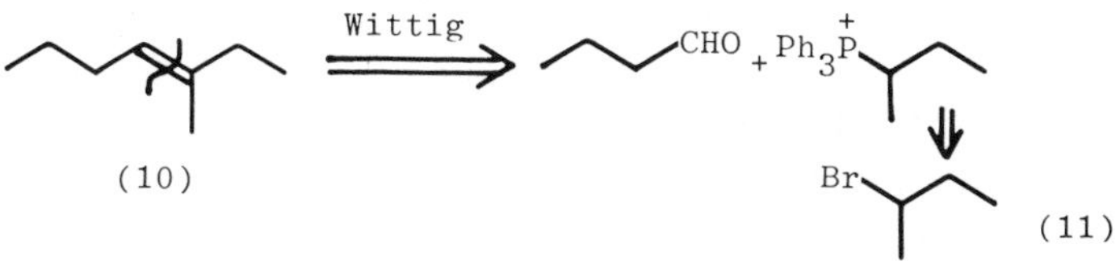

Synthesis[159]

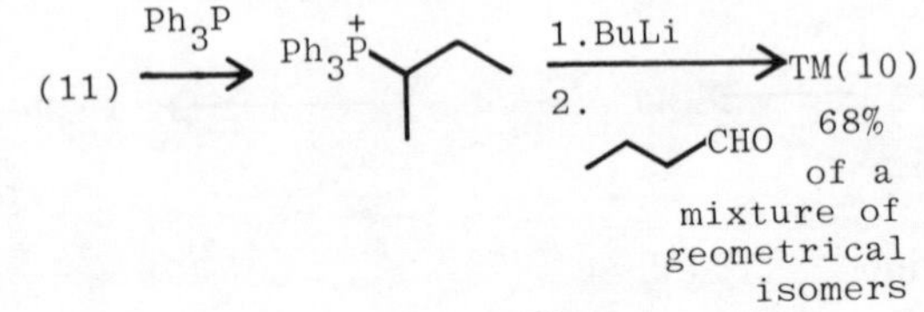

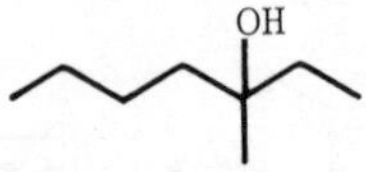

(12)

Substituted alkenes, e.g. (13), can be made by the Wittig reaction: the more reactive phosphonate ester (14) is often used when there is a stabilising group present.

Analysis

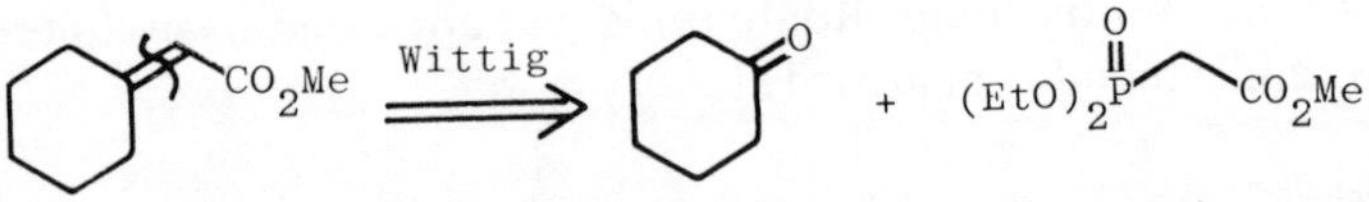

(13)

(14)

Synthesis[160]

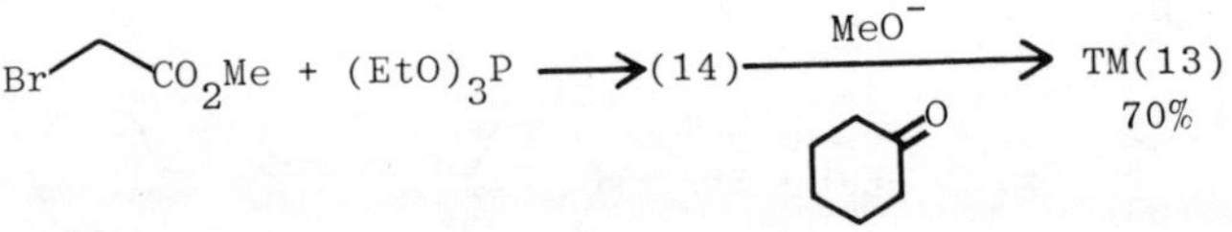

Stereoselectivity in Wittig reactions

The general rule[157] is that stabilised ylids (from **14**, or **15**, R^1 = Ar, COR, C=C etc.) react with aldehydes to give mainly *E* (*trans*) alkenes whilst unstabilised ylids (**15**, R^1 = alkyl) give mainly *Z* (*cis*) alkenes.

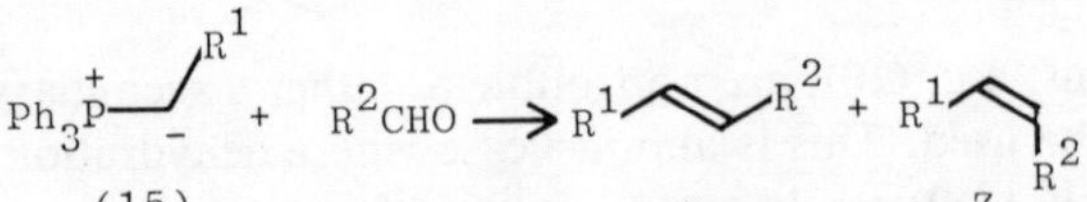

(15)

The optical brightener Palanil (16)—'whiter than white' washing powders contain these brighteners—is manufactured[2] by the Wittig reaction. Disconnection is bound to give an aryl-substituted ylid but the availability of aldehyde (17) (the corresponding acid is used to make terylene) makes this route the better of the two.

Palanil: *Analysis*

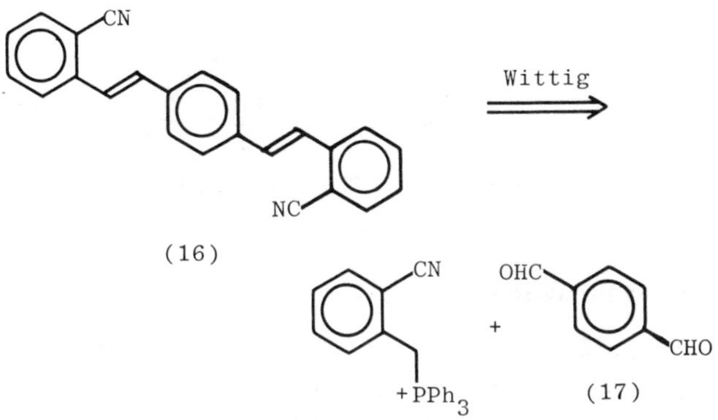

(16)

Wittig

(17)

The ylid is stabilised by CN as well as by aryl so the phosphonate ester (18) is used instead and the reaction is strongly *trans* selective.

Synthesis[2]

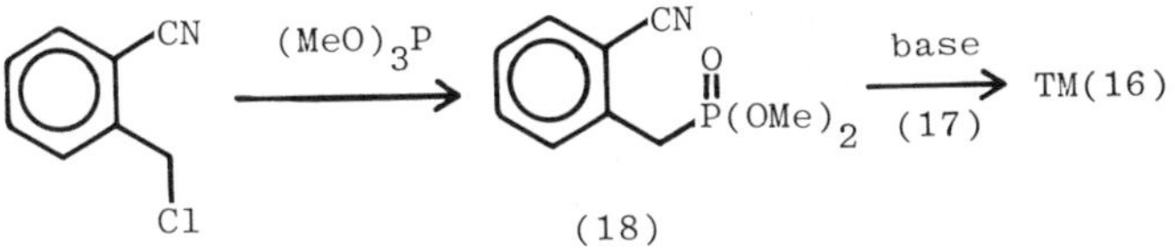

(18)

Many insect pheromones are derivatives of simple alkenes. Disparlure (19), an attractant for the gypsy moth, is an epoxide derived by stereospecific epoxidation (see Table 12.1) from a *cis* alkene (20). Disconnection in either sense gives an unstabilised ylid so *cis* selectivity is expected in the Wittig reaction.

Analysis

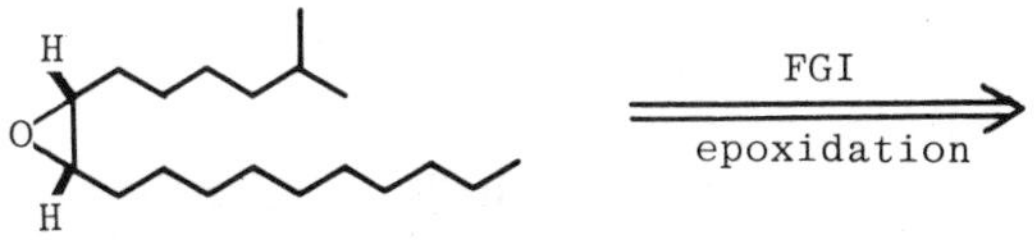

(19)

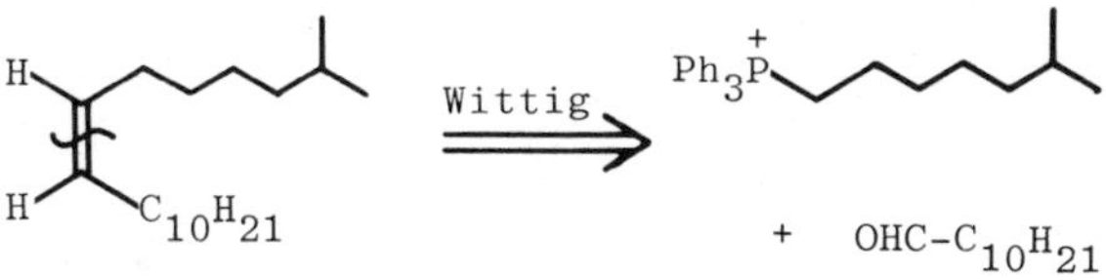

(20)

(21)

The synthesis has in fact been carried out with the reagents shown (21) in high yield and high stereoselectivity. No doubt the alternative combination would give equally good results. The synthetic material is as attractive to the moth as the natural product.

124

Synthesis[161]

$$Br\diagdown\diagup\diagdown\diagup\diagdown\diagup\diagdown\diagup \quad \xrightarrow{Ph_3P} \quad Ph_3\overset{+}{P}\diagdown\diagup\diagdown\diagup\diagdown\diagup\diagdown\diagup$$

86%

$$\xrightarrow[\text{2.C}_{10}\text{H}_{21}\text{CHO}]{\text{1.BuLi}} \quad (20) \quad \xrightarrow{\text{MCPBA}} \quad \text{TM}(19)$$

91% 83%

Diene synthesis by the Wittig reaction

Conjugated dienes **(22)** are important intermediates in synthesis as they are used in Diels–Alder reactions (Chapter 17). Again, either disconnection is acceptable, giving allylic phosphonium salts **(23)** and simple aldehydes or unstabilised phosphonium salts **(25)** and enals **(24)** as starting materials. If R^1 and R^2 are simple alkyl groups, route (a) is likely to give more *trans* double bond, route (b) more *cis*. The geometry of the other double bond is unaffected by the reaction and is decided by the starting materials **(23)** and **(24)**. Choosing which double bond to disconnect is therefore a matter of stereochemistry as well as the usual strategic guidelines (Chapter 11).

Analysis

$$R^1\diagup\diagdown\diagup\diagdown\diagup\diagdown R^2 \quad \underset{\text{(22)}}{}$$

$$\overset{a}{\Longrightarrow} R^1\diagup\diagdown\diagup\overset{\overset{+}{PPh_3}}{} \quad + \quad R^2CHO$$
(23)

$$\overset{b}{\Longrightarrow} R^1\diagup\diagdown\diagup CHO \quad + \quad Ph_3\overset{+}{P}\diagup\diagdown R^2$$
(24) (25)

The simple 1-alkyl butadiene **(26)** was needed for a Diels–Alder reaction[159] (see Chapter 17) and the central double bond had to be *trans*. The best disconnection is (a), giving two nearly equally small readily available starting materials. The direction of disconnection is chosen to give a stabilised ylid **(27)** and hence a *trans* double bond.

Analysis

$$\diagup\diagdown\diagup\diagdown\diagup \overset{a}{\underset{\text{(26)}}{\Longrightarrow}} \xrightarrow{\text{Wittig}} \diagdown\diagup\diagdown CHO + \diagup\diagdown\overset{}{+PPh_3} \overset{C-P}{\Longrightarrow} \diagup\diagdown\diagdown_{Br} + PPh_3$$

(26)

Synthesis[159]

$$Br\diagdown\diagup\diagdown \quad \xrightarrow{PPh_3} \quad (27) \quad \xrightarrow[\text{2.PrCHO}]{\text{1.BuLi}} \quad \text{TM}(26)$$

52%

The alternative direction ((b) on page 124) is suitable for 1,4-diaryl butadienes **(28)**. These were needed with a variety of substituents for a systematic study of electronic effects on the Diels–Alder reaction. Disconnection to give benzyl phosphonium salts **(30)** and easily made cinnamaldehydes **(29)** (see Chapter 18) is best.

Analysis

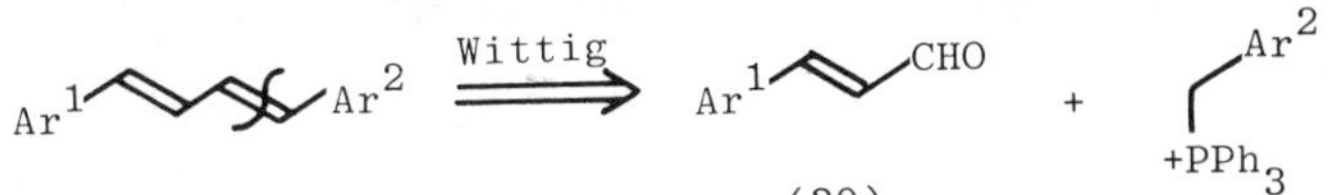

Synthesis[162]

CHAPTER 16

Strategy VII: Use of Acetylenes

This is a chapter different in kind from any before. We shall look at one class of compound — the acetylenes (alkynes) — and see what special jobs they can do in synthesis, particularly how they solve some problems already raised.

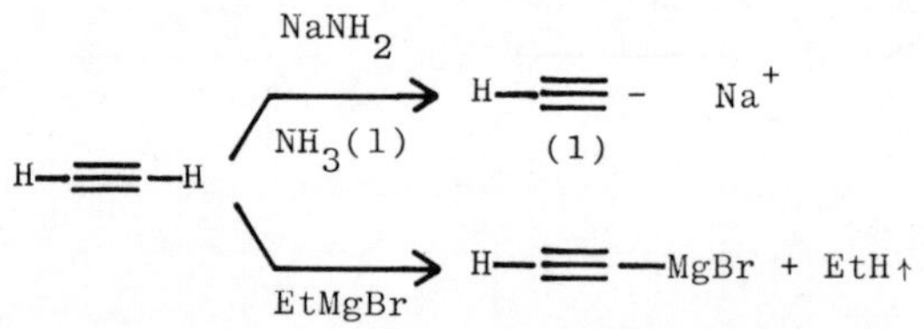

Acetylene itself is readily available and its first important property is its ability to form an anion (1), usually with sodamide in liquid ammonia, but also by Grignard exchange. This is one of the few carbanions available as a genuine intermediate. It reacts with alkyl halides, carbonyl compounds, and epoxides to give intermediates (2)-(4).

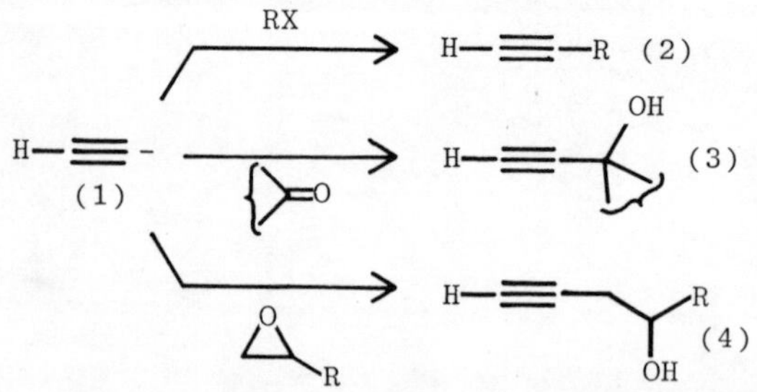

These products still have an acetylenic proton so they can react again with base and an electrophile, e.g. to give (5). Disconnection of any bond next to a triple bond is therefore reasonable.

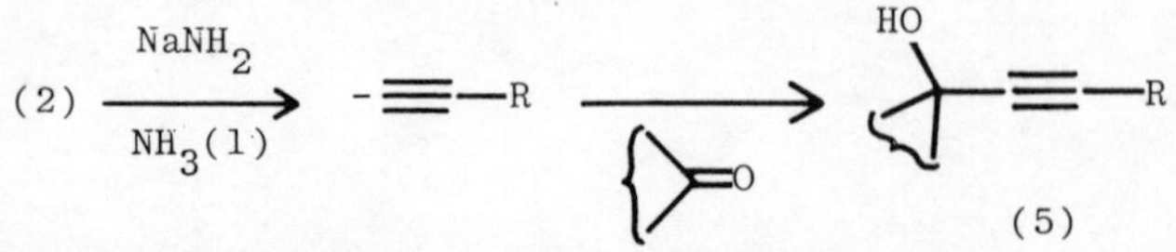

The tranquilliser Oblivon (6) is clearly an acetylene adduct[163] and the defoaming surfactant[164] Surfynol (7) is clearly an acetylene diadduct. The syntheses are straightforward.

Oblivon: *Analysis*   *Synthesis*[163]

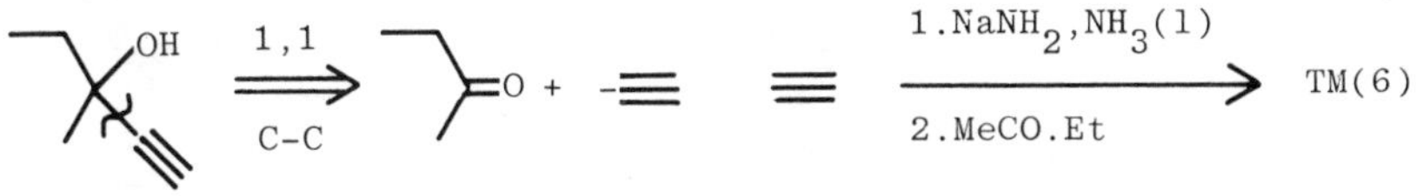

(6)

Surfynol: *Analysis*

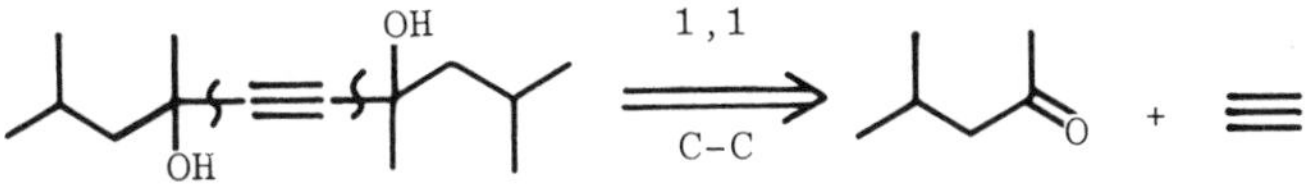

(7)

Synthesis[164]

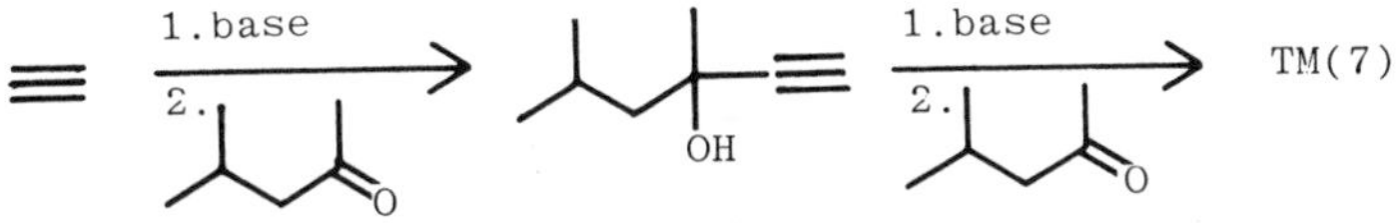

Otherwise few of these acetylene adducts are important in their own right, but they are valuable intermediates because disubstituted acetylenes (8) can be reduced to *cis* or *trans* alkenes at will and because monosubstituted acetylenes can be hydrated to methyl ketones (9).

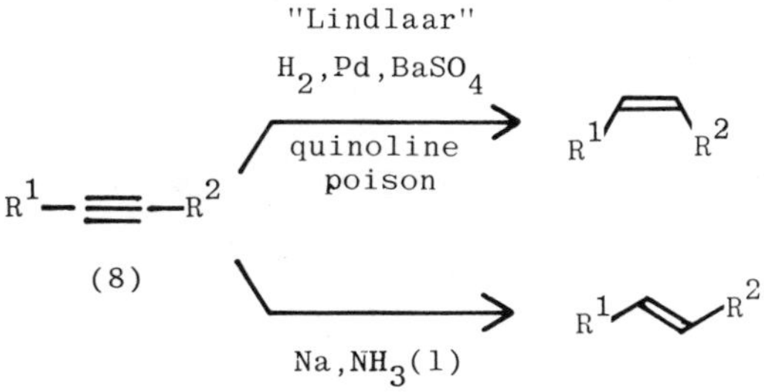

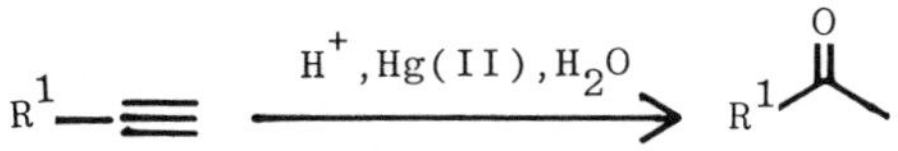

The important intermediate *cis*-butene diol **(10)** can be made by hydrogenation of the corresponding acetylene **(11)**. The butyne diol **(11)** is made commercially by a catalytic process using the strategy we have developed here.*

Analysis

$$ HO\diagdown\diagup OH \xrightarrow[\text{reduction}]{FGI} HO \diagdown\!\!=\!\!\diagup OH \Rightarrow \begin{array}{c} 2CH_2O \\ + \\ \equiv \end{array} $$

(10)   (11)

Synthesis[166]

$$ \equiv \xrightarrow[\substack{\text{metal} \\ \text{catalyst}}]{CH_2O} (11) \xrightarrow[\substack{H_2,Pd,BaSO_4 \\ \text{poison}}]{\text{Lindlaar}} TM(10) $$

(12)
cis-jasmone

The important perfumery constituent *cis*-jasmone **(12)** can be synthesised if a source of the *cis*-C5 side chain is available. Bromide **(13)** is often used for this purpose as, after it has been added to the rest of the molecule, Lindlaar reduction gives the *cis* double bond. The bromide **(13)** is clearly derived from alcohol **(14)**. Either disconnection (a) or (b) could be used here: we shall follow (a) as an important point of chemoselectivity emerges from it.

Analysis

(13)   (14)

The synthesis[167] will require the alkylation of dianion **(15)** on carbon. This is where alkylation does occur as the carbanion (pK_a of acetylene $\sim$ 25) is much more reactive than the oxyanion (pK_a of an alcohol $\sim$ 16) and protection of the alcohol is unnecessary.

*The fascinating story of the reluctant Reppe not divulging this process to the allies after World War II is told in ref. 165. We used **(10)** in Chapter 12.

Synthesis[167]

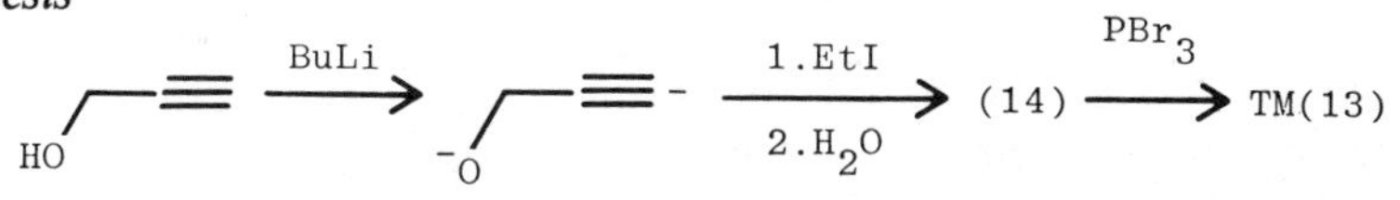

(15)

The *trans* acetate (16) is the pheromone used to trap pea moths[161, 168] to tell the farmer exactly when to spray to eliminate these destructive pests. It can be made from *trans* alcohol (17) and hence from acetylene (18) by reduction. Disconnection of (18) eventually requires an unsymmetrical compound (19). Diol (20) is available in large quantities so the statistical method (Chapter 5) can be used as very high yields are not so important when the starting material is cheap and the TM required in only small amounts.

Pea-moth pheromone: *Analysis*

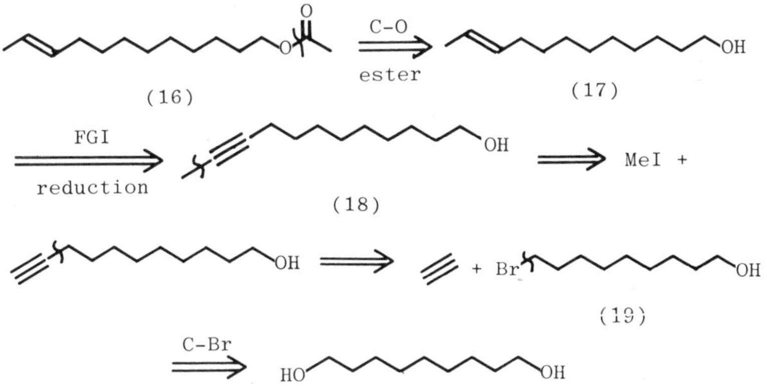

Experiments[168] showed that protecting one end of the diol (20) as its THP derivative (Chapter 9) gave good yields and the mono-bromo compound (21) could then be made. The protecting group was retained nearly to the end of the synthesis.

Synthesis

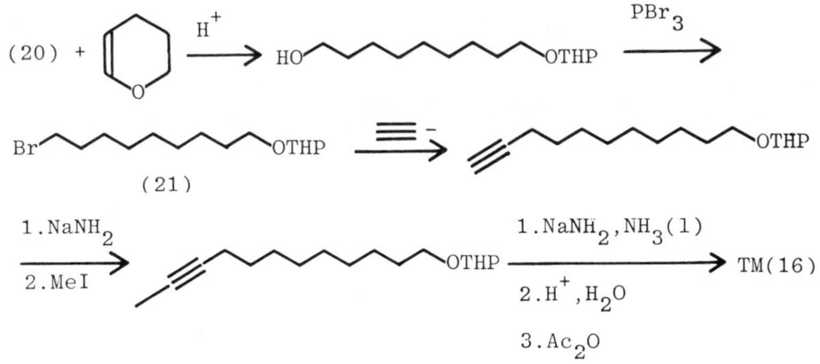

Diene Synthesis

Acetylenes are particularly useful for making dienes of type **(22)**. The double bond outside the ring comes from the acetylene and that inside the ring from the dehydration of an alcohol. This is very similar to the strategy outlined in Chapter 15 — but here the acetylene is the reagent for the vinyl anion synthon.

Analysis

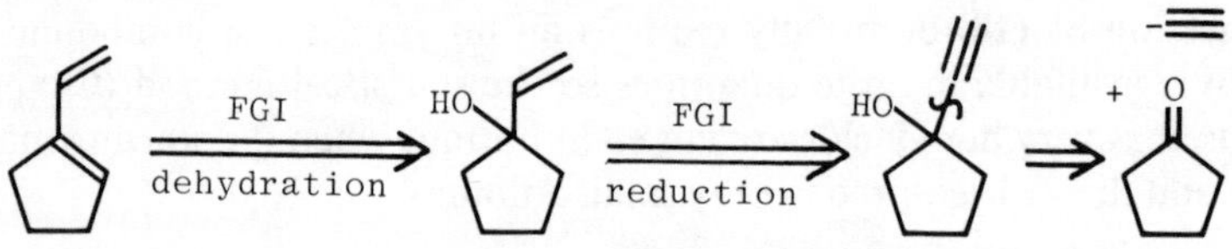

Synthesis[169]

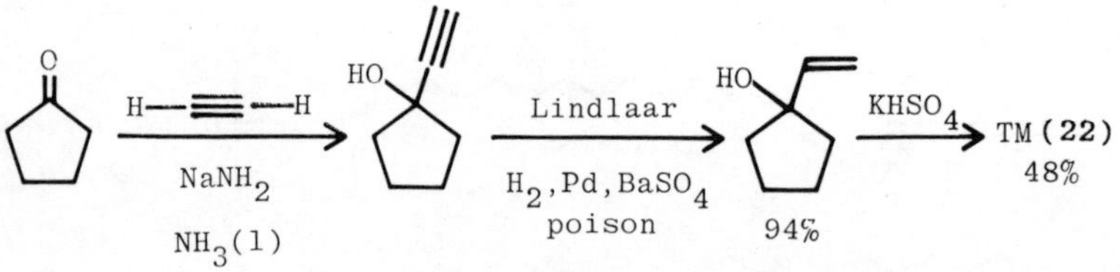

Hydration

Hydration of a triple bond with Hg(II) catalysis depends on forming the more stable vinyl cation **(23)**. Hence terminal acetylenes always give methyl ketones. The carnation perfume **(24)** can be made this way.[170]

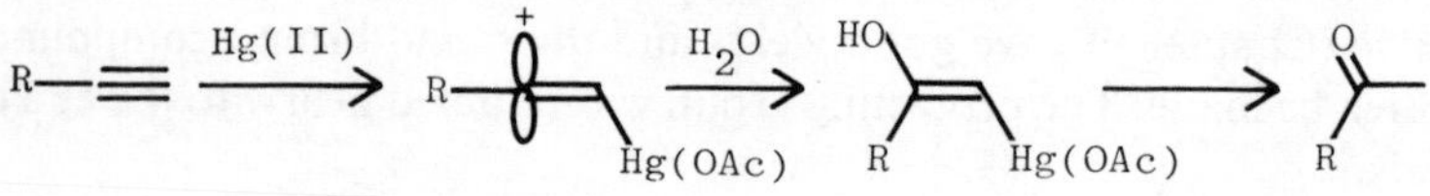

Analysis

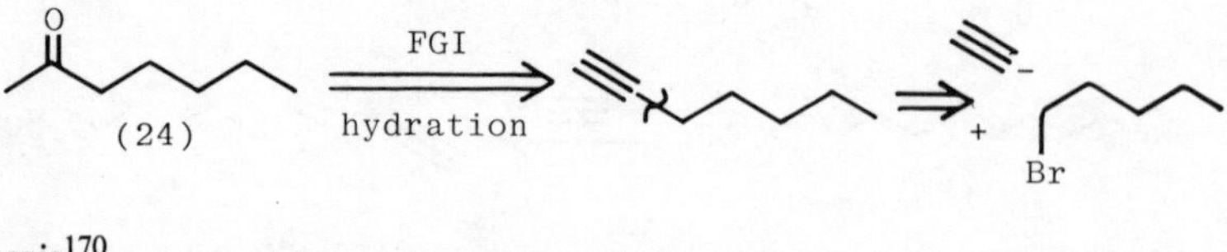

Synthesis[170]

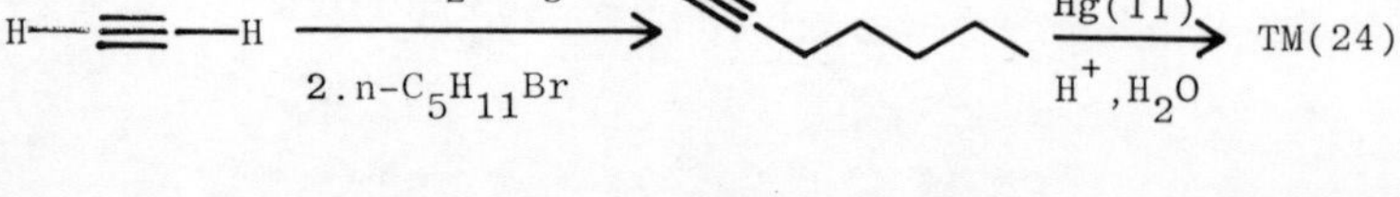

Conjugated acetylenes are hydrated more easily and without metal catalysts[171] by Michael addition of water to give 1,3-diketones, e.g. (25).

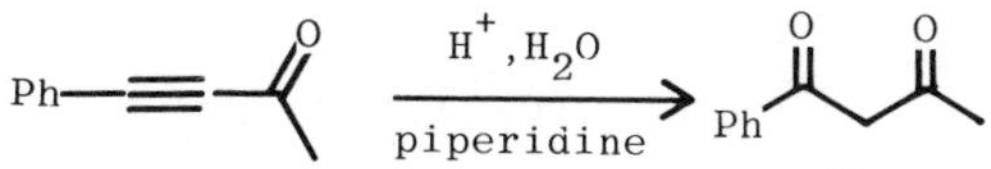

Other types of acetylene adduct (3) and (4) can also be hydrated but the ketone adducts (3) can give unexpected rearrangements. The simple adduct (26) hydrates as expected[172] but the cyclic compound (27) dehydrates[173] in the ring as well as being hydrated in the side chain. The useful product (28) is isomeric with (27).

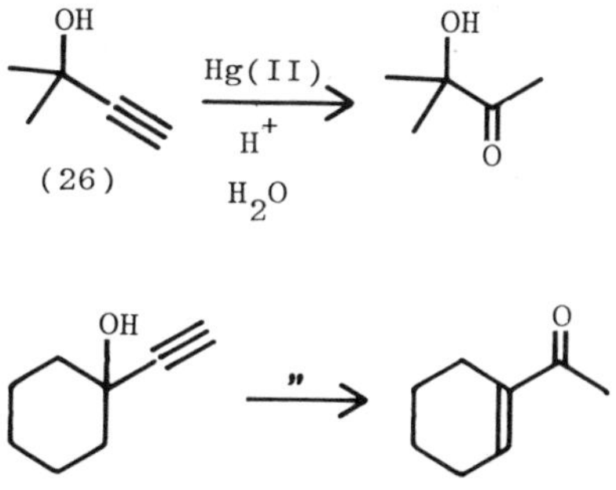

CHAPTER 17

Two-Group Disconnections I: Diels–Alder Reactions

The Diels–Alder[174] is one of the most important reactions in synthesis because it makes two C–C bonds in one step and because it is regio- and stereoselective. It is a pericyclic cycloaddition between a conjugated diene (1) and a conjugated alkene (2) (the dienophile), forming a cyclohexene.

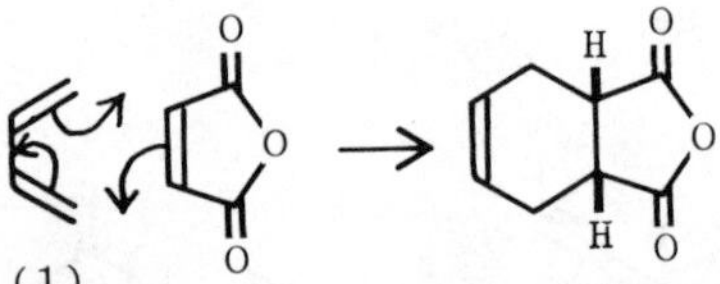

The corresponding disconnection is often best found by drawing the reverse reaction mechanism. Whenever you want to make a cyclohexene (4) with at least one electron-withdrawing group (Z) on the far side of the ring, draw three arrows round the ring, in either direction, starting from the double bond.

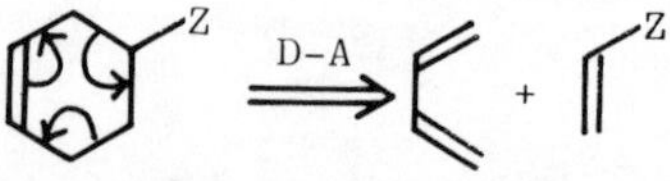

Aldehyde (5) can be made by a Diels–Alder reaction, easily discovered by this method. No special solvents or conditions are needed as no ionic intermediates are involved. The two components are simply mixed together and heated.

Analysis

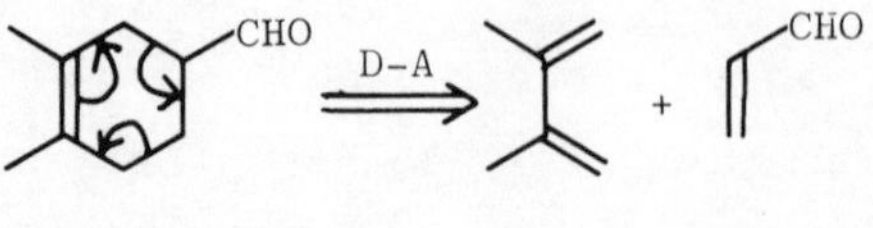

132

Synthesis[175]

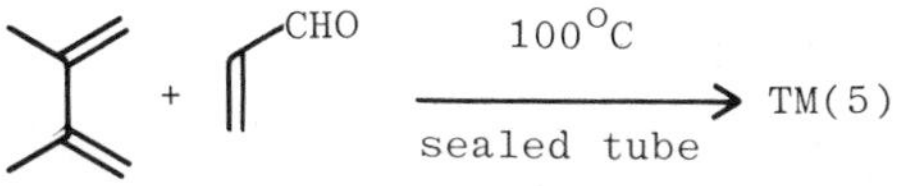

This is a two-group disconnection because it can be carried out only when both features — cyclohexene and electron-withdrawing group — are present and the relationship between them recognised. No matter how complicated the target molecule may be, e.g. **(6)**, if it contains a cyclohexene and an electron-withdrawing group in the right relationship, a Diels–Alder disconnection is worth trying. Other features, such as the four-membered rings in **(6)** shouldn't distract you.

Analysis

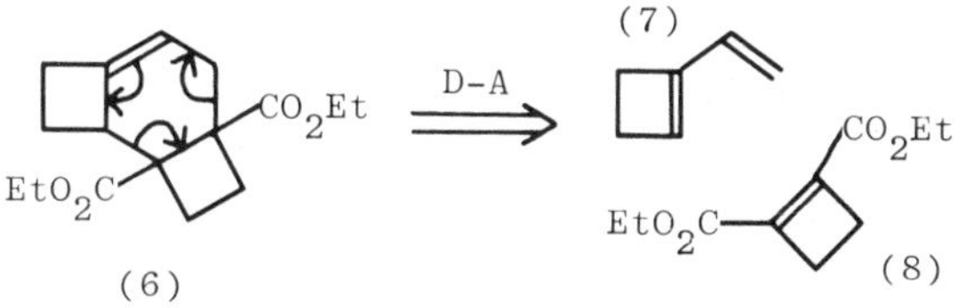

We made diene **(7)** in Chapter 15 and the dienophile **(8)** will be discussed later. The synthesis is straightforward and **(6)** was used[176] to make the curious benzene **(9)** with two fused four-membered rings.

Synthesis[176]

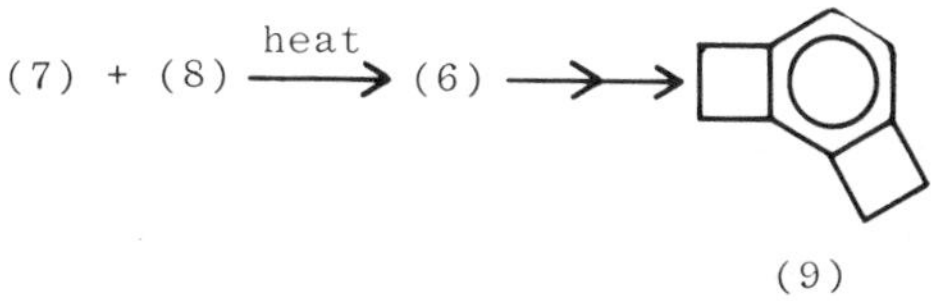

Stereospecificity and Stereoselectivity

The reaction occurs in one step so that neither diene nor dienophile has time to rotate and the stereochemistry of each must be faithfully reproduced in the product. *Cis* dienophiles give *cis* products (e.g. **3** and **6**) and *trans* dienophiles (e.g. **10**) give *trans* products.[177]

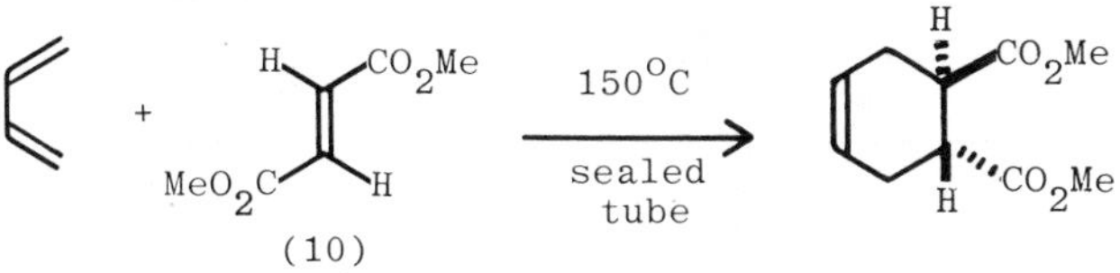

The synthetic attractant Siglure **(11)**, used as bait for the Mediterranean fruit fly,[178] has just such a *trans* relationship and can be made from a *trans* dienophile **(12)**.

Siglure: *Analysis*

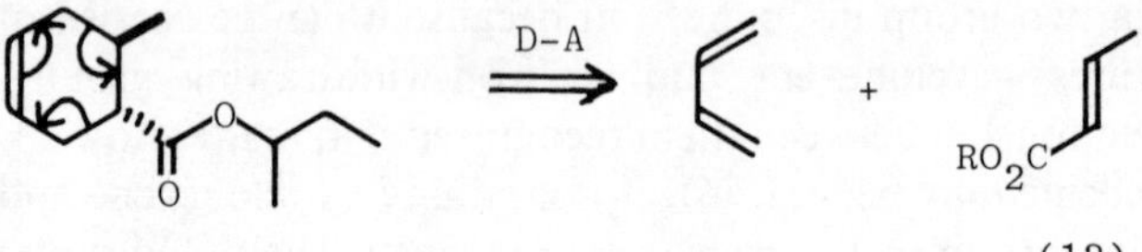

(11)   (12)

In manufacture it is easier to use the cheap ethyl ester **(12, R = Et)** and exchange with the more complex alcohol after the Diels–Alder reaction.[178]

Synthesis[178]

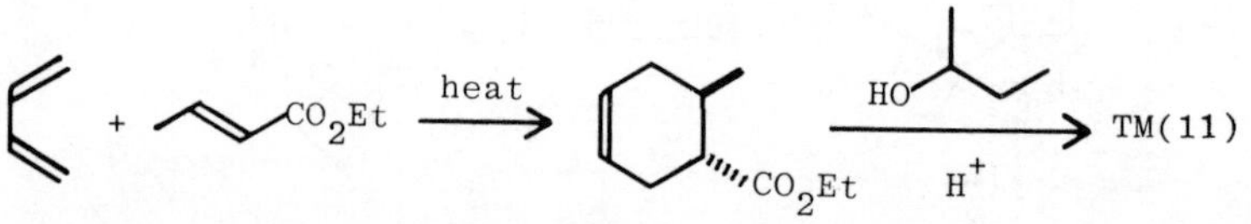

The stereochemistry of the diene is also faithfully transmitted to the product. Diene **(13)** adds to an acetylenic dienophile to give the *cis* product **(14)**. This is because the two reagents approach each other in parallel planes **(15)** so that the p-orbitals overlapping to form the new σ-bonds are as nearly coaxial as possible.

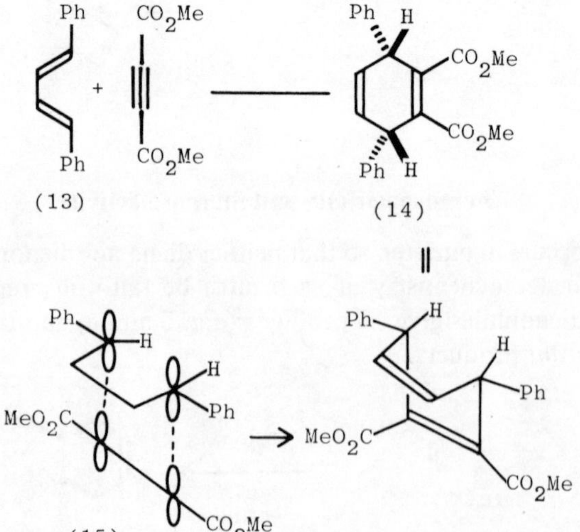

(13)   (14)

(15)

These two aspects of the Diels–Alder are both stereo*specific* (Chapter 12) in that the stereochemistry of the product is determined simply by the stereo-chemistry of the starting materials and not at all by how favourable one reaction pathway may be. There is one more stereochemical aspect of the Diels–Alder reaction — endo selectivity — and that is a stereo*selective* aspect.

Endo Selectivity

I deliberately used an acetylenic dienophile in the last example to avoid the question of the relationship between the stereochemistry of the diene and that of the dienophile. Though each stereochemistry must be preserved, in many cases two products can still be formed. This is easy to see in cyclic systems, e.g. (16).

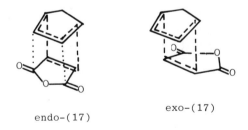

exo-(16)   endo-(16)

The two products are called *exo* and *endo*. These terms refer to the relation-ship between the Z groups of the dienophile (here CO) and the double bond in the new cyclohexene ring. In practice, the *endo* form is favoured as it is the kinetic product, though the *exo* is usually more stable. The role of the electron-withdrawing groups Z in the dienophile is to attract the diene through space in the endo transition state (17). This is a *secondary* orbital interaction which does not lead to bonding but which does help to hold the transition state together.*

endo-(17)

exo-(17)

---interaction leading to bonding
...secondary orbital interactions

---

*The reasons behind this are explained in Fleming's *Orbitals*, p.106.

136

With open chain compounds it is easier to work out the stereochemistry of the *endo* product if you draw the molecules one on top of the other. A simple case would be the synthesis of **(18)**. Three new chiral centres (• in **18**) are introduced in the reaction. Drawing the diene on top of the dienophile **(19)** with the hydrogen atoms at the developing chiral centres marked, and the carbonyl group arranged so that the secondary orbital interactions can occur **(20)**, gives the right stereochemistry for *endo*-**(18)**.

(18)

(19)   (20)   endo–(18)

The sequence of drawings **(19)** to *endo*-**(18)** should show clearly how to draw the stereochemistry of the *endo* adduct. All four drawings are not usually necessary. The imide **(21)**, used in Weinreb's cytochalasin synthesis,[179] will have all four chiral centres correct, that is with all four hydrogen atoms *cis*, if the correct geometrical isomer of diene **(22)** is used. One diagram **(23)** should be enough to show all four hydrogen atoms *cis* in the *endo* transition state if the *E,E*-diene **(24)** is used.*

Analysis

(21)   (22)

(23)   (24)

*The synthesis of this diene is discussed in Chapter 14 and in the workbook for Chapter 15.

Regioselectivity of Diels–Alder Reactions

Reactions between unsymmetrical dienes and unsymmetrical dienophiles are also regioselective. An explanation of this is beyond the scope of this book,* but to use the reaction in synthesis we need only a quick way to work out the answer. The easiest mnemonic is that the Diels–Alder reaction is 'ortho–para' directing. Thus 1-substituted butadienes give 'ortho' products (25) and 2-substituted butadienes give 'para' products (27), particularly under Lewis acid catalysis.[180] The 'meta' products (26) and (28) cannot be made this way.

(25)   (26)

(27)   (28)

Stereospecificity, stereoselectivity, and regioselectivity combined give an unprecedented degree of control over the Diels–Alder reaction and you can now see why it is so popular. The analgesic Tilidine (29), effective in cases of severe pain, is an obvious Diels–Alder product. The regioselectivity is correctly 'ortho' and the endo transition state (30) shows that the trans enamine (31) is needed. This is the geometry we get if we make the enamine in the usual way.

Tilidine: *Analysis*

(29)

(30)H and Ph cis   (31)

*It is fully explained in Fleming's *Orbitals*, p.132.

Synthesis[181]

$$\text{\raisebox{0pt}{\rlap{}}}\quad\text{CHO} \xrightarrow[\text{H}^+]{\text{Me}_2\text{NH}} \text{E-(31)} \quad + \quad \text{Ph}\underset{\text{}}{\overset{\text{}}{\diagup}}\text{CO}_2\text{H} \longrightarrow \text{TM(29)}$$

FGI on Diels–Alder Products

The Diels–Alder is such a powerful reaction that it is good tactics to take advantage of it wherever possible. Vig[180] chose to disconnect limonene **(32)** (a natural odour principle found in most citrus fruits) by a one carbon Wittig step because that revealed a *'para'* Diels–Alder product **(27)**.

Limonene: *Analysis*

$$(32) \xRightarrow{\text{Wittig}} (27) \xRightarrow{\text{D-A}} \quad + \quad$$

Synthesis[180]

$$\diagdown \quad + \quad \xrightarrow[\text{heat}]{\text{SnCl}_4} (27) \xrightarrow{\text{Ph}_3\overset{+}{\text{P}}\text{-}\bar{\text{C}}\text{H}_2} \text{TM(32)}$$

The cyclic ether **(33)** obviously comes from the diol **(34)** which can be made from a number of Diels–Alder adducts by reduction. The anhydride **(35)** is satisfactory.[182]

Analysis

$$(33) \xRightarrow[\text{ether}]{\text{C-O}} (34) \xRightarrow[\text{reduction}]{\text{FGI}}$$

$$(35) \xRightarrow{\text{D-A}} \quad + \quad$$

Synthesis[182]

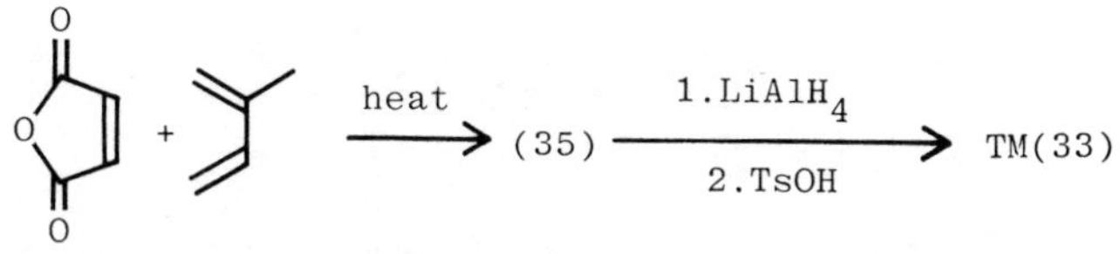

$$\text{maleic anhydride} + \text{2,3-dimethylbutadiene} \xrightarrow{\text{heat}} (35) \xrightarrow[\text{2.TsOH}]{\text{1.LiAlH}_4} \text{TM(33)}$$

CHAPTER 18

Strategy VIII: Introduction to Carbonyl Condensations

The next ten chapters are about the synthesis of carbon skeletons with two functional groups. Compounds such as **(1)**, **(2)**, and **(3)** will be grouped together as 1,3-difunctionalised compounds since the important thing is not the type of FG but its position. Our logic is that FGs can be derived from alcohols, ketones (or aldehydes), or acids by substitution reactions, and that these three can be interconverted by oxidation or reduction.

Analysis will mean using FGI and C–X disconnections to go back to the basic skeleton with only oxygen FGs at the right oxidation level and then disconnecting C–C bonds by the *two-group* methods we are about to learn. Though we have already met methods of C–X disconnection for difunction-alised compounds (Chapter 6), the C–C disconnections we shall now meet are strategically more fundamental and will allow us to synthesise far more complex TMs. This section of the book completes the reactions needed for basic synthesis design.

The order of events will be:

Chapter 19: 1,3-difunctionalised compounds and α,β-unsaturated carbonyl compounds.
Chapter 21: 1,5-dicarbonyl compounds.
Chapter 23: 1,2-difunctionalised compounds.
Chapter 25: 1,4-difunctionalised compounds.
Chapter 27: 1,6-difunctionalised compounds.

This curious order comes about because I want to start with natural or logical synthons. With two-group disconnections, e.g. **(4)**, each synthon is functionalised and we shall start with synthons in which the functional groups

140

help to stabilise the nucleophile and electrophile (Table 18.1). The nucleophile will usually be the enolate (5). In combination with 'direct' electrophiles (Table 18.1) a 1,3-relationship, e.g. (4), must be formed and this is the subject of Chapter 19. Then we move to conjugate electrophiles and hence the 1,5-relationship in Chapter 21.

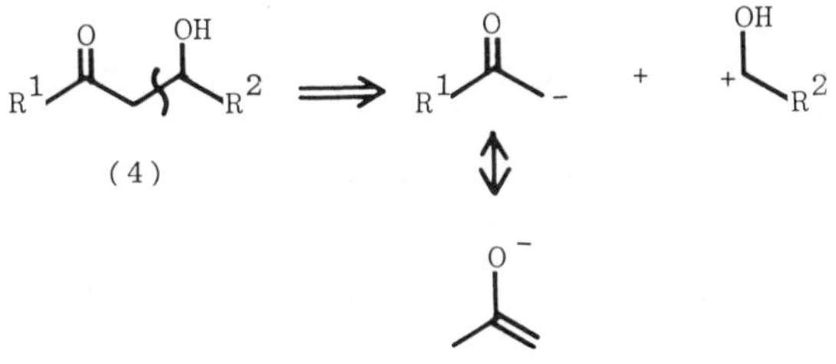

The 1,2- and 1,4-relationships require one synthon of unnatural polarity, such as (6), (7), or (8), in combination with a natural synthon from Table 18.1, and these relationships follow in Chapters 23 and 25. The 1,6-relationship is left until last (Chapter 27) as it requires a new logic—that of reconnection instead of disconnection.

Table 18.1 Natural or logical synthons

Nucleophile	Electrophile	
	Synthon	Reagent
Enolate ions (or enols)	(a) Direct	
	$\overset{+}{R}CHOH$	$RCHO$
	R^1-$\overset{+}{C}(-OH)$-R^2	R^1-$C(=O)$-R^2
	$R\overset{+}{C}O$	$RCOX$ ($X = Cl, OR^1$)
	(b) Conjugate (Michael)	

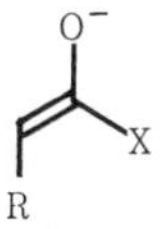

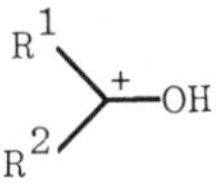

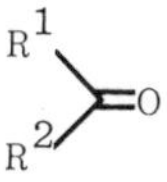

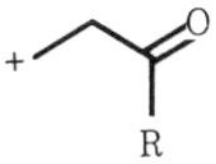

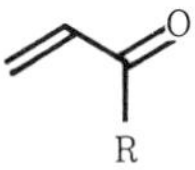

$$\underset{(6)}{\overset{O}{\underset{R}{\overset{\|}{\bigwedge}}}}^{-} \qquad \underset{(7)}{\overset{O}{\underset{R}{\overset{\|}{\bigwedge}}}}_{+} \qquad \underset{(8)}{\overset{O}{\underset{R}{\overset{\|}{\bigwedge}}}}^{-}$$

You will notice that all these methods depend on the carbonyl group. We have already used this group a great deal but this excursion into carbonyl condensations—reactions of one carbonyl compound with another—requires some more sophisticated strategic thinking and the strategy chapters will develop these ideas culminating in Chapter 28 with a general discussion of strategy in carbonyl synthesis. The strategy chapters are:

Chapter 20: Strategy IX: Control in Carbonyl Condensations.
Chapter 22: Strategy X: Use of Aliphatic Nitro Compounds in Synthesis.
Chapter 24: Strategy XI: Radical Reactions in Synthesis. FGA and its Reverse.
Chapter 26: Strategy XII: Reconnections.
Chapter 28: General Strategy B: Strategy of Carbonyl Disconnections.

Carbonyl chemistry began in the nineteenth century when it was the practice to give their discoverers names to new chemical reactions. I shall refer to many reactions by name for two reasons:

1. Many people like to learn a name to go with a new reaction—it helps them to remember the reaction with the name.
2. The chemical literature often employs the names as shorthand for the reactions, and I shall do so too later in the book.

There is no need to remember all the names: it is far more important to understand the reactions. Some named reactions—the Wittig and Diels–Alder are examples—are so important that all organic chemists should know their names and you will detect these because they are mentioned so often.

We shall be needing a variety of bases to form the anions from carbonyl compounds in the next ten chapters. Table 18.2 gives a guide to the strength of the various bases. Any base may be used to form the anion of a compound lower in the table: the conjugate acid of the base should have a higher pK_a than the carbon acid.

Table 18.2 Carbon acids and the bases used to ionise them

Carbon acid (H is the acidic proton)	pK_a	Base	(pK_a of the conjugate acid)	
Alk-H	42	BuLi	42	Available
		RMgBr	42	RBr + Mg
PhH	40	PhLi	40	
CH_2=CHCH_3	38			
PhCH_3	37	NaH	37?	Available
		R_2NLi	36	R_2NH + BuLi
MeSO.CH_3	35	NH_2^-	35	Na + NH$_3$(l)
(DMSO)		MeSO.CH_2^-	35	NaH + DMSO
Ph$_3$CH	30	Ph$_3$C$^-$	30	
HC≡CH	25			
CH_3CN	25			
CH_3CO$_2$Et	25			
CH_3COMe	20			
CH_3COPh	19	t-BuOK	19	Available
CH_3PPh$_3^+$	18	EtO$^-$, MeO$^-$	18	ROH + Na
ClCH_2COMe	17			
PhCH_2COPh	16	HO$^-$	16	Available
CH_2(CO$_2$Et)$_2$	13			
MeCOCH_2CO$_2$Et	11			
		PhO$^-$	10	PhOH + HO$^-$
CH_3NO$_2$	10	Na$_2$CO$_3$	10	Available
		R$_3$N etc.	10	
NCCH_2CO$_2$Et	9			
Ph$_3$P$^+$CH_2CO$_2$Et	6	NaHCO$_3$	6	Available
		AcO$^-$	5	Available
		Pyridine	5	Available

CHAPTER 19

Two-Group Disconnections II: 1,3-Difunctionalised Compounds and α,β-unsaturated Carbonyl Compounds

Direct disconnection of this group of compounds is possible at two oxidation levels—dicarbonyl **(1)** and β-hydroxy carbonyl **(3)**. Enones **(4)** come into this chapter since they are usually made by dehydration of **(3)**.

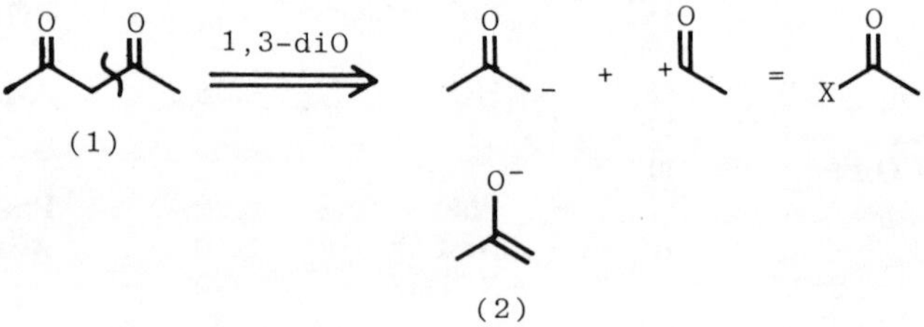

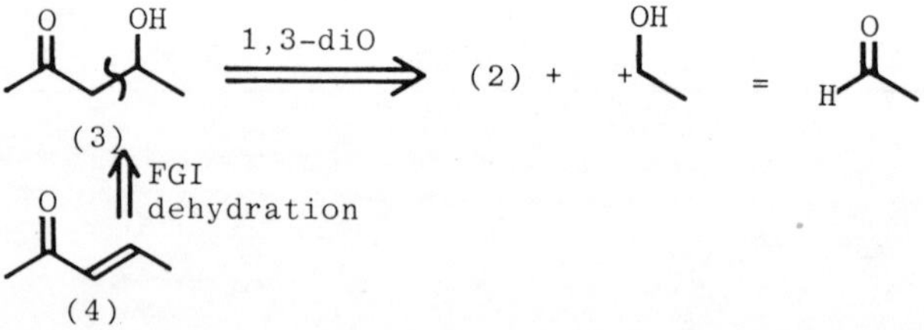

1,3-Dicarbonyl Compounds[183]

Disconnection **(1)** means that we are looking for a reaction which is the acylation of an enolate anion **(2)**. This is possible with esters (X = OR) or acid chlorides (X = Cl). The perfumery compound **(5)** (which has a 'fragrant balsamic odour of great power and tenacity'[184]) can be disconnected to the enolate of a ketone and an ester.

Analysis

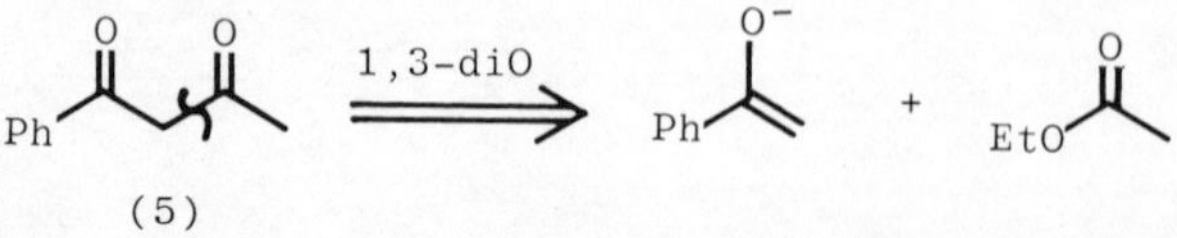

144

The reaction can be carried out by combining ketone (6) with the ester and a base strong enough (Table 18.2) to produce only a small concentration of the enolate, often EtO⁻. The reaction is therefore an equilibrium and it is driven over by formation of the stable delocalised enolate (7) of the product. Acid work-up then releases TM(5). This last step applies to all reactions of this sort but is not usually written down.

Synthesis[185]

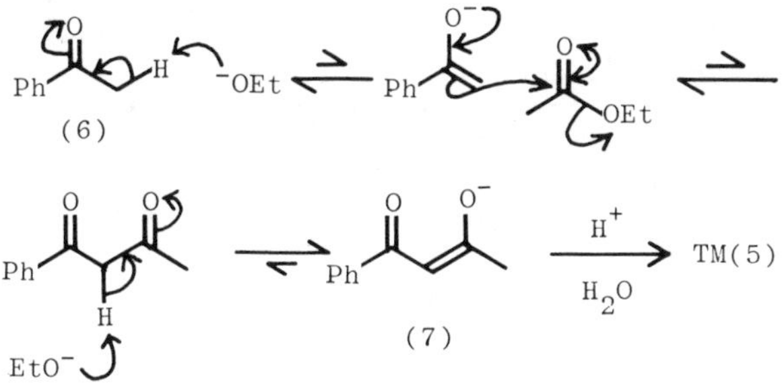

This synthesis was carried out by Claisen[185] and the reaction is known as the Claisen condensation. The acetoacetate (8) we used in Chapter 13 is also made this way. This time the starting materials are two molecules of the same compound. The synthesis,[186] known as the Claisen ester condensation, simply involves treating ethyl acetate with base.

Analysis

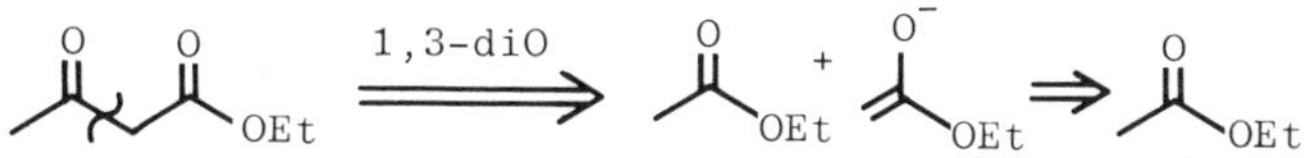

Synthesis[186]

$$MeCO_2Et \xrightarrow{\quad EtO^- \quad} TM(8)$$
$$60\%$$

Pival (9) is a rat poison[187] with three ketone groups each having a 1,3-relationship to the others. Of the two possible disconnections, (b) quickly leads back to available starting materials.

146

Pival: *Analysis*

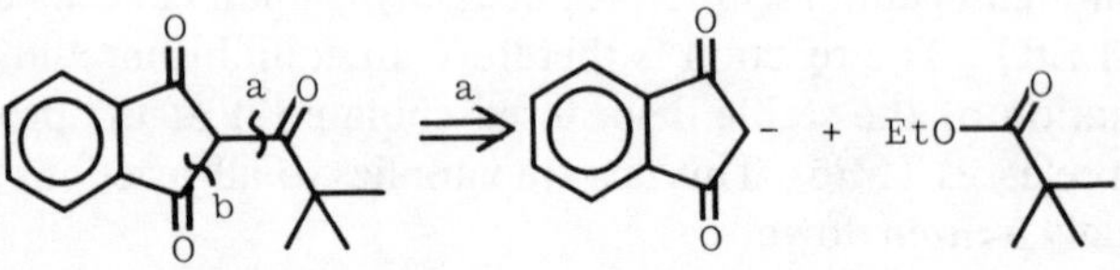

(9)

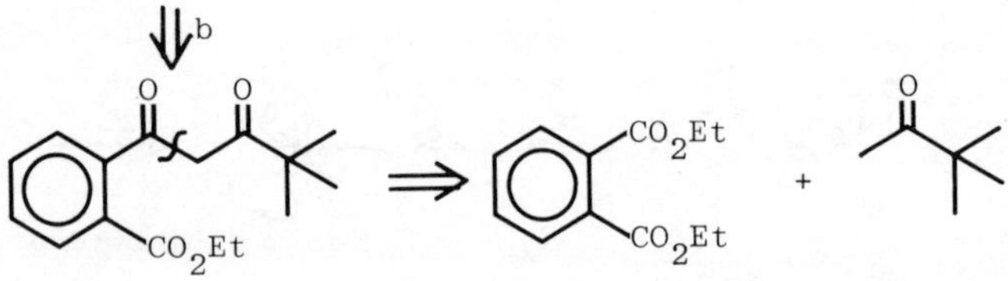

(10)

The synthesis,[188] as so often when a cyclisation is involved, is easier than expected as **(10)** cyclises to **(9)** under the conditions of its formation. The synthesis of unsymmetrical 1,3-diketones will be discussed more fully in Chapter 20.

Synthesis[188]

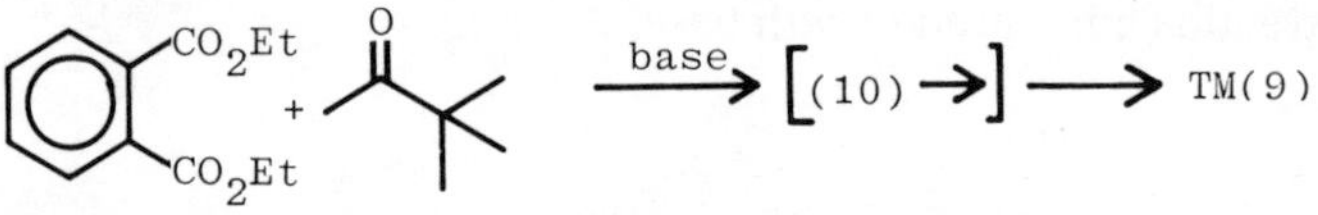

The important keto ester **(11)** can be disconnected in two ways. One (a) removes only one carbon atom and is bad strategy but the other gives the symmetrical and readily available diester **(12)** (Chapter 27) as starting material.

Analysis

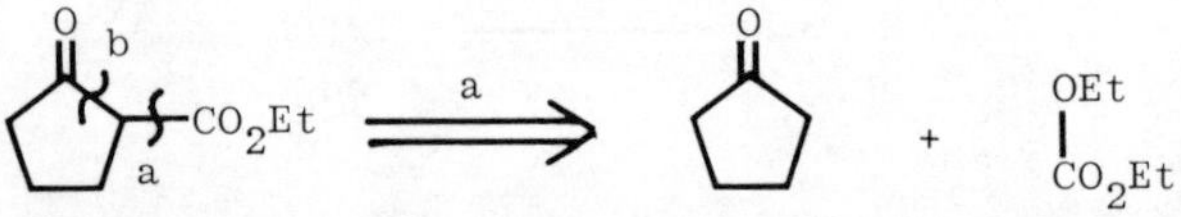

(11)

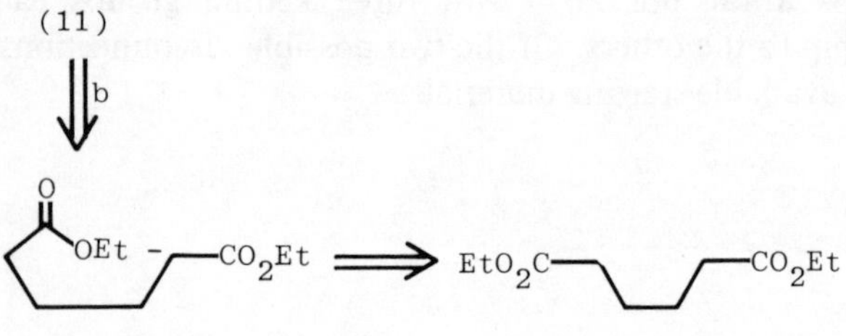

(12)

The reaction is intramolecular and hence fast and clean. Compound **(11)** is thus readily available and is used as a starting material for other five-membered rings.

Synthesis[189]

$$(12) \xrightarrow{\text{EtO}^-} \text{TM}(11)$$

A useful heterocyclic version of this reaction gives ketone **(14)** used in Chapter 11 in the synthesis of **(13)**. Direct disconnection of **(14)** by the 1,3-diX methods (Chapter 6) would require unstable ketone **(15)** as a starting material. However, if we insert a CO_2Et group β to the ketone (as in **11**) to get **(16)**, a 1,3-dicarbonyl disconnection gives symmetrical diester **(17)**. Two 1,3-diX disconnections now make excellent sense.

Analysis

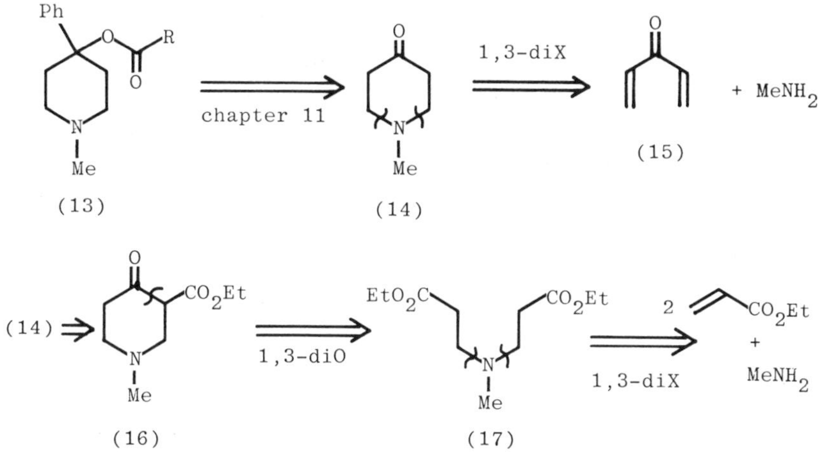

Synthesis[190]

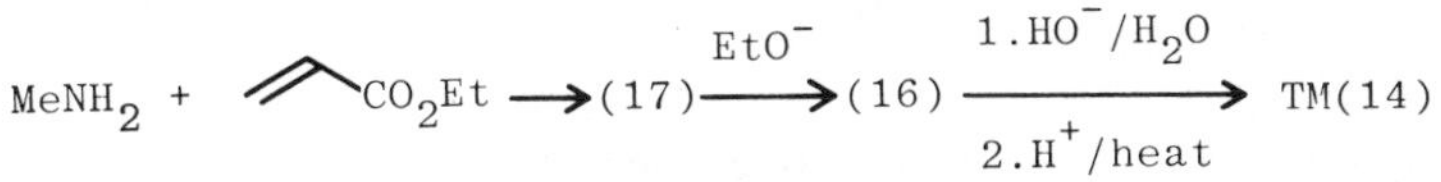

$$\text{MeNH}_2 + \underset{\text{CO}_2\text{Et}}{\diagup} \longrightarrow (17) \xrightarrow{\text{EtO}^-} (16) \xrightarrow[\text{2.H}^+/\text{heat}]{1.\text{HO}^-/\text{H}_2\text{O}} \text{TM}(14)$$

This route to a cyclic ketone is in fact a general route to symmetrical ketones, whether cyclic or not, e.g. **(18)** and we shall use it later in the book.

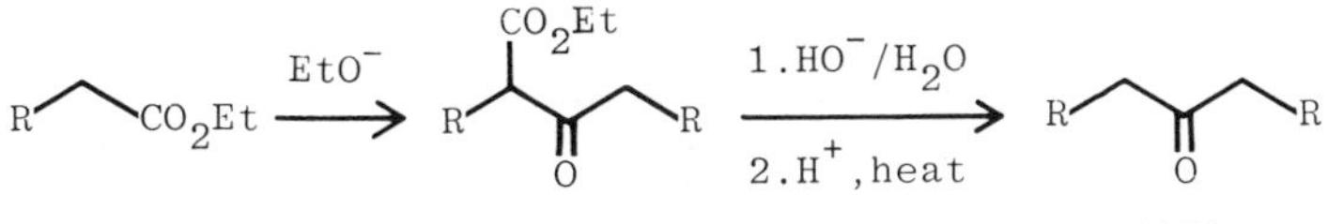

148

β-Hydroxy Carbonyl Compounds[191]

This is the same disconnection at a lower oxidation level, the ester being replaced by an aldehyde or ketone. Compound **(19)** may look complicated but only one disconnection is possible and the starting materials are two molecules of the same compound.

Analysis

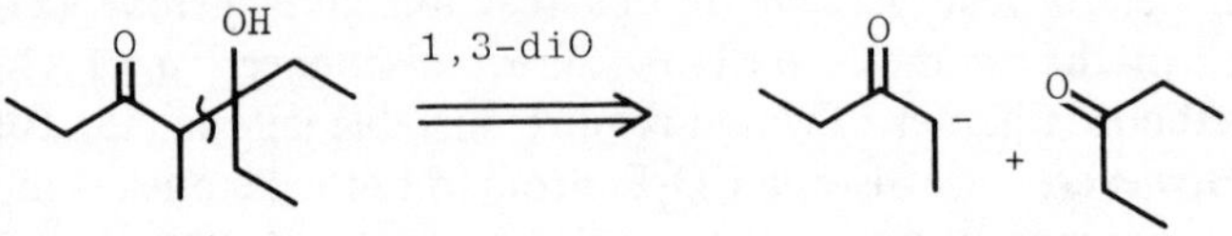

(19)

Synthesis[192]

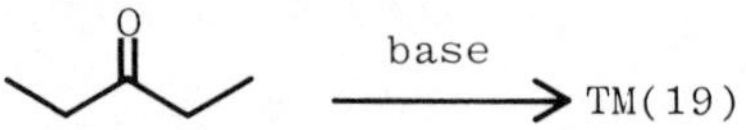

base
⟶ TM(19)

Other compounds may need FGI before disconnection. Meyer's heterocyclic reagent* **(21)** is made from diol **(20)** in a Ritter reaction (Chapter 8) and this comes from a β-hydroxy ketone by reduction. Disconnection again reveals two molecules of the same compound.

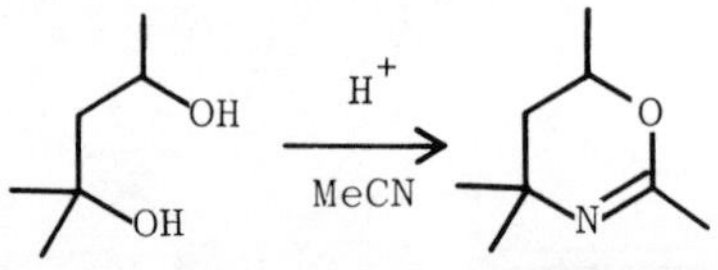

(20)   (21)

Analysis

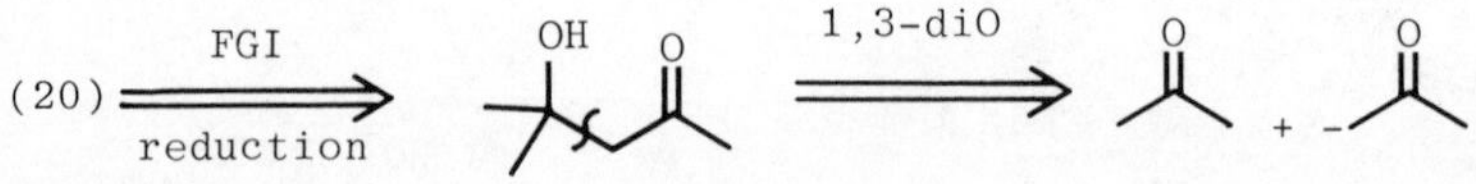

The condensation is conveniently carried out with barium hydroxide as an insoluble basic catalyst.

Synthesis[193]

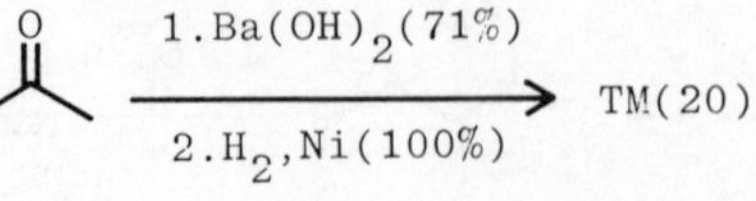

$1.Ba(OH)_2(71\%)$
⟶ TM(20)
$2.H_2,Ni(100\%)$

*Compound **(21)** is a versatile reagent for the synthesis of aldehydes, ketones, and acids.[193]

Sometimes it is easier to make 1,3-dicarbonyl compounds by oxidation of β-hydroxy carbonyl compounds. In 1970, chemists[194] wanted to synthesise (22) to study the photochemical reactions of a non-conjugated enone. A Wittig disconnection makes sense as it gives a 1,3-dicarbonyl compound (23) as starting material. Direct disconnection of (23) suggests a reaction between aldehyde (25) and ester (24): these have the same carbon skeleton so any ambiguity will be avoided if we use two molecules of aldehyde and oxidise later.

Analysis

(22)   Wittig ⟹   (23)   +   +PPh₃

(24) + (25)

Synthesis[194]

(25)   NaOH   (26) 85%   CrO₃ pyr   (23)   Ph₃P⁺   TM(22)

Note that the Wittig reaction has a chemoselective aspect—reaction is required at the aldehyde and not at the ketone—and a stereoselective aspect —*trans* (22) is required. Both these selectivities operate in our favour as the aldehyde is more reactive than the ketone and stabilised ylids (Chapter 15) give *trans* alkenes.

The condensation of an aldehyde with itself, as in (25) to (26), is an 'aldol' reaction as the product (26) is a hydroxy aldehyde or aldol. Some people refer to all these condensations, whether aldehydes or ketones are involved, as aldol reactions.

α,β-Unsaturated Carbonyl Compounds

Dehydration of β-hydroxy carbonyl compounds is very easy because the proton to be removed (H in 27) is enolic and the product (28) is conjugated.

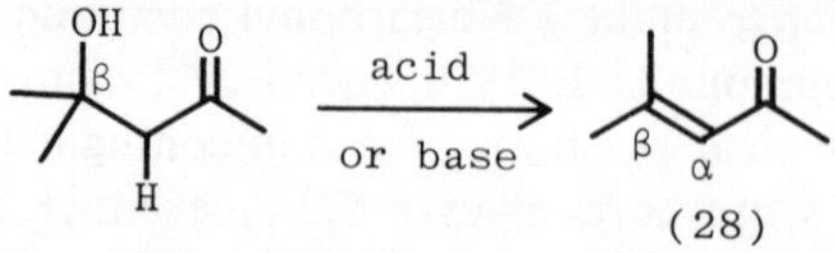

(27) → acid or base → (28)

The full analysis of an enone or other α,β-unsaturated carbonyl compound should be an FGI followed by a 1,3-diO disconnection. In practice, the dehydration often occurs during the condensation so that the intermediate (27) need not be isolated. In this example, acid catalysis gave a good yield.[195]

Analysis

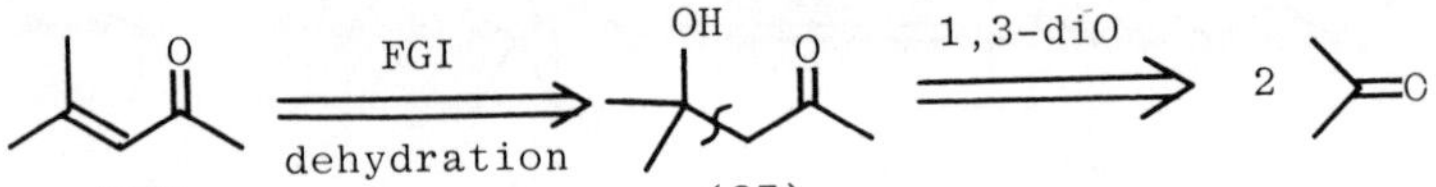

(28) ⟸ FGI / dehydration ⟸ (27) ⟸ 1,3-diO ⟸ 2

Synthesis[195]

→ H⁺ → TM(28)

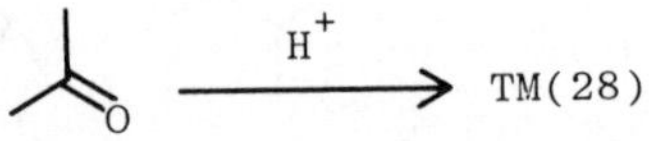

We therefore usually write a shorthand disconnection of the enone directly to the two carbonyl components:

Analysis

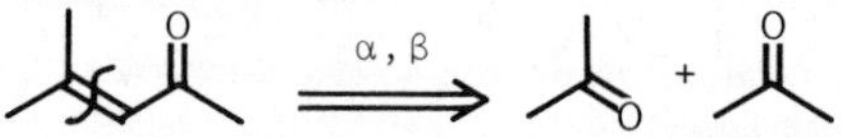

⟹ α,β ⟹ +

This is a very important disconnection and you should learn to look for it in unlikely places, e.g. (29). Simply disconnect the double bond and write a carbonyl group at the β atom. In this case we get two identical lactones and the reaction can be carried out in base.[196]

Analysis

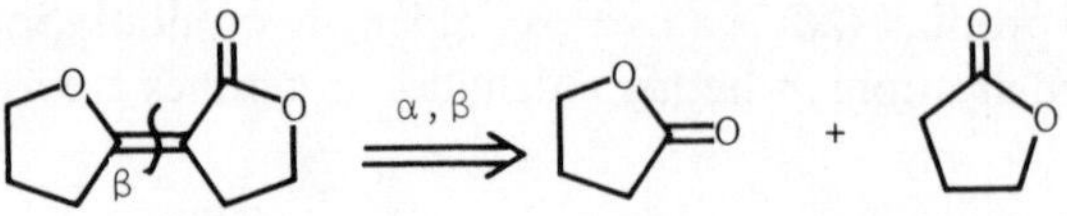

(29) ⟹ α,β +

Synthesis[196]

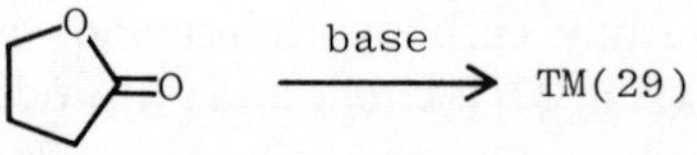

→ base → TM(29)

The famous French perfume ingredient Flosal[197] **(30)** is an enal and can be disconnected in this way.

Flosal: *Analysis*

(30)

The synthesis is carried out in base with a large excess of benzaldehyde to minimise self-condensation of the aliphatic aldehyde. Ambiguities of this kind will be discussed in Chapter 20.

Synthesis[197]

$$3PhCHO + n\text{-}HexCHO \xrightarrow[\text{H}_2\text{O},\text{EtOH}]{\text{NaOH}} TM(30) \quad 80\%$$

The synthesis of the minor tranquilliser Oxanamide[198] **(31)** illustrates how C–X disconnections and FGI can be added to this general plan. The molecule has amide and epoxide FGs. We know of one way to put in each of these groups—by C–N and C–O disconnection—so we must start here and decide the order of events later. This gives us an α,β-unsaturated acid, so we can disconnect at the double bond.

Oxanamide: *Analysis*

(31)

The two starting materials have the same skeleton so it will be better to use two molecules of the aldehyde **(32)** and oxidise later, as we did on page 149, to avoid ambiguities. Trial and error revealed that it was best to put in the amide before the epoxide.

Synthesis[198]

CHAPTER 20

Strategy IX: Control in Carbonyl Condensations

The last chapter introduced a series of good disconnections based on carbonyl chemistry but avoided all questions of chemo- or regioselectivity. These reactions are so important that it is worthwhile learning some methods to control both types of selectivity. All the chief difficulties crop up in the synthesis of the enone (1).

Analysis

(1)

(2)   (3)

Synthesis (no good)

(2)   (4)   TM(1)

The analysis is very simple but when we come to the synthesis all is not well. We want ketone (2) to enolise, but why shouldn't aldehyde (3) enolise too? We want (2) to enolise on the methyl side to give (4), but the benzyl side might be preferred. We want the enolate (4) to attack the aldehyde, but might it not also attack another molecule of (2)?

For an unambiguous condensation, there must be no doubt as to which compound enolises and which compound acts as the electrophile (chemoselectivity) nor any doubt about the regioselectivity of the enolisation. We need definite and favourable answers to three questions:

1. Which compound enolises?
2. On which side does it enolise?
3. Which compound acts as the electrophile?

Fortunately, it is rare that all three are serious questions, but even if they are, enough methods now exist to solve all but the most intractable of cases. In this chapter we shall look at methods of control using reactions from Chapter 19 as examples. The problems arise because the most easily enolised carbonyl compounds are also the most electrophilic (Table 20.1).

Table 20.1 Reactivity of carbonyl compounds

──────────────────────────────▶ Most enolisable			
──────────────────────────────▶ Most electrophilic			
RCONR$_2$'	RCO$_2$R'	(RCO)$_2$O	RCOCl
	RCOR'	RCHO	

Self-condensations

Questions of chemoselectivity (questions 1 and 3 above) can be avoided if the two compounds are the same. There were examples of these self-condensations in Chapter 19, where we went out of our way, by preliminary FGI, to ensure that other cases became self-condensations. Aldehydes (**5**),[199] symmetrical ketones (**6**),[200] and esters (**7**)[201] all provide unambiguous examples.

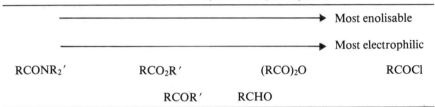

154

Unsymmetrical ketones retain the question of regioselectivity in enolisation (see Chapter 14 for an earlier discussion). Some selectivity can be achieved by varying the conditions. In base, kinetic control ensures that the more acidic proton, usually on the less substituted carbon atom, is removed, e.g. (8) gives (9). In acid, rapid equilibration between ketone and enol means that the more stable enol, usually the more heavily substituted enol, is formed, e.g. (10). Hence different products (11) and (12) are formed in basic[202] and acidic[203] conditions from ketone (8).

These are rather delicately balanced factors and cannot be relied upon. Experiment is needed in any new case to see if this method of control is adequate. In any case, the range of compounds available from self-condensation is limited and methods for controlling cross-condensation are necessary.

Intramolecular Reactions

These are a halfway house between self-condensations and cross-condensations. Although only one molecule is involved, all three questions may still arise; but since these carbonyl condensations are usually reversible, a route leading to a stable five- or six-membered ring is thermodynamically preferred. Intramolecular reactions are therefore easier to control than the bimolecular equivalent.

Cyclisation of symmetrical diketone (13) could occur in two ways, depending on which side of the ketone enolises. In practice, in acid or base, only route (a) is followed to give stable six-membered (14) rather than less stable eight-membered (15). In acid solution, 85% of (14) is formed.[204]

(13) (14) (15)

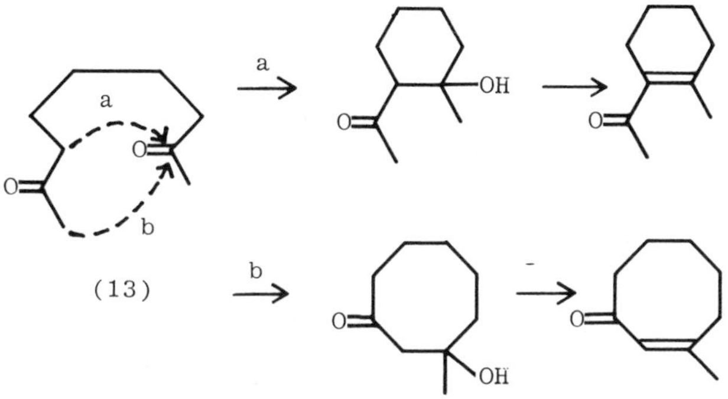

Discrimination against a small ring is equally effective. Unsymmetrical diketone **(16)** could enolise in four different ways (a–d) and each enol could cyclise to give four different products **(17–20)**. Products **(18)** and **(19)** both contain three-membered rings and will instantly revert to **(16)**. Product **(17)** contains only five-membered rings but it is a crowded bridged compound and it cannot dehydrate (see under Cross-condensations III, page 167). Product **(20)** is stable and it dehydrates to **(21)** easily and this is the only product.[205]

(16) (17) (18) (19) (20) (21) 85%

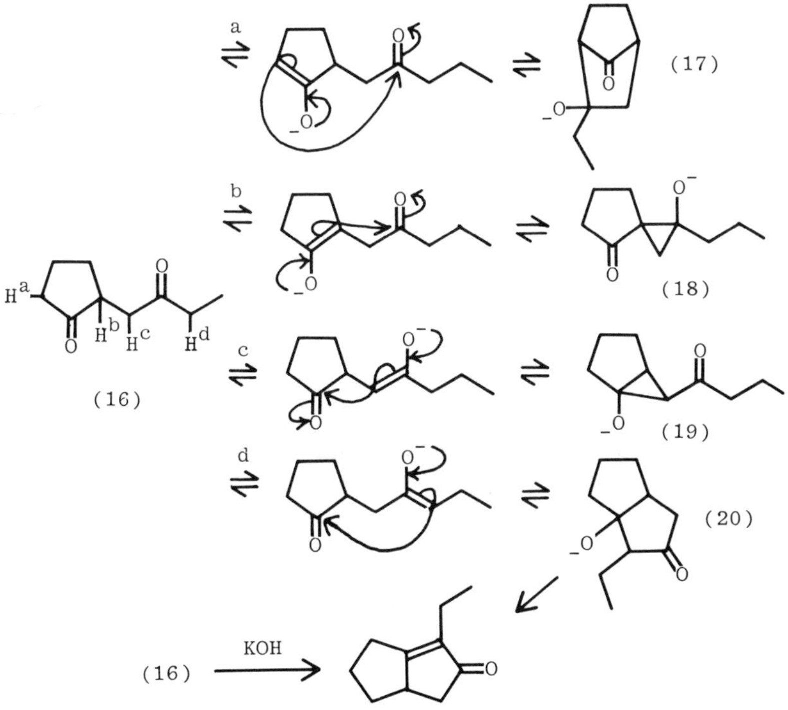

Cross-condensations I: Use of Compounds which cannot Enolise

If a compound cannot enolise, it can react only as the electrophilic component in the condensation. Such compounds are summarised by formula (22) in which neither substituent has any α-protons. Some examples are (23)–(25).

$$R^1, R^2 = H, OEt, Cl,$$

$$Ar, t\text{-}Alk, CO_2Et$$

(22)

(23) (24) (25)

One useful case is the addition of the activating group CO_2Et (see Chapter 19) in such compounds as (26). The disconnection requires a carbonic acid derivative: ester $CO(OEt)_2$ (diethylcarbonate) and the half acid chloride (24) (ethyl chloroformate) are good reagents for this task.

Analysis

$$\text{(26)} \xrightarrow{1,3\text{-diO}} + \begin{array}{c} OEt \\ | \\ CO_2Et \end{array}$$

(26)

Synthesis[206]

$$\xrightarrow[\text{(EtO)}_2\text{CO}]{EtO^-} \text{TM(26)}$$

Aryl substituted malonates (27) are usually made this way as the alternative disconnection (a) requires the unknown S_N2 reaction at an aryl halide.

Analysis

$$\xrightarrow[\text{1,2 C-C}]{a} ArBr + \begin{array}{c} CO_2Et \\ < \\ CO_2Et \end{array} \quad \text{no good}$$

$$Ar\text{-}\begin{array}{c} a \\ b \end{array}\begin{array}{c} CO_2Et \\ \\ CO_2Et \end{array}$$

(27)

$$\xrightarrow[\text{1,3-diO}]{b} Ar\text{-}\begin{array}{c} - \\ | \\ CO_2Et \end{array} + CO(OEt)_2$$

Synthesis[207]

$$Ph\diagup\diagdown CO_2Et \xrightarrow[\text{NaH}]{\text{CO(OEt)}_2} TM(27) \quad (Ar=Ph) \\ 86\%$$

Compounds such as $ArCHO$, $ArCO_2Et$, and HCO_2Et are all useful in this type of control, but there is still one word of warning. Although the electrophiles cannot enolise, their reaction partners may also be elctrophilic and may self-condense. Thus Ph_2CO, a poor electrophile, cannot be used in the synthesis of **(28)** by condensation with acetaldehyde as the self-condensation product **(29)** is formed instead.

The electrophilic component must therefore be *more* electrophilic than the enol component. Chalcones **(30)** may be synthesised by the obvious route from **(31)** and **(32)** as only **(31)** can enolise and aldehyde **(32)** is more electrophilic than ketone **(31)**.

Analysis

Synthesis[208]

$$Ar^1COMe + Ar^2CHO \xrightarrow{\text{base}} TM(30)$$

Regioselectivity of enolisation is still important. Acid **(33)** was used in a (very early!) stage of Woodward and Eschenmoser's vitamin B_{12} synthesis.[209] Disconnection gives an unsymmetrical ketone **(8)** and reactive, non-enolisable **(34)**. Reaction in acid (page 154) ensured enolisation on the more substituted side of **(8)**.

Analysis

Synthesis[209]

$$(8) + (34) \xrightarrow{\text{H}_3\text{PO}_4} \text{TM}(33) \ 82\%$$

By contrast, in Woodward's synthesis[210] of the supposed structure **(35)** of the antibiotic patulin, analysis of intermediate **(36)** by the standard 1,3-dicarbonyl disconnection reveals reactive non-enolisable **(37)** and unsymmetrical ketone **(38)**, required to enolise at the *methyl* group. Condensation in base ensured kinetic control (page 154).

Alleged patulin intermediate: *Analysis*

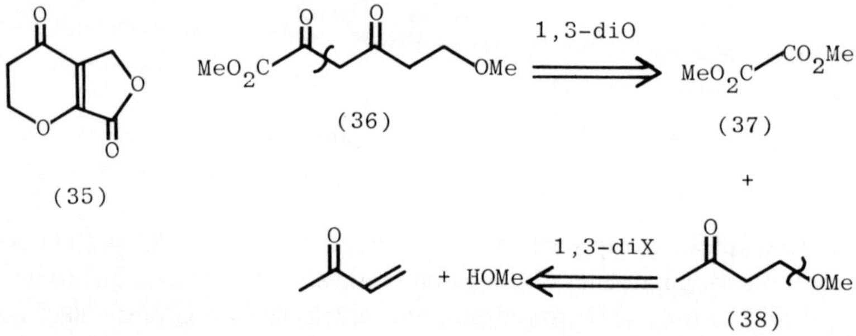

(35)

(36)

1,3-diO

(37)

+

+ HOMe

1,3-diX

(38)

Synthesis[210]

$$\xrightarrow{\text{MeOH}} (38) \xrightarrow[\;(37)\;]{\text{base}} \text{TM}(36)$$

This synthesis showed that **(35)** was not the correct structure for patulin, and Woodward then made **(39)**, the correct structure.

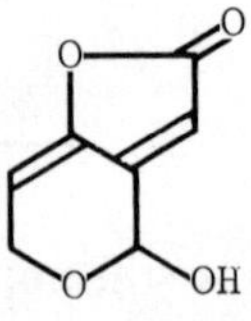

(39)

Formaldehyde: the Mannich reaction

One obvious candidate for a reactive non-enolisable carbonyl compound is formaldehyde, CH_2O. The trouble with this compound is that it is rather too reactive, adding repeatedly and transforming the product by Cannizarro reactions to give such compounds as **(40)**.

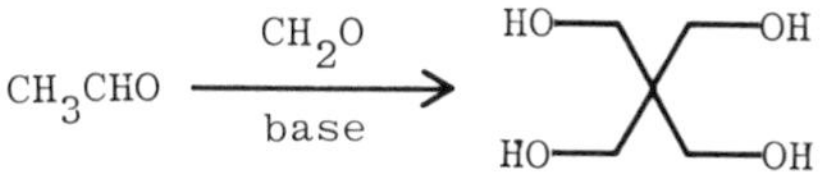

(40)
pentaerythritol

We need a formaldehyde equivalent which is less reactive than form-aldehyde itself. The most popular method is the Mannich reaction[211] in which formaldehyde reacts with the enolic component and a secondary amine. The intermediate (41) is first formed: this adds to the enol to form the Mannich base (42).

(41)

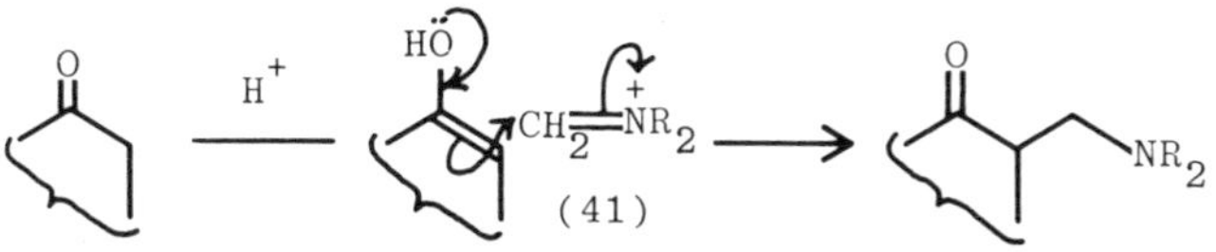

(42)

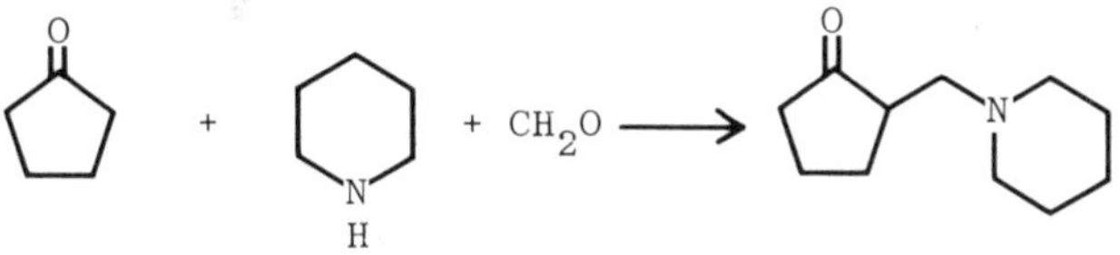

(43) 90%

These are not very interesting compounds in themselves but can be isolated and purified, e.g. (43),[212] and then by alkylation and elimination converted into (44), the products of formaldehyde condensation. Methylene ketones (44) are unstable and an advantage of the Mannich method is that they may be stored as the Mannich base (42) and released only when they are needed.

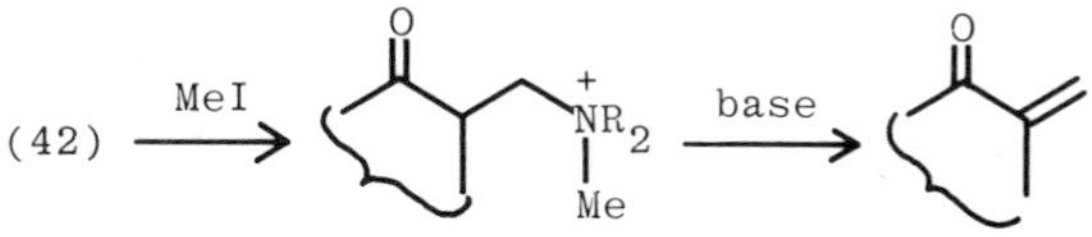

(44)

Whiting needed acetal (45) in his synthesis[213] of compounds from the heartwood of conifers. Disconnection of the acetal gives diol (46) which can be made from enone (47). This leads into the Mannich routine and hence to the Friedel–Crafts product (48).

Analysis

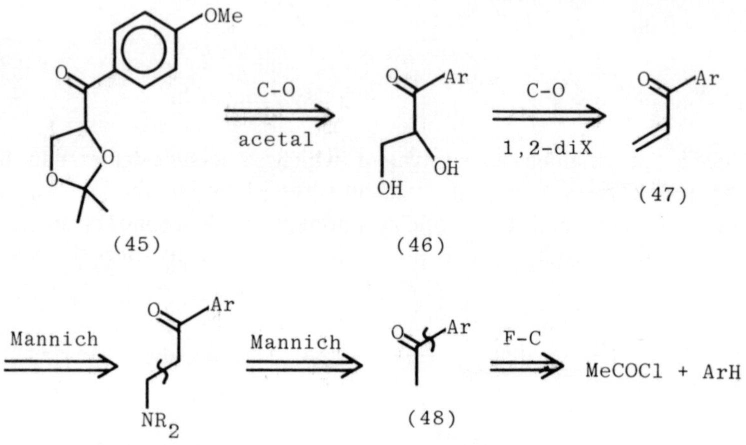

Whiting used Me₂NH for the Mannich reaction and found that the elimination step could be carried out with bicarbonate as the very mild base. The hydroxylation was done by epoxidation and hydrolysis.

Synthesis[213]

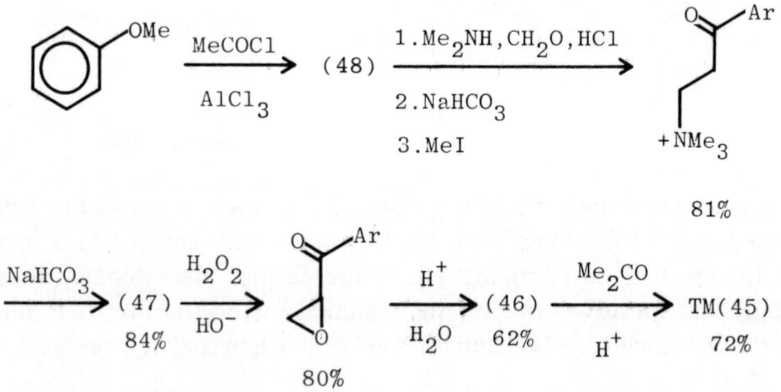

Cross-condensations II: Use of Specific Enol Equivalents

Problems of regio- and chemoselectivity are all answered by the use of specific enol equivalents—that is reagents which behave as the regiospecific enol of one particular carbonyl compound.

Activating groups

We have already seen (Chapters 13, 14) how the regioselective alkylation of ketones can be carried out by adding an activating group, usually CO_2Et, at the position we wish to be enolised. The same method can be used for carbonyl condensations. Condensation of the enol of an ester with an aldehyde as electrophile will not normally work as the aldehyde is more reactive in both senses and condenses with itself to give (50) instead.

If the ester is replaced by malonate (51), the specific enol equivalent, the condensation works very well. Malonate (51) enolises completely under the reaction conditions whilst the aldehyde is only slightly enolised and the most electrophilic carbonyl group is still the aldehyde. A mixture of weak acid and weak base is often used as conditions should be as mild as possible to encourage only the fastest reaction (kinetic control). This is known as the Knoevenagel reaction.[214] The product can be hydrolysed and decarboxylated in the usual way to give TM(49).

Any combination of two carbonyl groups or similar anion-stabilising groups is suitable for this reaction. The amide and cyanide groups in (52) point the way to a disconnection to a ketone (53) and the malonate derivative (54).

Analysis

We shall discuss the synthesis of (53) in Chapter 30. Otherwise the synthesis is straightforward. TM(52) was converted into a non-toxic, rapid acting barbiturate.[215]

162

Synthesis

$$(53) + (54) \xrightarrow[\substack{\text{reflux} \\ \text{separate water}}]{\text{NH}_4\text{OAc, HOAc}} \begin{array}{c} \text{TM}(52) \\ 74\% \end{array}$$

If the decarboxylated product, e.g. **(49)**, is wanted, it is a short cut to use the free acid, e.g. malonic acid, instead of the ester. The acid enolises nearly as well as the ester and decarboxylation occurs under the reaction conditions. This is one of the best ways[216] to make the important cinnamic acids **(55)**.

Analysis

$$\text{Ar} \overset{\alpha,\beta}{\diagdown}\text{CO}_2\text{H} \overset{\alpha,\beta}{\Longrightarrow} \text{ArCHO} + \underset{\substack{\text{activation} \\ \text{needed}}}{\text{CH}_3\text{CO}_2\text{H}}$$

(55)

Synthesis[216]

$$\text{ArCHO} + \text{CH}_2(\text{CO}_2\text{H})_2 \xrightarrow[\substack{\text{DMSO} \\ 80°C}]{\text{piperidine}} \text{TM}(55)$$

Dienophiles for the Diels–Alder reaction (Chapter 17) can be made this way. The obvious Diels–Alder disconnection on **(56)** leads to a dienophile **(57)** which is clearly a malonate condensation product.

Analysis

(56)   (57)

Synthesis[217]

$$\text{PhCHO} + \text{CH}_2(\text{CO}_2\text{Et})_2 \rightarrow (57) + \overset{\text{heat}}{\diagup\!\!\!\diagdown} \xrightarrow{\text{heat}} \text{TM}(56)$$

Synthesis of 1,3-dicarbonyl compounds by this approach requires the acylation of malonates or other simple 1,3-dicarbonyl compounds as a first step. Sodium or potassium enolates **(58)** acylate on oxygen but in the corresponding magnesium enolates **(59)** the oxygen atoms are chelated by the metal, leaving the carbon free to react.

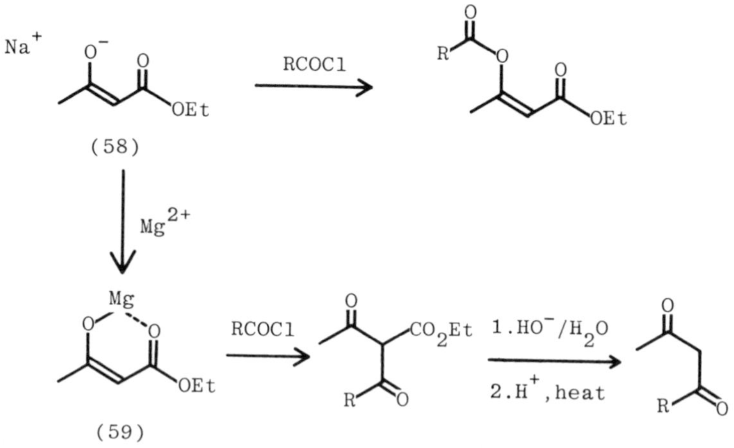

The natural product bullatenone was thought to have structure **(60)** from spectroscopic analysis of the very small amounts that could be isolated. Authentic **(60)** had to be synthesised for comparison with the natural product. Disconnection of the C–O bond reveals an enol **(61)** and hence a 1,3-diketone **(62)** disconnected in the usual way.

Alleged bullatenone: *Analysis*

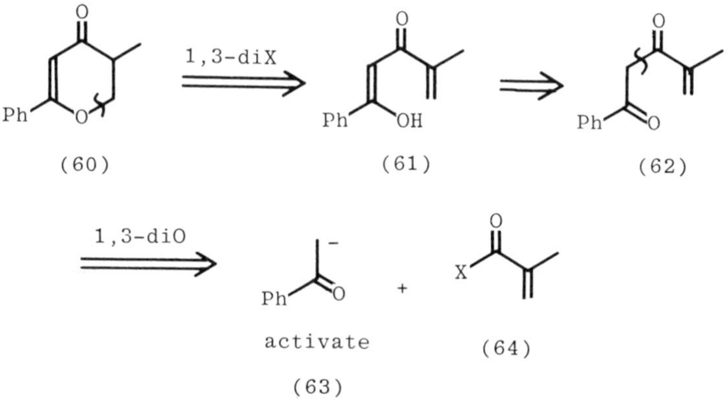

It was decided to use the activated form **(65)** as the specific enol equivalent of **(63)** and acylate the magnesium derivative with the acid chloride **(64, X = Cl)**.[218]

164

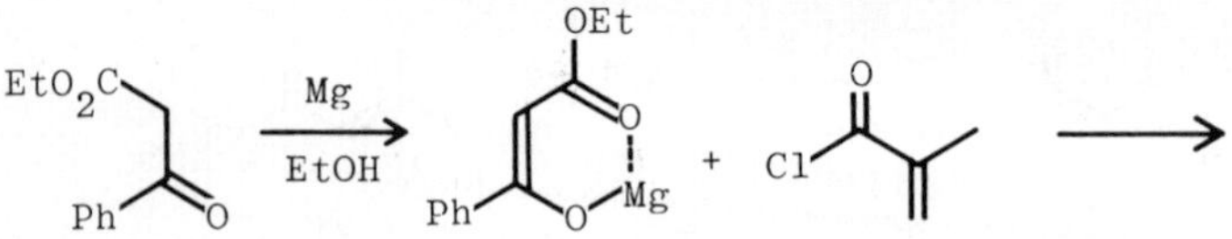

(65)

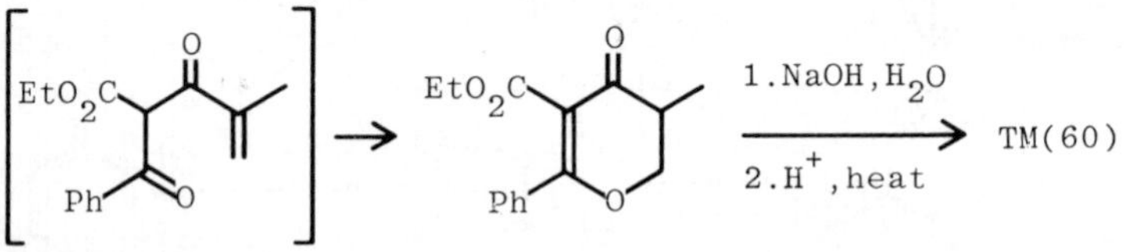

The cyclisation was spontaneous. Hydrolysis and decarboxylation gave **(60)** which proved not to be identical with natural bullatenone.*

Wittig and Reformatsky reagents

We have already met one series of specific enol equivalents as Wittig reagents **(66)** may be used in this way (see Chapter 15).

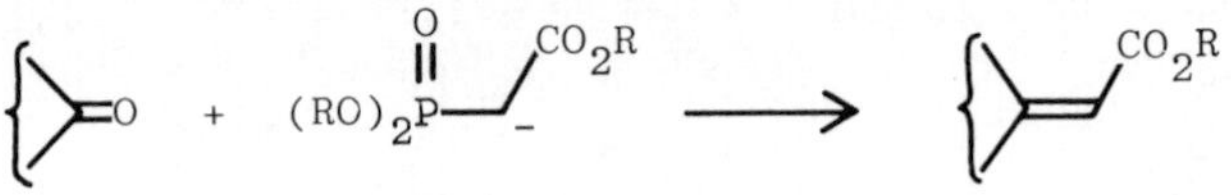

(66)

Alternatives which are more useful for β-hydroxy carbonyl compounds are organometallic reagents. Organozinc reagents **(67)** react with aldehydes and ketones but not with esters so they can be made from α-halo esters. This, the Reformatsky reaction,[219] has the advantage that the α-halo esters are easily made (see Chapter 7).

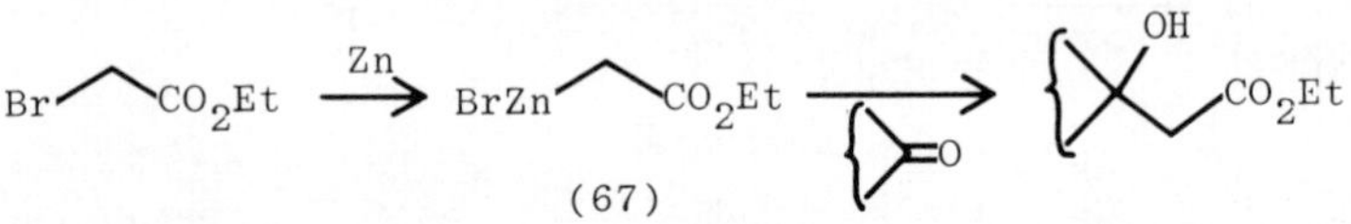

(67)

The alcohol **(68)** is best disconnected as a β-hydroxy carbonyl compound. The synthesis gives good yields by the Reformatsky reaction.[220]

Analysis

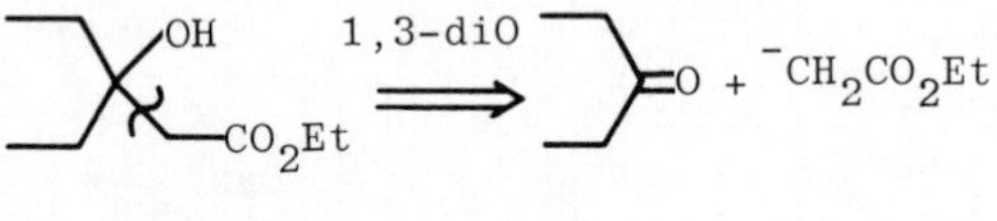

(68)

*The correct structure is given on page 195.

Synthesis[220]

$$BrCH_2CO_2Et \xrightarrow[\text{2.Et}_2\text{CO}]{\text{1.Zn}} TM(68)$$

Mevalonic acid (MVA) **(69)** is a very important biosynthetic intermediate used by us to make steroids and by plants to make terpenes. Tracing these biosynthetic pathways required radioactively labelled MVA and J. W. Cornforth[221] has published many such syntheses. MVA has two 1,3-dicarbonyl relationships **(70)**, one of which can be disconnected to **(71)** and **(72)**. If the OH and CO_2H groups are both protected as esters, we can use the Reformatsky method to activate **(72)** and finally remove both protecting groups in one step.

MVA: *Analysis*

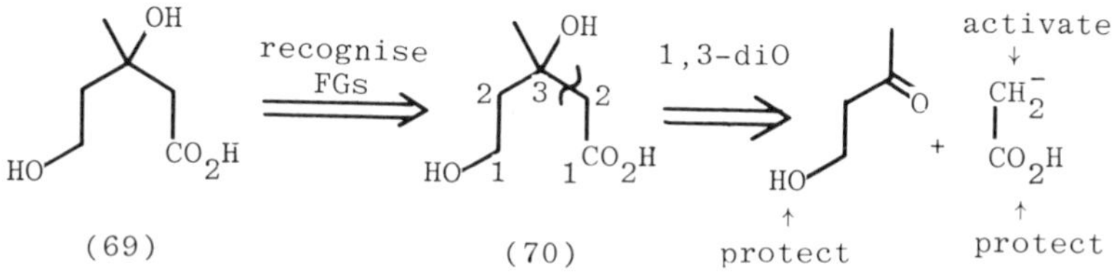

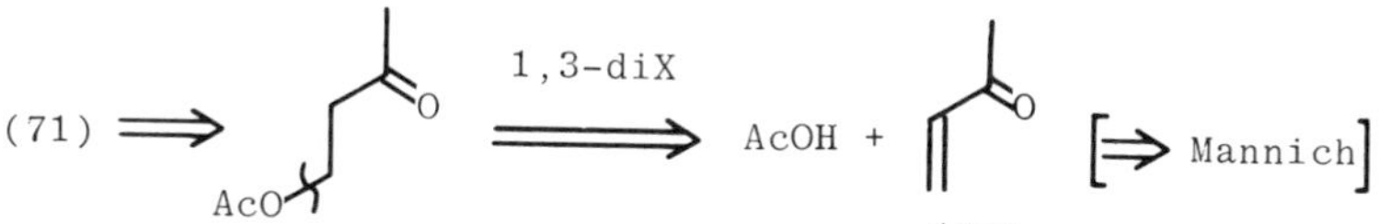

Methyl vinyl ketone **(73)** can be released[222] from the Mannich base when required and addition of acetic acid then occurs under very mild conditions. MVA is normally released as the lactone **(74)** and Cornforth[223] used this synthesis to make ^{14}C-labelled (•) MVA **(74)**.

Synthesis[223]

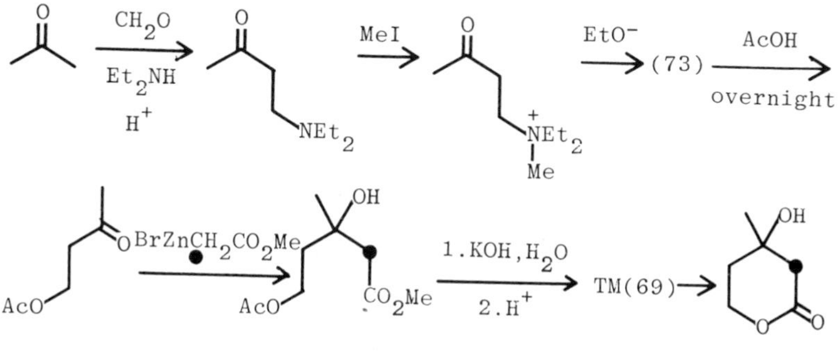

Enamines

The best specific enol equivalents for aldehydes are enamines[224] **(75)** and these are also very useful for ketones. They are easily made from the carbonyl compound and a secondary amine, are stable isolable compounds, and react **(76)** in the same way as enols.

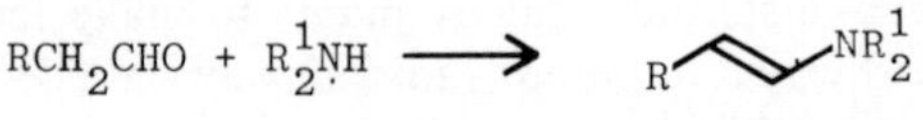

$$RCH_2CHO \ + \ R_2^1NH \ \longrightarrow$$

(75)

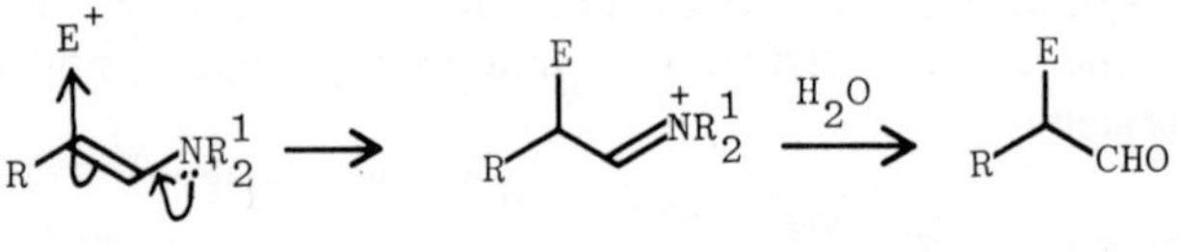

(76) E^+=electrophile

Analysis of enone **(77)** reveals a condensation between an aldehyde and the enol of a less reactive ketone, easily achieved[225] by first making the enamine of the ketone—the cyclic secondary amine morpholine **(78)** is often used.

Analysis

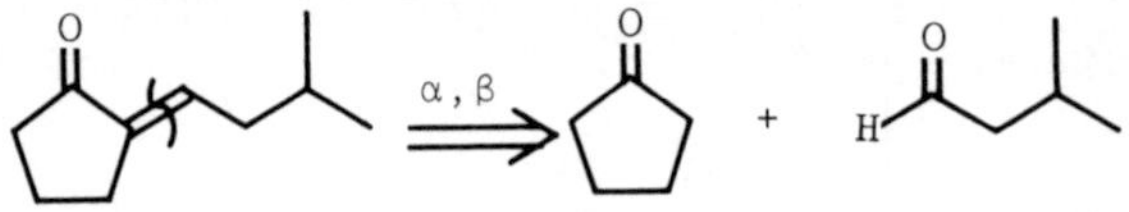

(77)

Synthesis[225]

(78)   (79)

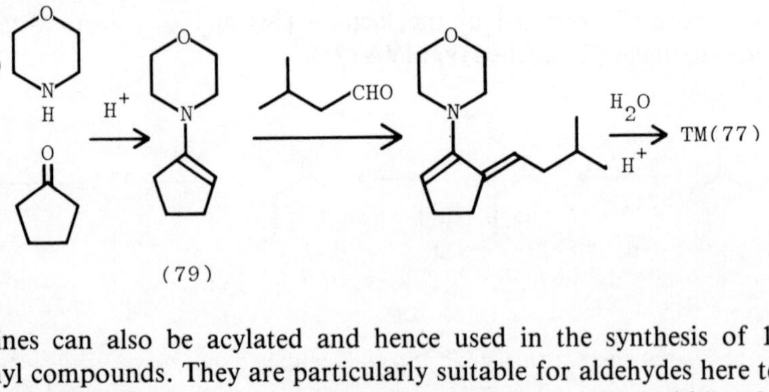

Enamines can also be acylated and hence used in the synthesis of 1,3-dicarbonyl compounds. They are particularly suitable for aldehydes here too. Disconnection of the keto aldehyde **(80)** is best carried out in the middle of the molecule to give two readily available starting materials. The synthesis[226] uses the enamine and the acid chloride.

Analysis

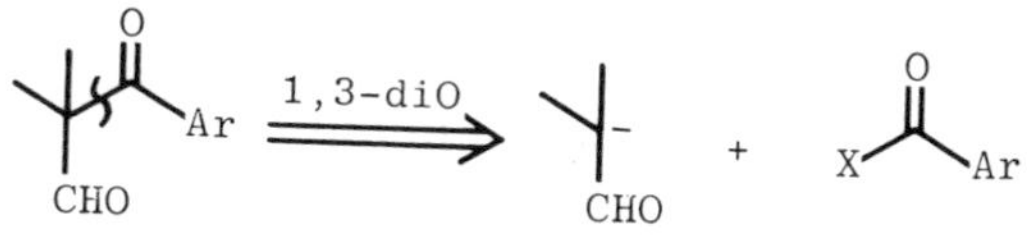

(80)

Synthesis[226]

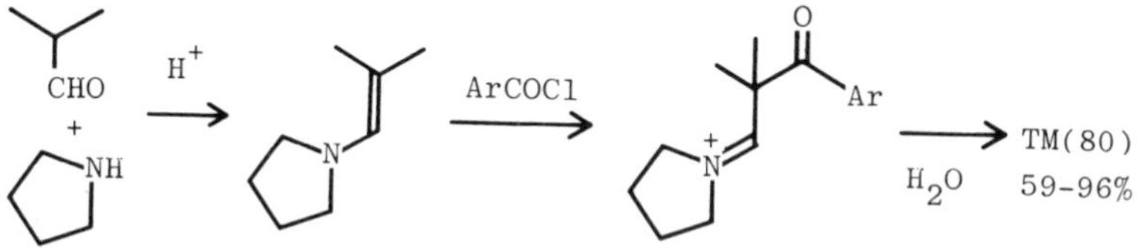

Cross-condensation III: Removal of One Product

The irreversible removal of one product is a powerful method of thermo-dynamic control when all products are in equilibrium. This type of control is commonly exercised through dehydration to give enones, e.g. (81)[227], product ionisation to give stable β-dicarbonyl anions, e.g. (82),[228] or decarboxylation (page 108), e.g. to give (83).[229]

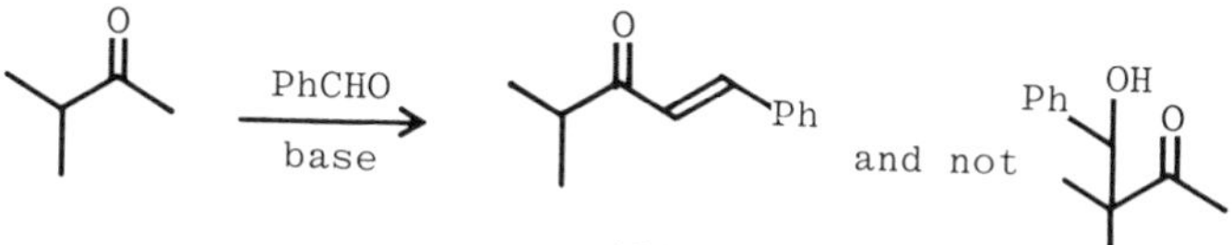

(81)

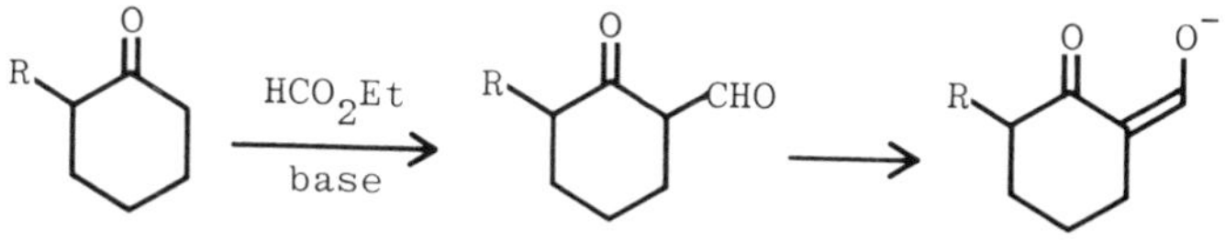

(82) 74%, R=Me

$$RCHO + CH_2(CO_2H)_2 \longrightarrow R \diagup\diagdown CO_2H$$

(83)

The first two examples allow for control over the regioselectivity of enolisation of an unsymmetrical ketone if only one side can dehydrate or form an anion. Formaldehyde reacts with (84) at its most enolisable site,[230] but the product (85) reacts at its methyl group as the alternative, and no doubt favoured, product (86) cannot dehydrate. The final product (87) was needed in an alternative approach to patulin (see page 158).

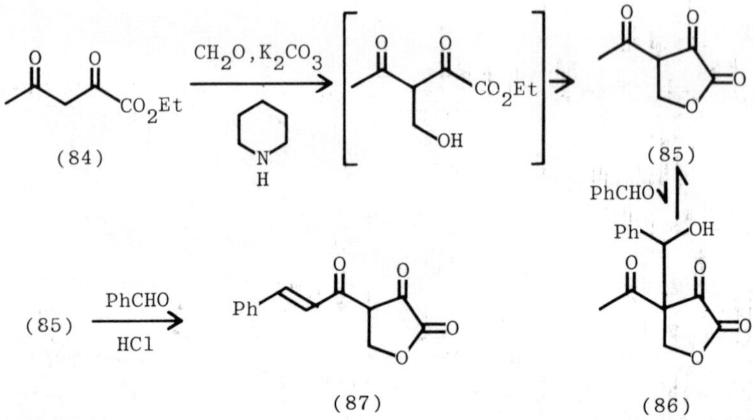

168

(84)

(85)

PhCHO

(85) $\xrightarrow[\text{HCl}]{\text{PhCHO}}$

(87)

(86)

The regioselective reaction of HCO_2Et or $CO(OEt)_2$ with an unsymmetrical ketone is important as it activates that side of the ketone towards enolisation (Chapters 14 and 20). Thus ketone (89) gives (90) and not (88) on condensation[231] with $CO(OEt)_2$ as only (90) can form a stable enolate ion (91). The product was needed for a synthesis of the ant pheromone (92) (see Chapter 29).

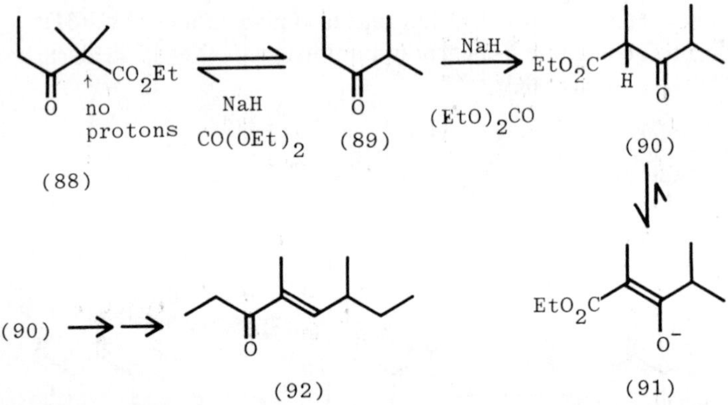

(88)

(89)

(90)

(90) $\rightarrow \rightarrow$

(92)

(91)

The decarboxylation method was discussed on page 162 as it is part of the Knoevenagel method of control. The European cockchafer, a root-eating pest, is lured to traps by butyl sorbate[232] (93), a synthetic pheromone. The ester is obviously made from the acid (94) which can be disconnected at the α,β bond to give aldehyde (95), again disconnected at the α,β bond.

Analysis

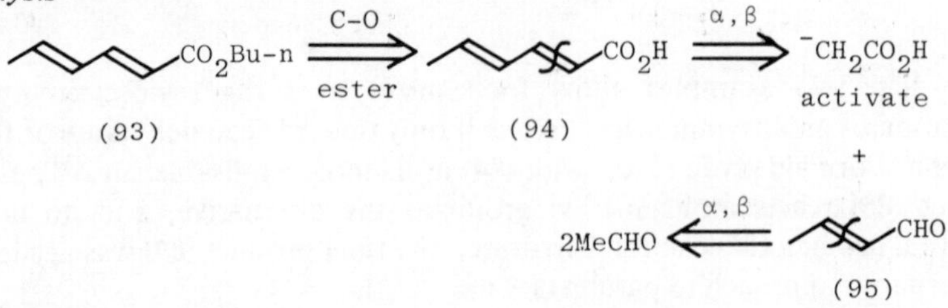

(93)

(94)

(95)

The self-condensation to give **(95)** will need no control, and the decarboxylation method is ideally suited to the second step.

Synthesis[232]

$$MeCHO \xrightarrow{base} (95) \xrightarrow[pyr,pip]{CH_2(CO_2H)_2} (94) \xrightarrow[H^+]{BuOH} TM(93)$$

CHAPTER 21

Two-Group Disconnections III: 1,5-Difunctionalised Compounds, Michael Addition and Robinson Annelation

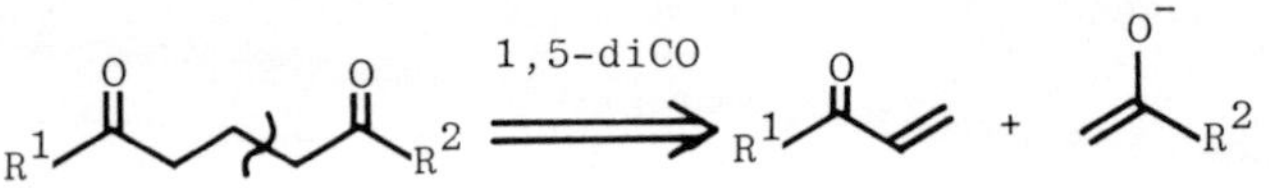

1,5-Dicarbonyl compounds (1) can be disconnected at either α,β bond in a reverse Michael reaction. All the previous questions of control remain so we should be well advised to have an activating group on enolate (3) to ensure both enolisation at this site (Chapters 14 and 20) and Michael rather than direct addition to (2) (Chapter 14). The normal CO_2Et group serves very well.[233]

Synthesis

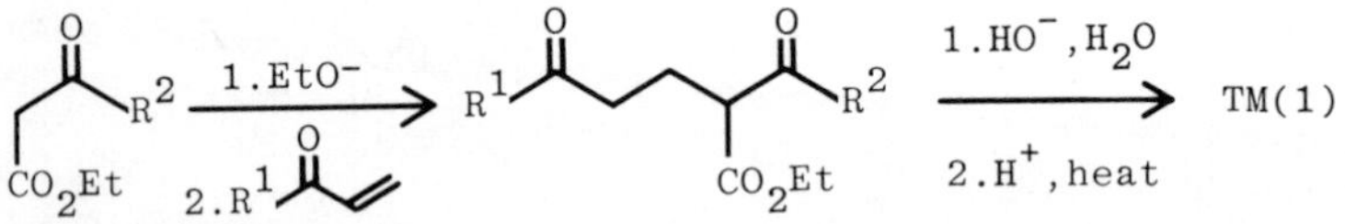

The cyclic ketone (4) is best disconnected where ring and chain meet and synthon (5) is best represented by malonate.

Analysis

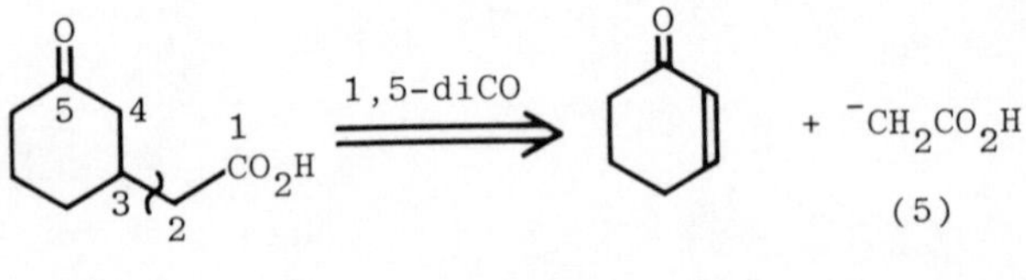

170

Synthesis[234]

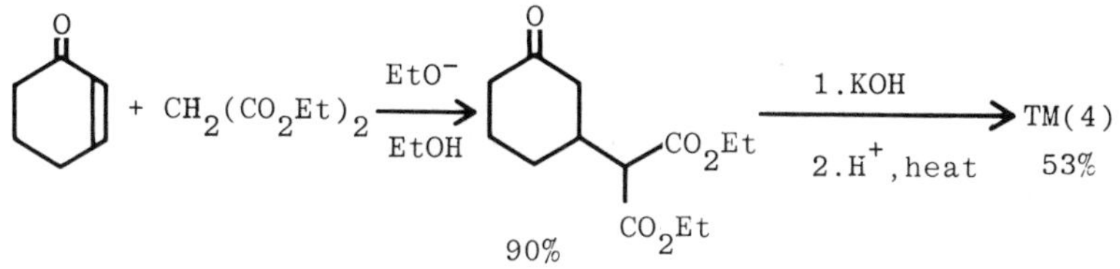

Michael reactions of this kind work best when they follow a catalytic cycle in which the first formed enolate anion of the product **(7)** is a strong enough base to regenerate the anion **(6)** of the starting material.[235] Then only a catalytic amount of base is needed unless a mole is consumed in converting the product into the stable anion **(8)**, as in 1,3-dicarbonyl synthesis (Chapter 19).

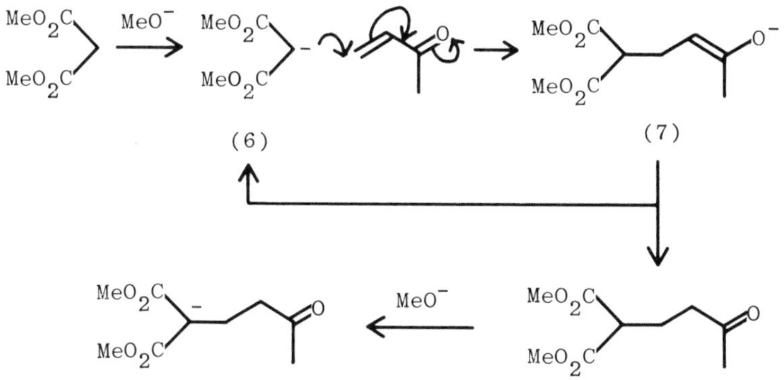

Stevens needed amine **(10)** in his synthesis of coccinelline[236] **(9)** — the defence compound ladybirds exude from their knees. The branched primary amine must come from a ketone **(11)** (Chapter 8) and it is probably better to leave the acetal protecting groups in place during these manipulations.

Analysis 1

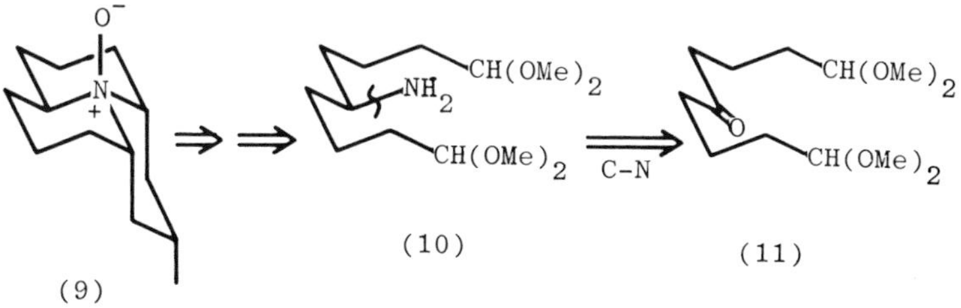

This (11) is a symmetrical ketone so we can use the 1,3-diCO disconnection (12) which follows the addition of a CO_2Et activating group (Chapter 19). This strategy is good because it leads back through a self-condensation to a single starting material (13). If we remove the acetal groups, we now have a 1,5-dicarbonyl compound (14) and disconnection by reverse Michael can give malonate as one starting material.

Analysis 2

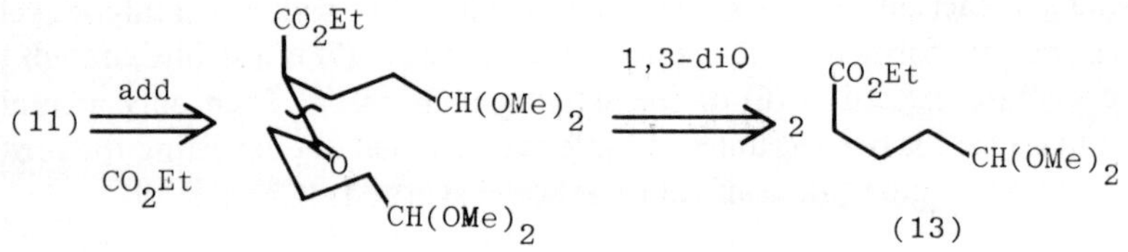

(12)

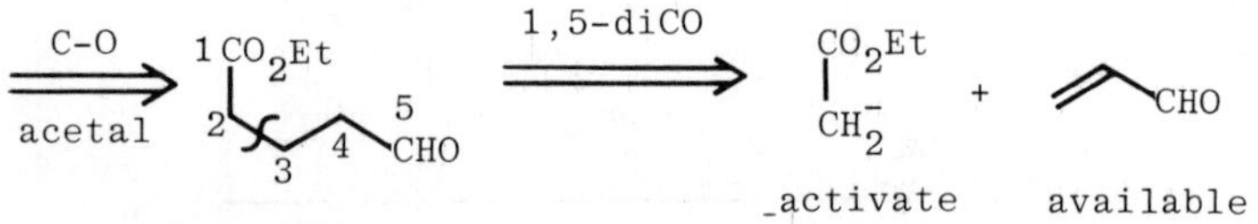

(13)

(14)

Stevens chose[236] to protect the aldehyde immediately after the Michael reaction to prevent side reactions on this reactive group, and to put in the amine by reductive amination (Chapter 8) using sodium cyanoborohydride as the reducing agent. Note the short cut of decarboxylation with NaCl in wet solvent (DMF or DMSO) when the ester, e.g. (13) is needed.

Synthesis[236]

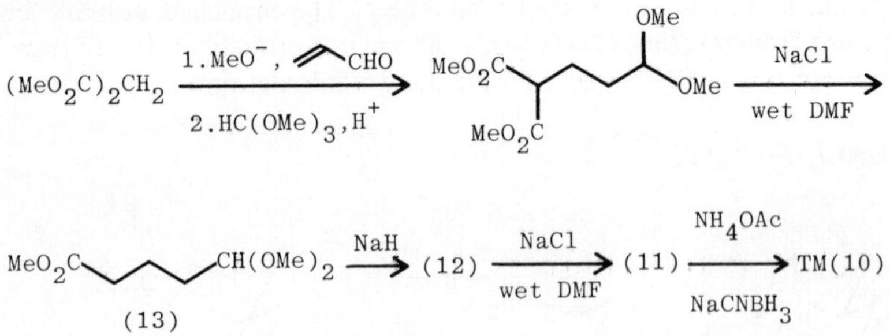

(13)

Activation by Enamine Formation

We discussed the use of enamines as specific enol equivalents in Chapter 20. They are equally useful in Michael reactions.[224] Enamine (15) adds cleanly to acrylic ester **(16)** in Michael fashion and the first-formed product **(17)** is in equilibrium with **(18)**. Hydrolysis of **(18)** in aqueous acid releases the 1,5-dicarbonyl product **(19)**.

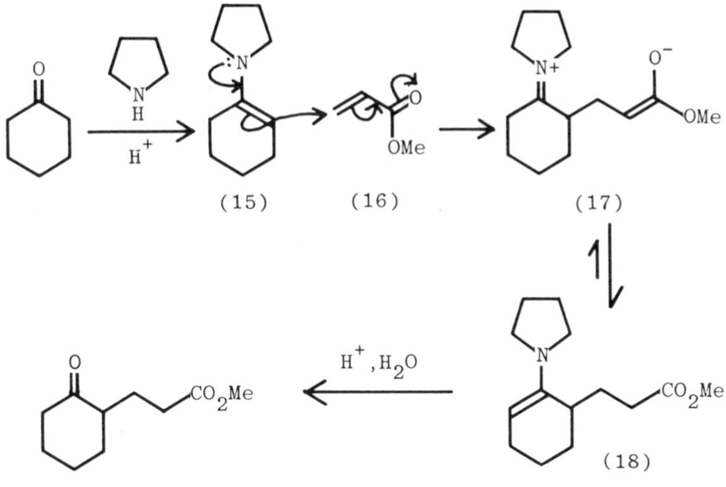

(15)   (16)   (17)

(18)

(19)   60%

The unsymmetrical 1,5-diketones **(21)**, needed for photochemical experiments,[237] were first synthesised from acid chloride **(20)** by a poor statistical strategy. The chemists really wanted compounds **(22)** with an extra methyl group and so were forced to consider the much better 1,5-dicarbonyl disconnection at the branch point (C2).

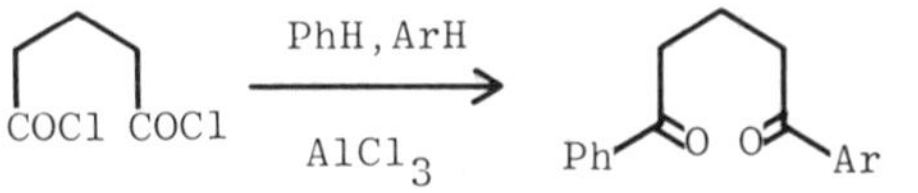

(20)

(21) low yields

Analysis

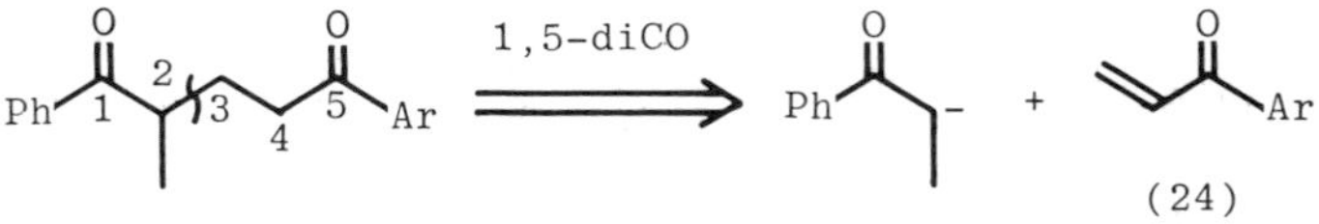

(22)   (23)   (24)

Synthon **(23)** was represented by an enamine and the vinyl ketone **(24)** could be made by the Mannich reaction (see Chapter 20) if required, as discussed in the next section.

Synthesis[237]

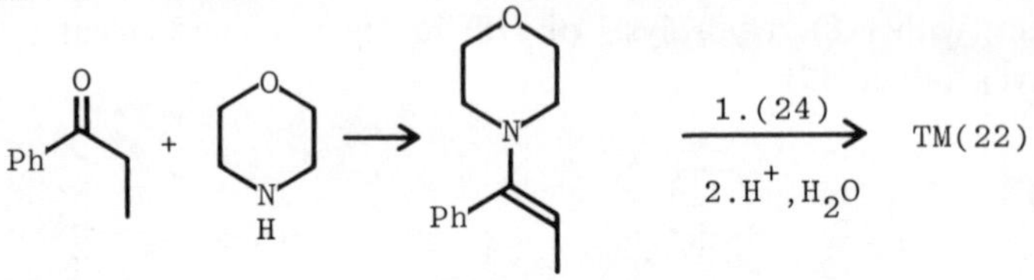

$$Ph \overset{O}{\underset{}{\parallel}} + \quad \text{morpholine} \quad \rightarrow \quad \text{enamine} \quad \xrightarrow[2.H^+,H_2O]{1.(24)} \quad TM(22)$$

Michael Acceptors by the Mannich Reaction

Vinyl ketones are rather reactive and dimerise readily by Diels–Alder reactions to give **(25)**. It is often better not to make them until they are needed. This can be done (see Chapter 20) by adding instead an alkylated Mannich base, which decomposes under the basic conditions used for the Michael reaction to release the vinyl ketone into the reaction mixture.

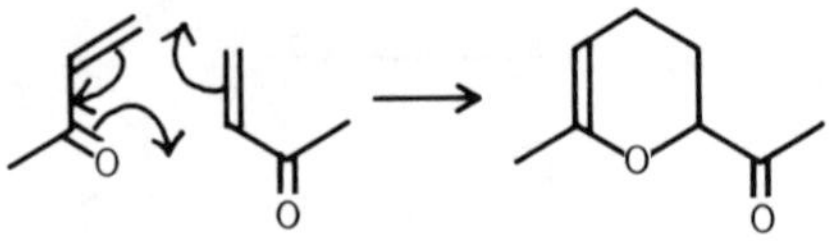

(25)

Cyclic ketones with exo-methylene groups, e.g. **(27)**, are usually protected in this way as the exposed CH_2 group is very electrophilic. In keto ester **(26)** the extra ester group tells us which disconnection to make to avoid the need for control.

Analysis

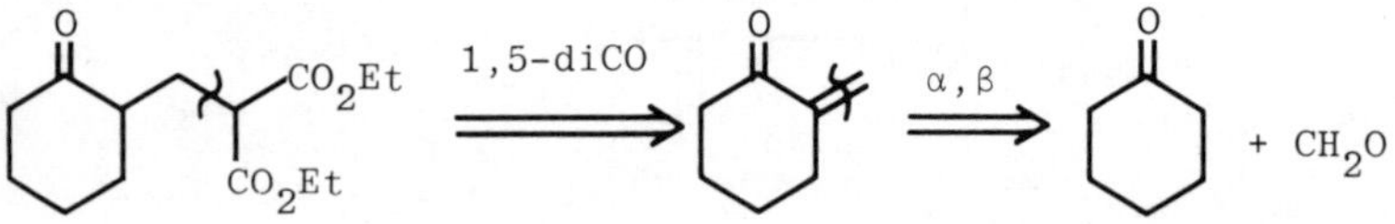

| (26) | (27) | use Mannich |

Synthesis[238]

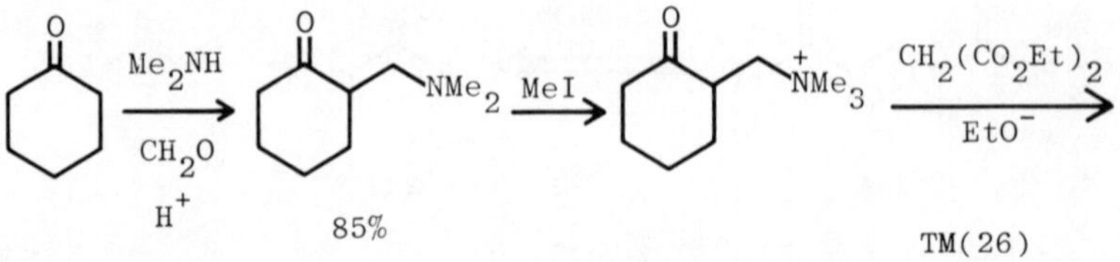

TM(26)
66%

An alternative route to TM(19) would be via TM(26) — both 1,5-diCO disconnections are useful.

The Robinson Annelation

One extension of the Michael reaction is an important way to make six-membered rings. At first glance you might think of a Diels–Alder disconnection for (28) as it is a cyclohexene with carbonyl groups, but the relationship is wrong. Disconnection of the α,β double bond, however, reveals a simple 1,5-dicarbonyl disconnection (29).

Analysis

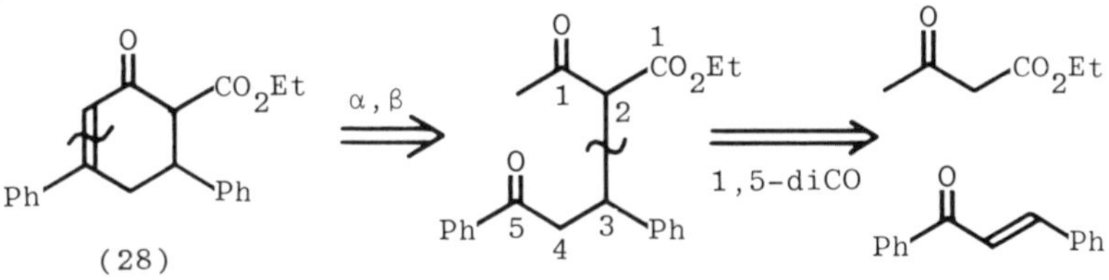

The second step of the synthesis, the cyclisation of (29) to (28), often goes spontaneously and the whole process of addition and cyclisation, being a ring formation or annelation, is known as the Robinson annelation.[239]

Synthesis[240]

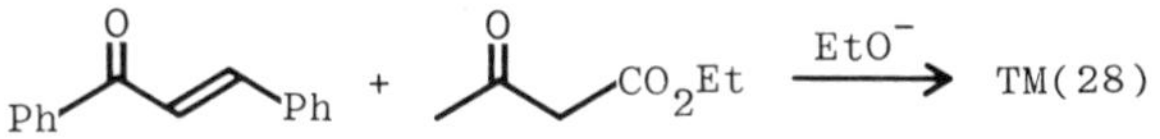

The bicyclic ketone (31) is an ideal intermediate in the synthesis of steroids such as (30) as the unconjugated ketone is correctly placed to add ring C. The α,β-disconnection on (31) gives a 1,5-diketone (32) with an excellent disconnection at the ring–chain junction to give the symmetrical diketone (33).

Analysis

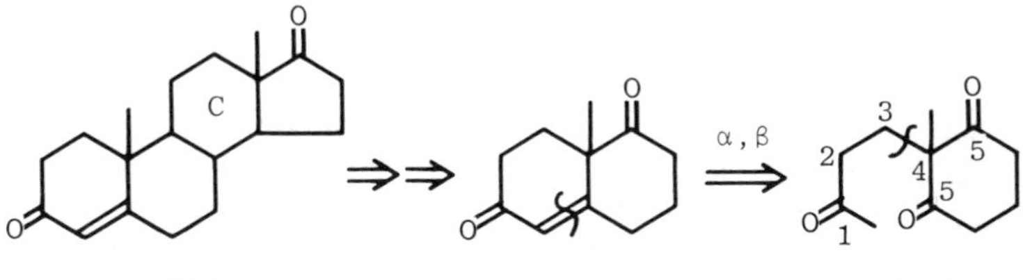

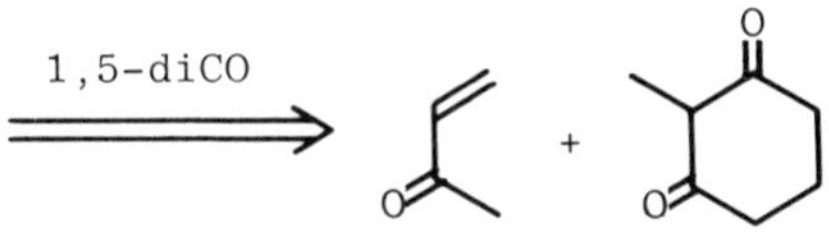

176

Activation is built into (33) and cyclisation of symmetrical (32) is unambiguous. In this case, the intermediate (32) is usually isolated[241] and the cyclisation carried out as a separate stage. Only a catalytic amount of base is needed for the Michael reaction as position C4 in (32) is blocked and cannot enolise (page 171).

Synthesis[241]

(33) $\xrightarrow[\text{MeOH}]{\text{KOH(cat.)}}$ (32) $\xrightarrow[\text{H}^+]{}$ TM(30)
65%

Simple cyclohexenones can be made by Robinson annelation, if necessary after the addition of an activating group. The α,β-disconnection on (34) reveals an unsymmetrical open chain 1,5-diketone (35) and so an activating group such as CO_2Et is necessary to control the Michael reaction leading to (36). The activating group will add to the methyl group of ketone (37) as the alternative product cannot enolise (Chapter 20)

Analysis

(34)   (35)   (36)

(37)

The Mannich method of control (Chapter 20) can be used here—the vinyl ketone being released in the reaction mixture. Cyclisation of (36) is unambiguous as the alternative product (38) cannot dehydrate to a conjugated enone. Hydrolysis and decarboxylation occur under the reaction conditions.

(38)

Synthesis[242]

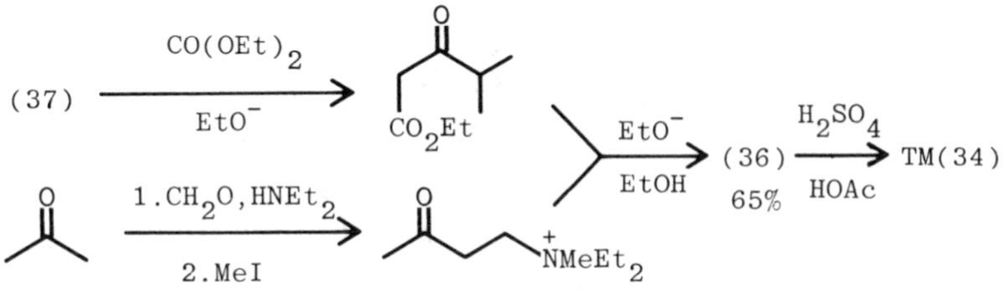

Cyclic 1,3-Diketones

In the same way, cyclisation of keto esters often occurs under the conditions of their synthesis. A famous example is dimedone (39). Either disconnection (a) or (b) would no doubt lead to a good synthesis but (b) is used because both starting materials are readily available, the enone (40) being an acetone dimer (Chapter 19).

Dimedone: *Analysis*

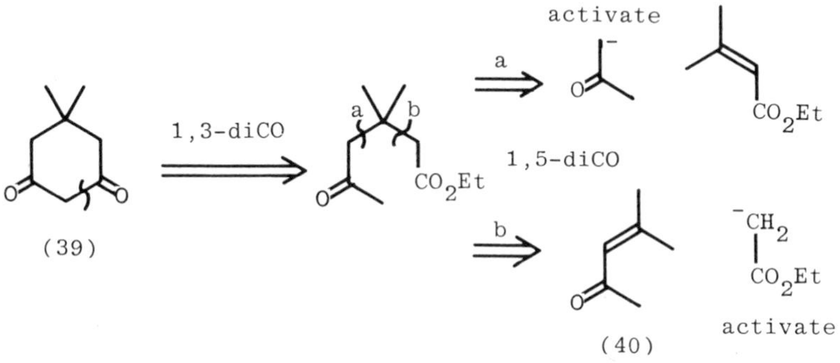

Synthesis[243]

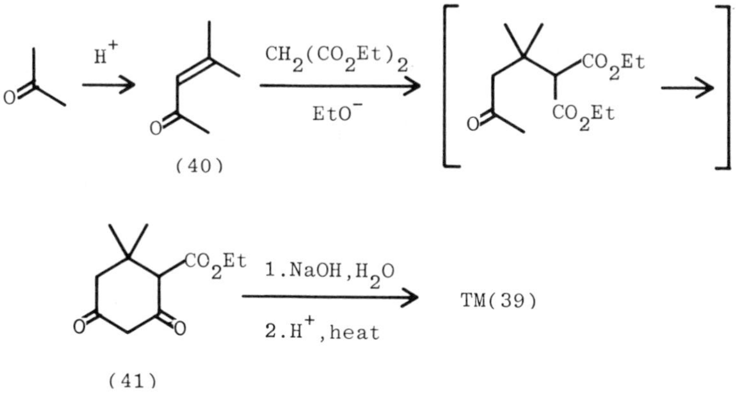

Activation by CO$_2$Et addition is easiest. The intermediate **(41)** can be isolated in 85% yield if desired, but it can be hydrolysed and decarboxylated directly to dimedone. Notice that cyclisation gives the stable six-membered **(41)** (thermodynamic control) as cyclisation of the most stable enolate **(42)** can give only a four-membered ring.

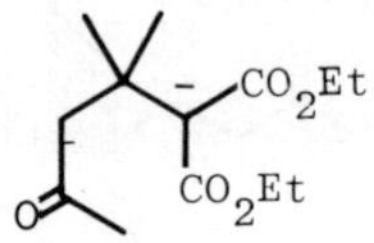

(42)

CHAPTER 22

Strategy X:
Use of Aliphatic Nitro Compounds in Synthesis

In controlling carbonyl condensations (Chapter 20) we ignored one possible device—using a compound which could enolise but not act as the electrophilic component—because it is unreasonable to expect any carbonyl compound to do this. This role can be filled by aliphatic nitro compounds such as nitromethane (1). The nitro group is not susceptible to nucleophilic attack but is very powerfully anion-stabilising. One nitro group is at least as anion-stabilising as *two* carbonyl groups: the pK_a of nitromethane is less than that of malonate esters (see Table 18.2). Even weak bases will 'enolise' nitromethane —it dissolves in aqueous sodium hydroxide.

$$CH_3 \overset{+}{N}\!\!\diagdown^{O}_{O^-} \quad \xrightarrow[\text{H}_2\text{O}]{\text{HO}^-} \quad CH_2 = \overset{+}{N}\!\!\diagdown^{O^-}_{O^-}$$

$$(1) \qquad\qquad\qquad (2)$$

Few nitro compounds are wanted as target molecules in their own right and the importance of nitro compounds in synthesis is that the nitro group is easily converted into two FGs in great demand: amines, e.g. (3), by reduction and ketones, e.g. (4), by $TiCl_3$ catalysed hydrolysis.[244] These reactions suggest two new concepts in synthetic design: latent functionally and acyl anion equivalence. They are the topics of this chapter and will be important too in the next few chapters.

$$RNO_2 \xrightarrow[\text{or H}_2,\text{Pd-C}]{\text{LiAlH}_4} RNH_2$$

$$(3)$$

$$\underset{R^2}{\overset{R^1}{>}}\!\!-NO_2 \xrightarrow{\text{TiCl}_3} \underset{R^2}{\overset{R^1}{>}}\!\!=O$$

$$(4)$$

179

The only readily available nitro compounds are nitromethane (1) and 2-nitropropane. However, nitromethane so readily forms an anion (2) that it can be alkylated in malonate fashion to give other nitro compounds (5). If the product is reduced to a primary amine (6) then the anion (2) is a reagent for the synthon $^-CH_2NH_2$. The amino group was not present at the time of the alkylation: it was *latent*. This is what we mean by latent functionality.

$$CH_3NO_2 \xrightarrow[\text{2.RHal}]{\text{1.base}} RCH_2NO_2 \longrightarrow RCH_2NH_2$$

$$(1) \qquad\qquad (5) \qquad\qquad (6)$$

This approach is particularly useful for primary amines bearing *t*-alkyl groups as few, if any, of our previous methods (Chapter 8) can be used for these compounds. The appetite suppressant Chlorphentermine[245] (7) is therefore best made from nitro compound (8). The best disconnection is then to the benzyl halide (9) and 2-nitropropane (10): these are both available and it is good strategy to break off as large a fragment as possible.

Chlorphentermine: *Analysis*

Synthesis[245]

$$(10) \xrightarrow[\text{2.(9)}]{\text{1.base}} (8) \xrightarrow[\text{cat}]{H_2} TM(7)$$

Nitroalkanes RCH_2NO_2 also condense with aldehydes and ketones and both double bond and nitro group in the products (11) can be reduced in one step to give amines (12). The condensation is unambiguous as nitromethane forms an anion so easily and does not react with itself. The reduction can be carried out catalytically or with 'Red-Al',[246] $NaAlH_2(OCH_2CH_2OMe)_2$

Diamine (13) was needed[247] as a monomer for polyamide manufacture.*
Disconnection of the shorter side chain by the scheme we have just described
takes us back to amino aldehyde (14). The remaining NH_2 group could come
from NO_2 or CN—we prefer CN because it gives us a 1,5-relationship (15)
which we can disconnect at the branch point by the usual Michael method
(Chapter 21).

Analysis

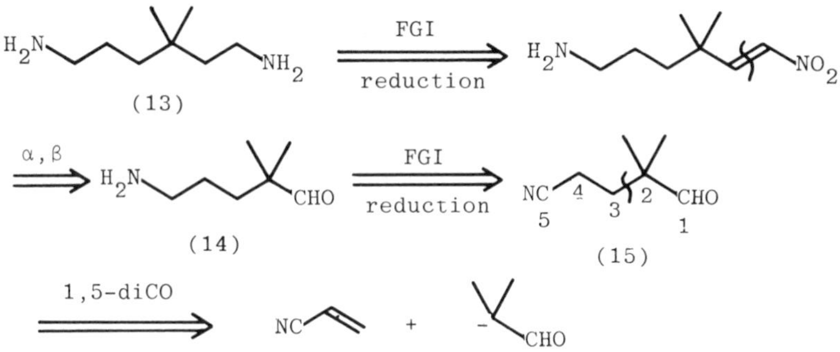

We shall not want to liberate (14) as it would cyclise. In any case, it is shorter
to reduce CN in the same step as reduction of C=C and NO_2 at the end.
Experiments showed that it was better to do the nitromethane condensation in
two steps.

Synthesis[248]

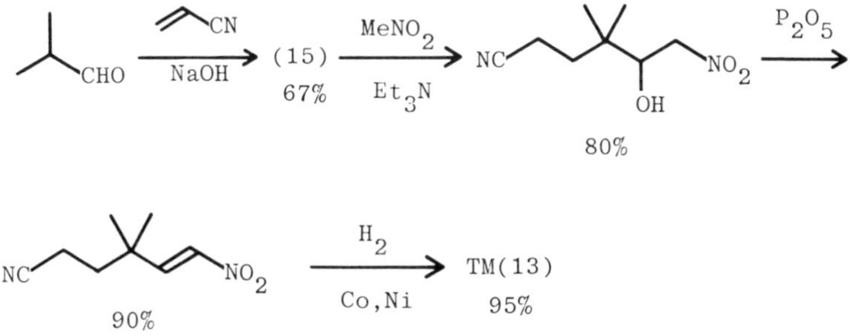

α,β-Unsaturated nitro compounds are good dienophiles so that cyclo-
hexylamines (17) can be made from Diels–Alder adducts (16) by reduction.

*Polyamides, such as nylon, are made from diamines and dicarboxylic acids:

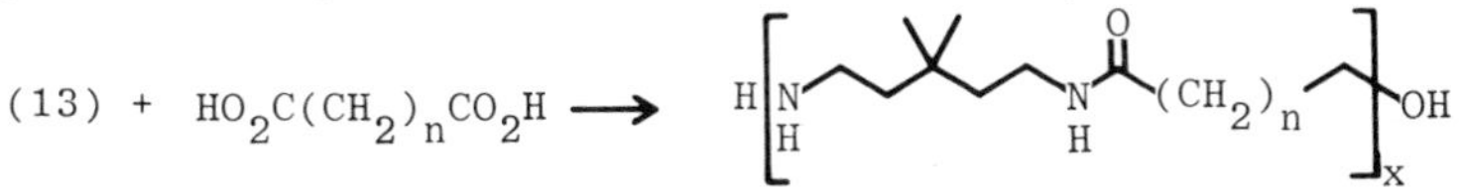

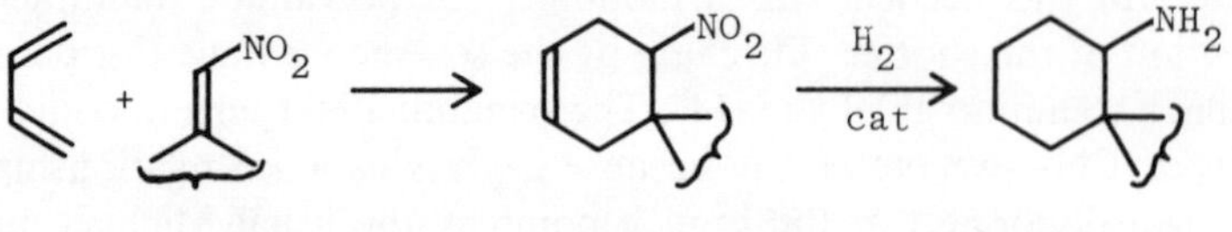

(16)   (17)

The stimulant Fencamfamin **(18)** is a candidate for this approach (once the ethyl group has been removed by the methods of Chapter 8) as it has the cage structure of cyclopentadiene **(19)** Diels–Alder adducts (Chapter 17).

Fencamfamin: *Analysis*

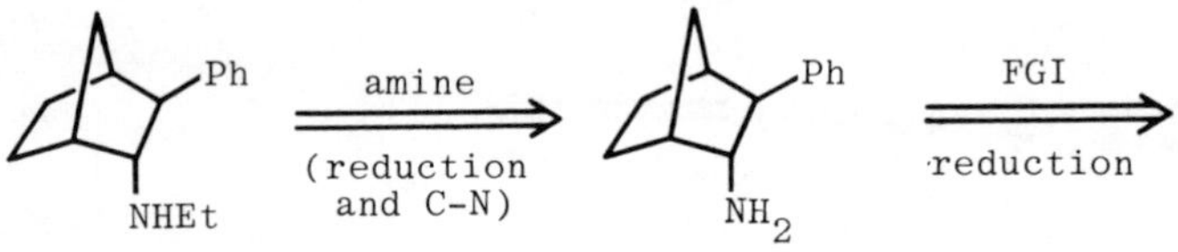

The ethyl group was added by imine reduction (Chapter 8). The stereochemistry will be as shown since the condensation selectively gives *trans* $PhCH:CHNO_2$ and the nitro group goes endo (Chapter 17).

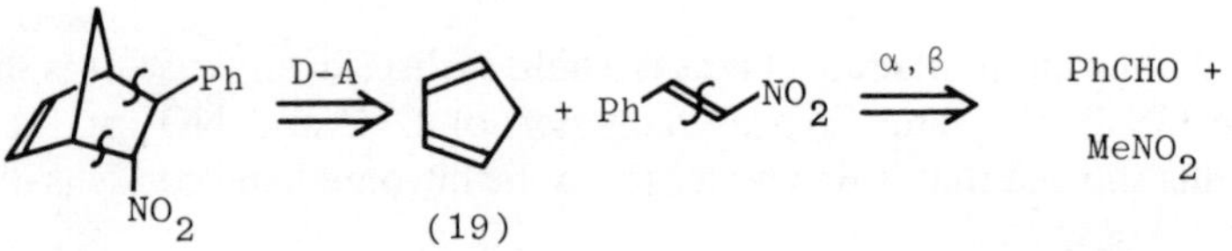

Synthesis[249]

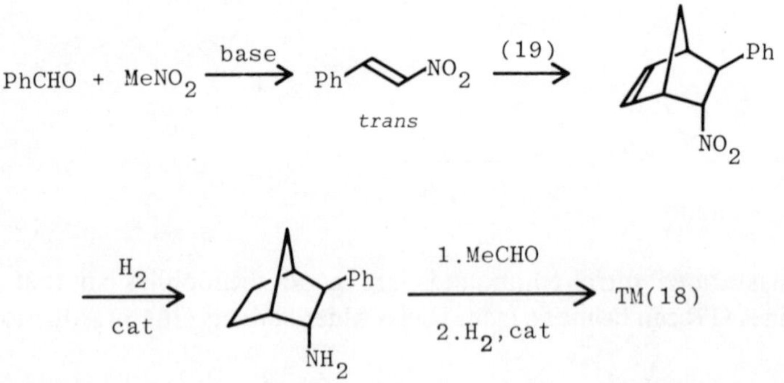

Nitro compounds also react cleanly in Michael additions. Disconnections of cyclic amide **(20)** gives a 1,5-relationship if we write NO_2 for NH_2—the tertiary centre next to NH_2 in **(21)** is a strong hint that we should do this. Only one Michael disconnection is then possible.

Analysis

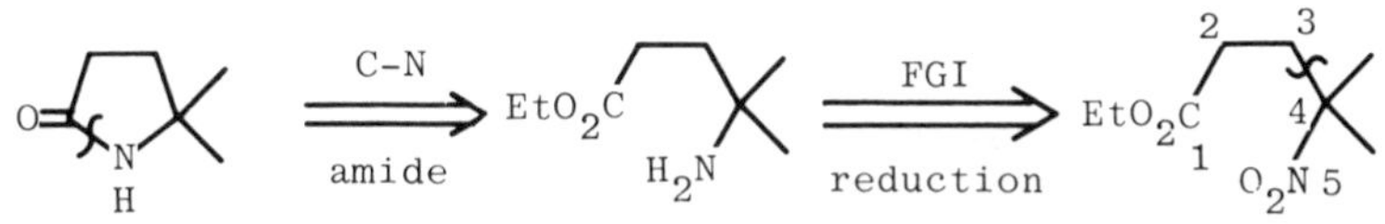

No control is needed in the Michael reaction as the nitro compound 'enolises' so readily. Catalytic reduction attacks NO_2 before carbonyls and cyclisation of (21) is spontaneous.

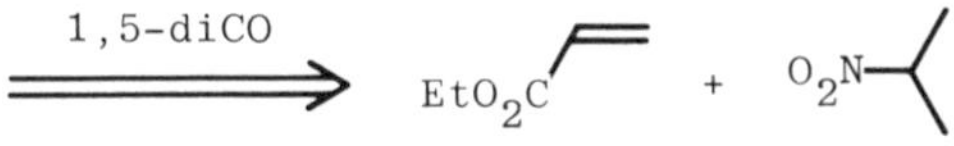

Synthesis[250]

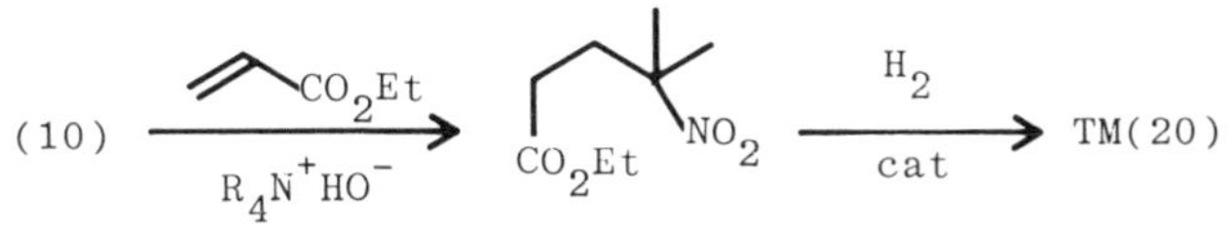

Ketones from Nitro Compounds

The conversion of secondary nitro compounds (22) into ketones used to be attempted by the rather violent and unsatisfactory Nef reaction with conc. sulphuric acid. McMurry's discovery[244] that $TiCl_3$ will catalyse the reaction under much milder conditions has made it a useful reaction.

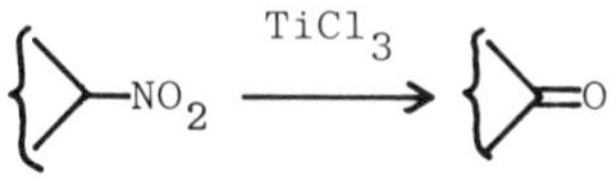

Alkylation of primary nitro compounds (23) is a simple way to make secondary nitro compounds. In this sequence, the starting material (23) has latent carbonyl functionality so that its anion is a reagent for the synthon R^1CO^-—it is an *acyl anion equivalent*. We shall need this type of synthon in the next few chapters.

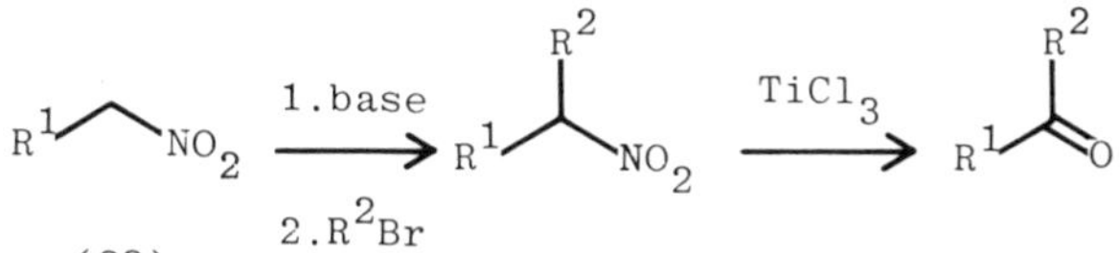

The unconjugated enone **(24)** is an example of a ketone which can be made from a nitro compound. Direct Diels–Alder disconnection is impossible, but replacement of carbonyl by NO_2 **(25)** reveals a simple Diels–Alder disconnection. The condensation will again give a *trans* double bond, but this is irrelevant to the final product **(24)**.

Analysis

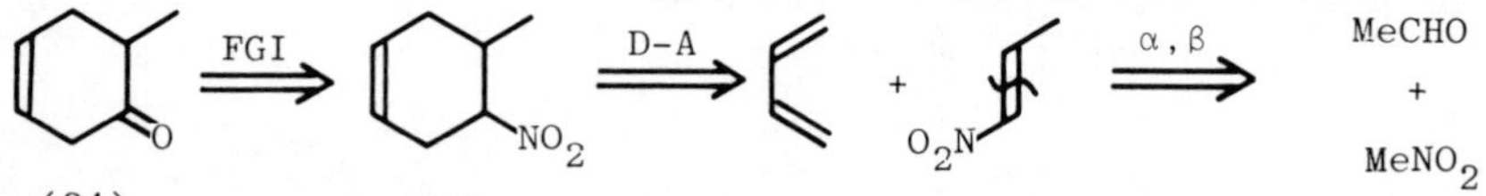

FGI
NO2
D–A
+
O2N
α, β
MeCHO
+
MeNO2

(24) (25)

Synthesis[244]

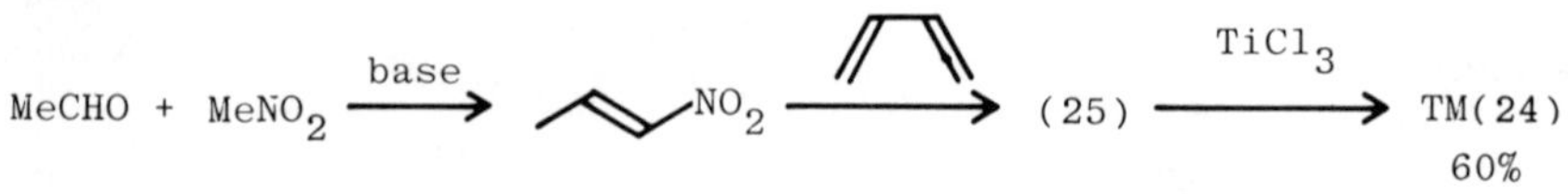

MeCHO + MeNO2
base
NO2
(25)
TiCl3
TM(24)
60%

CHAPTER 23

Two-Group Disconnections IV:
1,2-Difunctionalised Compounds

There are no grand unifying strategies to make 1,2-disubstituted compounds. Rather there are many diverse methods which I shall try to classify in a helpful fashion. Here, even more than elsewhere, it is important to judge each case on its merits. One common theme is a preference for the disconnection (1) of the bond joining the two substituted atoms.

$$(1) \qquad (2) \qquad (3)$$

The nature of the problem is now revealed. We have many candidates for the synthon (3), the most obvious being an aldehyde (Y = OH). But what of synthon (2)? This synthon has unnatural polarity (illogical) and the usual strategy will be to devise a reagent for (2) or to avoid it altogether by some alternative strategy.

Methods Using Acyl Anion Equivalents[251]

Disconnection of α-hydroxy ketones such as (4) requires acyl anion (5) equivalents. We have already met two of these, the nitro compounds in the last chapter and the acetylenes in Chapter 16. The acetylide ion is satisfactory here, adduct (6) being hydrated to TM(4) with Hg(II) catalysis.

Analysis

$$(4) \qquad\qquad (5)$$

Synthesis[172]

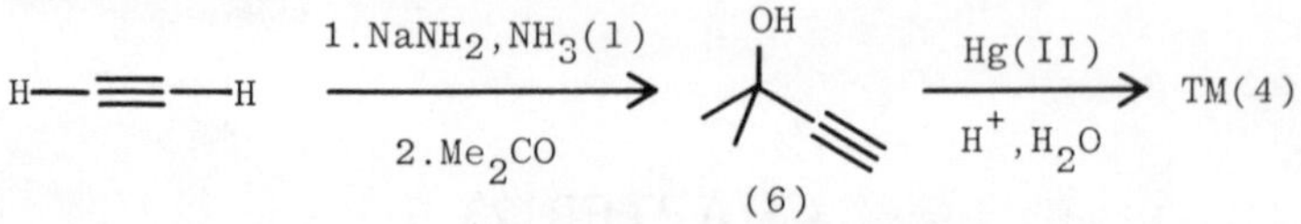

$$H \!\!-\!\!\equiv\!\!-\!\!H \quad \xrightarrow[\text{2.Me}_2\text{CO}]{\text{1.NaNH}_2,\text{NH}_3(1)} \quad (6) \quad \xrightarrow[\text{H}^+,\text{H}_2\text{O}]{\text{Hg(II)}} \quad TM(4)$$

Ketone **(8)** was an intermediate in a synthesis of the boll weevil hormone grandisol **(7)**. This cyclohexenone is a Robinson annelation product: disconnection by the method of Chapter 21 leads back to the available aldehyde **(9)** and the curious enone **(10)**.

Analysis 1

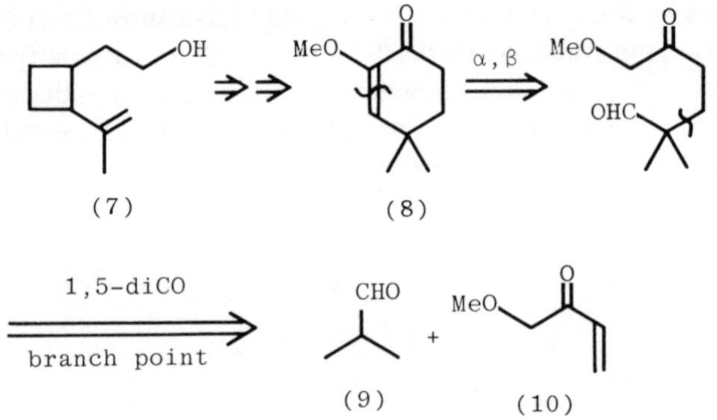

(7)   (8)

(9)   (10)

It might be possible to make **(10)** by a regioselective Mannich reaction but a more cunning approach comes from the realisation that any β-substituted compound **(11)** will do as well. If we put X = OMe, symmetry can be introduced by FGI to an acetylene **(12)** and readily available butyne diol **(13)** becomes the starting material (see Chapter 16).

Analysis 2

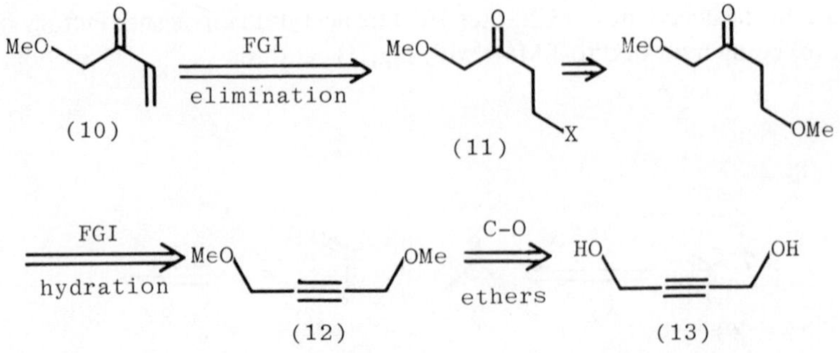

(10)   (11)

(12)   (13)

Synthesis 1 [252]

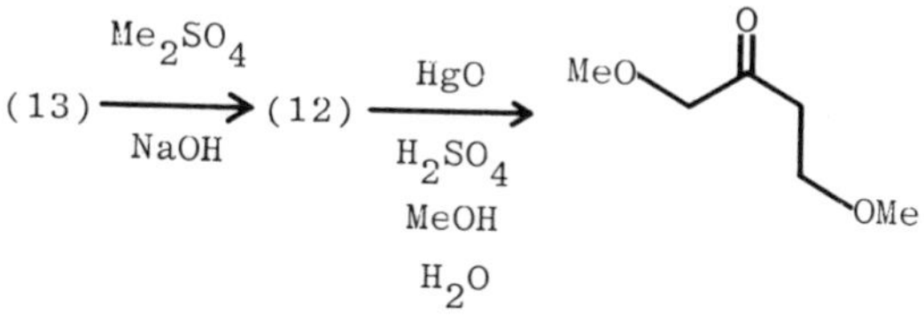

$$(13) \xrightarrow[\text{NaOH}]{\text{Me}_2\text{SO}_4} (12) \xrightarrow[\substack{\text{H}_2\text{SO}_4 \\ \text{MeOH} \\ \text{H}_2\text{O}}]{\text{HgO}}$$

The synthesis can be completed either by activating aldehyde **(9)** as an enamine and eliminating to give **(10)** before combining the two in a Robinson annelation (synthesis 2), or by the lazy man's method of simply using no control at all[254]—with good results in this case (synthesis 3).

Synthesis 2 [253]

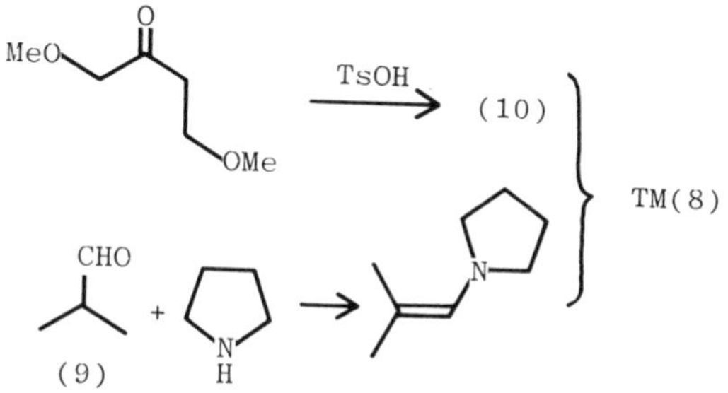

Synthesis 3 [254]

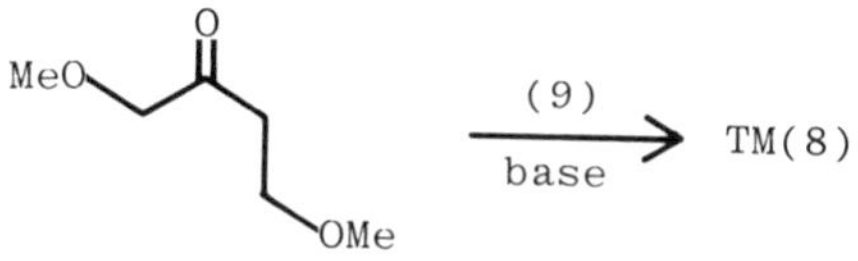

Another acyl anion equivalent we have already met is CN⁻, a reagent for ⁻CO₂H. We discussed its addition to carbonyl compounds **(14)** (Chapter 6) and the variation which leads to α-amino acids **(16)** (Chapter 6).

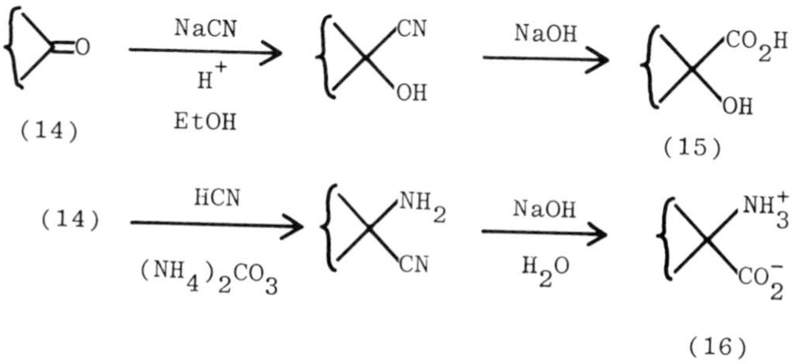

$$(14) \xrightarrow[\substack{\text{H}^+ \\ \text{EtOH}}]{\text{NaCN}} \quad \xrightarrow{\text{NaOH}} \quad (15)$$

$$(14) \xrightarrow[(\text{NH}_4)_2\text{CO}_3]{\text{HCN}} \quad \xrightarrow[\text{H}_2\text{O}]{\text{NaOH}} \quad (16)$$

In this chapter we shall take these methods a stage further by looking at TMs derived from these 1,2-disubstituted compounds (15) and (16). The minor tranquilliser Phenaglycodol[255] (17), used for treating petit mal, has two adjacent tertiary alcohol groupings. Disconnection of both methyl groups from one of these (Grignard, Chapter 10) gives an α-hydroxy ester (18), a compound of type (15).

Phenaglycodol: *Analysis*

In practice,[255] there may have been some difficulty in the conversion of cyanide to ester as the amide (19) was isolated as an intermediate. One molecule of Grignard neutralises the OH group in (18) so an excess is needed. An alternative approach would have been to use the SeO_2 oxidation described on page 191.

Synthesis[255]

The benzoin condensation[256]

Cyanide also plays a part in the benzoin condensation. The simplest example is the synthesis of benzoin (21) from benzaldehyde. The disconnection and logic are the same as in the synthesis of (4), as the cyanide transforms one molecule of benzaldehyde into the acyl anion equivalent (20). The reaction is successful only for non-enolisable aldehydes.

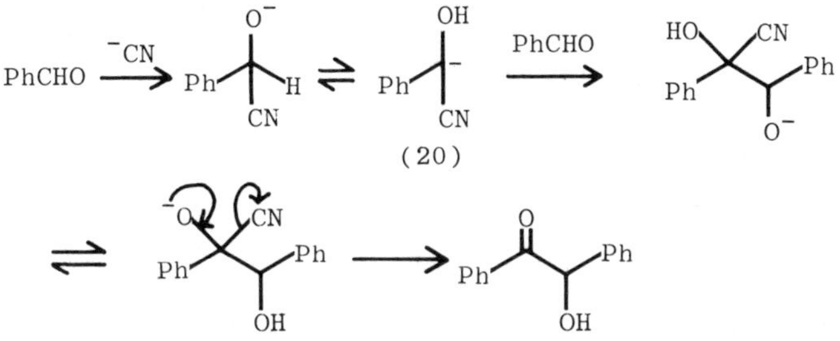

(20)

(21)

Analysis   *Synthesis*[256]

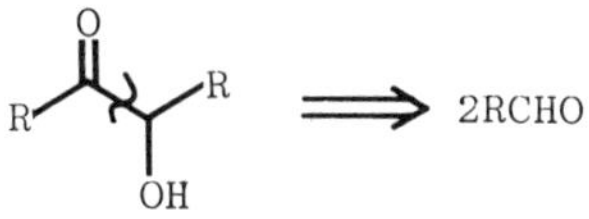

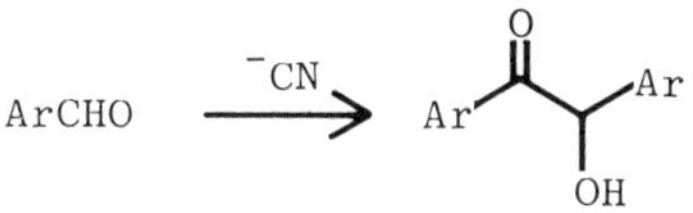

Methods from Alkenes

Alkenes are easily made (Chapter 15), have adjacent functionalised atoms, and can be converted into 1,2-disubstituted compounds by epoxidation, halogenation, or hydroxylation (Table 23.1). The Wittig reaction is the most general approach to alkenes so the disconnection is again between the two functionalised atoms.

Table 23.1 Alkenes as sources of 1,2-difunctionalised compounds

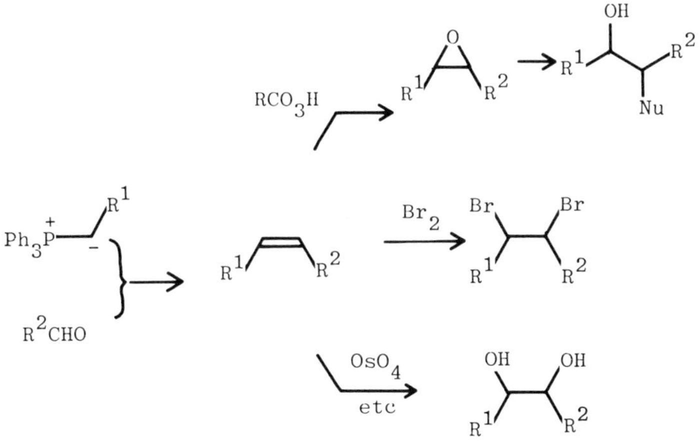

We have already discussed epoxides as sources of 1,2-difunctionalised compounds by C–X (Chapter 6) and C–C (Chapter 10) disconnections so we shall concentrate here on hydroxylation. The stereochemistry of these reactions has also been discussed (Chapter 12).

Lambert[257] wished to study the effect of an adjacent leaving group on the performance of another and so he wanted single diastereoisomers of ditosylates (22) with a variety of aryl groups. These are esters of 1,2-diols (23), which can be made stereospecifically from alkenes (24). Wittig disconnection gives (25) and (26) both made from commercially available aryl acetic acids (27).

Analysis

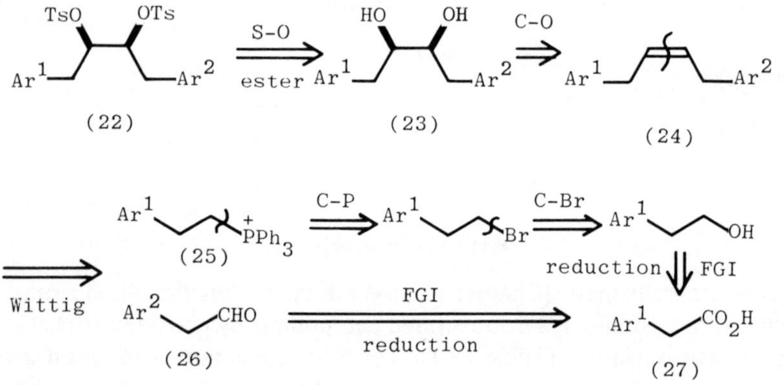

The hindered reducing agent (28) converts acids (27) into aldehydes (26) whilst LiAlH$_4$ gives the alcohols. The Wittig reaction stereoselectively gives *cis* alkenes (24) (Chapter 15), which Lambert chose to hydroxylate *cis*.

Synthesis[257]

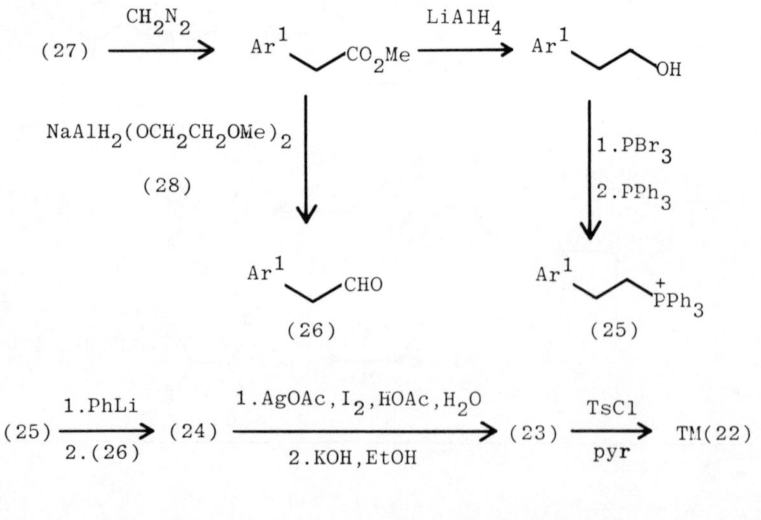

α-Functionalisation of Carbonyl Compounds

This approach was discussed in Chapter 6. The synthesis of the spasmolytic Diphepanol (28) is an example. Disconnection of one or both phenyl groups gives α-amino carbonyl compounds which can be made from the corresponding α-halo compounds (see Chapter 8). Halogenation of both (29) and (30) is regioselective as the other side of the carbonyl group cannot enolise.

Diphepanol: *Analysis*

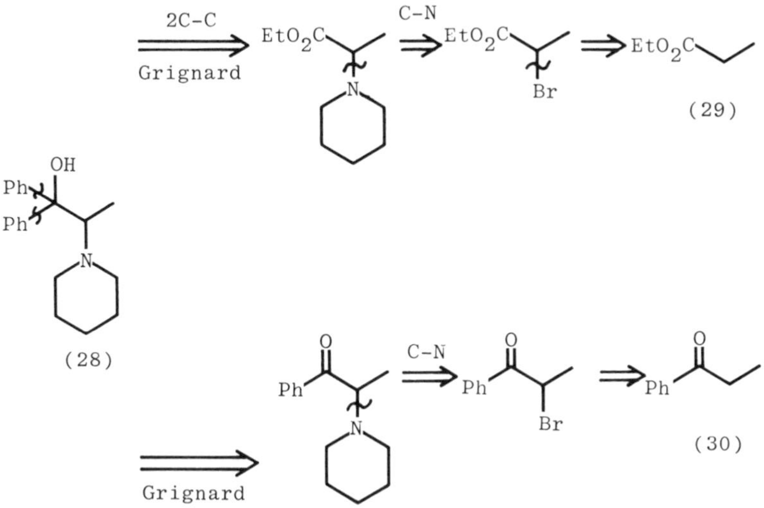

One published synthesis[258] uses the Friedel–Crafts route *via* (30), though no doubt the other would work well too. There are obvious alternatives *via* epoxides.

Synthesis[258]

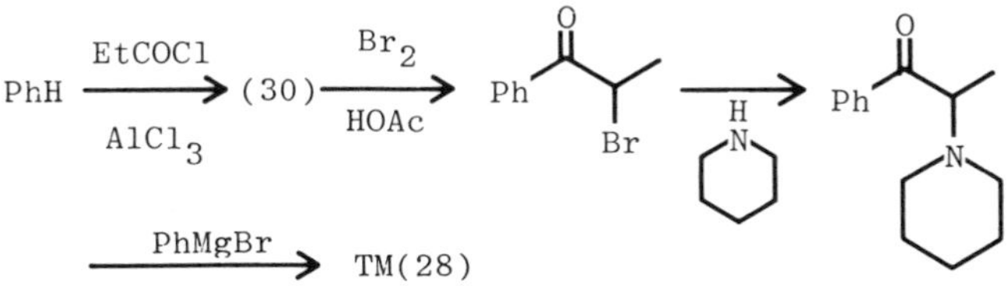

A related approach is the construction of a new carbonyl group next to the old. Such reactions occur *via* enols and are unambiguous only if enolisation is so too. Selenium dioxide (SeO_2) converts ketones directly to α-diketones[259] (31), whilst nitrosation[260] gives (32) which is in tautomeric equilibrium with oxime (33). Hydrolysis gives the same α-diketone (31).

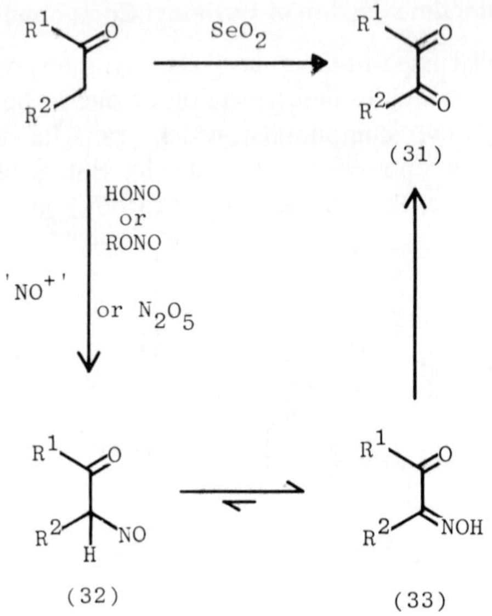

(31)

(32)   (33)

Both methods have been used to make adrenaline analogues. The broncho-dilator Metaproterenol (34) can be made from the α-keto aldehyde (35) by familiar methods. Our new α-oxidation makes aryl ketone (36) a suitable starting material, though the hydroxyl groups will need protection during the oxidation.[261] The strategy of α-oxidation appeals here because of the easy synthesis of ArCOMe by the Friedel–Crafts reaction.

Metaproterenol: *Analysis*

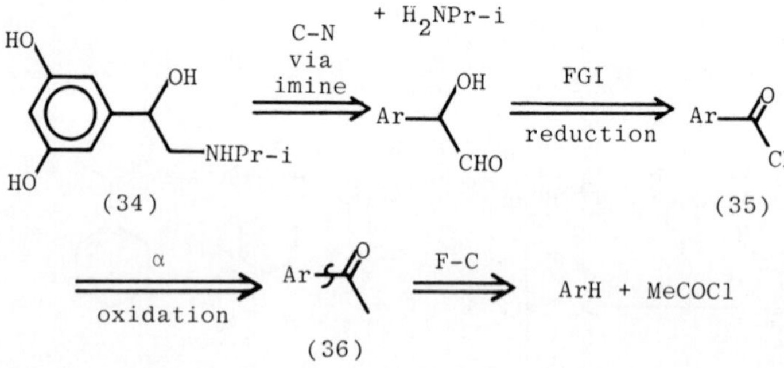

(34)   (35)

(36)

The synthesis has been carried out with methyl ether protecting groups, using SeO_2 for the α-oxidation. Both reductions (ketone and imine) were done in the same step.

Synthesis[261]

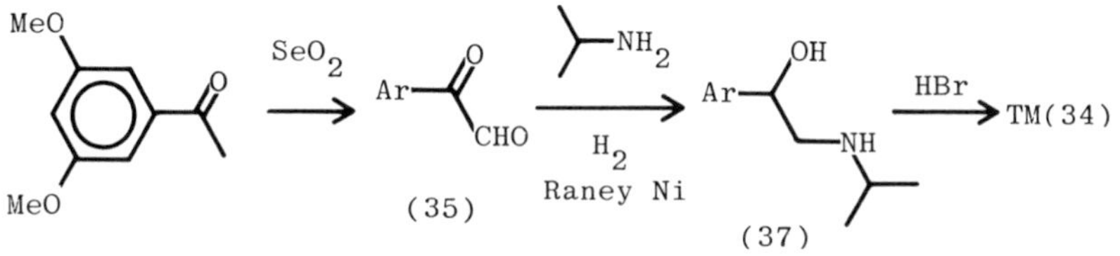

(35)

(37)

TM(34)

Notice that α-dicarbonyl compounds are very electrophilic and that the aldehyde in (35) reacts with *i*-PrNH$_2$ much more rapidly than does the conjugated ketone.

The triester (38) was needed for an investigation into intramolecular pericyclic reactions between electron-rich (a) and electron-poor (b) double bonds.[262] A Wittig disconnection on double bond (b) (nearer the centre of the molecule) demands an α-dicarbonyl compound whichever way we write it. The keto diester (39) can easily be made from a malonate ester by α-oxidation, so this route is preferred. Further disconnection of phosphonium salt (40) suggests allylic alcohol (41) as intermediate and hence regioselective reduction (Chapter 14) of α,β-unsaturated aldehyde (42) (Chapter 19).

Analysis

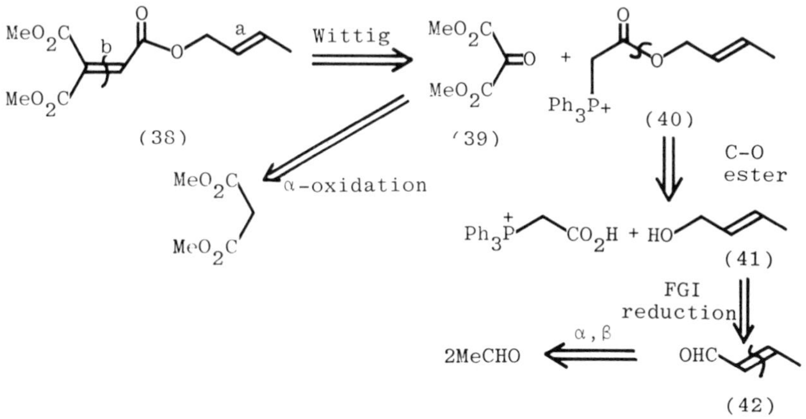

(38)

(39)

(40)

(41)

(42)

Since chloracetyl chloride is available, the synthesis of (40) followed a slightly different order of events from the analysis. This strategy of being guided by the availability of compounds is the subject of the next section.

Synthesis[262]

$$MeCHO \xrightarrow{\text{base}} (42) \xrightarrow{\text{NaBH}_4} \text{(41)} \xrightarrow{\text{Cl} \frown \text{Cl}}$$

(41)

$$Cl \frown O \frown \frown \xrightarrow{\text{PPh}_3} (40)$$

$$(MeO_2C)_2CH_2 \xrightarrow{\text{N}_2\text{O}_4} (39) \xrightarrow[\text{base}]{(40)} TM(38)$$

Strategy of Available Starting Materials

Since 1,2-difunctionalised skeletons are awkward to construct, one sensible strategy is to disconnect to a readily available compound (Table 23.2) instead of disconnecting the 1,2-relationship.

One way to disconnect diol (43) is to remove both phenyl groups (Grignard) and leave the skeleton (44) of lactic acid (45). In the synthesis, esterification is unnecessary as lactic acid (45) forms a dimeric lactone (46) which gives (43) with the Grignard reagent.[263]

Analysis

$$\text{(43)} \xleftrightarrow[\text{Grignard}]{\text{2C-C}} \text{EtO}_2C \text{ (44)} \xleftrightarrow[\text{ester}]{\text{C-O}} \text{HO}_2C \text{ (45)}$$

(43)   (44)   (45)

Synthesis[263]

$$(45) \xrightarrow{\text{heat}} \text{(46)} \xrightarrow{\text{PhMgBr}} TM(43)$$

(46)

The cyclic unsaturated ketone (48) has been used in a synthesis of bullatenone (47) (see Chapter 20 and workbook for Chapter 20) and as a dienophile in Diels–Alder reactions.[264] Disconnection of the enol ether reveals both 1,2- and 1,3-relationships in (49) and one of the 1,3-disconnections (a) gives available (4) as starting material.

Table 23.2 Some available 1,2-Difunctionalised compounds

$\begin{matrix}CO_2H\\|\\CO_2H\end{matrix}$ Oxalic acid (also esters and acid chloride)	$\begin{matrix}CHO\\|\\CHO\end{matrix}$ Glyoxal (as aqueous solution)	$\begin{matrix}CH_3-CO\\|\\CH_3-CO\end{matrix}$ Diacetyl
$\begin{matrix}CO_2H\\|\\CHO\end{matrix}$ Glyoxylic acid (.H$_2$O)	$\begin{matrix}CO_2H\\|\\COCH_3\end{matrix}$ Pyruvic acid	Chloracetyl chloride
Glycollic acid (.H$_2$O) $\begin{matrix}CO_2H\\\\OH\end{matrix}$	Lactic acid $\begin{matrix}CO_2H\\\\OH\end{matrix}$	Tartaric acid ($\pm$) and meso

$\begin{matrix}R\\|\\H_2N\quad CO_2H\end{matrix}$ Wide range of naturally occurring α-amino acids, R = Alk, Ar etc.

(See Chapter 16) Acetoin

Phenylglyoxal $Ph-CO-CHO.H_2O$

Benzoin (see Chapter 23) $\xrightarrow{HNO_3}$ Benzil $\longrightarrow$ Benzilic acid

Glycol $HO\frown OH$

Ethanolamine $HO\frown NH_2$

Ethylenediamine $H_2N\frown NH_2$

Analysis

$$(47) \Longrightarrow (48) \xrightarrow[\text{ether}]{C-O} (49) \xrightarrow[\text{diCO}]{(a)\ 1,3-} (4) + HCO_2Et$$

No control is needed in the first step as enolisation of **(4)** is possible only on one side and HCO$_2$Et cannot enolise (Chapter 20). Cyclisation of **(49)** occurs spontaneously but distillation is needed to dehydrate the hemiacetal **(50)**.

Synthesis[265]

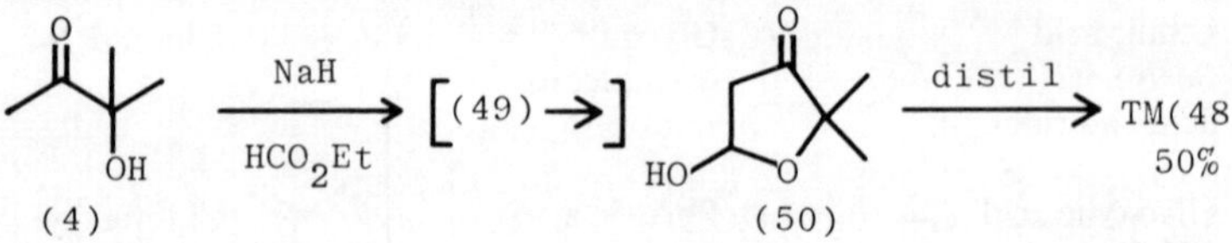

This completes a selection of methods for 1,2-difunctionalised compounds. More methods, based on radical reactions, appear in the next chapter.

CHAPTER 24

Strategy XI: Radical Reactions in Synthesis. FGA and its Reverse

We have so far discussed only ionic and pericyclic reactions and rightly so as they are more important in synthesis than radical reactions. However, some radical reactions are useful and it is convenient to group them in this chapter as many of them lead to 1,2-difunctionalised compounds.

Functionalisation of Allylic and Benzylic Positions[266]

Ionic routes to allylic (2) and benzylic (4) alcohols use reduction of carbonyl compounds (1) and (3) since these are readily available by condensation or Friedel–Crafts reaction.

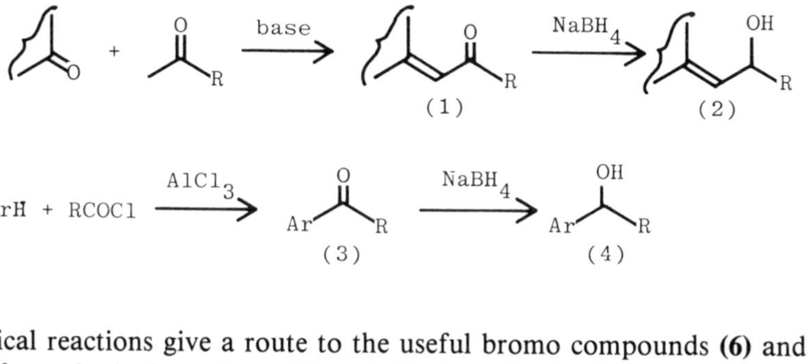

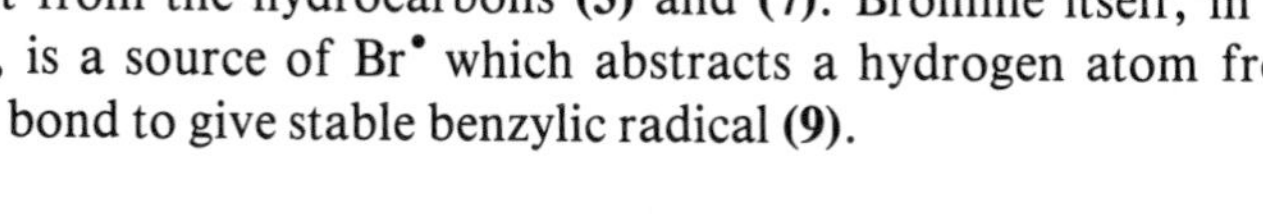

$$ArH + RCOCl \xrightarrow{AlCl_3} (3) \xrightarrow{NaBH_4} (4)$$

Radical reactions give a route to the useful bromo compounds (6) and (8) direct from the hydrocarbons (5) and (7). Bromine itself, in the presence of light, is a source of Br$^\bullet$ which abstracts a hydrogen atom from the weakest C–H bond to give stable benzylic radical (9).

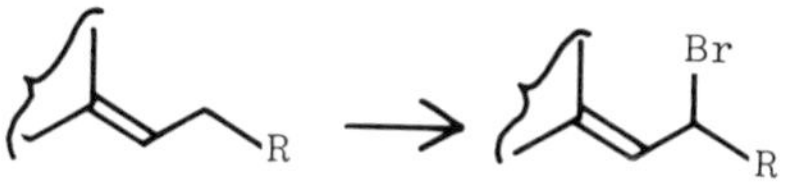

197

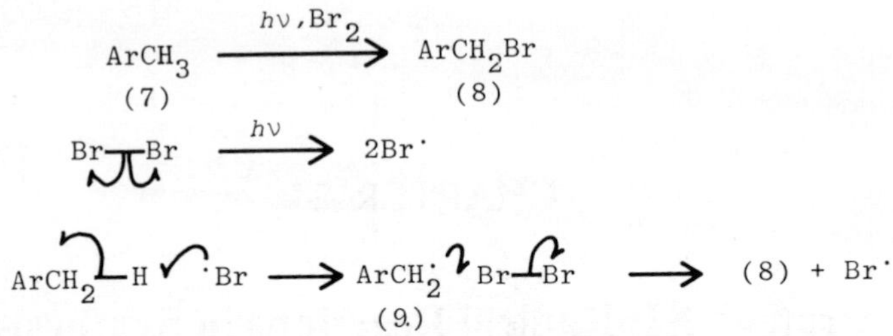

$$ArCH_3 \xrightarrow{h\nu,Br_2} ArCH_2Br$$
$$(7) \qquad\qquad (8)$$

$$Br\text{—}Br \xrightarrow{h\nu} 2Br\cdot$$

$$ArCH_2\text{—}H \quad \cdot Br \longrightarrow ArCH_2\cdot \quad Br\text{—}Br \longrightarrow (8) + Br\cdot$$
$$(9)$$

For allylic bromination, e.g. **(10)** → **(11)**, 'NBS', *N*-bromsuccinimide **(12)**, is often used. This acts as a radical generator and a source of bromine.*

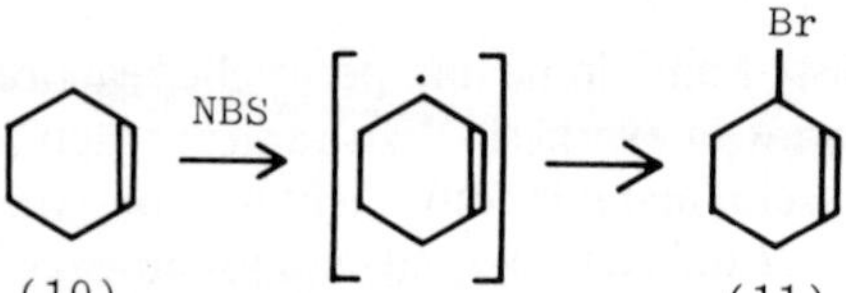

$$(10) \xrightarrow{NBS} \left[\ \cdot\ \right] \longrightarrow \overset{Br}{}$$

$$(10) \qquad\qquad\qquad (11)$$

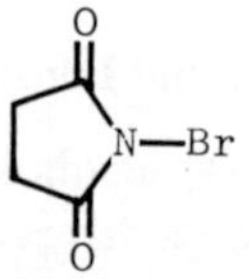

(12) NBS

When Confalone[267] was planning his synthesis of biotin **(13)**, he chose intermediate **(14)** to supply all the carbon atoms as well as one nitrogen and the sulphur atom. A Michael disconnection gives allylic thiol **(15)** which can be made from allylic bromide **(16)** and hence from cycloheptene.

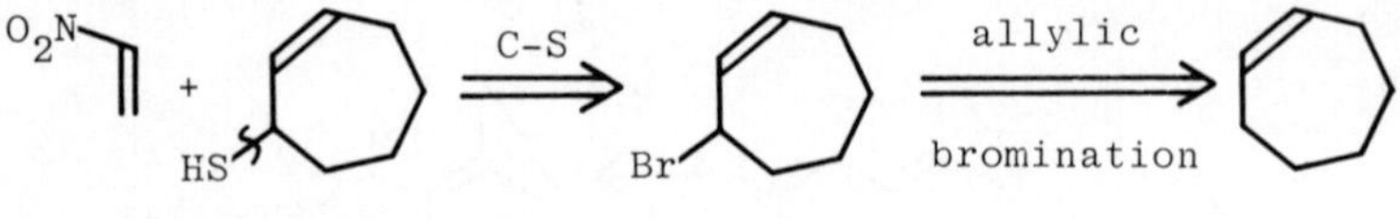

Analysis

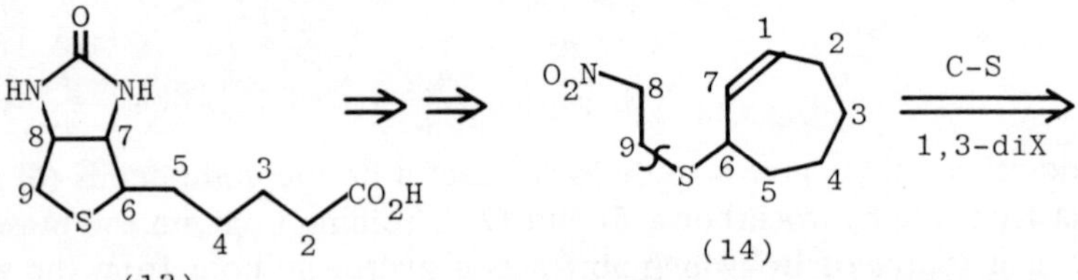

*The mechanism is given in the workbook.

NBS **(12)** was used for the allylic bromination. Thiol synthesis needs protection for sulphur (see Chapter 5) and an ester can be used. The nitro compound was also protected as an ester so that nitroethylene was released into the reaction mixture.

Synthesis[267]

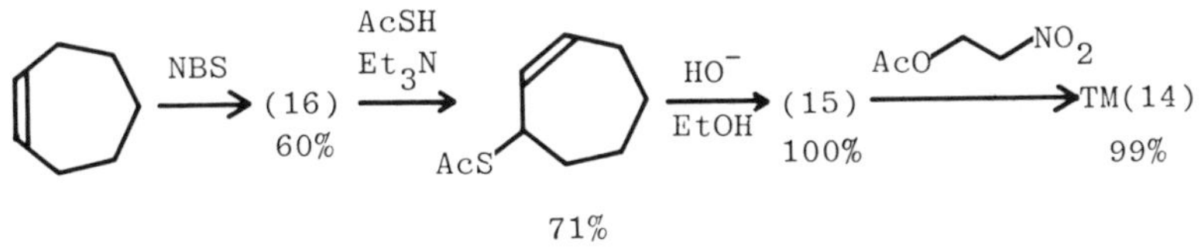

The *bis*-benzylic bromide **(17)** was needed[268] to study its cyclisation reaction in base: it gives the remarkable product **(18)**. Reversing the benzylic bromination gives symmetrical α-diketone **(19)** which can be made by a benzoin condensation (Chapter 23) after FGI. NBS was used here too as the radical generator in a short synthesis of this complicated TM.

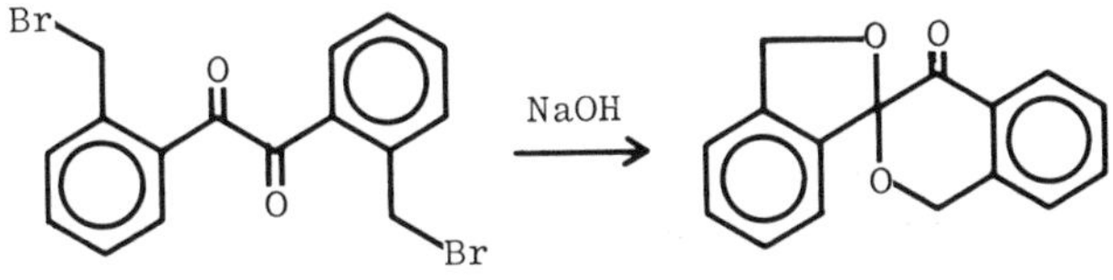

(17)   (18)

Analysis

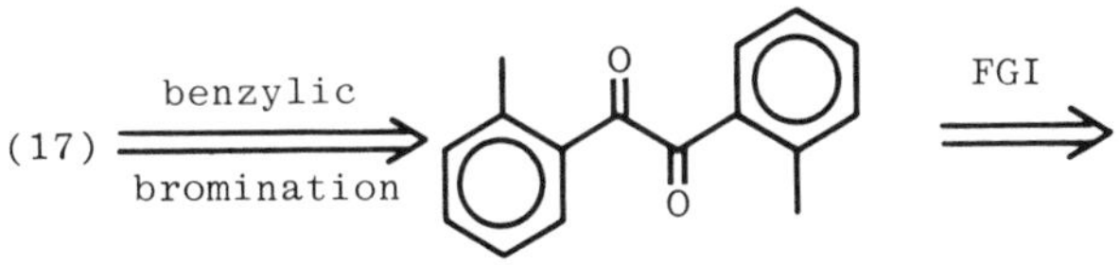

(19)

Synthesis[269]

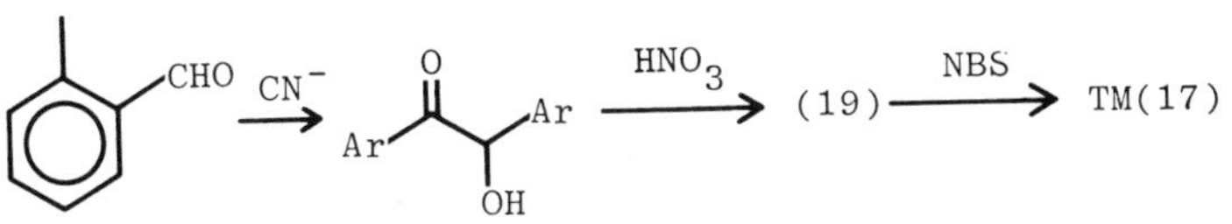

Though not radical reactions, the oxidation of allylic and benzylic positions by SeO_2 (see Chapter 23) is best mentioned here. Analysis of the cyclohexenone (20) by conventional (Chapter 21) and allylic oxidation strategies shows how different they are.

Analysis 1: Robinson annelation

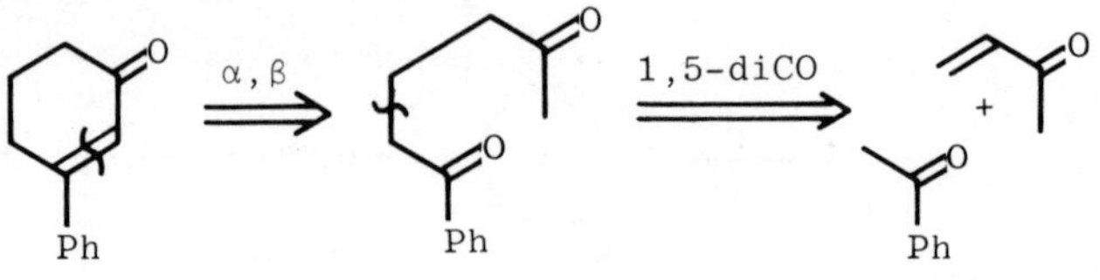

Analysis 2: allylic functionalisation

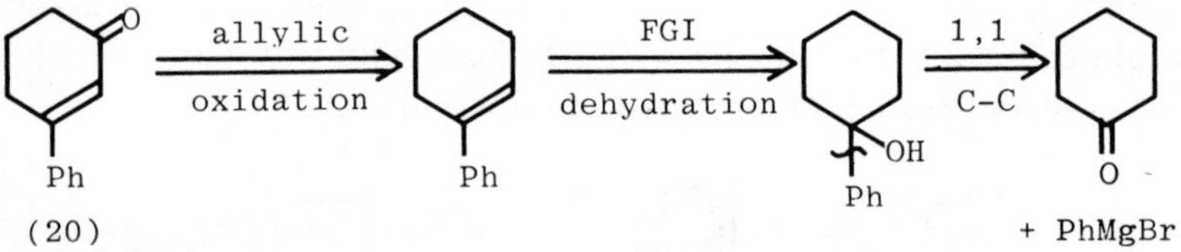

This synthesis has been carried out[270] by the new strategy. The first few stages are simple Grignard chemistry from Chapters 10 and 15, and CrO_3 was used instead of SeO_2.

Synthesis[270]

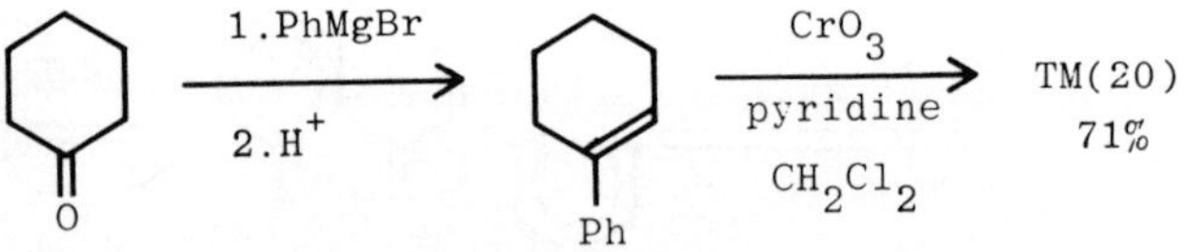

C–C Bond Formation

Radical reactions are used industrially[271] in many C–C bond-forming reactions, particularly the polymerisation of olefins. Such methods are beyond the scope of this book. Some simpler molecules are made by radical dimerisation. Diene (21), used in the manufacture of pyrethroid insecticides, is a dimer of the stable allylic radical (22) and is made industrially[272] at ICI from allylic halide (23).

Analysis

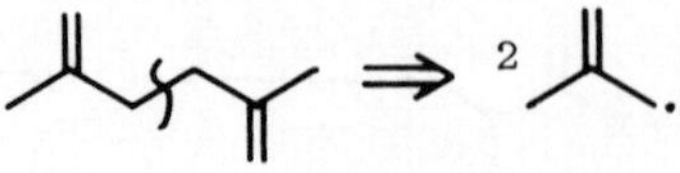

Synthesis[272]

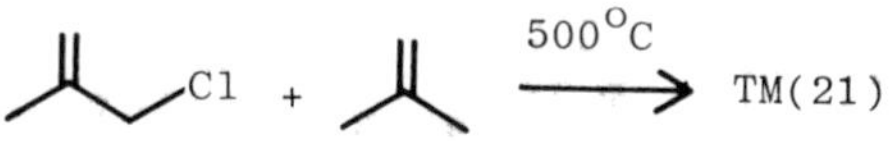

(23)

This section is concerned with radical dimerisations of this kind (22-23), mostly to make 1,2-difunctionalised compounds. Thus a 1,2-diol (24) could be made from two radicals (25) and this is indeed how it is made. The radicals (25) are generated by metal catalysed reduction of acetone. The product (24) is known as pinacol and the reaction is sometimes called pinacol reduction. The reaction works for most ketones, including enolisable ketones.

Pinacol: *Analysis*

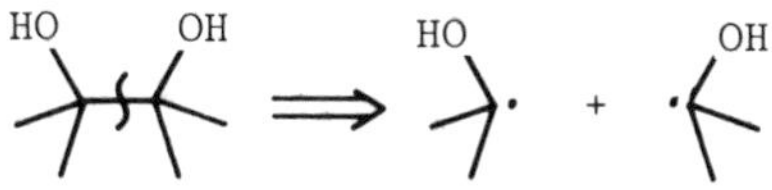

(24)   (25)   (25)

Synthesis

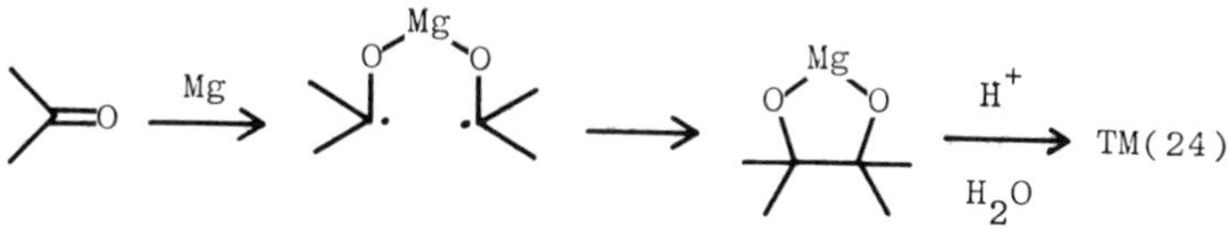

The synthetic oestrogen[273] dienoestrol (26) is a dehydration product of symmetrical diol (27) which we can disconnect in this way. Dehydration was most easily accomplished by acetyl chloride (perhaps *via* the acetate).

Dienoestrol: *Analysis*

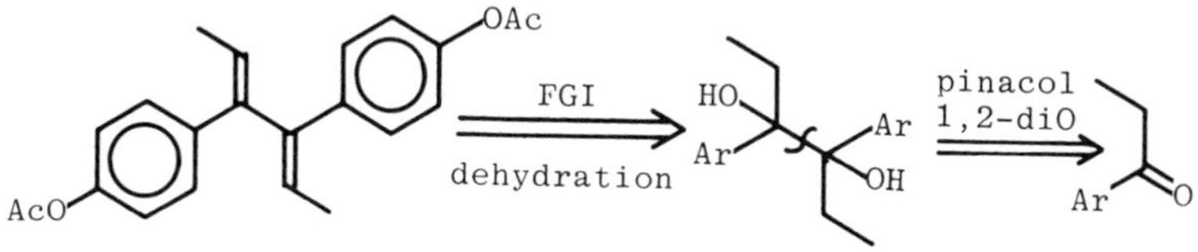

(26)   (27)

Synthesis[273]

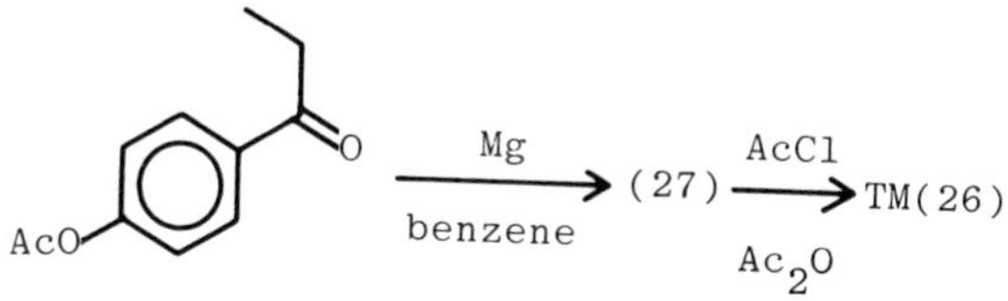

The acyloin reaction[274] is a similar dimerisation at the ester oxidation level. Metals again act as electron donors to give diradicals (28). The reaction was traditionally used only to make large rings, and these diradicals do cyclise to diketones (29) even when n is as large as 42. Further reduction to (30) is inevitable as the diketone (29) is more electrophilic than the original ester. On aqueous work-up the α-hydroxy ketone or acyloin (31) is released.

(28)

(29)   (30)   (31)

The acyloin reaction is still particularly useful for large rings but it is now available for other ring sizes and for open chain compounds by the addition of Me_3SiCl which traps dianion (30) and removes the EtO^- byproduct, preventing side reactions.[275] The silylated product (32) is easily hydrolysed to the acyloin (33).

(32)   (33)

α-Diketones (35) were needed to synthesise tetronic acids (34) and hence the natural products pulvinones.[276] FGI on (35) gives α-hydroxy ketones (36). These cannot be made by a benzoin condensation (Chapter 23) as the aldehyde required would enolise. The acyloin reaction in the presence of Me_3SiCl avoids this difficulty.

Analysis

(34)   (35)   (36)

Synthesis[276, 277]

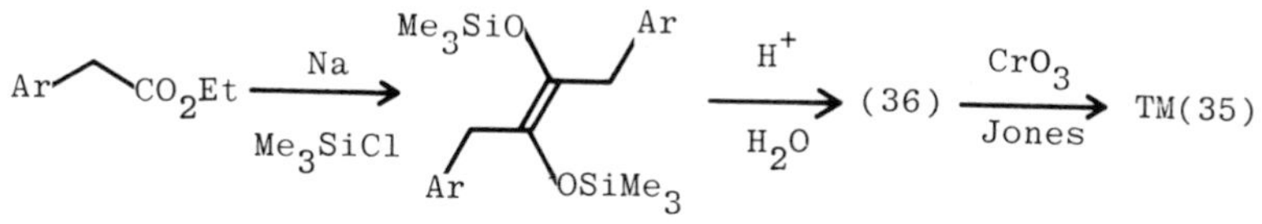

The α-diketones, e.g. **(37)**, made by oxidation of benzoin or acyloin products, can be converted into acetylenes[278] either by removing both oxygens with P(OR)$_3$ or via the *bis*-hydrazone **(38)** and oxidation with Hg(II). This method is useful for acetylenes such as **(39)** which cannot be made by displacement on alkyl halides by acetylide ion (Chapter 16). In this case, the starting material **(37)** is made from benzoin (Chapter 23) by nitric acid oxidation.

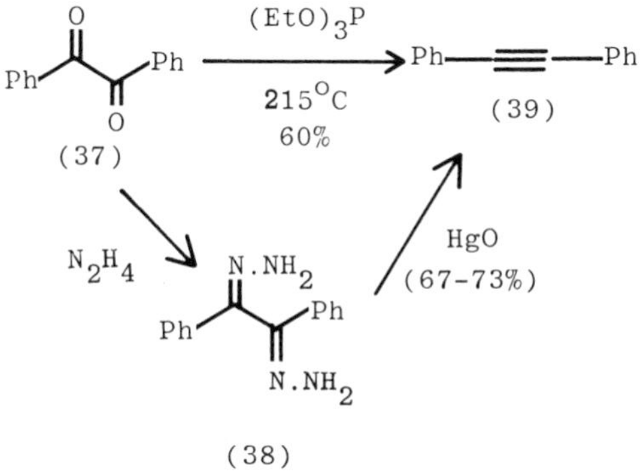

This method was used to make the smallest cyclic acetylenes — even an eight-membered ring **(40)** can have a triple bond in it. The α-diketone **(41)** is best made by an acyloin reaction from diester **(42)**.[279]

Analysis 1

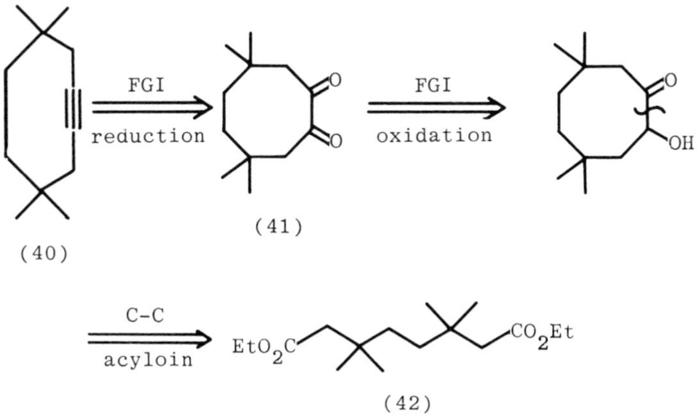

204

Symmetrical **(42)** is an ideal target for synthesis from two radicals (page 200). These could be generated by electrolytic decarboxylation of half ester **(43)**, available from diacid **(44)** *via* the cyclic anhydride (Chapter 5). The acid **(44)** is made by oxidative cleavage of dimedone (Chapter 21).

Analysis 2

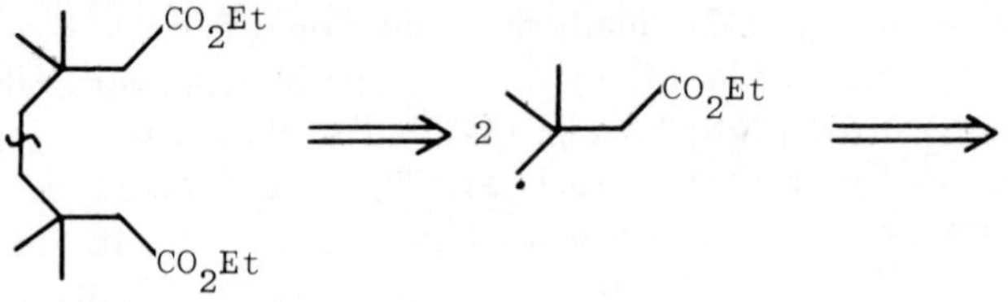

(42a)

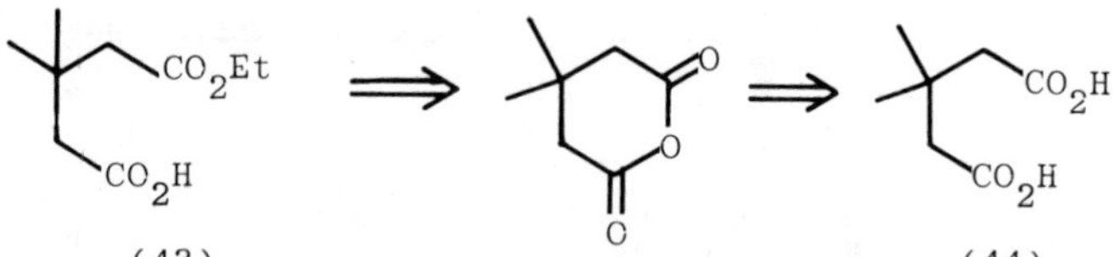

(43)   (44)

Synthesis[279]

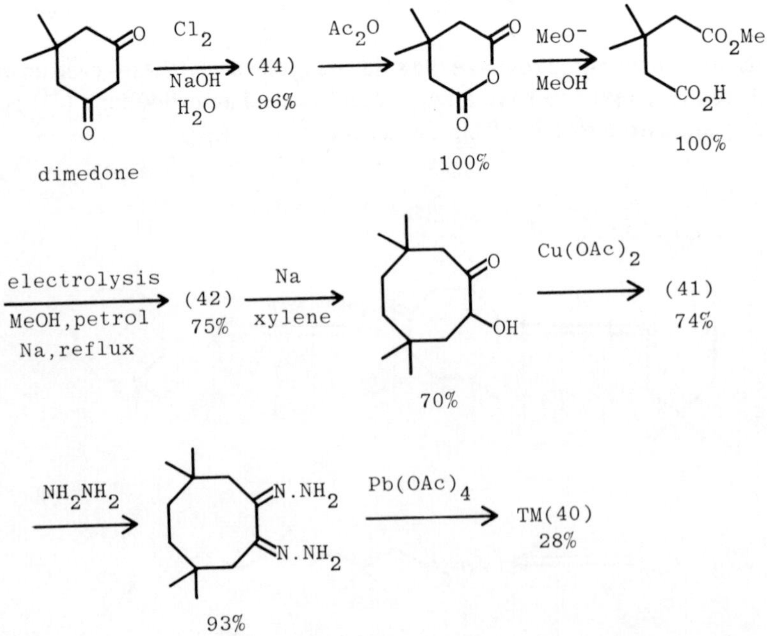

dimedone   100%   100%

electrolysis
MeOH,petrol
Na,reflux   (42)
75%   Na
xylene   Cu(OAc)₂   (41)
74%

70%

NH₂NH₂   Pb(OAc)₄   TM(40)
28%

93%

Functional Group Addition (FGA)

The synthesis of hydrocarbons

The FGA of the title is strategic FGA: that is we add a functional group during the analysis so that we can disconnect the molecule. During the synthesis* we shall remove the FG. The most obvious application is to the synthesis of hydrocarbons: they have no FGs so we must do FGA before any disconnection is possible.

Simple branched hydrocarbons such as **(45)** were needed to study petrol performance.[280] A helpful FGA is the addition of a hydroxyl group at the branch point, allowing a simple Grignard disconnection.

Analysis

There are many methods for replacing OH and other FGs by hydrogen (Table 24.1). One way with alcohols is to dehydrate (which olefin is formed is unimportant) and hydrogenate.

Synthesis[280]

FGA has an important part to play in the synthesis of functionalised compounds too. In the synthesis of tetralone **(47)** it solves a problem of regioselectivity. Friedel–Crafts disconnection of both Ar–C bonds is unacceptable as acylation will occur *para* to the MeO group and the wrong isomer **(48)** will be formed.[284]

Analysis 1

*This is the *reverse* of the first part of this chapter where the FG was added during the synthesis and removed during the analysis.

Table 24.1 Removal of functional groups

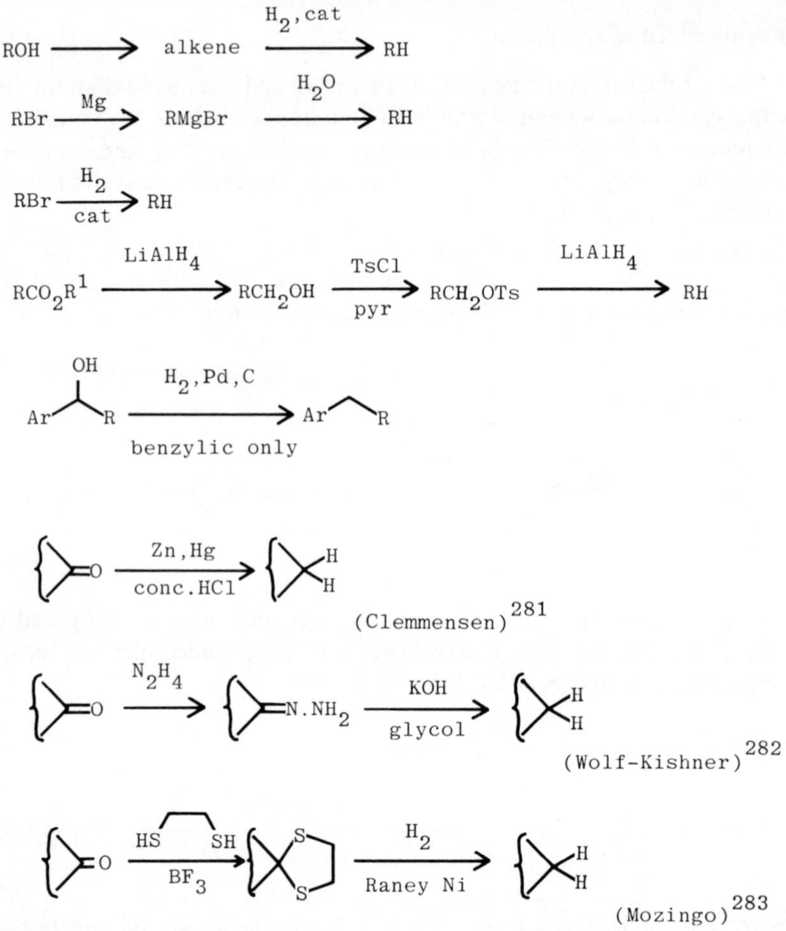

ROH $\longrightarrow$ alkene $\xrightarrow{\text{H}_2,\text{cat}}$ RH

RBr $\xrightarrow{\text{Mg}}$ RMgBr $\xrightarrow{\text{H}_2\text{O}}$ RH

RBr $\xrightarrow[\text{cat}]{\text{H}_2}$ RH

$RCO_2R^1 \xrightarrow{\text{LiAlH}_4} RCH_2OH \xrightarrow[\text{pyr}]{\text{TsCl}} RCH_2OTs \xrightarrow{\text{LiAlH}_4} RH$

benzylic only

(Clemmensen)[281]

(Wolf-Kishner)[282]

(Mozingo)[283]

It is better to follow methods from Chapter 3 and disconnect the acyl group first. The intermediate **(49)** can be made by a Friedel–Crafts reaction if we introduce a carbonyl group (FGA) to make the starting material the symmetrical anhydride **(50)**.

Analysis 2

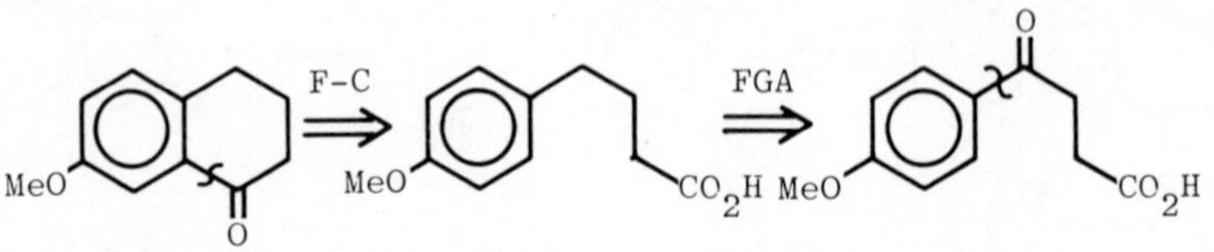

(47)   (49)

Any of the methods (Table 24.1) for removing a carbonyl group might work as the molecule is tough enough to survive all three. The Clemmensen method has been used[285] and must be carried out while the two carbonyl groups are still different, before the cyclisation.

Synthesis[285]

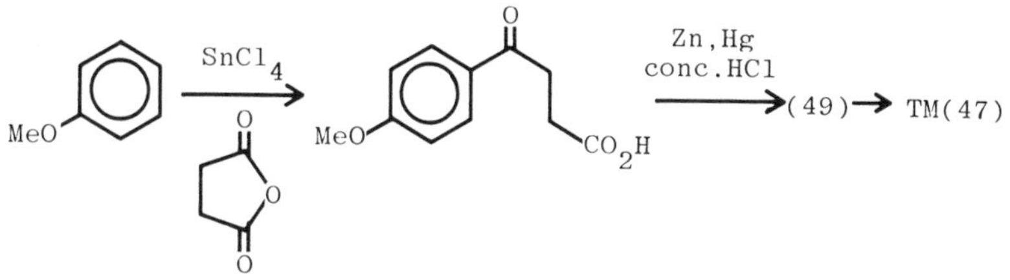

(50)

FGA of a C=C double bond is important in bringing two distant FGs into a helpful relationship. Raspberry ketone **(51)** is widely used to imitate the taste and smell of real raspberries. There is no obvious disconnection until we add a double bond **(52)** when a carbonyl condensation is revealed.

Raspberry ketone: *Analysis*

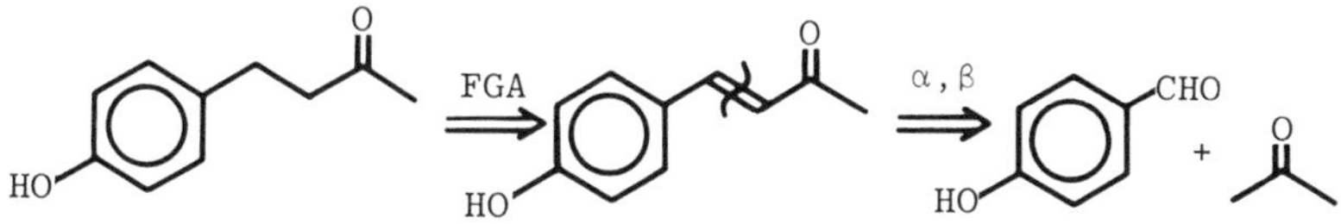

(51)   (52)

Hydrogenation will remove even a conjugated double bond without affecting the carbonyl group (Chapter 14) or an aromatic ring.

Synthesis[286]

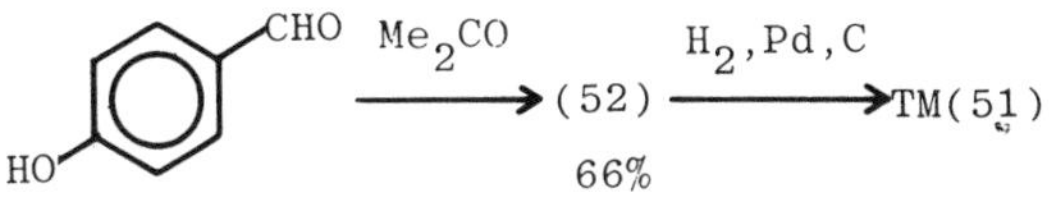

66%

Extended versions of this FGA can bring even very distant groups into helpful relationship. The first disconnection of cyclic ketone **(53)** is easy, but how are we to continue? One solution is a double FGA to reveal two successive carbonyl condensations.

208

Analysis

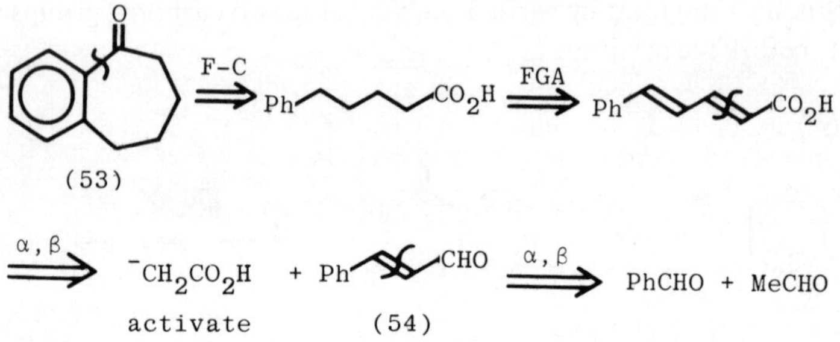

(53)

Cinnamaldehyde (54) is readily available so we need do only the remaining stages.[287]

Synthesis[287]

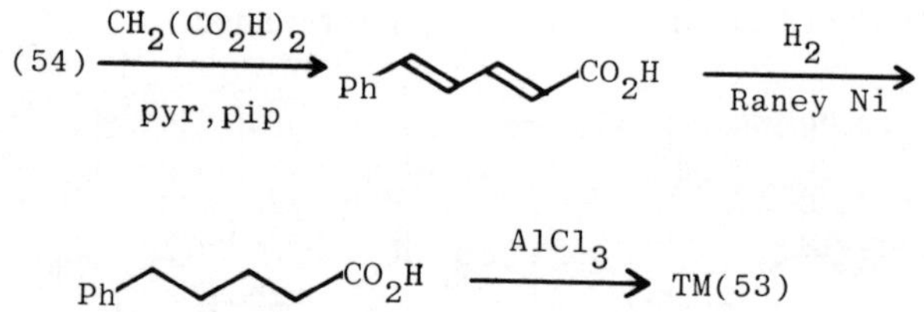

CHAPTER 25

Two-Group Disconnections V: 1,4-Difunctionalised Compounds

The problem of an unnatural (illogical) synthon arises here too (cf. Chapter 23). A 1,4-diketone (1) can be disconnected at its central bond into the natural enolate (2) but that requires also an unnatural synthon, the α-carbonyl cation (3). We shall need reagents for this synthon as well as for related synthons at different oxidation levels. We met some of these reagents—α-halo carbonyl compounds and epoxides—in Chapter 6.

Analysis

Methods using Unnatural Electrophilic Synthons

α-Halo carbonyl compounds (4) are useful reagents for synthon (3). In this reaction control is important as the halo substituent in (4) also enhances the acidity of the α-protons (H in 4) (see Table 18.2). The usual method is to use a specific enol equivalent (Chapter 20) for synthon (2).

Synthesis

Disconnection of the central bond (a) in (5) is good strategy as it separates the ring from the chain. The most popular specific enol equivalents are enamines (6) and compounds (7) activated with a CO_2Et group: both have been used in the synthesis of (5). The synthesis of (7) is discussed in Chapter 19.

Analysis

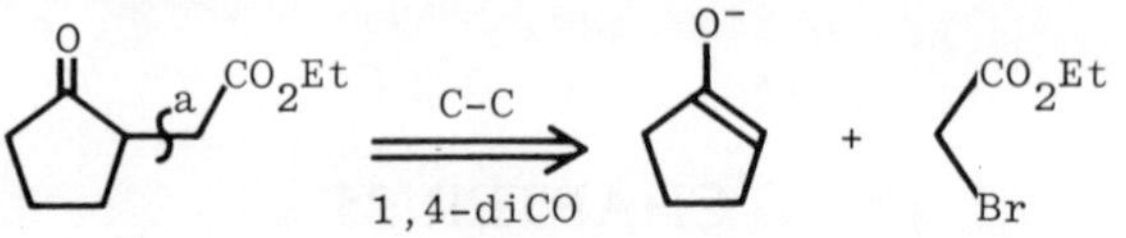

(5)

Synthesis using an enamine[288]

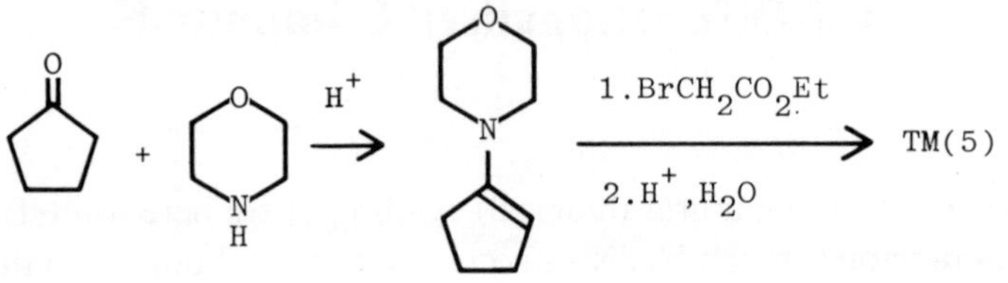

(6)

Synthesis with CO_2Et activating group[289]

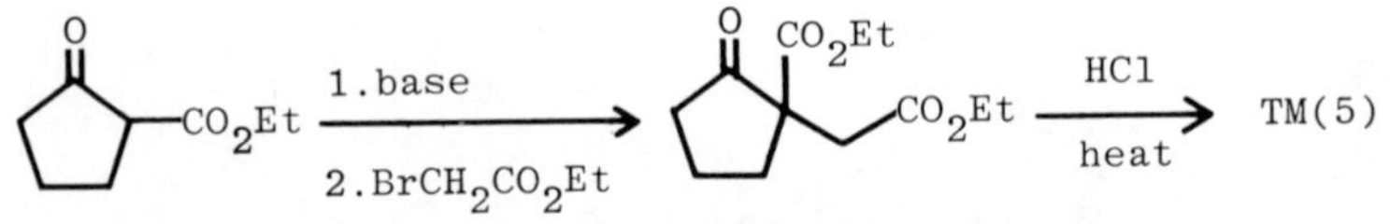

(7)

Keto ester (9) was needed for a synthesis of the antibiotic methylenomycin (8). The first disconnection should be at the α,β double bond and this reveals an obvious 1,4-disconnection (10) with the activating group CO_2Et already present.

Analysis

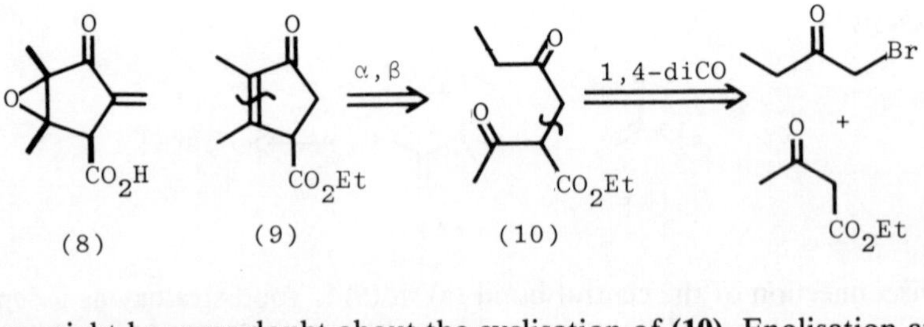

(8)　　　(9)　　　(10)

There might be some doubt about the cyclisation of (10). Enolisation at C3 is no problem (Chapter 20) as it could give only a three-membered ring. The alternative cyclisation to a five-membered ring (11) does not occur:[290] thermodynamic control favours the more substituted double bond in the TM(9).

Synthesis[290]

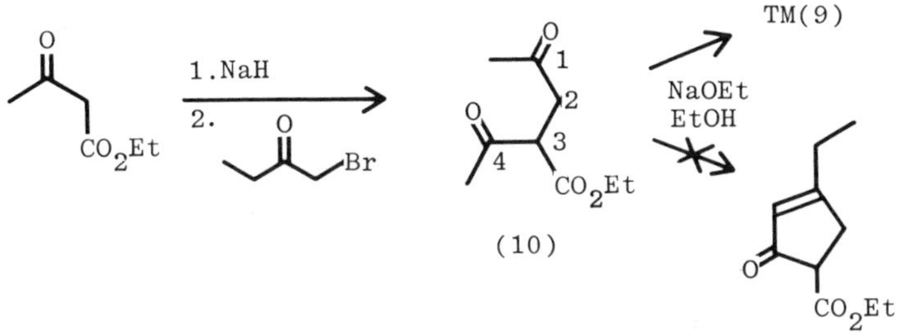
O
CO2Et
1.NaH
2.
O
Br
O
1
2
O
3
4
CO2Et
TM(9)
NaOEt
EtOH
O
CO2Et

(10) (11)

At the alcohol oxidation level **(12)**, epoxides **(14)** are the obvious reagents for the unnatural synthon **(13)**. The *trans* hydroxy acid **(15)** was needed for conformational studies. Disconnection at the ring–chain junction is sensible as regio- and stereospecificity are assured with symmetrical epoxide **(16)**.

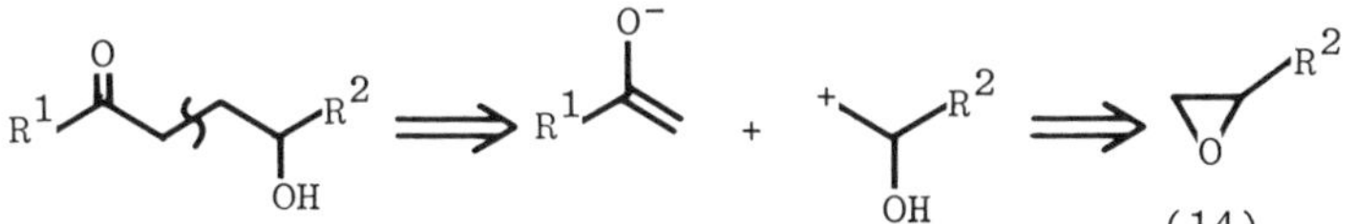
O
R1
R2
OH
O−
R1
+
+
R2
OH
R2
O

(12) (13) (14)

Analysis

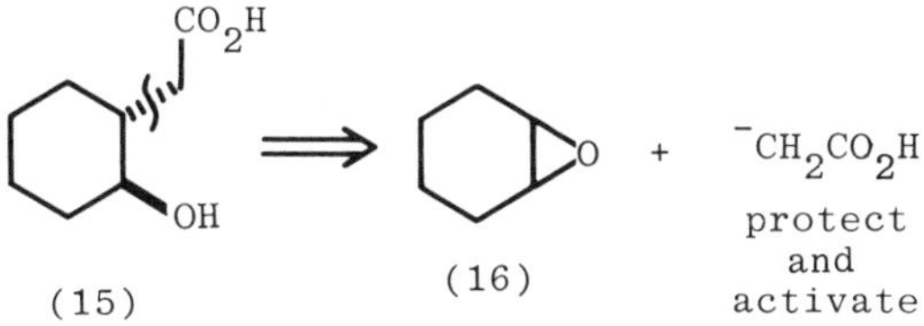
CO2H
OH
O
+
−CH2CO2H
protect
and
activate

(15) (16)

Synthesis[291]

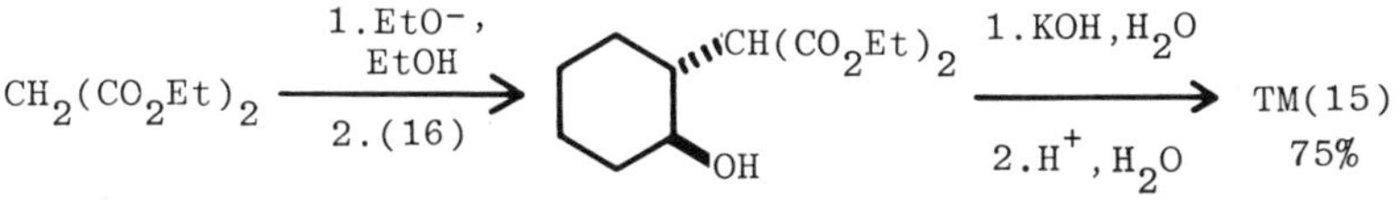
CH2(CO2Et)2
1.EtO−,
EtOH
2.(16)
CH(CO2Et)2
OH
1.KOH,H2O
2.H+,H2O
TM(15)
75%

The useful synthetic intermediate **(17)** is another compound of this sort. Disconnection gives acetoacetate and ethylene oxide itself.

Analysis

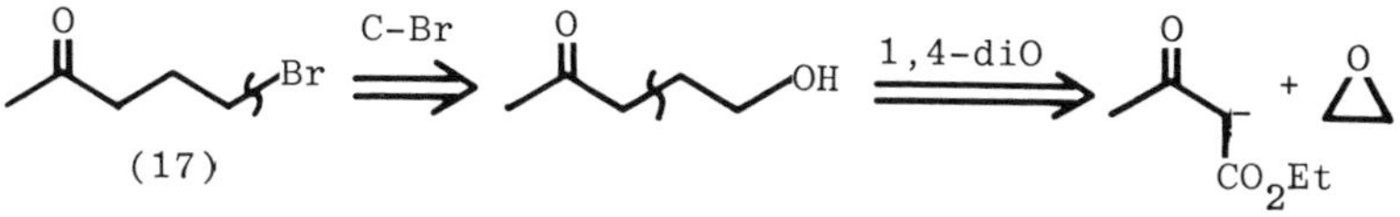
O
Br
C–Br
O
OH
1,4-diO
O
CO2Et
+
O

(17)

212

The intermediate (18) cyclises to the lactone (19) but this is by no means a nuisance as (19) can be converted into TM(17) in one step. Substitution by Br⁻, hydrolysis, and decarboxylation all occur under the same conditions.

Synthesis[292]

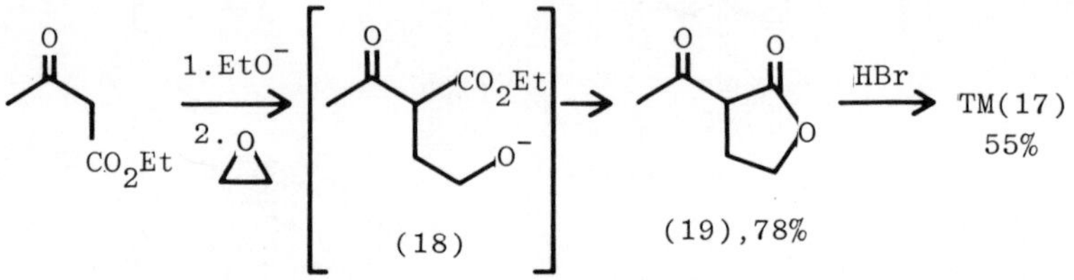

(18)   (19),78%

Unnatural Nucleophilic Synthons

The alternative disconnection (20) requires the Michael addition of an acyl anion equivalent. Cyanide ion, a reagent for $^-CO_2H$ and nitro alkane anions (Chapter 22) are both good at Michael additions, so that addition of cyanide ion to (21) is a route to γ-keto acids (22).

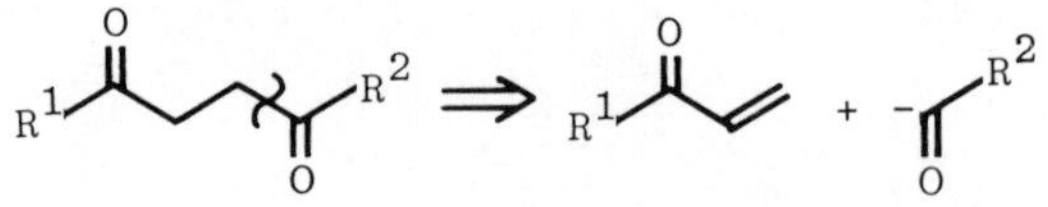

(20)   (21)

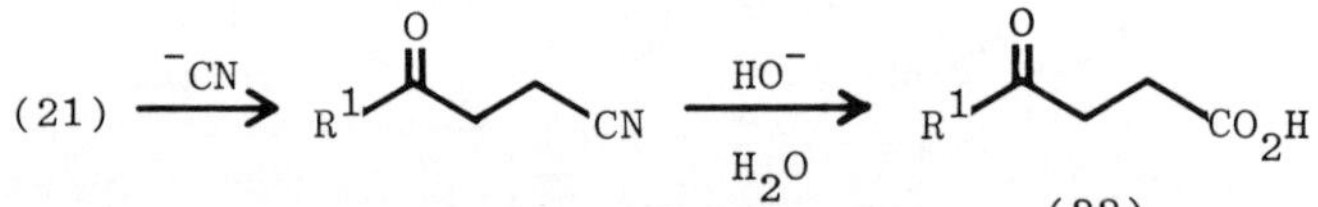

(22)

The anticonvulsant Phensuximide (23), being an imide, is made from diacid (24). We can disconnect either CO_2H group but prefer the branch point disconnection (a).

Phensuximide: *Analysis*

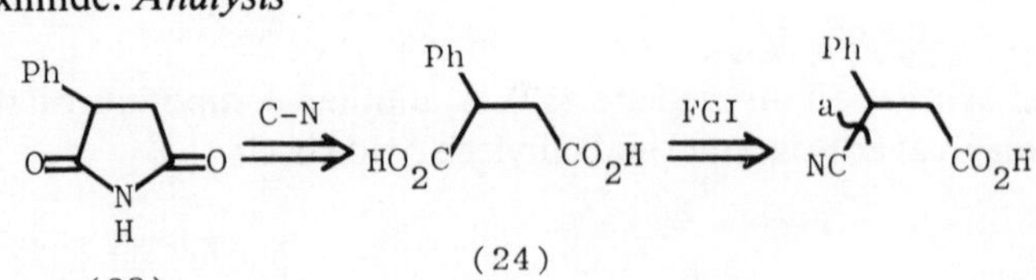

(23)   (24)

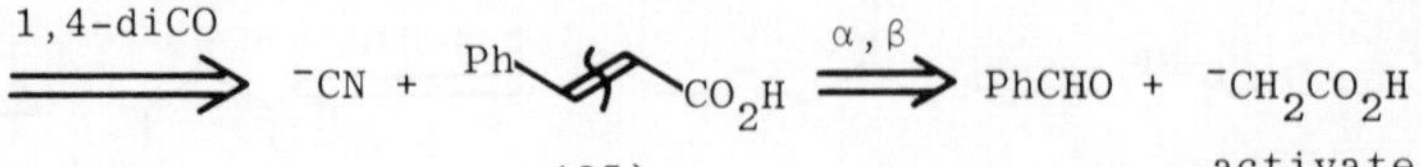

(25)   activate

In practice, cyanide added only slowly to cinnamic acid (25) so an activating group (CO₂Et) was used to help both the first condensation and the Michael addition. Any malonate derivative will do—the cyanide (26) has been used successfully.

Synthesis[293]

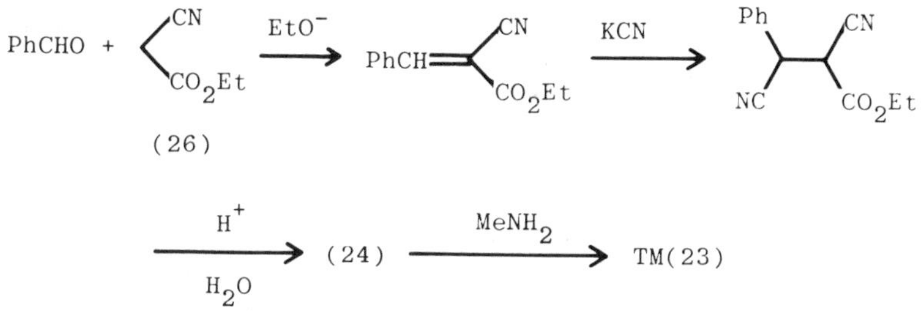

Nitro alkane anions (Chapter 22) will add to enones in Michael reactions and therefore behave as acyl anion equivalents in the synthesis of 1,4-diketones (27).[244]

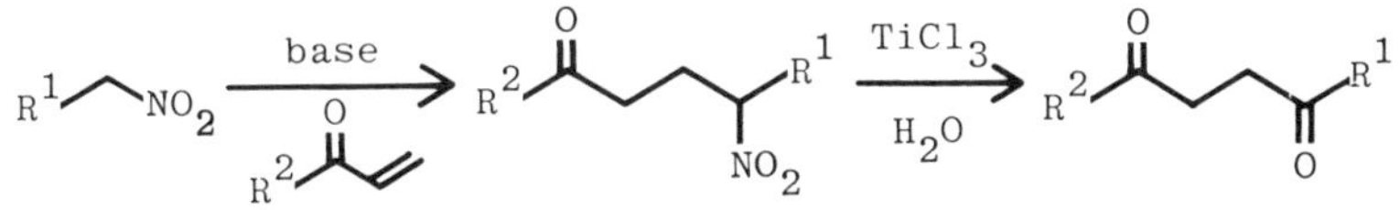

The cyclopentenone (28) has an obvious α,β-disconnection, giving 1,4-diketone (29). Either 1,4-disconnection will do, so we choose available ketone (30) as our guide.

Analysis

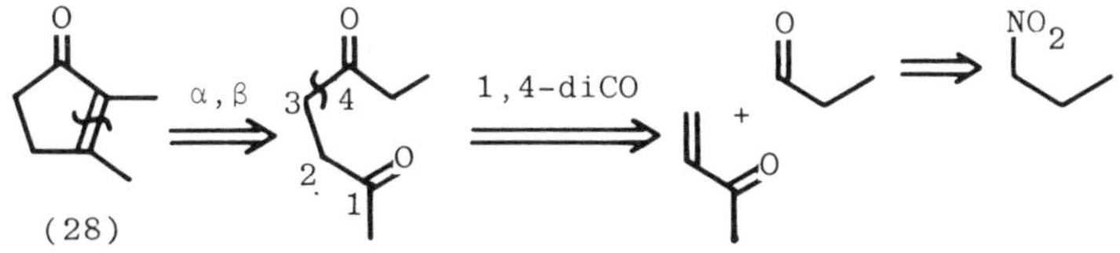

Nitro alkanes are so acidic that only a weak base is needed for the Michael reaction. The cyclisation gives the more substituted double bond.

Synthesis[244]

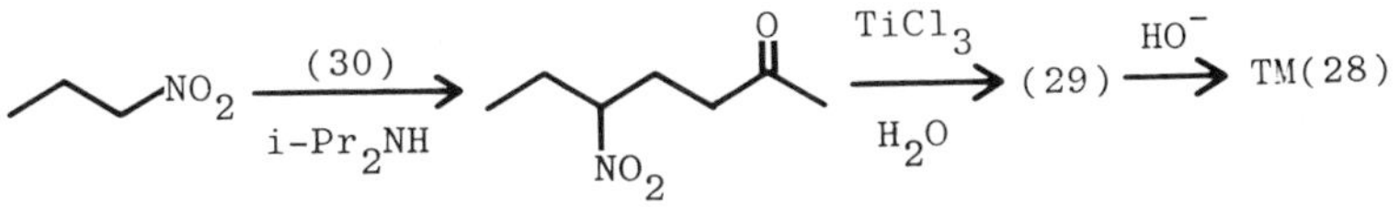

Strategy of Available Starting Materials

One way of avoiding the problem of unnatural polarity is to start with a cheap and readily available 1,4-difunctionalised starting material. A selection is given in Table 25.1. Examples of Friedel–Crafts reactions with some of these compounds were discussed in Chapter 24, and Diels–Alder reactions with maleic anhydride in Chapter 17.

Amino acid (31) is an intermediate in the synthesis of some β-blockers. It contains 1,4-dicarbonyl groups and an amino group 1,3 to the ketone. If the amino group comes from a nitro group (32) then FGA (Chapter 24) gives a good disconnection to (33).

Table 25.1 Some available 1,4-Difunctionalised Compounds

Butane-1,4-diol	cis-Butene diol (Chapter 16)	Butyne diol

Amines and halides		

Putrescine

X; $X=Br$, Cl

γ-Substituted ketones (Chapter 25)	Butyrolactone	Glutamic acid

X $X=Cl$, OH

Succinic acid and derivatives	Succinic anhydride	Maleic anhydride

Various furans	Especially furfuraldehyde (Chapter 40)	Its reduction products, e.g.:

Fumaric acid and derivatives	Levulinic acid	Acetonyl acetone

Dibenzoyl ethylene	3-Benzoyl propionic acid	

Analysis 1

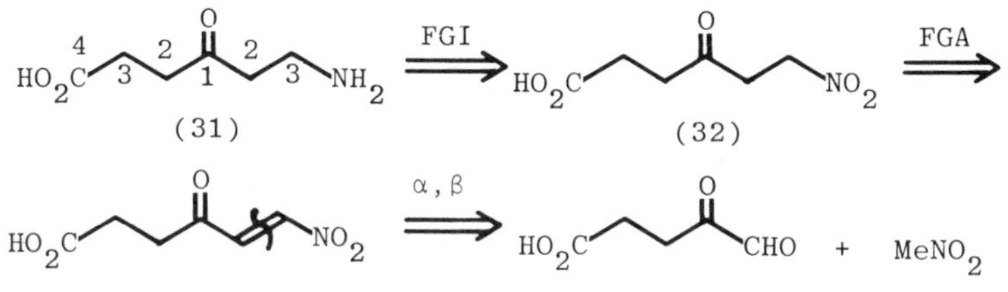

We now have both 1,2- and 1,4-dicarbonyl relationships, exactly the ones present in the very cheap starting material furfural **(34)** (a byproduct from the manufacture of Quaker oats).

Analysis 2

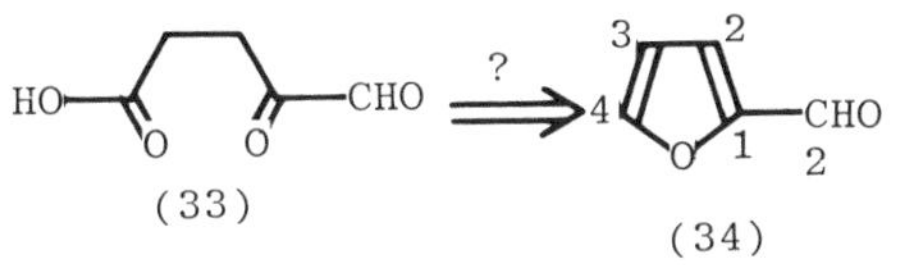

The condensation with nitromethane is best done first[294] to give **(35)** which is at the same oxidation level as **(32)**, surprisingly enough. Hydrolysis in acid solution gives a good yield of **(32)**.

Synthesis[294]

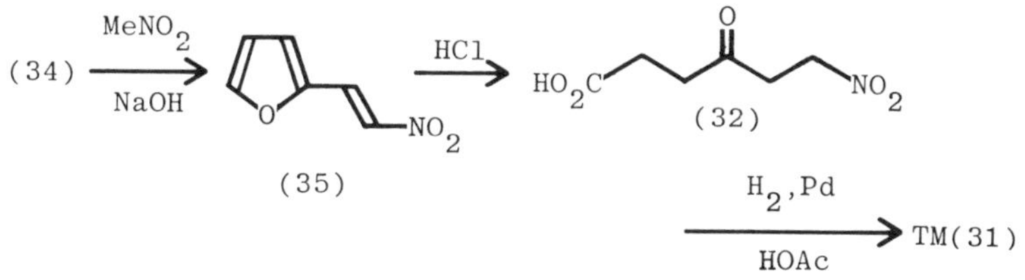

FGA Strategy

FGA of a triple bond* gives a simple strategy for assembling a 1,4-difunctionalised skeleton. Diadducts **(36)** of acetylene and carbonyl compounds can be hydrogenated (no catalyst poison) to remove the triple bond.

Analysis

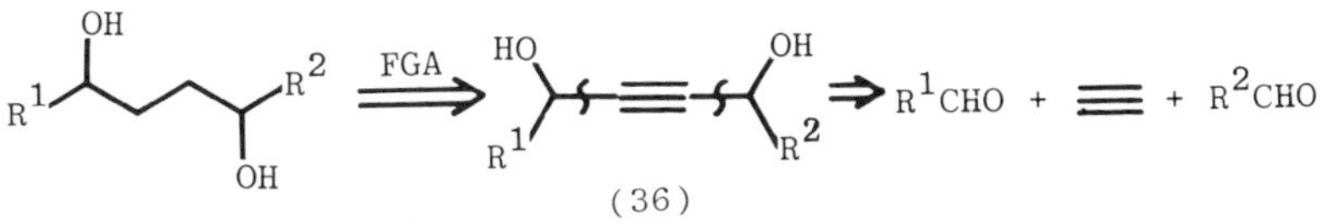

*See Chapter 24 for FGA and Chapter 16 for the use of acetylenes in synthesis.

Synthesis

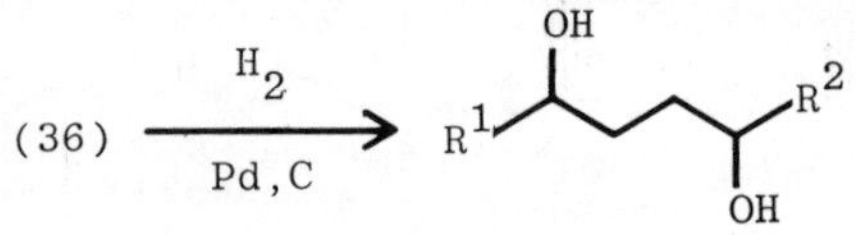

$$(36) \xrightarrow[\text{Pd,C}]{\text{H}_2} \quad R^1\text{-(OH)-(OH)-}R^2$$

γ-Lactones **(37)** can be made similarly by disconnecting a carbonyl compound from one end of **(38)** and CO_2 from the other.

Analysis

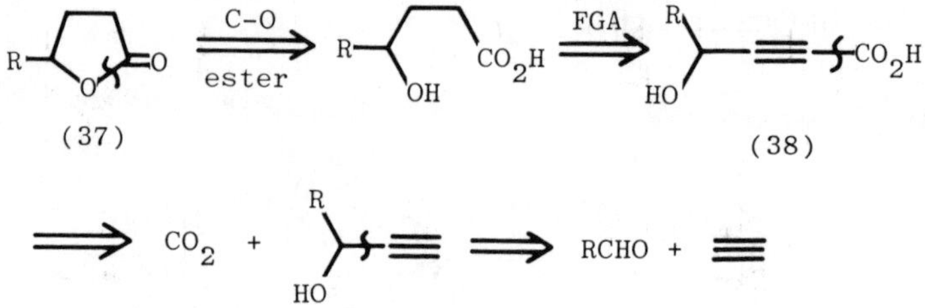

$$R\text{-lactone (37)} \xRightarrow[\text{ester}]{\text{C-O}} R\text{-(OH)-}CO_2H \xRightarrow{\text{FGA}} R\text{-(HO)-}\!\equiv\!\text{-}CO_2H \text{ (38)}$$

$$\Rightarrow \quad CO_2 \; + \; R\text{-(HO)-}\!\equiv \Rightarrow RCHO \; + \; \equiv$$

This three step synthesis[295] (the cyclisation occurs spontaneously on hydrogenation) makes an interesting contrast with the synthesis on page 96 where the available glutamic acid was used as starting material in a stereo-specific synthesis.

Synthesis[295]

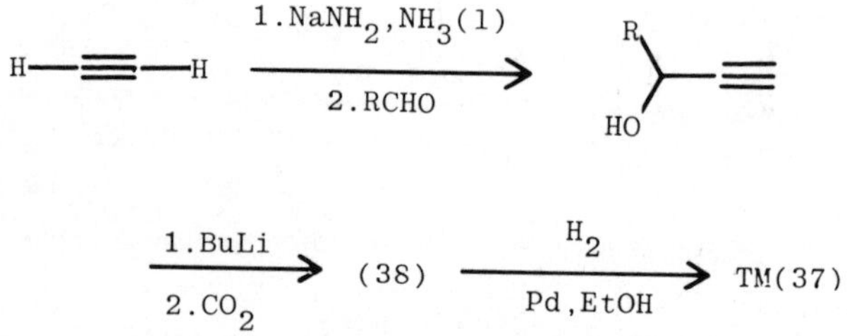

$$H\text{—}\!\equiv\!\text{—}H \xrightarrow[\text{2.RCHO}]{\text{1.NaNH}_2, \text{NH}_3(1)} R\text{-(HO)-}\!\equiv$$

$$\xrightarrow[\text{2.CO}_2]{\text{1.BuLi}} (38) \xrightarrow[\text{Pd,EtOH}]{\text{H}_2} TM(37)$$

CHAPTER 26

Strategy XII: Reconnections

Synthesis of 1,2- and 1,4-Difunctionalised Compounds by C=C Cleavage

While we must use polarity inversion to add the α-carbonyl cation synthon (1) to an enolate ion, no such tricks are needed to add synthon (3): allyl halides are easily made (Chapter 24) and are reactive in S_N2 reactions. The conversion of (4) to (2) by oxidative cleavage of the C=C double bond provides another route to 1,4-dicarbonyl compounds. The various methods of double bond cleavage are summarised in Table 26.1.

Table 26.1 Double bond cleavage methods[296]

Ozonolysis with reductive work-up

$$R^1 \diagdown \diagup R^2 \xrightarrow{1.O_3 \quad 2.Me_2S} R^1CHO + R^2CHO$$

Ozonolysis with oxidative work-up

$$R^1 \diagdown \diagup R^2 \xrightarrow{1.O_3 \quad 2.H_2O_2} R^1CO_2H + R^2CO_2H$$

Hydroxylation and cleavage of diol

$$R^1 \diagdown \diagup R^2 \xrightarrow[\text{or KMnO}_4]{OsO_4} R^1 \diagup^{OH}_{OH} R^2 \xrightarrow[\text{or Pb(OAc)}_4]{NaIO_4} R^1CHO + R^2CHO$$

Hydroxylation and cleavage combined

$$R^1 \diagdown \diagup R^2 \xrightarrow[\text{excess of NaIO}_4]{KMnO_4 \text{ or } OsO_4(cat)} R^1CHO + R^2CHO$$

217

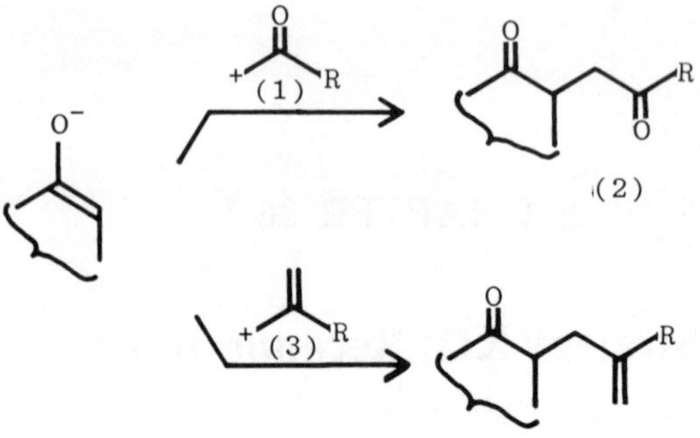

Disconnection of ester aldehyde (5) might suggest an α-halo aldehyde (6) as reagent. These are highly reactive compounds and very difficult to handle without protection. A simple allyl halide (7) is an attractive alternative.

Analysis

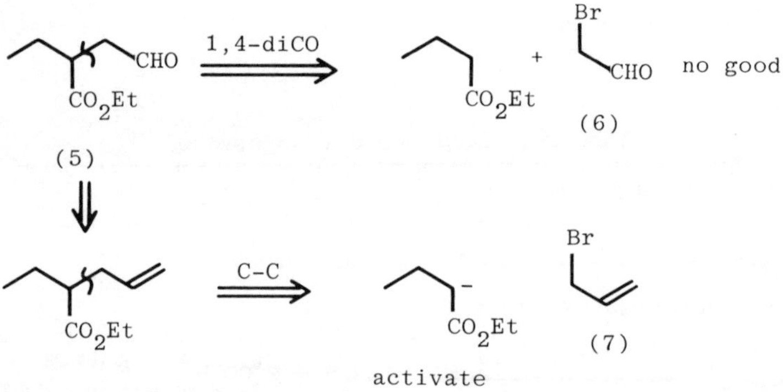

This synthesis has been carried out[297] with ozone as the reagent for double bond cleavage, and malonate for the activated enolate.

Synthesis[297]

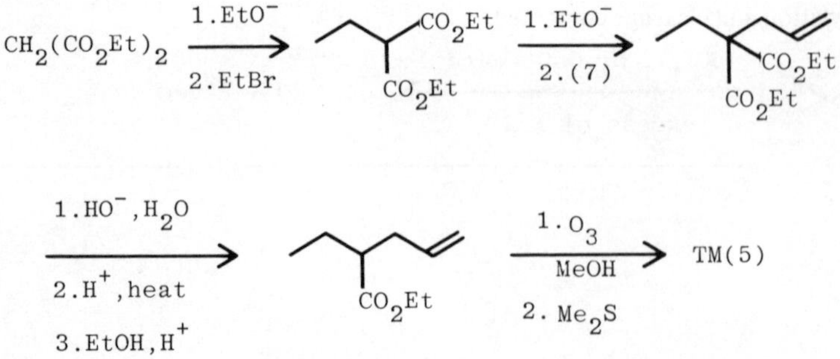

When analysing this approach to dicarbonyl compounds, we use not a disconnection but the reverse, we must join up a bond in the molecule which will be broken during the synthesis. This operation is called *reconnection*.

Consider the problem of making *cis* enone **(8)**, a structure found in insect pheromones, flavourings, and perfumes. A Wittig reaction would give the right stereochemistry (Chapter 15) but would require the selectively protected keto aldehyde **(9)**, as a Wittig reagent is probably too reactive to show the required chemoselectivity.

Analysis 1

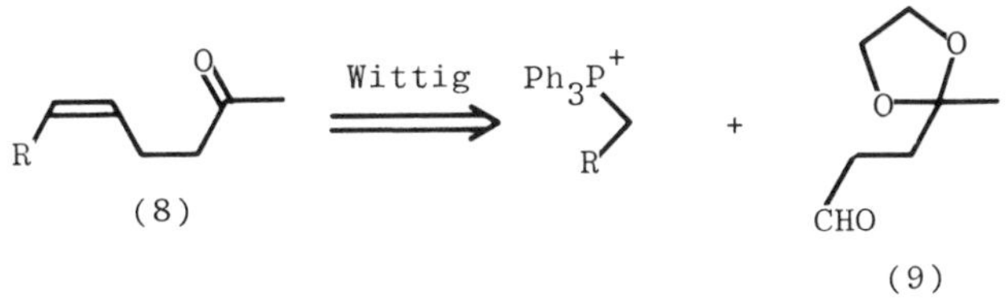

This is a 1,4-dicarbonyl problem and can be solved by reconnection to **(10)**. Thus the aldehyde can be introduced after the ketone is protected and problems of chemoselectivity are avoided.

Analysis 2

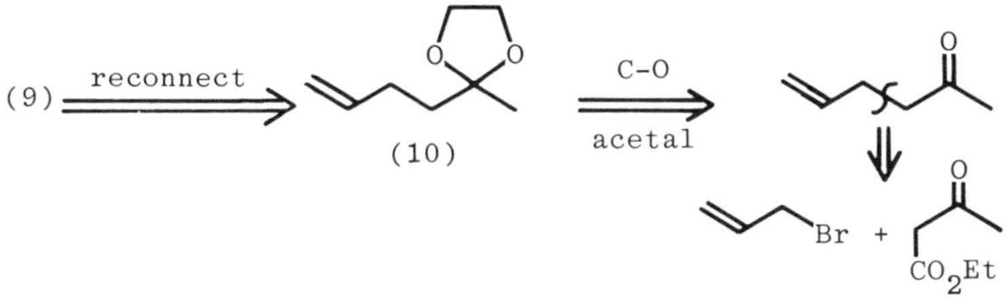

Synthesis[298]

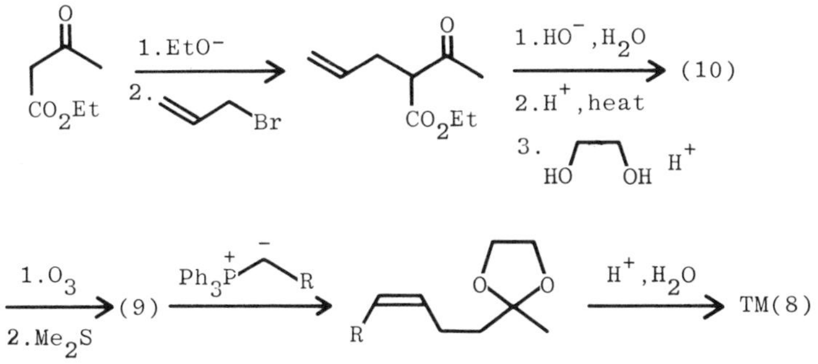

Reconnection can also be used in the synthesis of 1,2-dicarbonyl compounds since ozonolysis of enones **(11)** gives α-keto aldehydes **(12)** or acids **(13)** depending on the work-up. Ozonolysis of enones of type **(14)** gives α-diketones.

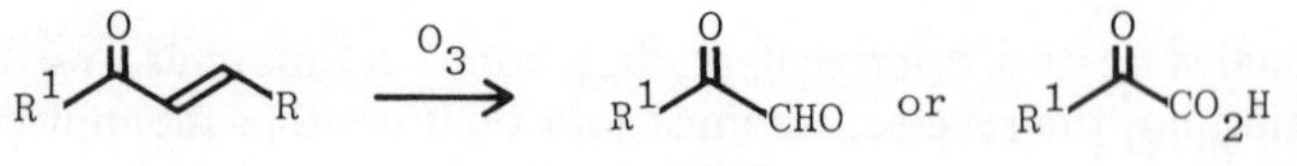

220

(11) → (12) or (13)

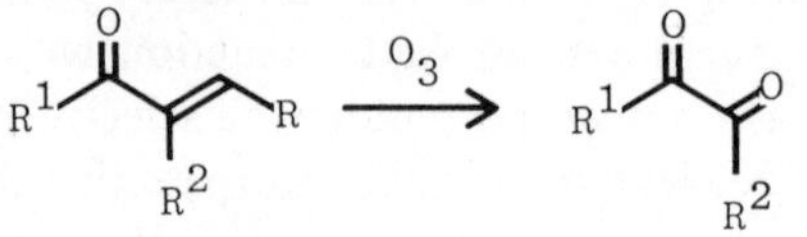

(14) →

The group R in **(11)** and **(14)** is unimportant as it is lost during ozonolysis and R=Ph is often used so that enones **(11)** and **(14)** may be made by unambiguous condensations with benzaldehyde.

Compound **(15)** was used in the synthesis of the extraordinary polycyclic staurone **(16)**. Reconnection of the aldehyde to **(17)** is obvious, and α,β-disconnection to benzaldehyde and **(18)** follows.

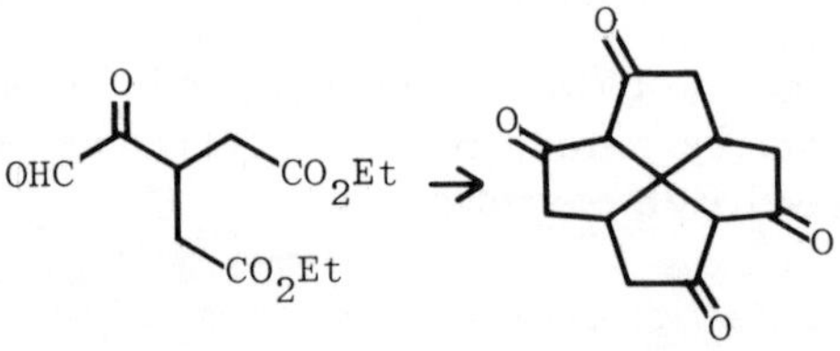

(15)   (16)

Analysis 1

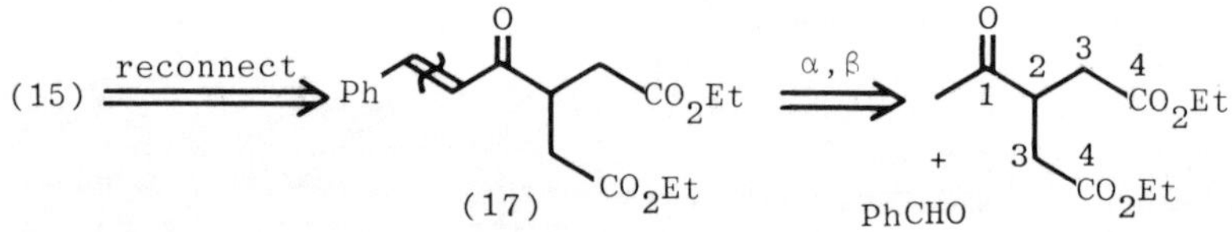

(15) $\xrightarrow{\text{reconnect}}$ (17) $\xrightarrow{\alpha,\beta}$ (18) + PhCHO

The two 1,4-relationships in symmetrical **(18)** can both be disconnected to bromacetate **(19)** if an activating group is first added.

Analysis 2

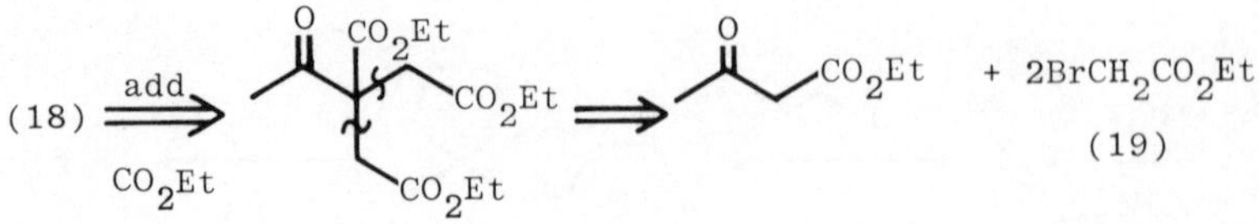

(18) $\xrightarrow{\text{add}}$ → + 2BrCH$_2$CO$_2$Et (19)

In practice,[299] benzyl acetoacetate **(20)** was used so that the free acid could be released and decarboxylated by hydrogenolysis. The benzaldehyde condensation is unambiguous as dehydration of the alternative product is impossible (Chapter 20).

Synthesis[299]

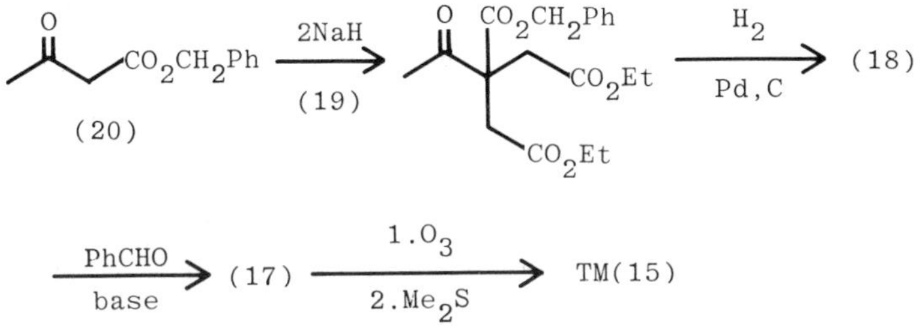

$$\text{(20)} \xrightarrow[\text{(19)}]{2\text{NaH}} \text{[structure]} \xrightarrow[\text{Pd,C}]{H_2} \text{(18)}$$

$$\xrightarrow[\text{base}]{\text{PhCHO}} \text{(17)} \xrightarrow[2.\text{Me}_2\text{S}]{1.\text{O}_3} \text{TM(15)}$$

1,4-Functionalisation without Reconnection

Allyl **(7)** and propargyl **(22)** halides can act as latent unnatural synthons **(24)** and **(25)** respectively by electrophilic addition to the double bond in **(21)** or hydration of the triple bond in **(23)**. If the nucleophile is an enolate ion, a 1,4-difunctionalised compound must be formed.

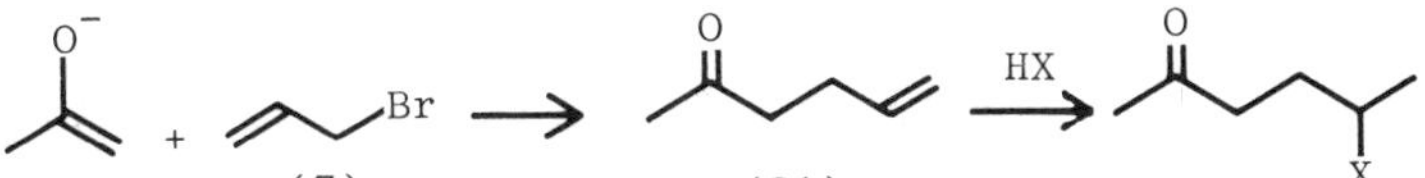

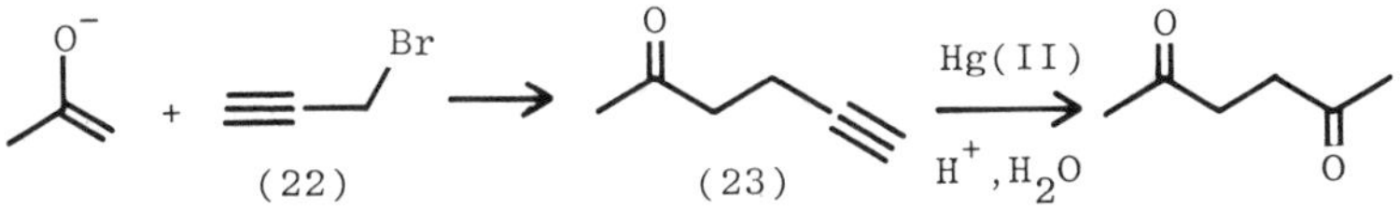

An interesting application of propargyl halides **(22)** is the formation of cyclopentenones from 1,4-diketones in a five-ring version of the Robinson annelation (Chapter 21). Disconnection of **(26)** at the α,β bond reveals a 1,4-relationship in **(27)** which needs synthon **(25)**.

Analysis

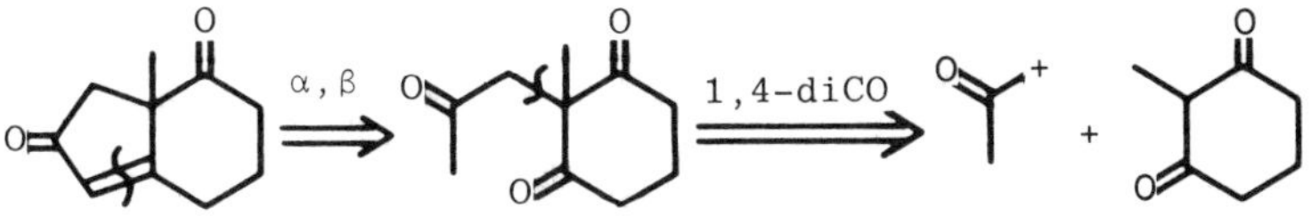

An allylic halide could be used for (25) with ozonolysis to generate the carbonyl group, or propargyl bromide (22) with hydration[300] of (29).

Synthesis 1

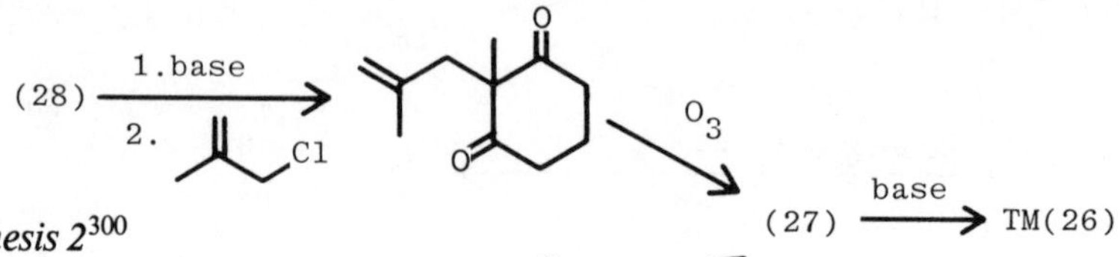

Synthesis 2[300]

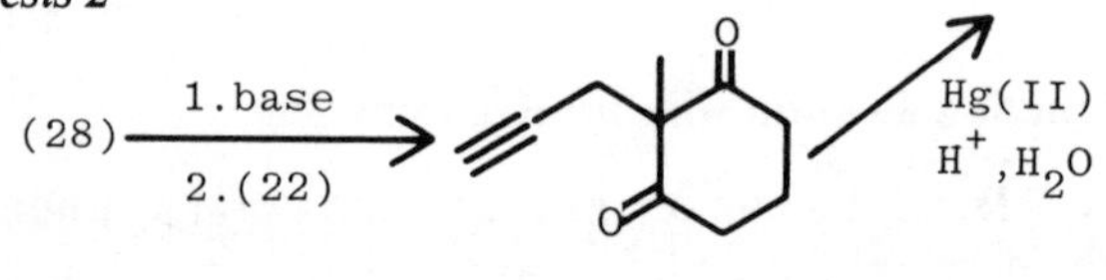

(29)

Addition of HBr to an allyl group gives the alkyl bromide—the most useful derivative for substitution. The lactone (30) can be made from bromo acid (31) and the bromide comes from the allyl group in (32). In order to add the allyl group, the CO_2H group is best replaced by CN to make a more stable carbanion.

Analysis

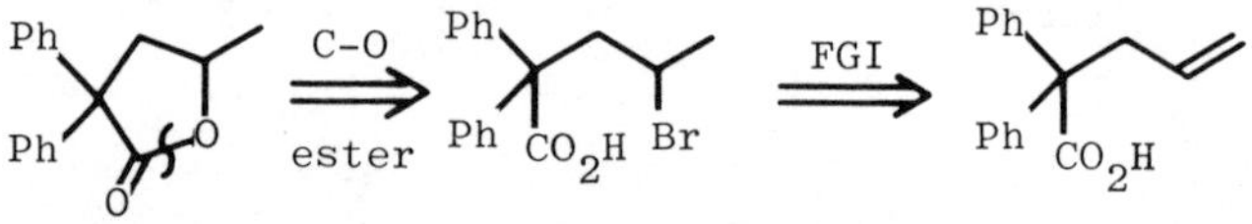

(30)   (31)   (32)

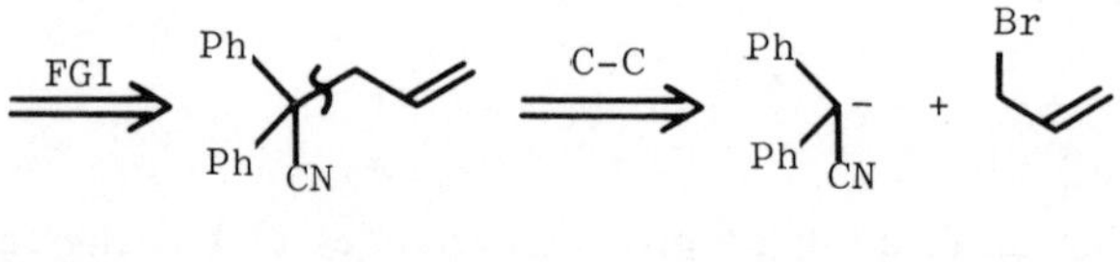

Hydrolysis of cyanide to CO_2H and of bromide to OH can both be accomplished in one step.[301]

Synthesis[301]

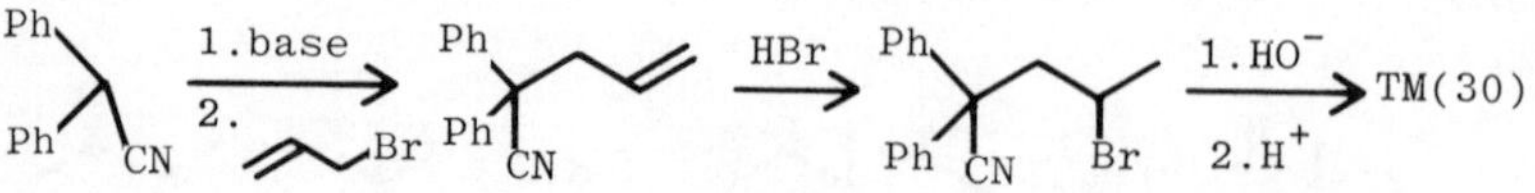

CHAPTER 27

Two-Group Disconnections VI:
1,6-Difunctionalised Compounds

Reconnection is the usual strategy for synthesising 1,6-difunctionalised compounds since the cyclohexenes required for the oxidative cleavage are easily made. Adipic acid (1) is available from cyclohexene itself and is a source of five-membered rings by condensation reactions (Chapter 19).

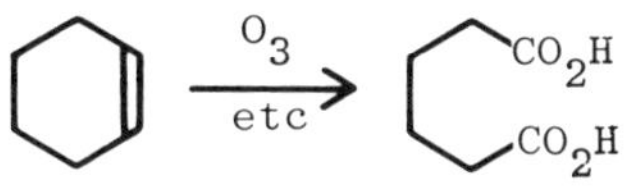

(1)

Cyclohexenes with substituents in the 1-position (2), made from cyclohexanone and Grignard reagents, cleave to give keto acids (3) or keto aldehydes according to conditions.

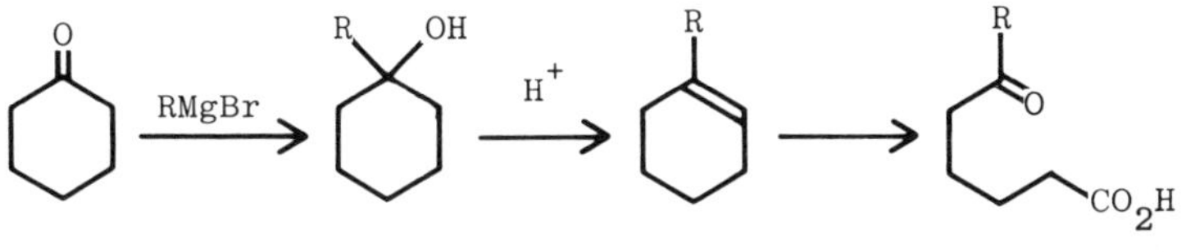

(2)   (3)

Ketone (5) was needed in a synthesis of the bicyclic ketone (4).[302] The α,β-disconnection reveals a 1,6-dicarbonyl compound (6) and reconnection gives cyclohexene (7) made by the Grignard route from ketone (8).

223

(4)   (5)   (6)

(7)   (8)

The synthesis of compound (8) is discussed in Chapter 36. Cleavage of (7) by ozone with reductive work-up (Table 26.1) gives the keto aldehyde (6) and cyclisation gives the most substituted double bond.

Synthesis[302, 303]

$$(8) \xrightarrow[\text{2.H}^+]{\text{1.MeLi}} (7) \xrightarrow[\text{2.Me}_2\text{S}]{\text{1.O}_3} (6) \xrightarrow[\text{MeOH}]{\text{KOH}} TM(5)$$

Perhaps the most important way to make cyclohexenes is by the Diels–Alder reaction and this opens up a wide field of 1,6-dicarbonyl products. Diels–Alder adducts (9) have at least one other carbonyl group so that the cleavage products have 1,4- and 1,5-dicarbonyl relationships as well as 1,6.

(9)   (10)

Heathcock[304] required diester (11) for his synthesis of the antibiotic pentalenolactone. Reconnection gives a symmetrical cyclohexene (12) with the right substitution pattern for a Diels–Alder adduct. Minor adjustments of the oxidation level suggest anhydride (13) as a suitable starting material and the stereochemistry will be correct if we use maleic anhydride in the Diels–Alder reaction.

Analysis

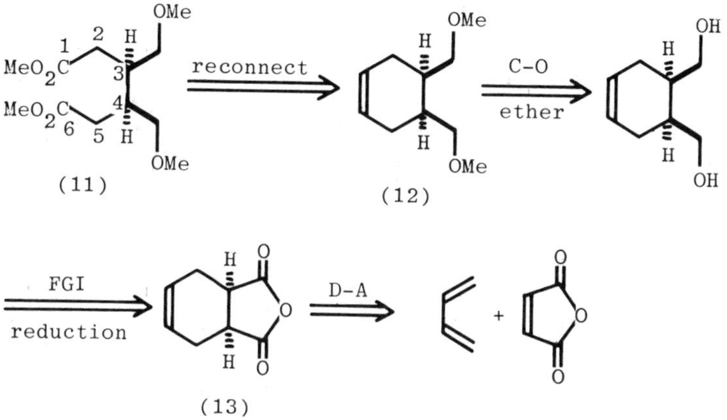

(11)   reconnect ⟹   (12)   C–O ether ⟹

(13)

⟸ FGI reduction ⟸ D–A ⟸

Reduction of the anhydride **(13)** to the diol is possible with LiAlH₄. The cleavage step was carried out with ozone and oxidative work-up and the diester **(11)** formed in the same step with diazomethane (CH₂N₂).

Synthesis[304]

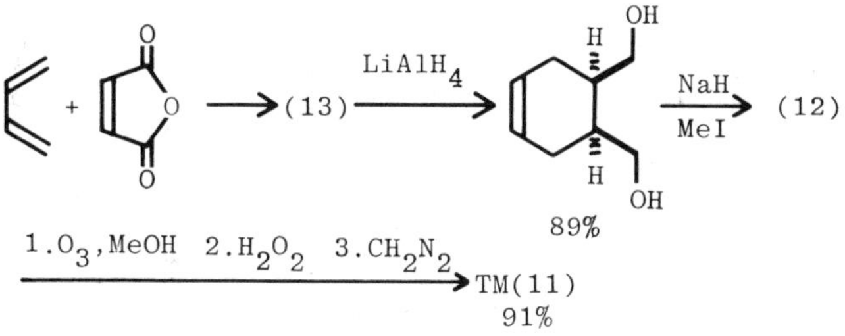

→ (13) —LiAlH₄→ —NaH/MeI→ (12)

89%

1.O₃,MeOH 2.H₂O₂ 3.CH₂N₂ ⟶ TM(11)

91%

The bicyclic double lactone **(14)** was used by Eschenmoser[305] as the precursor for all four heterocyclic rings in his vitamin B₁₂ synthesis. Disconnecting both lactones reveals a ketone **(15)**.

Analysis 1

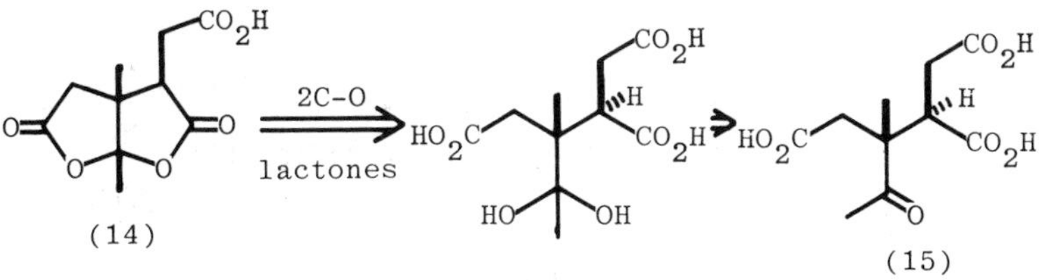

(14)   2C–O lactones ⟹   ⟹ (15)

Keto triacid **(15)** contains 1,4-, 1,5- and a single 1,6-dicarbonyl relationship **(15a)**. Reconnecting the 1,6 carefully to avoid losing the stereochemistry we find an obvious Diels–Alder adduct **(16)**.

Analysis 2

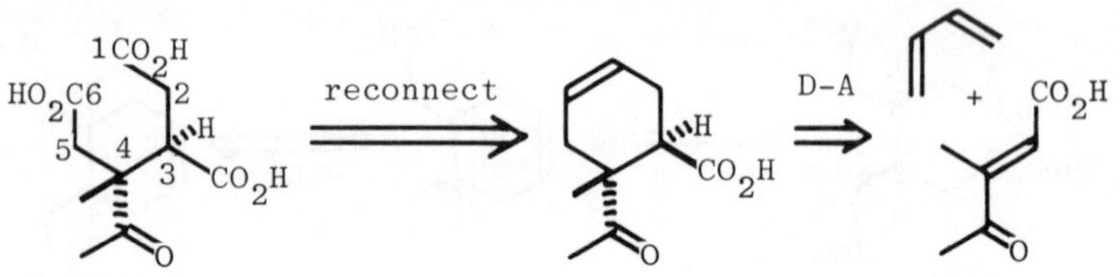

(15a)   (16)   (17)

The dienophile disconnects in two α,β ways. Controlling (a) to react once only might be difficult, but (b) is unambiguous in all but the regiochemistry of ketone enolisation.

Analysis 3

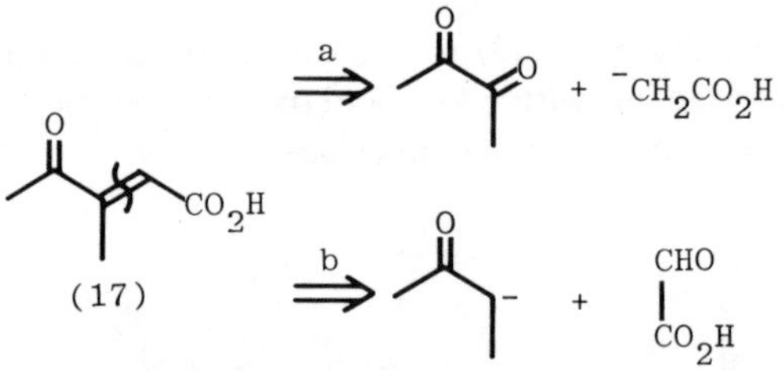

(17)

Acid conditions ensure enolisation on the more substituted side (Chapter 20) and thermodynamic control gives the *E*-isomer of **(17)** — the one we want. Eschenmoser found that oxidative cleavage of **(16)** under acidic conditions led to spontaneous lactone formation without the isolation of **(15)**.

Synthesis[305]

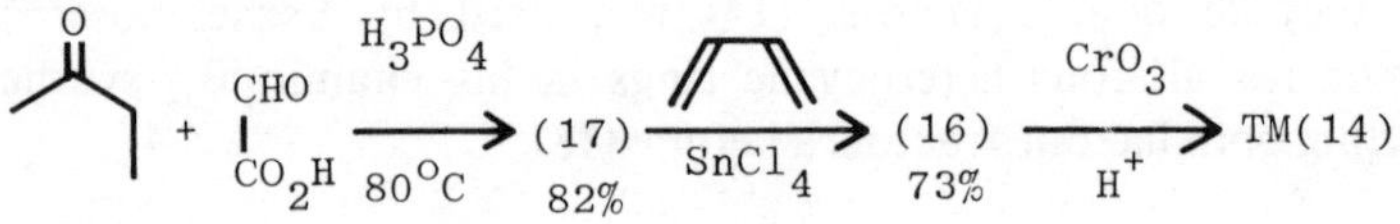

Oxidative Cleavage by the Baeyer–Villiger Reaction

Cyclohexanones may also be cleaved oxidatively by peracids **(18)**, a rearrangement **(19)** which has the effect of inserting an oxygen atom into the ring to give a lactone.[306]

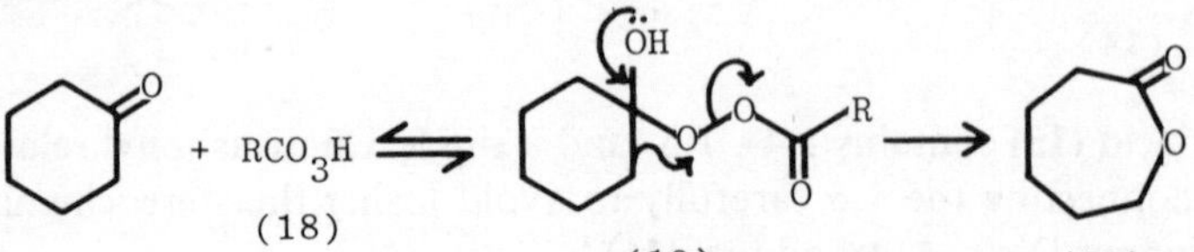

(18)

(19)   (20)

The aldehyde ester (21) was needed[307] for a biotin synthesis (cf. Chapter 24) and looks a formidable problem of chemoselectivity. Reconnection to lactone (20) after FGI preserves the difference between the carbonyl groups and the synthesis becomes straightforward.

Analysis

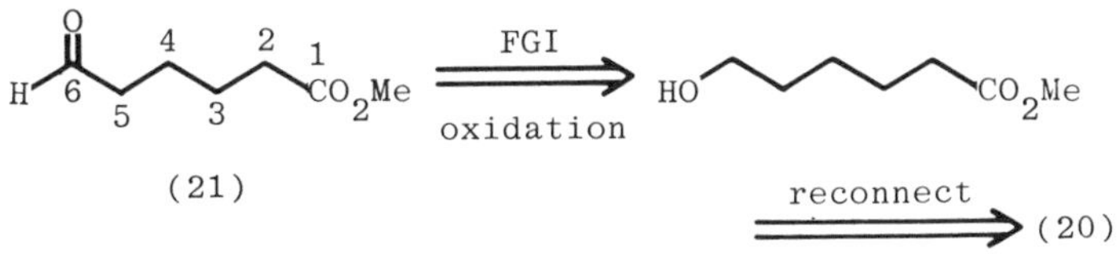

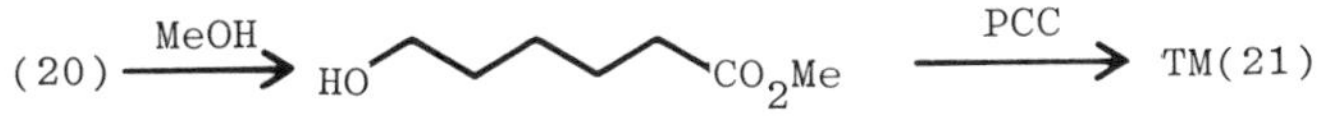

The Baeyer–Villiger reaction is regioselective—the more substituted group migrates—and stereospecific—it does so with retention. Hydroxy ketone (22) was needed for insect pheromone synthesis[308] and is a 1,6-difunctionalised compound. It could be made by nucleophilic displacement by an organo-metallic reagent (R⁻) on lactone (23), the Baeyer–Villiger product from (24) which can be made from aromatic (26).

Analysis

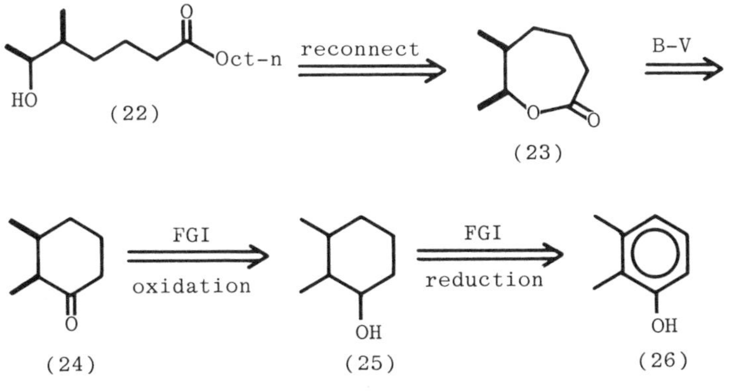

Catalytic reduction of (26) gave a mixture of isomers of (25) from which the *cis* compounds (27) could be separated by chromatography. The Baeyer–Villiger reaction involves migration of the more substituted group with retention of configuration and an organolithium compound was found by experiment to convert (23) into (22).

228

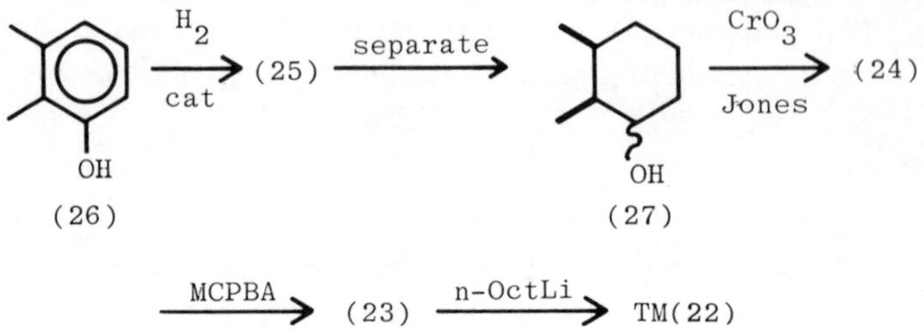

Other Approaches

There is no reason why 1,6-difunctionalised compounds should not be made by conventional methods, essentially ignoring the 1,6-relationship. The symmetrical spiro ketone **(28)** disconnects to 1,6-dicarbonyl compound **(29)** which could no doubt be made by cleavage of **(30)**. An alternative approach is to disconnect the ring from the chain to give **(31)** easily made from butyrolactone **(32)** (Table 25.1).

Analysis

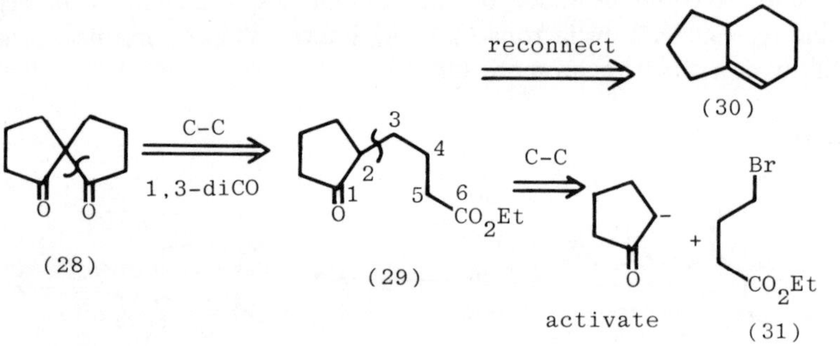

Synthesis[309]

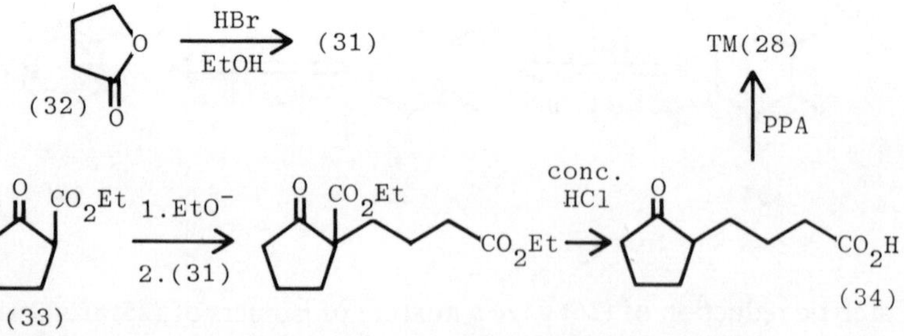

Keto ester **(33)** was discussed in Chapter 19. The final cyclisation was carried out on the free acid **(34)** using polyphosphoric acid (PPA), a powerful dehydrating agent.

CHAPTER 28

General Strategy B:
Strategy of Carbonyl Disconnections

This chapter links the carbonyl disconnections of the last ten chapters with the general principles for disconnections established in Chapter 11. We shall discover some new principles but the main purpose of this chapter is to use the principles of Chapter 11 to decide why some carbonyl disconnections are better than others.

We could look at all carbonyl relationships in the TM, consider all possible disconnections based on them, and decide which we prefer. This can be a very long job, and I shall do it for just one example. Thereafter we shall use our established guidelines to make choices as we disconnect.

Pratt and Raphael needed the cyclohexenone (1) for a synthesis of the anti-tumour compound vernolepin.[310] Disconnection of the α,β bond reveals the basic skeleton (2): no FGI is needed as all the FGs are carbonyl groups.

Analysis 1

(1)   (2)

Compound (2) has 1,3-, 1,4-, 1,5-, and 1,6-dicarbonyl relationships (2a–2d). The next stage is to disconnect each of these (reconnect the 1,6) as in analysis 2.

Analysis 2

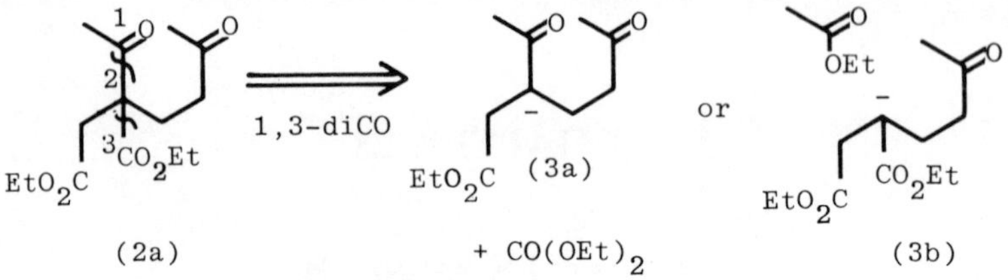

1,3-diCO

(2a)

(3a)

or

+ CO(OEt)$_2$

(3b)

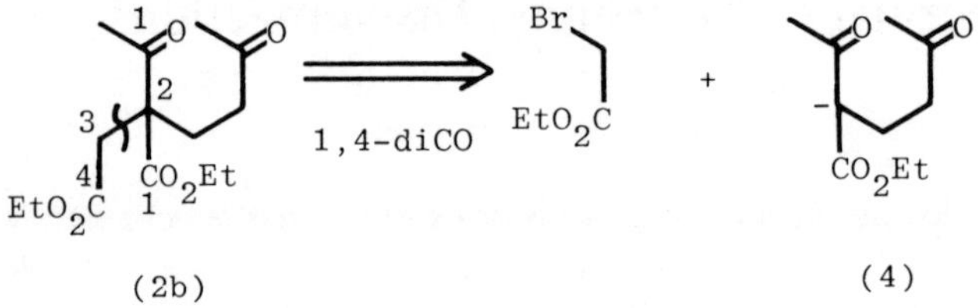

1,4-diCO

(2b)

+

(4)

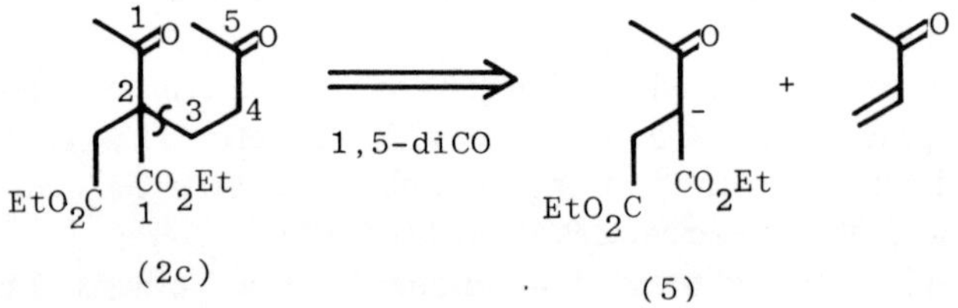

1,5-diCO

(2c)

(5)

1,6-diCO
reconnect

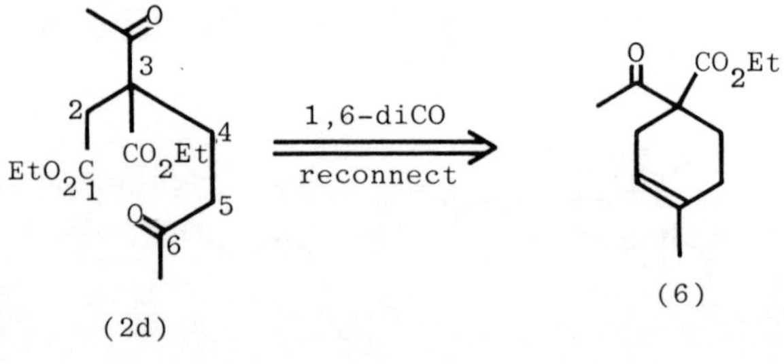

(2d)

(6)

We can forget the 1,3-disconnection as it will be nearly impossible to make specific enolates for (3a) or (3b)—there are at least four roughly equivalent sites for enolisation in each molecule. The 1,4- and 1,5-disconnections look promising as (4) and (5) are stable enolates, and the 1,6 is also promising as (6) looks like a Diels–Alder adduct. We can continue the analysis for (4), (5), and (6).

Analysis 3

(4a) → 1,5-diCO ⟹ (7) + (8)

CO$_2$Et
(4a)

(7) CO$_2$Et

(8)

(5a) → 1,4-diCO ⟹ (9) + (7.)

EtO$_2$C   CO$_2$Et
(5a)

Br
CO$_2$Et
(9)

CO$_2$Et
(7.)

(6a) → D–A ⟹ (10) +

O   CO$_2$Et
(6a)

O   CO$_2$Et
(10)

The starting materials for the 1,4 and the 1,5 approaches are the same —
(7), **(8)**, and **(9)** — just the order of events is different. The Diels–Alder strategy
is good as the orientation is *para* (Chapter 17) and the dienophile **(10)** can be
made by the Mannich method from **(7)**. With three equally good-looking
prospects, the best strategy is to try the 1,4 or the 1,5 approaches since the
starting materials can be used for the other if the one fails. Pratt and
Raphael[310] found the 1,5 strategy *via* **(5)** to be successful — the others may well
be so too.

Synthesis

(7) → 1. EtO$^-$ 2. (9) → (5) → cat EtO$^-$ → (2) → HOAc → TM(1)

CO$_2$Et
(7)

CO$_2$Et
CO$_2$Et
(5)

CO$_2$Et
CO$_2$Et
(2)

N
H

We shall now choose disconnections as we go along and turn to the less
likely only when the 'obvious' fails. Don't forget that the syntheses I give are
not necessarily the only successful ones.

When there are C–X bonds in the TM, e.g. **(11)**, it is usually best to disconnect them first as we can then see the carbon skeleton displayed. The next stage is to count up the relationships between the various functionalised carbon atoms **(12)**.

Analysis 1

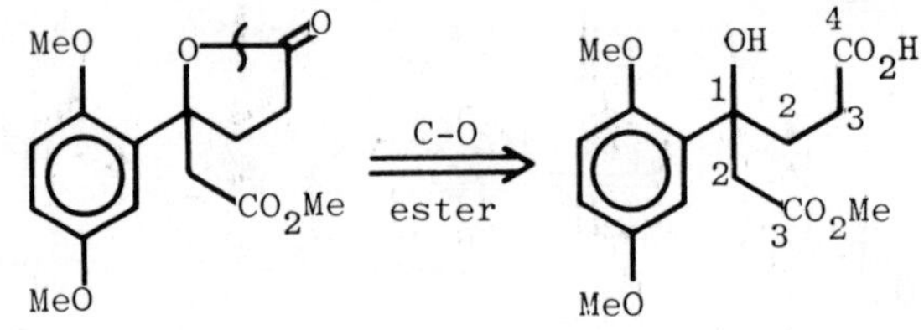

(11)   (12)

Here we have a 1,3- and a 1,4-relationship. Disconnection at the branch point (C1) is strategically best and this is easiest to do for the 1,3-disconnection as the 1,4 requires the difficult synthon **(14)**.

Analysis 2

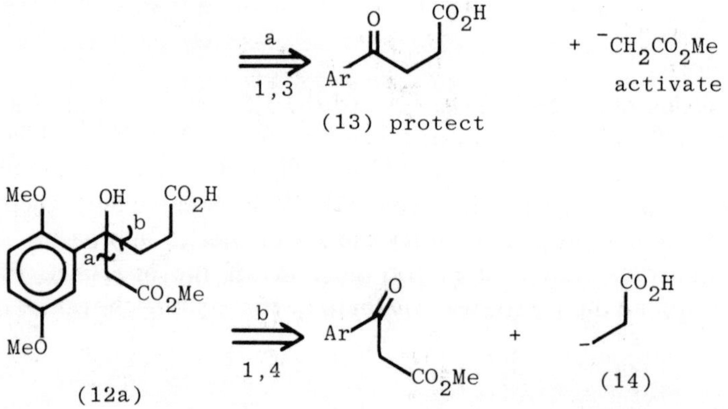

Continuing with **(13)**, all we have left is a 1,4-dicarbonyl relationship. The best strategy is to use the available 1,4 compound succinic anhydride in a Friedel–Crafts reaction (Table 25.1).

Analysis 3

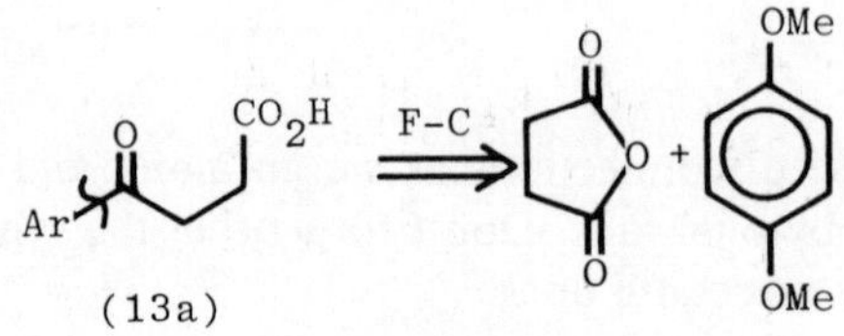

(13a)

There are no problems of orientation in the Friedel–Crafts reaction as all positions in the aromatic ring are the same. Reformatsky (Chapter 20) is usually the best method of control for β-hydroxy esters **(12)** so we shall need to protect the CO_2H group as an ester **(15)**. The final cyclisation occurs during work-up.

Synthesis[311]

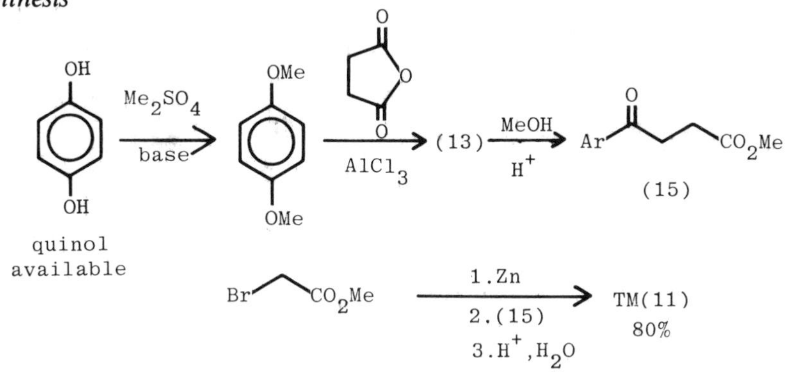

Clear thinking is required when none of the 'obvious' disconnections is promising. Bicyclic **(16)**, needed[312] for a prostaglandin synthesis, contains an acetal and disconnection of this reveals the carbon framework **(17)**. This is symmetrical and has 1,4- and 1,5-relationships.

Analysis 1

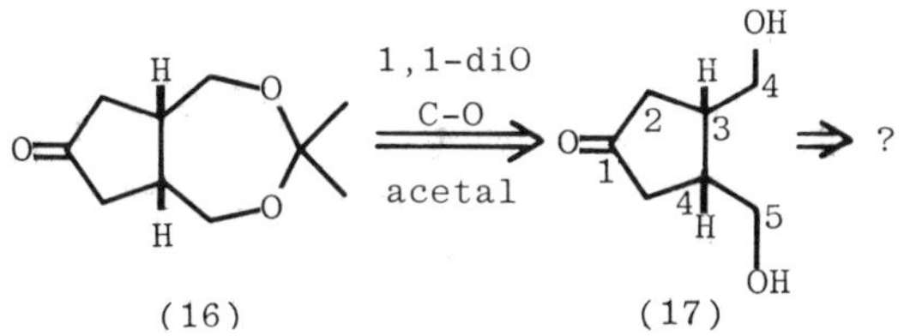

No 1,4- or 1,5-disconnections look very promising (e.g. (a), (b), or (c) and they would destroy the symmetry. An alternative strategy is to introduce an activating group **(18)** according to the method introduced in Chapter 19. A 1,3-dicarbonyl disconnection then restores the symmetry in **(19)**.

Analysis 2

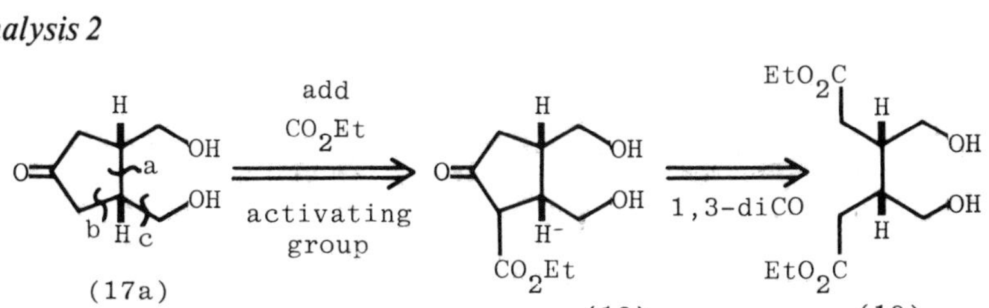

234

The earlier unhelpful 1,4- and 1,5-relationships are still there but a new 1,6-relationship has appeared. Reconnection preserves the symmetry **(20)** and reveals a Diels–Alder adduct **(21)** after the oxidation level is adjusted.*

Analysis 3

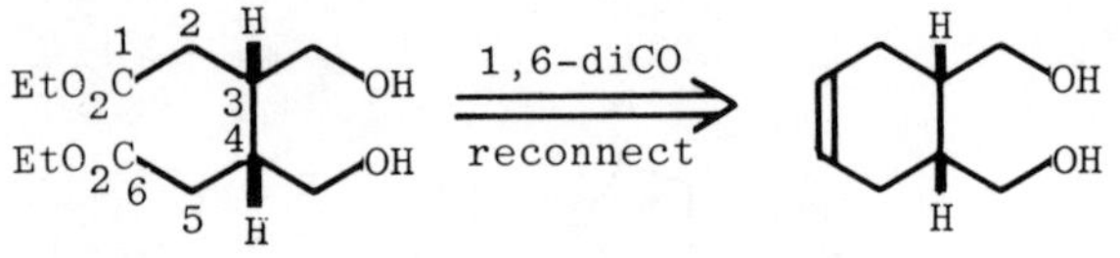

(19a)   (20)

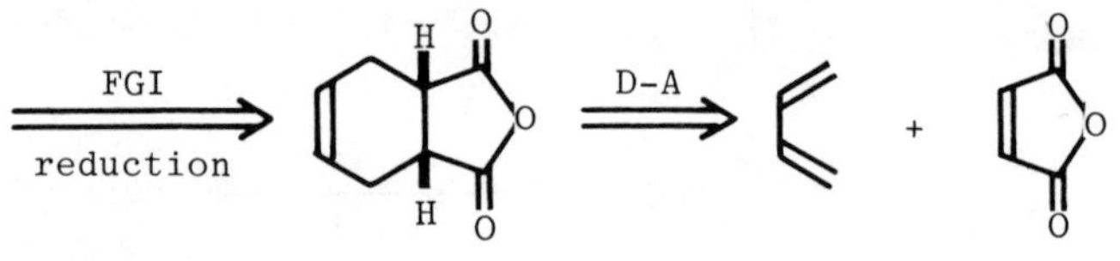

(21)

Maleic anhydride as dienophile ensures the correct relationship between the two chiral centres. It is better to reduce and protect before the oxidative cleavage so that the distinction between the right and left sides of the molecule is maintained.

Synthesis[312]

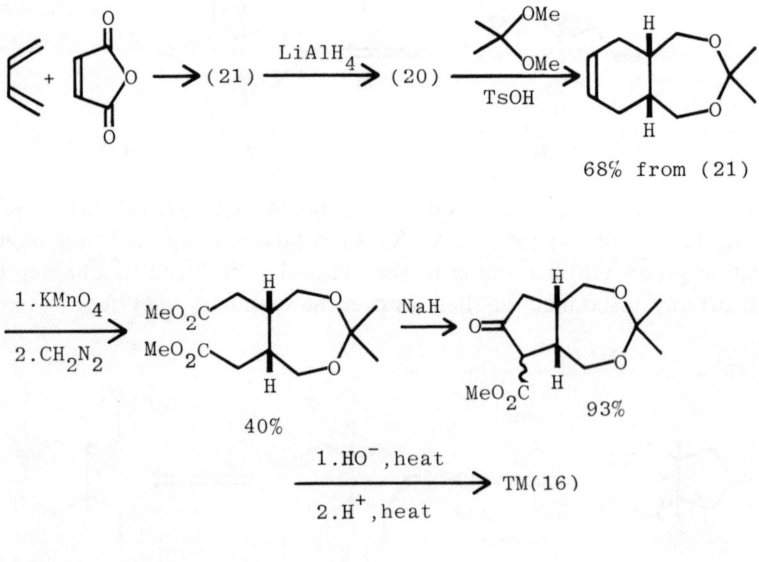

68% from (21)

$$\xrightarrow[\text{2.H}^+,\text{heat}]{\text{1.HO}^-,\text{heat}} \text{TM(16)}$$

*Cf. Chapter 27.

The process of abandoning bad strategy and trying again may have to be repeated several times. Two promising approaches to spiro enone **(22)**, needed by Corey[313] in his gibberellic acid synthesis, eventually peter out and a third approach has to be found. The first disconnection on **(22)** must surely be at the α,β bond to give **(23)**. This is a 1,4-dicarbonyl compound most obviously disconnected at the branch point (a) to **(24)**.

Analysis 1

| (22) | (23) | (24) |

Aldehyde **(24)** is a 1,5-difunctionalised compound but it cannot be made by a Michael reaction.* Disconnection (a) still looks the best so we might consider reconnecting the aldehyde in **(23)** and reversing the polarity of disconnection (a) according to the strategy outlined in Chapter 26.

Analysis 2

| (25) | (26) |

Unfortunately, allylic bromide **(26)** will almost certainly react at C3 with nucleophiles so this approach is not promising enough to try. The alternative branch point disconnection **(25b)** requires the addition of a vinyl copper derivative **(27)** to an enone **(28)** (see Chapter 14 for an example). Further disconnection on enone **(28)** by the usual method gives a simple cyclohexanone **(29)**.

*Though no doubt it could have been made by reduction of an aromatic compound, see Chapter 36.

Analysis 3

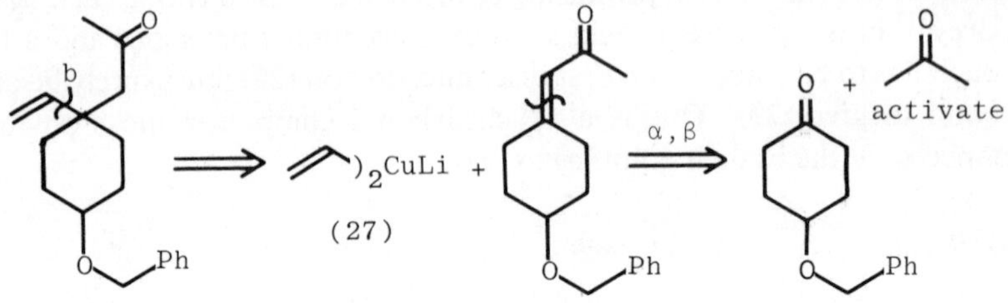

(25b)   (27)   (28)   (29)

The relationship between the ether and ketone in **(29)** is most easily set up by reduction of an aromatic compound: quinol, as used on page 233, is the obvious choice.

Analysis 4

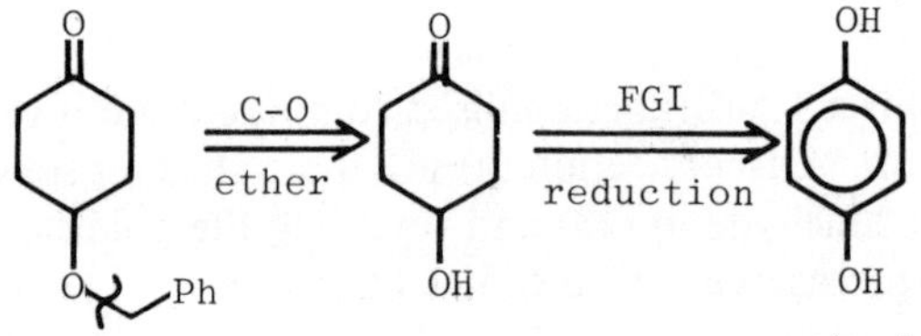

(29)   quinol

The first stage of the synthesis[314] requires chemoselectivity between two identical groups—the OH groups of quinol. This had to be statistical (Chapter 5) and experiments showed that monobenzylation of diol **(30)** gave the best results. The yield is only moderate (50%) but the byproducts **(30)** and **(32)** can be recycled and a poor step at the start of a synthesis is not so bad as it can be carried out on a very large scale with cheap starting materials.

Synthesis[314]

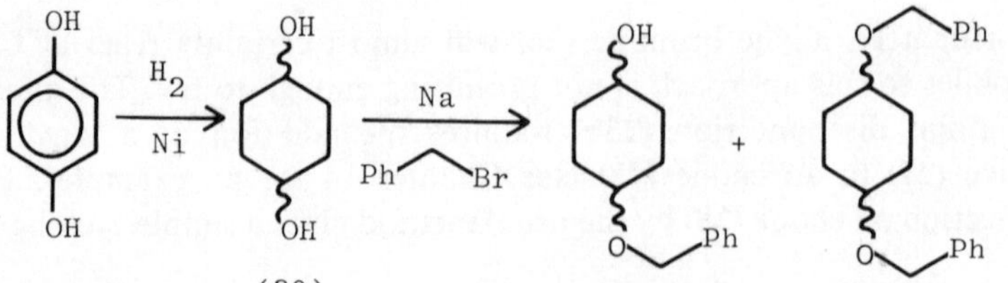

(30)   (31)   (32)

$$(31) \xrightarrow{CrO_3} (29)$$

Corey[313] chose the Wittig method to control the condensation to give **(28)**. The rest of the synthesis went according to plan.

Synthesis 2[313]

$$(EtO)_3P + Br\diagup\!\!\diagup\!\!\overset{O}{\diagdown} \longrightarrow (EtO)_2\overset{O}{\underset{}{P}}\diagup\!\!\overset{O}{\diagdown} \xrightarrow[\text{(29)}]{\text{NaOH}}$$

(28)

$$\xrightarrow[\text{Cu}_2\text{I}_2]{\diagup\!\!\diagdown\diagdown\text{MgBr}} \quad (25) \quad \xrightarrow[\text{NaIO}_4]{\text{OsO}_4} \quad (23) \quad \xrightarrow[\text{EtOH}]{\text{NaOH}} \quad \text{TM(22)}$$

Another strategy is to change the relationship between two substituents by chain lengthening* or shortening so that an unhelpful relationship becomes a helpful one. Bicyclic ketone **(33)** has a 1,3-diketone disconnection which reveals a 1,6-dicarbonyl compound **(34)**. The normal reconnection approach is impossible here as strained olefin **(35)** cannot be made. Reducing the chain length by one carbon atom would turn the 1,6- into a 1,5-relationship **(37)** which can be disconnected at the ring–chain junction by the usual Michael approach (Chapter 21) after FGI. In the synthesis, chain lengthening is most easily achieved by cyanide displacement of a halide.

Analysis 1

$$(33) \xRightarrow{\text{1,3-diCO}} (34) \xRightarrow{\text{reconnect}} \begin{matrix}(35): \\ \text{impossible}\end{matrix}$$

---

*Chain lengthening is treated more fully in Chapter 31.

Analysis 2

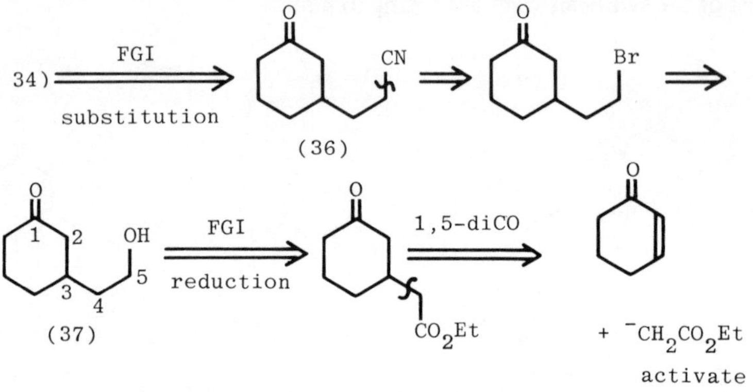

The Michael reaction is satisfactory with malonate as nucleophile, but the ketone must be protected while the ester is reduced.[315] The rest of the synthesis follows the plan.

Synthesis[315]

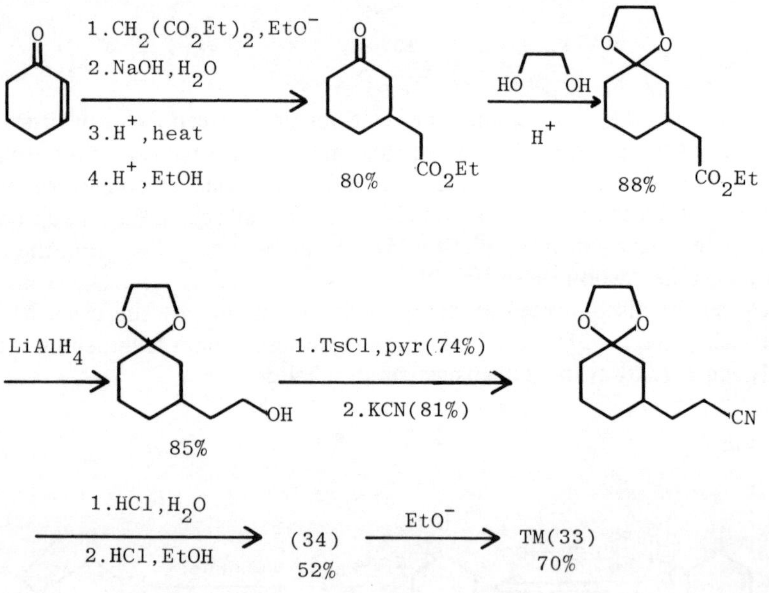

This is a long synthesis. Though the yields are good, the overall yield is only 13%. A second successful and much shorter synthesis comes from the FGA strategy (Chapter 24). Introducing unsaturation throughout the whole skeleton of (34) gives an aromatic compound (38) with an obvious α,β-disconnection.

Analysis 3

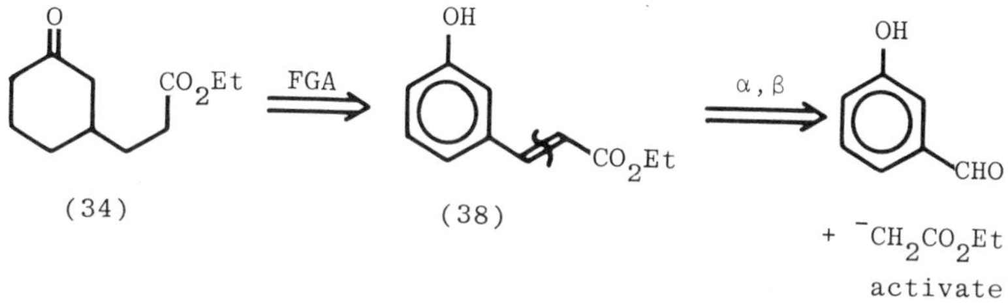

(34) (38)

+ $^-CH_2CO_2Et$
activate

Meta-hydroxybenzaldehyde is available and hydrogenation of **(38)** with Raney nickel catalyst reduces the double bond and the aromatic ring, leaving only reoxidation for a short synthesis of **(34)**. The overall yield is 25%.

Synthesis 2[315]

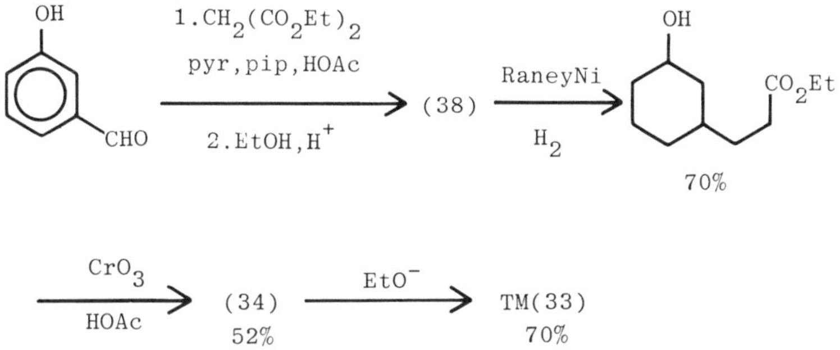

These two strategies are just examples of the many ways in which it is now possible to solve difficult problems. We shall meet more in the remaining chapters.

Summary of Approach

1. Convert all FGs to those based on oxygen (OH, CO, etc.) by FGI or C–X disconnections so that the basic carbon framework is displayed.
2. Identify the 1,*n*-relationships.
3. Adjust oxidation level (if necessary) and disconnect using reactions from Chapters 18–28 guided by principles set out in Table 11.2.
4. If necessary, look at all possible disconnections until a good synthesis emerges.
5. If necessary, add extra FGs (FGA) or activating groups, or change the 1,*n*-relationships by chain lengthening or shortening to get a good synthesis.
6. If a bad step must be included, put it as early as possible in the synthesis.

CHAPTER 29

Strategy XIII: Introduction to Ring Synthesis. Saturated Heterocycles

The exceptional ease of cyclisation reactions has been a theme running through the book. It appeared in Chapters 7 and 20 where I emphasised that controls may be relaxed for cyclisations as they usually take precedence over intermolecular reactions. The next nine chapters concern ring formation, taking ring sizes in order from three to six with the usual alternate strategy chapters on related topics. This chapter explores in more depth the reasons why ring formation is favourable using the synthesis of saturated heterocycles as examples.

Cyclisations

Intramolecular reactions are usually favoured kinetically over intermolecular reactions for reasons of entropy because the two reactive sites are part of the same molecule and there is no need for a bimolecular collision. This factor is greatest in three-membered ring formation where the two ends of the reagent (1) are always close, and for five-membered ring formation where natural thermal motion of (2) brings the reactive groups into bonding distance.

Six-membered ring formation is kinetically reasonable: the problem here, as for larger rings, is that rotation brings the reactive ends too close (3) and a folding of the chain is needed before bonding can occur. Four-membered ring formation is uniquely slow: the chain normally adopts a conformation (4) with the reactive groups far apart and, even in the best conformation (5) for cyclisation, they are still distant.

(3)

(4)   (5)

The six-membered ring is, however, uniquely favoured thermodynamically in its chair conformation (6) with all groups staggered, especially if large substituents can become equatorial, as in (6). Five-membered rings are also stable, but three- and four-membered rings are strained with angles of 60° and 90° instead of the usual 109° for sp^3 or 120° for sp^2 atoms.

(6)

All these conclusions are rough and ready and must be assessed in the context of the precise structure under consideration. Table 29.1 gives a summary.

Table 29.1   Factors affecting ring formation

Ring size	Kinetic factors	Thermodynamic factors
3	√√√	×
4	×	×
5	√√	√
6	√	√√
7	√	√

Taking both factors into account, five-, six-, and seven-membered rings are easy to make, three-membered rings are easy to make but often break down again under the conditions of their formation, whilst four-membered rings are uniquely awkward and often need special methods. In Perkin's original work[316] on ring formation by the double alkylation of malonate, three-, four-, five-, six-, and seven-membered rings were all formed in good yield, but the four-membered ring was formed very slowly (i). With acetoacetate, the four-membered was not formed, an enol ether being formed instead (ii).

242

(i)

(CH₂)ₙ₋₃ with two Br + CH₂(CO₂Et)₂ → EtO⁻ → (CH₂)ₙ₋₃ ring with CO₂Et, CO₂Et

n=3,4,5,6,7

(ii)

Except for four-membered rings, these factors are favourable to ring formation so that we can use the normal disconnections and all the normal guidelines to choose one disconnection rather than another. We have already analysed the formation of many cyclic compounds in this way (e.g. in Chapters 17, 20, 21, 27, and 28), but we shall now take each ring size in turn and discuss its synthesis systematically, starting with C–X disconnections in this chapter, and progressing to C–C disconnections in Chapters 30–37.

Saturated Heterocycles

Three-membered rings

Epoxides are usually made from alkenes—both C–O bonds being disconnected at once. The reagent is a peracid RCO_3H often MCPBA (Chapter 7).

Analysis

Synthesis

MCPBA

Disconnection of only one C–O bond in epoxide (7) suggests chloro alcohol (8) as an intermediate, and then by C–C disconnection, an α-chloro ketone (9) might be a good starting material. Cornforth[317] developed this method, using the attack of Grignard reagents on (9) to make (8). The ease of displacement to give the epoxide (7) from (8) shows that three-membered ring formation is kinetically favoured.

Analysis

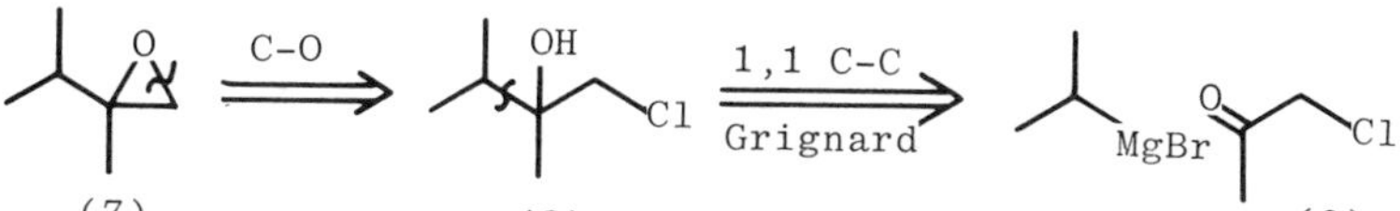

(7)   (8)   (9)

Synthesis[317]

Four-membered rings

Special methods for four-membered carbocyclic rings are given in Chapter 32. The normal cyclisation approach[318] often gives poor results but is sometimes successful. Straightforward C–N disconnection of amine **(10)** suggests 1,3-dibromide **(11)** as a starting material. Lactone **(12)** can be used to make **(11)** (strategy of Table 25.1) and it does indeed give TM**(10)** with a primary amine.[319]

Analysis

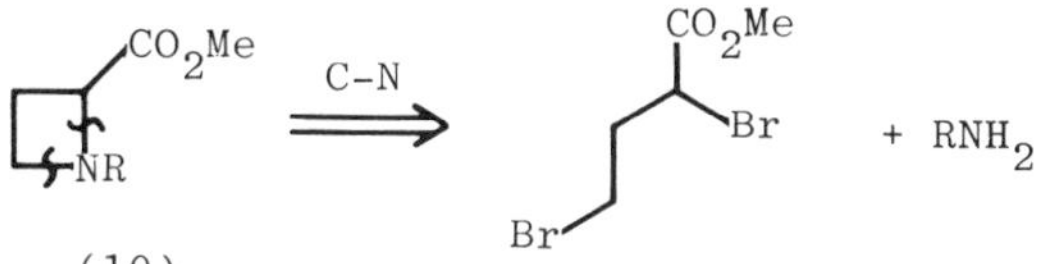

(10)   (11)

Synthesis[319]

(12)

In very favourable cases, four-membered cyclic ethers can also be made by cyclisation. We have already seen that *cis* **(13)** (but not *trans*) cyclises in base (Chapter 13). A similar C–O disconnection on cyclic acetal **(14)** takes us back to a β-hydroxy ketone **(15)** and hence to ketone **(16)**.

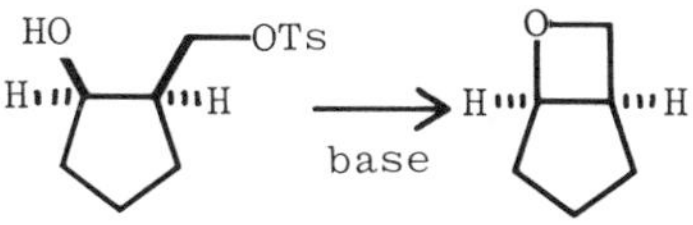

(13)

244

Analysis

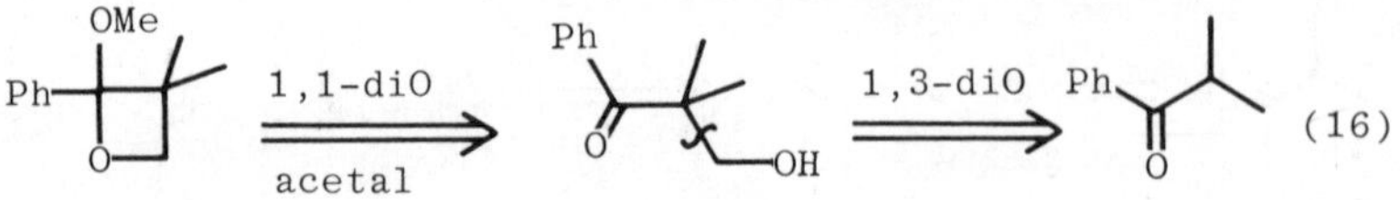

(14)   (15)   (16)

+ CH$_2$O

Mannich control is unnecessary here as **(16)** has only one enolisable proton, though a weak base is used[320] to prevent Cannizzaro reduction of **(15)**. We need to make conditions for cyclisation as favourable as possible so the OH group in **(15)** is converted into a better leaving group and MeOH is added as the more reactive MeO⁻.

Synthesis[320]

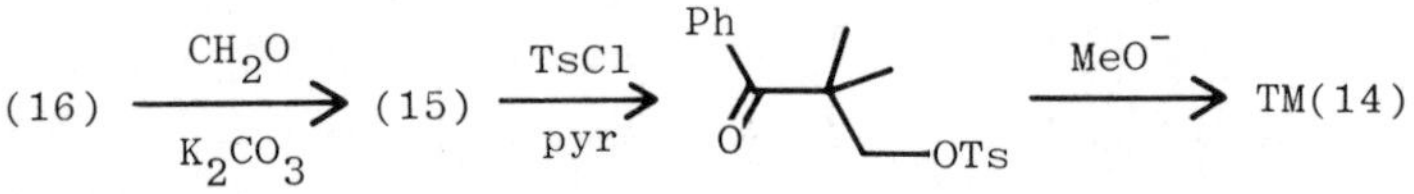

Five-membered rings

Cyclisation to five-membered rings is very favourable. Hydroxy acids **(17)** cannot usually be isolated: the anion **(18)** is stable but neutralisation causes instant cyclisation to lactones **(19)**. Syntheses of five-membered lactones have been discussed in Chapters 12, 25, 26, 27, and 28.

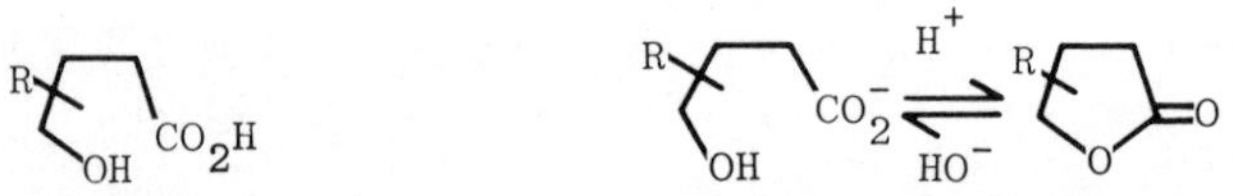

(17)   (18)   (19)

Tactics for the synthesis of five-membered heterocycles are to disconnect both C–X bonds and to identify the electrophilic carbon fragment needed to add to the nucleophilic heteroatom. Cyclic sulphide **(20)** requires *cis*-dihalide **(21)** which comes from *cis*-butenediol **(22)** (Chapter 16).

Analysis

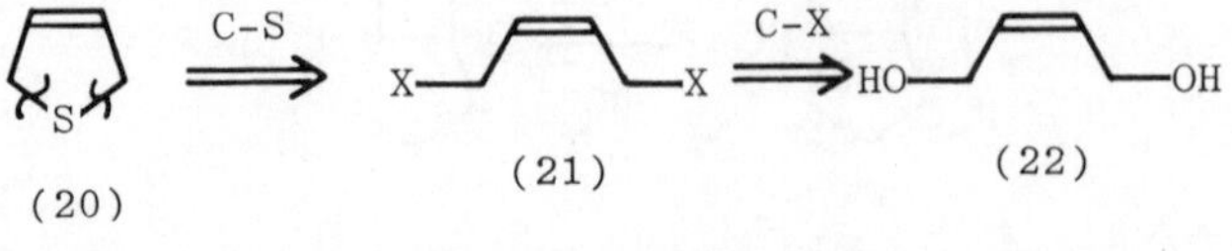

(20)   (21)   (22)

Synthesis[321]

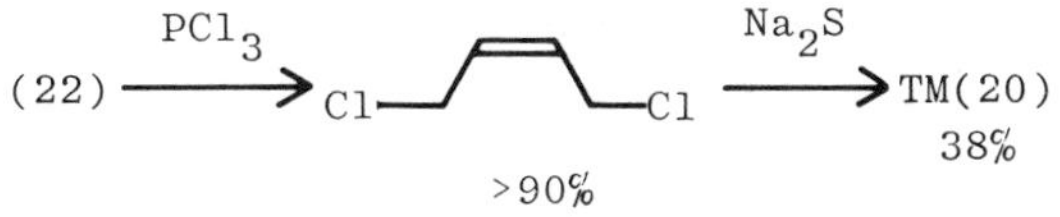

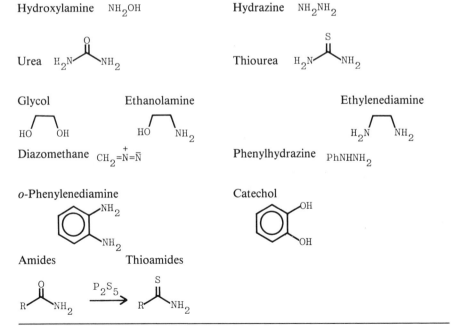

With two heteroatoms in the ring, it is often easiest to identify a readily available fragment (Table 29.2) containing both heteroatoms and disconnect it from a suitable electrophilic fragment. Hence **(24)**, needed for a synthesis[231] of the ant alarm pheromone mannicone **(23)**, clearly contains hydrazine, NH_2NH_2. Disconnection reveals keto ester **(25)**.

Analysis 1

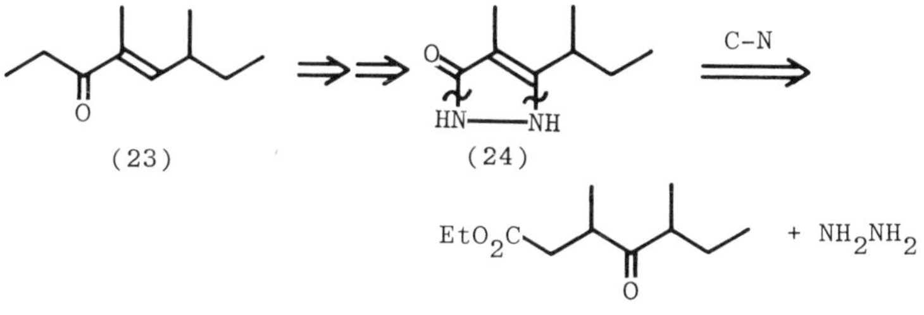

Table 29.2 Some available reagents containing two heteroatoms

Hydroxylamine NH_2OH	Hydrazine NH_2NH_2
Urea	Thiourea
Glycol Ethanolamine	Ethylenediamine
Diazomethane $CH_2=\overset{+}{N}=\bar{N}$	Phenylhydrazine $PhNHNH_2$
o-Phenylenediamine	Catechol
Amides Thioamides	

See also Table 23.3.

Disconnection of 1,3-dicarbonyl (25) can be at (a) or (b): (a) is ambiguous as (26) can self-condense, but (b) is not as only the required product (25) can form a stable enolate ion (Chapter 20).

Analysis 2

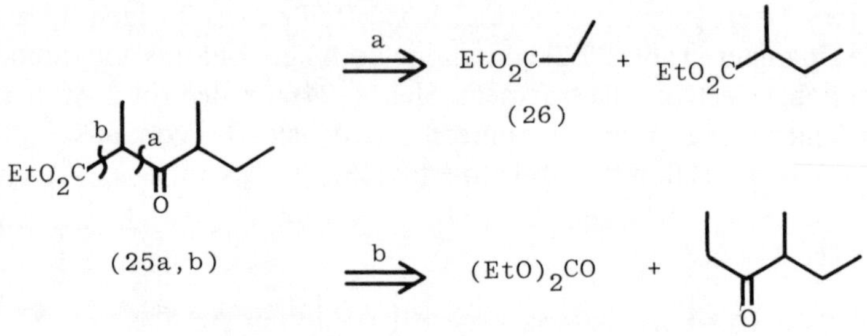

(25a,b)

(26)

(EtO)$_2$CO +

Synthesis[231]

$$\xrightarrow[\text{NaH}]{\text{(EtO)}_2\text{CO}} (25) \xrightarrow{\text{NH}_2\text{NH}_2} \text{TM(24)}$$

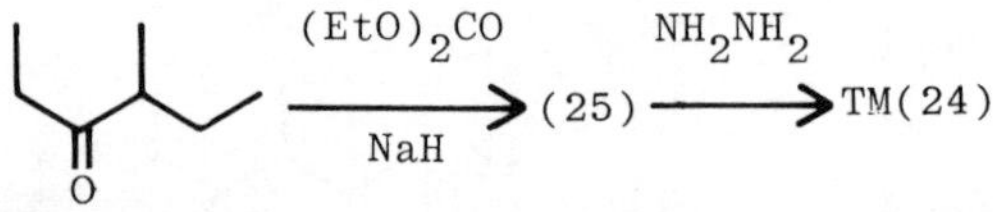

Disconnection of C–C bonds may help. The obvious C–X disconnections on (27) give a difficult starting material (28) whereas 1,3-dicarbonyl disconnection first to (29) allows more helpful C–S disconnections.

Analysis 1 (C–X disconnections)

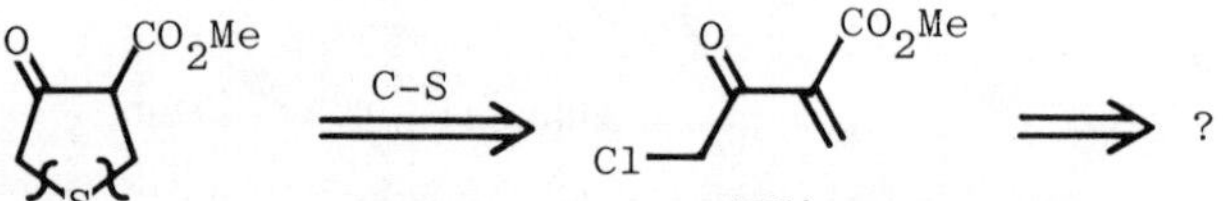

Analysis 2 (C–C disconnections)

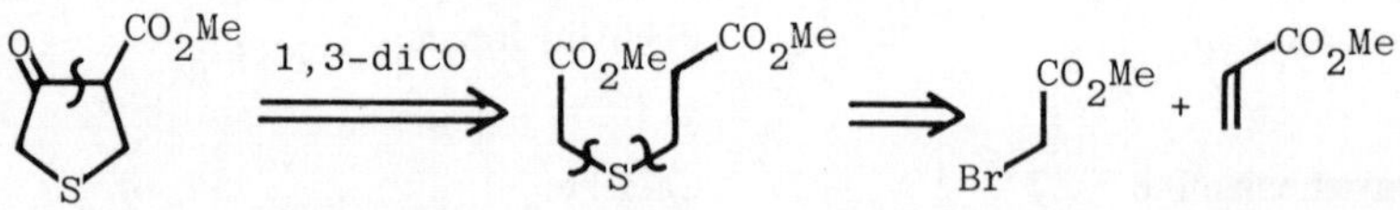

This sulphide was made by Woodward[322] who chose to do the Michael reaction second, because thiolacetic ester (30) is available.

Synthesis[322]

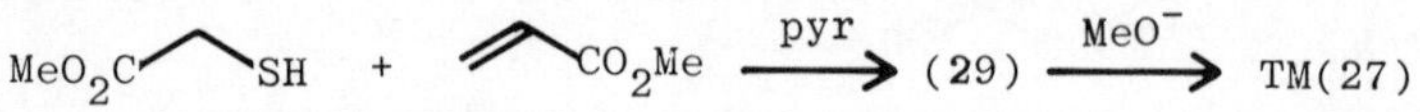

$$\text{MeO}_2\text{C}\diagup\text{SH} + \diagup\text{CO}_2\text{Me} \xrightarrow{\text{pyr}} (29) \xrightarrow{\text{MeO}^-} \text{TM(27)}$$

(30)

1,3-Dipolar cycloadditions

One special method for five-membered rings is 1,3-dipolar cycloaddition,[323] e.g. (31). This is like a Diels–Alder reaction with the diene replaced by a three atom, four electron unit — the 1,3-dipole.

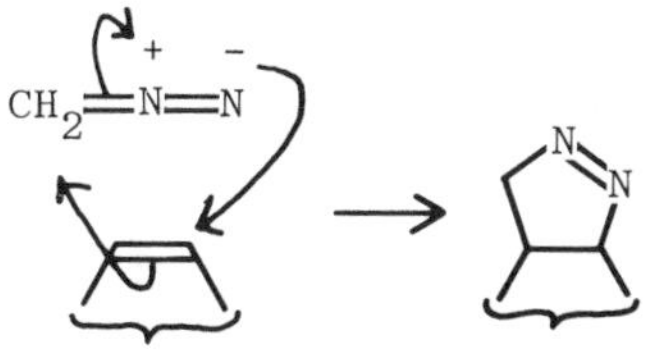

(31)

Conjugated alkenes react best, as in the Diels–Alder reaction, so that the heteroatoms and the ester group in (32) suggest a 1,3-dipolar disconnection. The orientation of such reactions is beyond the scope of this book, and is discussed in full by Fleming.[324]

Analysis

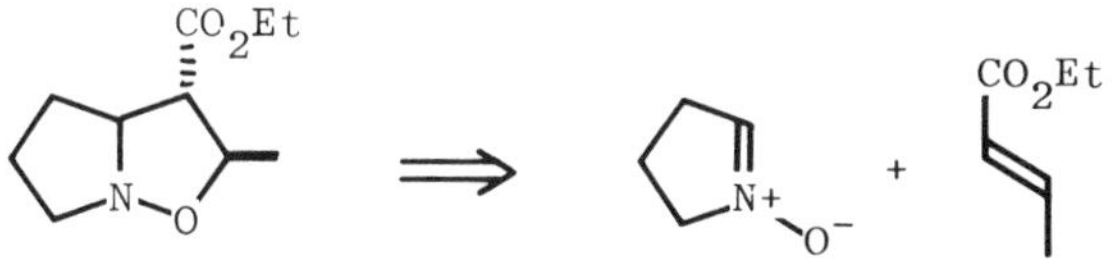

(32)   (33)

The 1,3-dipole (33) is a nitrone, made from a cyclic amine by a sequence of oxidation, elimination, and another oxidation. The cycloaddition gives only the isomer (32) required and the *trans* stereochemistry of the unsaturated ester must be preserved.

Synthesis[325]

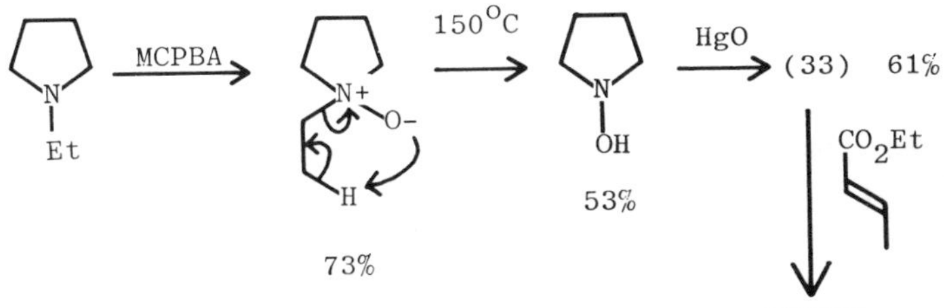

Six-membered rings

These are again simple and the same principles apply as for five-membered rings. Where the heteroatom is joined to carbon atoms at different oxidation levels, e.g. (34), (35), or (36), the best order of events is usually to disconnect (34) first and (36) last.

(34) 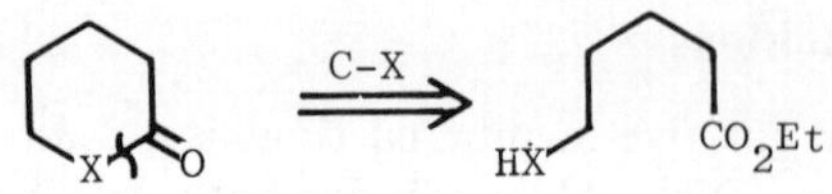 ⟹ᶜ⁻ˣ

(35) 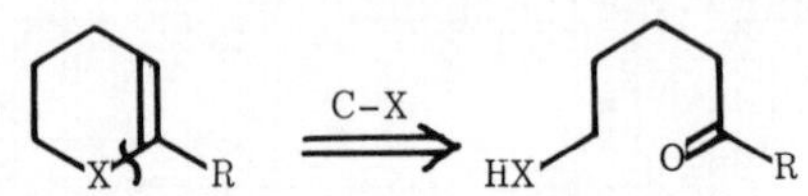 ⟹ᶜ⁻ˣ

(36) 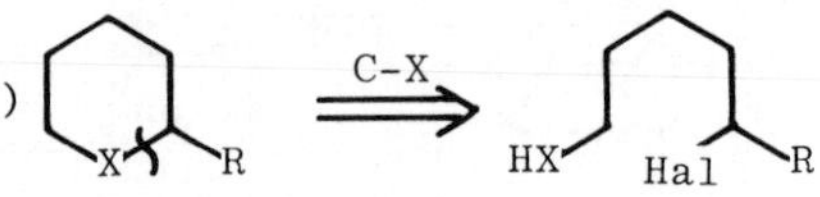 ⟹ᶜ⁻ˣ

Following this order of events, disconnection (a) on **(37)** should come first. This reveals a symmetrical amino ketone **(38)** with a 1,5-disconnection if the amino group comes from a cyanide.

Analysis

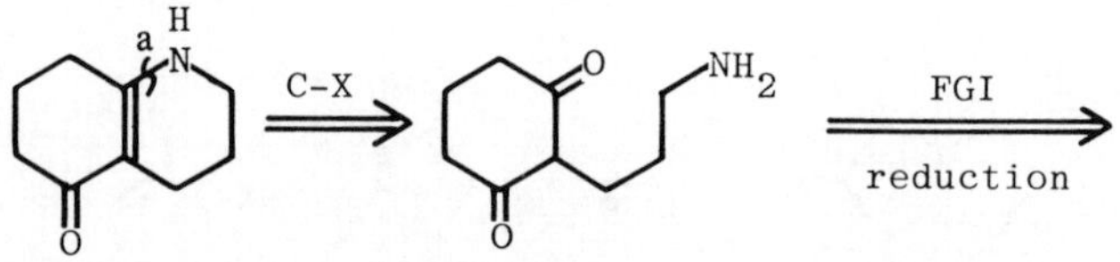

(37)   (38)

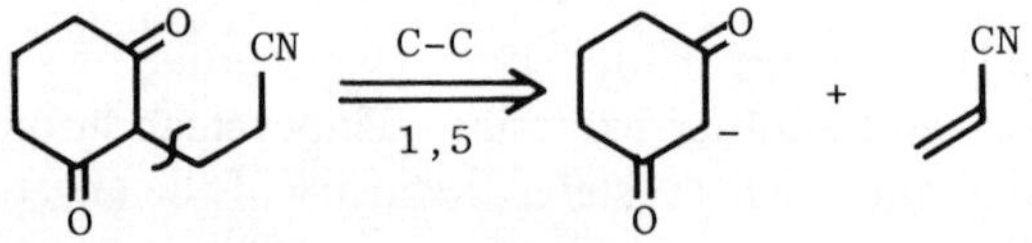

The diketone **(40)** can be made[326] by reduction of resorcinol **(39)** and the same catalyst (Raney Ni) promotes reduction of the cyanide in the presence of the carbonyl groups. Cyclisation of **(38)** is spontaneous.

Synthesis[326]

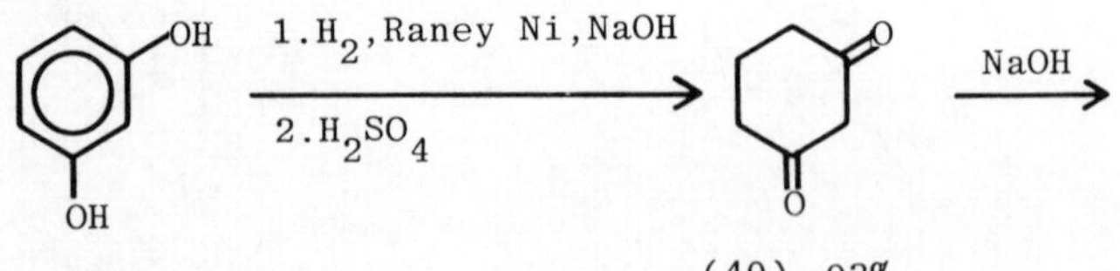

(39)   (40) 93%

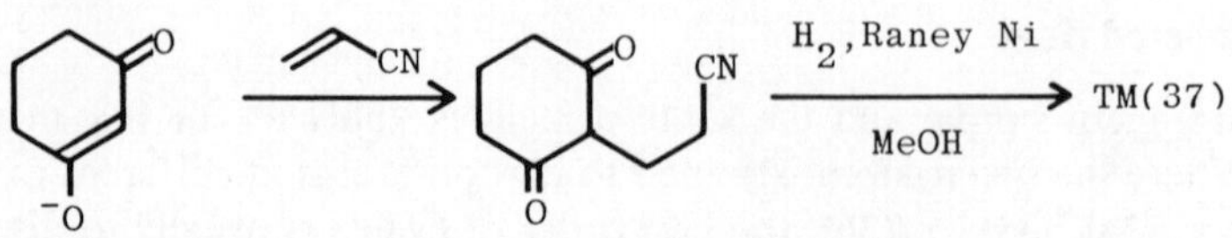

97%

4-Piperidones can be disconnected at C–N (41) or by the strategy used for other symmetrical ketones (Chapter 19) (43).

Analysis

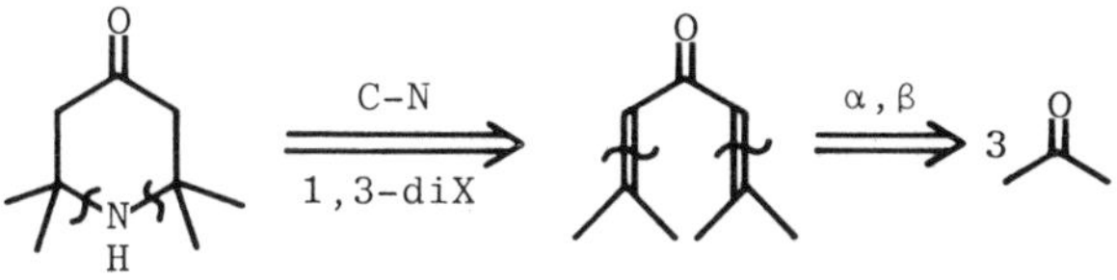

(41)   (42)

Since (42) is made from three molecules of acetone, the synthesis is easier than expected and (41) is a readily available compound.

Synthesis[327]

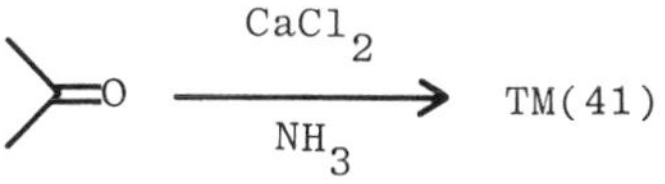

The C–C disconnection strategy is more general (44) and has been used to make many pharmaceuticals.[328]

Analysis

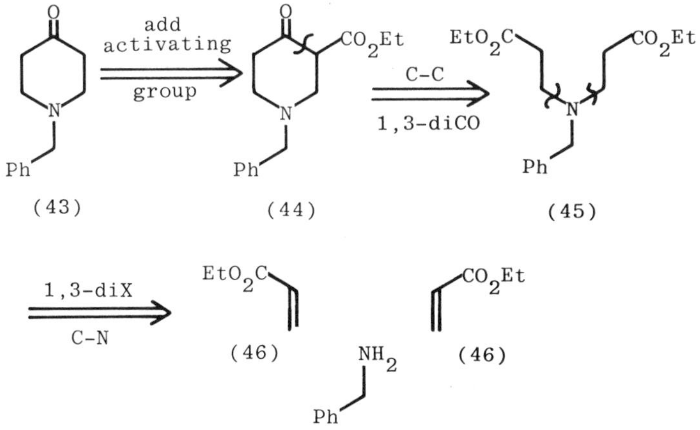

(43)   (44)   (45)

(46)   (46)

The synthesis is short and straightforward from readily available benzylamine and acrylate ester (46) (cf. Chapter 19).

Synthesis[329]

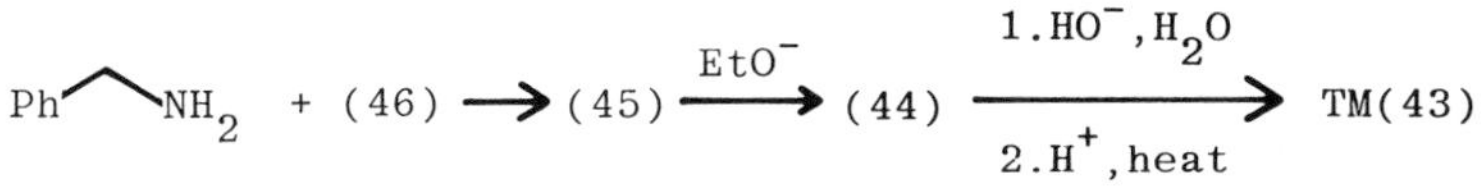

Ph⌒NH$_2$ + (46) → (45) $\xrightarrow{\text{EtO}^-}$ (44) $\xrightarrow[\text{2.H}^+,\text{heat}]{1.\text{HO}^-,\text{H}_2\text{O}}$ TM(43)

With two heteroatoms in the ring, it is again useful to look for a recognisable fragment containing both. Uracil (47), one of the bases in the nucleic acids, can be disconnected to urea and a suitable electrophilic fragment. Michael addition to an electrophilic acetylene is a suitable reaction.

Uracil: *Analysis*

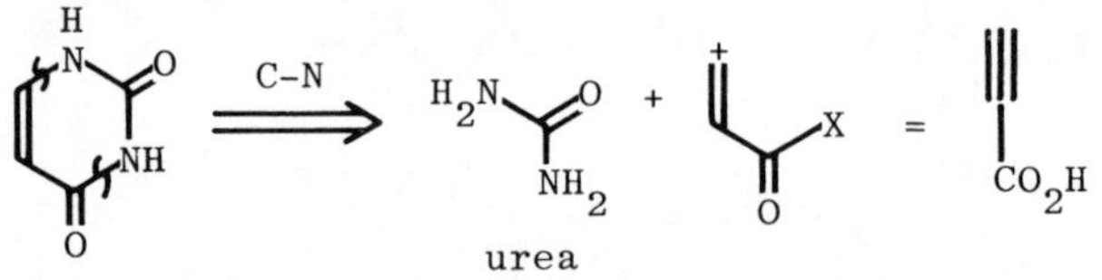

Synthesis[330]

$$\equiv\!\!-CO_2H + (NH_2)_2CO \xrightarrow[\text{heat}]{H^+} TM(47)$$
$$65\%$$

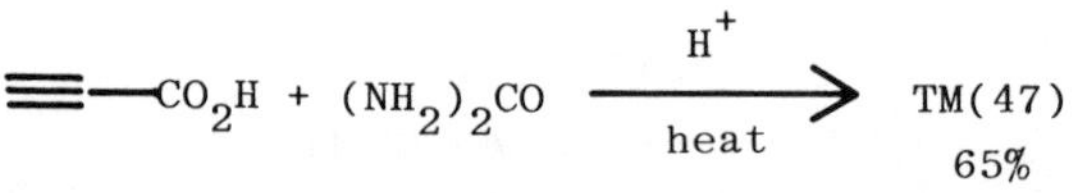

Seven-membered rings

Cyclisation to seven-membered rings may lack the great kinetic or thermodynamic advantages of cyclisation to five- or six-membered rings, but neither are there great difficulties and normal disconnections may be used. The tranquillisers Librium and Valium are based on structures like (48). Disconnection of the aliphatic fragment as chloracetyl chloride (Chapter 23) leaves the imine (49) of an aromatic ketone.

Analysis

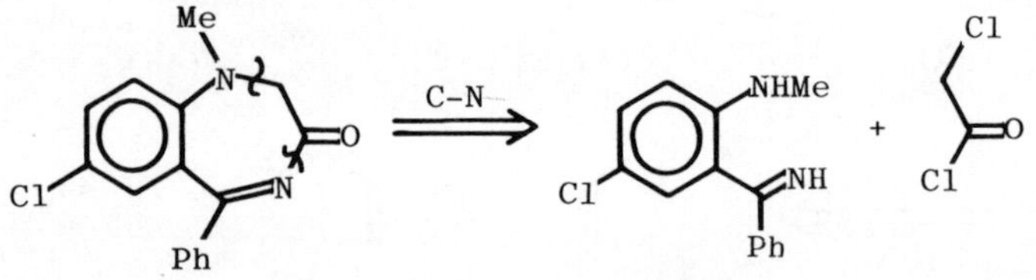

In practice, the oxime (50) can be used in the cyclisation reaction.

Synthesis[331]

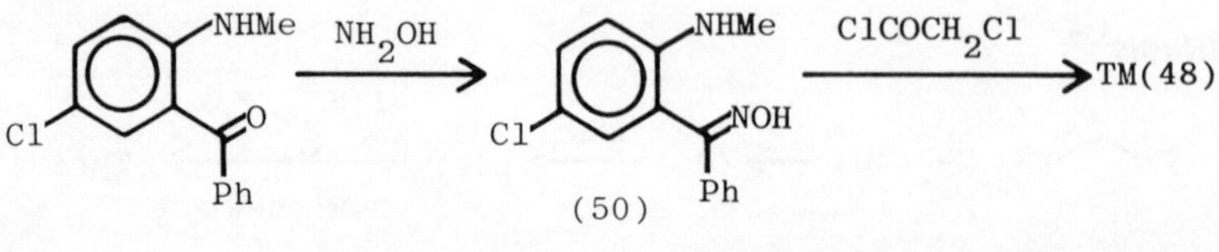

CHAPTER 30

Three-Membered Rings

Cyclisations

Three-membered ring formation is kinetically favourable but thermo-dynamically unfavourable so that three-membered rings are often destroyed under the conditions of their formation. Since most carbonyl condensations are reversible, they are not usually good routes to three-membered rings. Alkylation of carbonyl compounds is usually irreversible and, since the cyclisation is kinetically favoured, it gives three-membered rings without control.

Cyclopropyl ketones, e.g. (1), can be made from γ-halo ketones (2), whose synthesis from epoxides and activated ketones was discussed in Chapter 25.

Analysis

$$\text{(1)} \quad \xrightarrow{\text{C-C}} \quad \text{(2)} \quad \Rightarrow\Rightarrow$$

Synthesis[332]

$$\xrightarrow[\triangle]{\text{EtO}^-} \quad \xrightarrow{\text{HBr}} (2) \xrightarrow{\text{base}} \text{TM}(1)$$

The biologically patterned insecticide Permethrin,[333] developed by M. Elliott's team at Rothamstead, contains ester (3). Disconnection (a) gives a starting material with a tertiary halide (4), yet even this cyclises readily in base, ring formation being so favourable.

251

Analysis

$$(3) \quad\quad\quad \overset{C-C}{\Longrightarrow} \quad\quad\quad (4)$$

Synthesis[334]

$$(4) \quad \xrightarrow{\text{NaOEt}} \quad \text{TM}(3)$$

Insertion Reactions

Epoxides

Most three-membered rings are made by insertion reactions in which two bonds form in a single step. We have already seen (Chapters 7 and 29) that epoxides are usually made this way, the peracid acting as an electrophile.

$$R\diagdown\diagup \quad \xrightarrow{\text{RCO}_3\text{H}} \quad R\diagdown\triangleleft^{O}$$

The corresponding disconnection of a carbon atom suggests a carbene, e.g. **(6)**, as intermediate. This disconnection is important for epoxides **(5)** of α,β-unsaturated carbonyl compounds.

Analysis

$$\overset{a}{\Longrightarrow} \quad \diagup\hspace{-0.3em}=\hspace{-0.3em}\diagdown^{\text{CO}_2\text{Et}} \quad + \quad \text{'O'}$$

$$\overset{b}{\Longrightarrow} \quad \diagup\hspace{-0.3em}=\hspace{-0.3em}\text{O} \quad + \quad \overset{..}{\text{C}}\text{HCO}_2\text{Et}$$

$$(6)$$

In route (a) the reagent for the oxygen atom needs to be nucleophilic so HOO⁻, from H_2O_2 and base, is used.

Synthesis (a)

$$\diagup\hspace{-0.3em}=\hspace{-0.3em}\diagdown^{\text{CO}_2\text{Et}} \quad \xrightarrow[\text{base}]{\text{H}_2\text{O}_2} \quad \overset{O}{\diagdown\triangle\diagdown}^{\text{CO}_2\text{Et}}$$

$$\text{TM}(5)$$

Route (b) requires a reagent for the carbene **(6)** and the halo ester **(7)** is used in the Darzens reaction.[335] A carbene is not in fact an intermediate.

Synthesis (b)

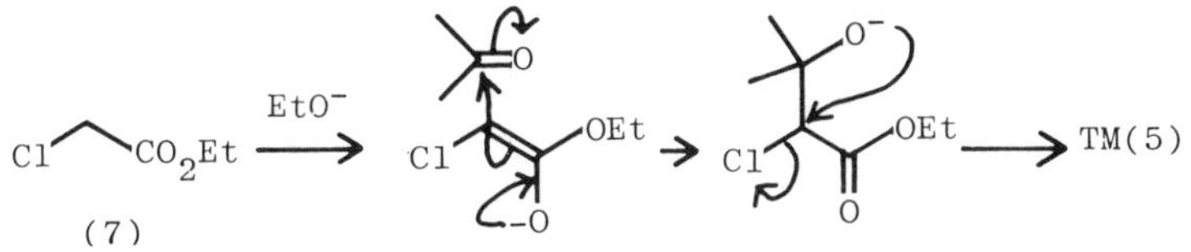

Epoxide **(9)** was used to supply the marked atoms in piperolide **(8)**, a natural product found in the plant *Piper sanctum*.[336] Darzens disconnection leads to cinnamaldehyde **(10)**, easily made by a carbonyl condensation.

Analysis

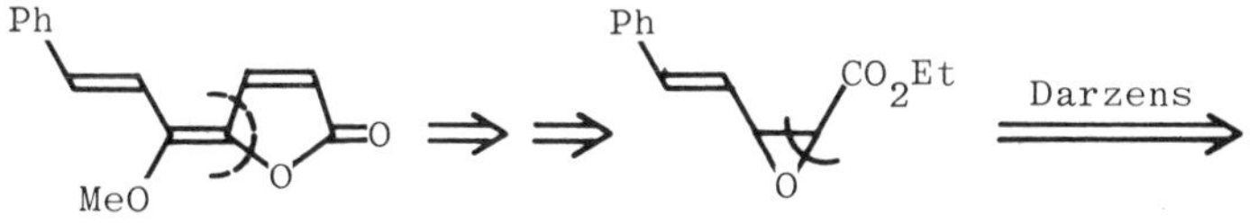

$$(7) + Ph \overset{CHO}{\longrightarrow} \xrightarrow{\alpha, \beta} PhCHO + MeCHO$$

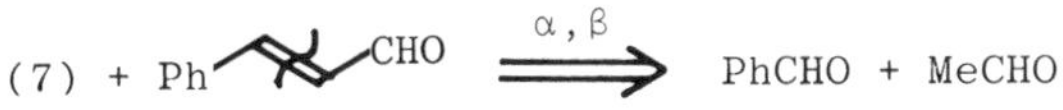

(10)

Synthesis[336]

$$PhCHO + MeCHO \xrightarrow{base} (10) \xrightarrow[base]{(7)} TM(9)$$

This same disconnection of a carbon atom is also helpful for epoxides **(11)** without carbonyl substituents. The reagent should be a nucleophilic carbene equivalent and a sulphur ylid **(12)** is the answer.[337] These can be made from the sulphide **(13)** by a similar process to phosphorus ylid synthesis (Chapter 15), though the reactions of the two ylids with carbonyl compounds are significantly different (there is a third type of reaction with the 'ylid' CH_2N_2 in Chapter 31).

Analysis

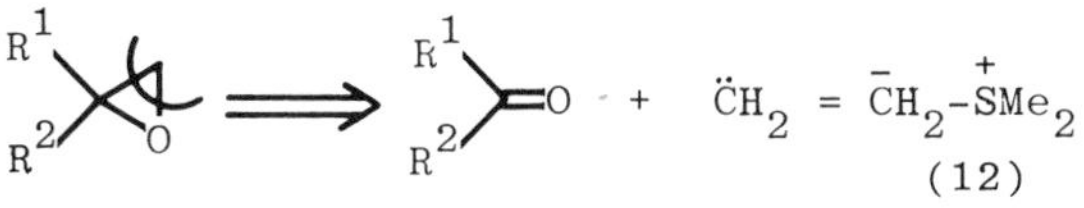

254

$$Me_2S \xrightarrow{MeI} Me_3S^+ \xrightarrow{base} (12) \xrightarrow[R^1CO.R^2]{} TM(11)$$
$$(13)$$

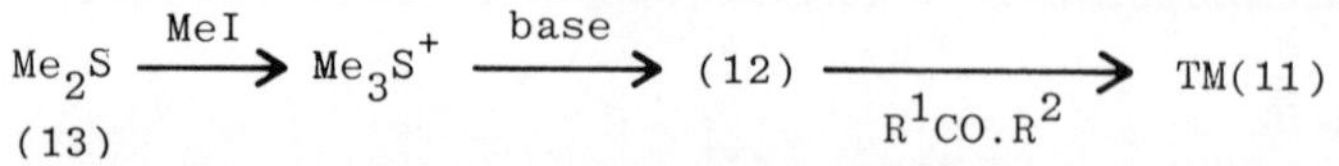

We have already analysed one approach (Chapter 7) to the asthma drug Salbutamol (14). An alternative is the 1,2-diX disconnection to epoxide (15) and 'carbene' disconnection from (15) to aldehyde (16) which can be made from (17) by addition of formaldehyde.

Salbutamol: *Analysis*

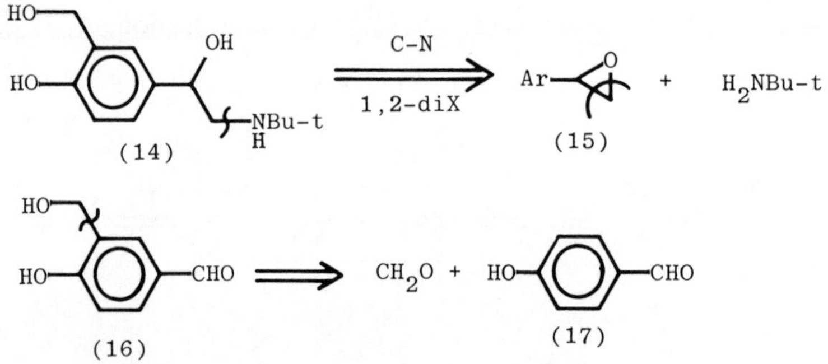

Protection of the two hydroxyl groups in (16) will be needed before the ylid (12) is added: an acetal is the easiest way.

Synthesis[88]

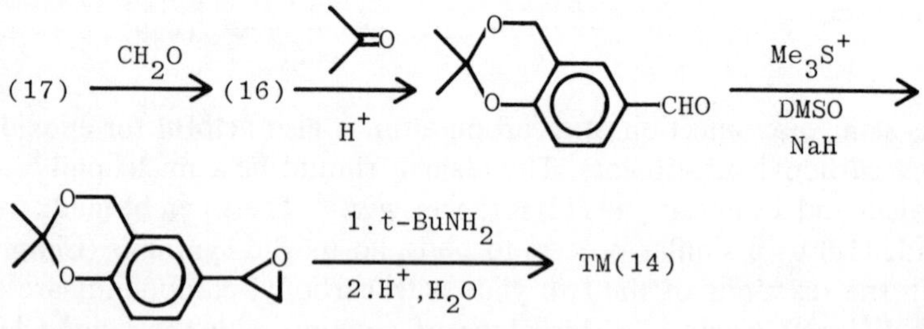

Cyclopropyl ketones[338]

A similar disconnection on cyclopropyl ketones (18) requires a reagent for carbene (19). This can be supplied as the diazoketone (20), made from acid chloride (21) and CH_2N_2, diazomethane. Carbene (19) is probably an intermediate in this reaction when (20) is either photolysed or heated with Ag(I) salts.

Analysis

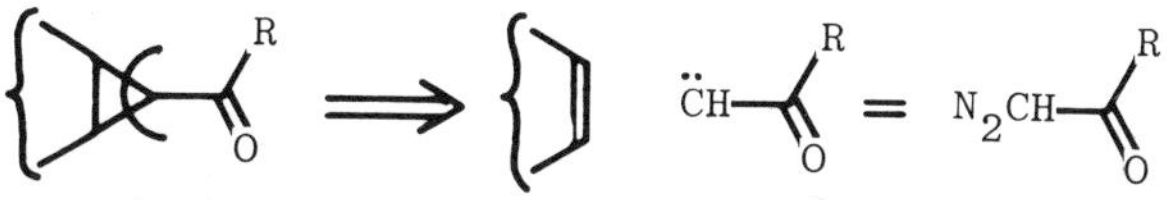

(18) (19) (20)

Synthesis

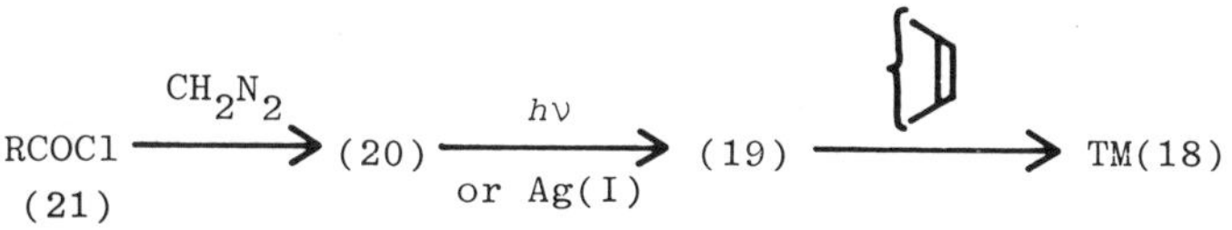

(21)

Tricyclic ketone **(22)** soon becomes a simple problem if we use this disconnection as the starting material is the monocyclic acid **(23)**.

Analysis 1

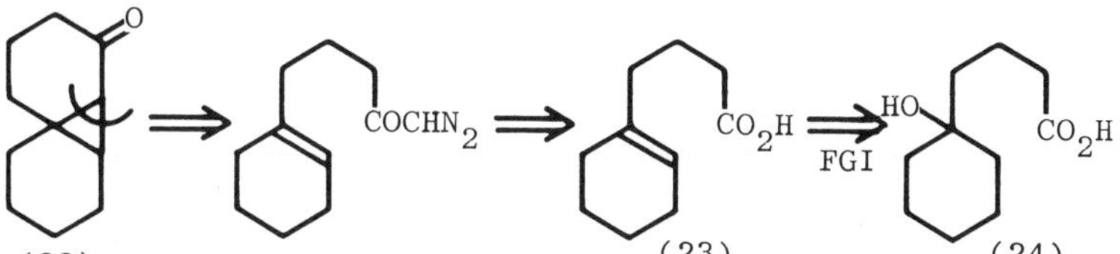

(22) (23) (24)

Acid **(23)** could be made by dehydration of alcohol **(24)**, but this 1,5-difunctionalised compound cannot be made by a Michael reaction because the alcohol is tertiary. Instead a malonate disconnection on **(23)** gives an alkyl halide which comes from **(25)** by FGI. The Reformatsky route gives **(25)**.

Analysis 2

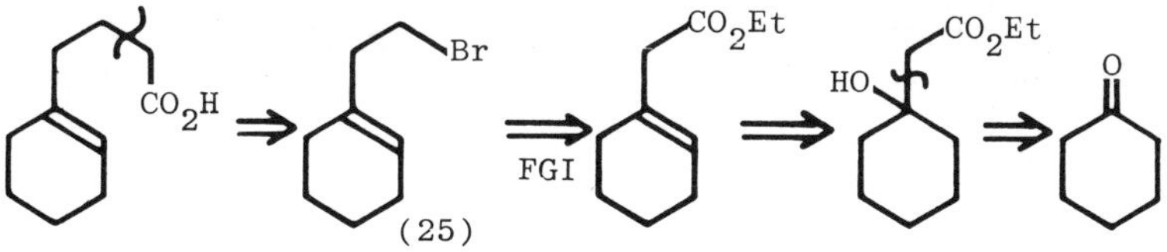

(25)

*See workbook for this chapter.

Synthesis[339]

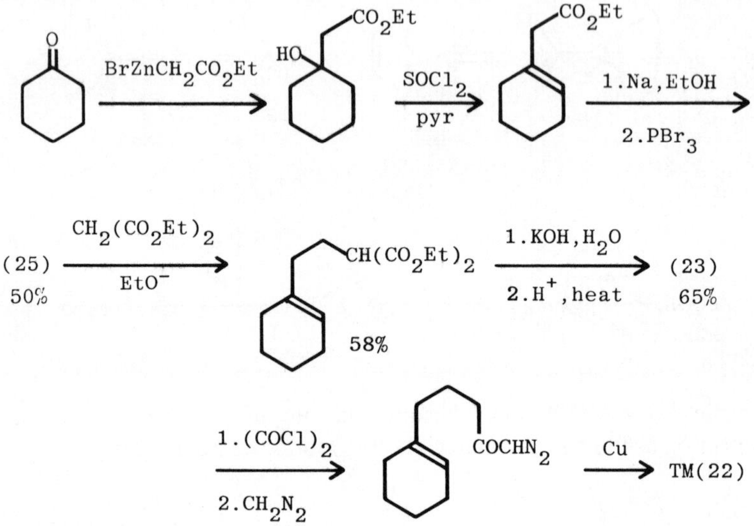

Diazoacetic esters **(27)** are available from diazotisation of glycine esters **(26)** and provide carbenes for three-membered ring synthesis.

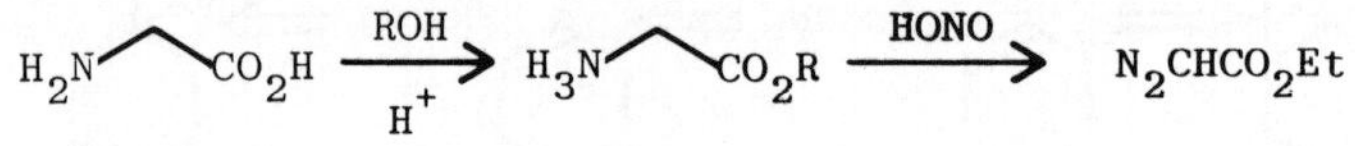

A possible synthesis of three-membered ring compound **(28)** is by intramolecular Friedel–Crafts reaction on acid chloride **(29)** and some chemists[340] wished to investigate this possibility. They first had to synthesise **(29)**. Preliminary FGI to ester **(30)** allows our diazoacetate disconnection to styrene **(31)**.

Analysis

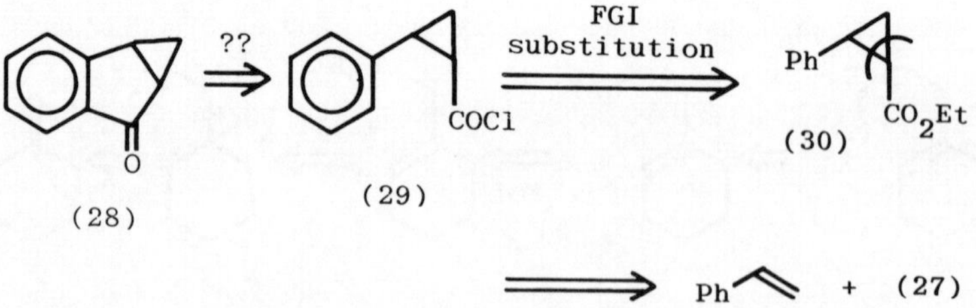

The diazoacetate addition went in excellent yield, acid chloride **(29)** was made this way,[340] and it did indeed cyclise to **(28)**.

Synthesis[340]

$$Ph \diagdown\!\!\!\diagdown \xrightarrow[\substack{77\% \\ 2.SOCl_2}]{N_2CHCO_2Et} (30) \xrightarrow[2.SOCl_2]{1.NaOH} (29) \xrightarrow{AlCl_3} TM(28)$$

Cyclopropanes

All disconnections are the same on cyclopropane, requiring a carbene equivalent which will add to an unactivated double bond. Diazomethane will do this, but one of the best carbene sources is CH_2I_2 with a zinc–copper couple (the Simmons–Smith reaction[341]). This works particularly well on allylic alcohols (31), no doubt because of hydrogen bonding between the OH group and the reagent. The reaction is then totally stereoselective.

Analysis

$$\text{cyclopropane} \Longrightarrow \text{cyclobutane} \quad + \quad \ddot{C}H_2 = CH_2I_2 + Cu/Zn$$

Synthesis

$$(31) \xrightarrow[Cu/Zn]{CH_2I_2}$$

When a study of the thermal rearrangement of ketone (32) was undertaken,[342] synthesis of (32) was completed *via* allylic alcohol (33) in order to ensure efficient cyclisation. Allylic alcohol (33) is best made by regioselective reduction (Chapter 14) of enone (34).

Analysis

$$(32) \overset{FGI}{\Longrightarrow} \overset{carbene}{\Longrightarrow} (33)$$

$$\overset{FGI}{\underset{reduction}{\Longrightarrow}} (34)$$

The Simmons–Smith reaction worked well here[342] and neither alcohol need be isolated.

258

Synthesis[342]

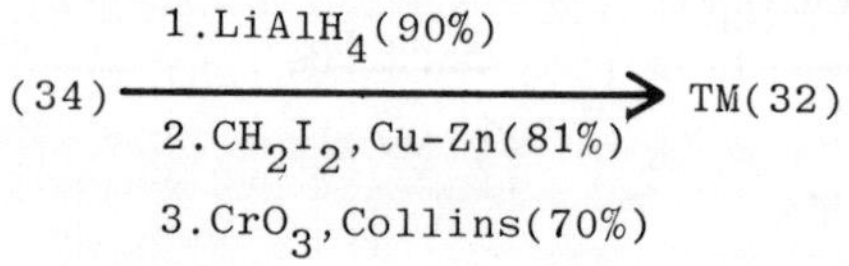

$$(34) \xrightarrow[\begin{array}{c} 2.\text{CH}_2\text{I}_2,\text{Cu-Zn}(81\%) \\ 3.\text{CrO}_3,\text{Collins}(70\%) \end{array}]{1.\text{LiAlH}_4(90\%)} \text{TM}(32)$$

Halocarbenes are relatively easy to use and can be made from haloforms with base or by decarboxylation of trihaloacetate ions.

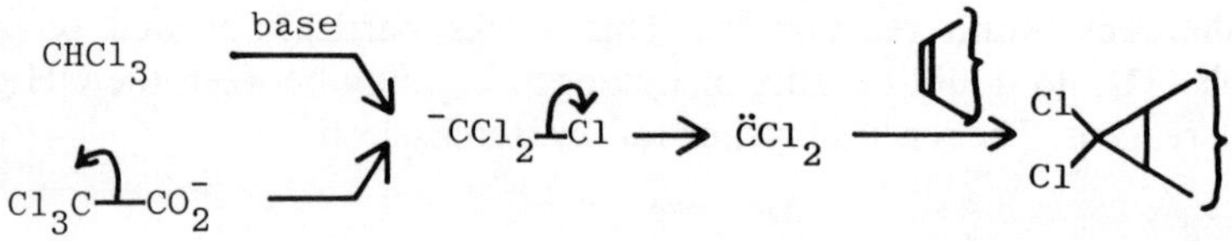

Compound **(35)** was used in a synthesis of the natural product himalchene.[343] Dihalocarbene disconnection reveals **(36)**, easily made from acrolein **(37)** and the glycol **(38)** made from malonate **(39)**.

Analysis

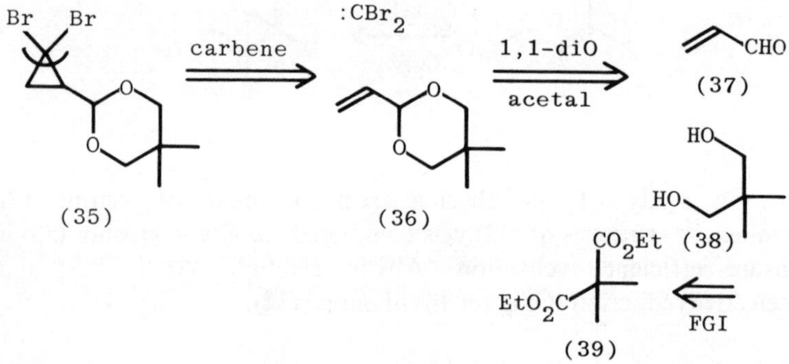

(35)   (36)

CO$_2$Et (38)

(39)

Synthesis[343]

$$\text{CH}_2(\text{CO}_2\text{Et})_2 \xrightarrow[\text{MeI}]{\text{base}} (39) \xrightarrow{\text{LiAlH}_4} (38) \xrightarrow[\substack{\text{TsOH} \\ \text{MgSO}_4}]{(37)}$$

$$(36) \xrightarrow[\text{NaOH}]{\text{CHBr}_3} \text{TM}(35)$$

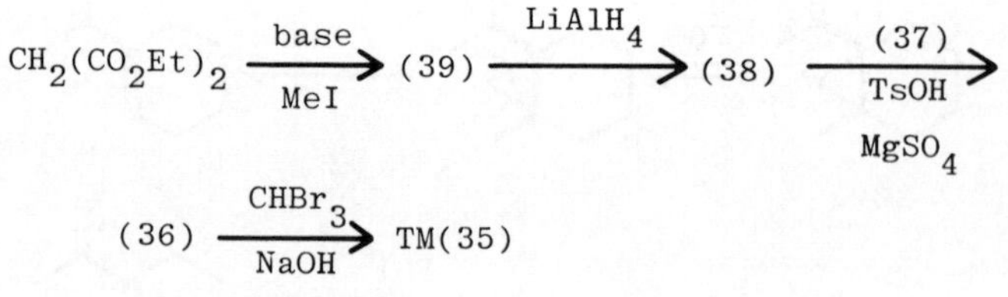

CHAPTER 31

Strategy XIV: Rearrangements in Synthesis

If the carbon framework of a TM is difficult to construct, one strategy is to construct a slightly different framework by conventional reactions and rearrange it to the target molecule. These methods range from simple chain extensions to deep-seated skeletal rearrangements very difficult to analyse.

Diazoalkanes

In the last chapter we met the use of diazoalkanes in the synthesis of three-membered rings. These same diazoalkanes are useful reagents for rearrangement *via* carbenes or carbonium ions.

Chain extension by diazomethane: the Arndt–Eistert procedure

If a diazoalkane (2), made from diazomethane and an acid chloride, is heated or photolysed in the absence of a carbene acceptor, the carbene (3) rearranges to an electrophilic ketene* (4) which captures a nucleophilic solvent, e.g. to give ester (5). The result is that the chain length of the original acid (1) has been increased by one CH_2 group.

$$RCO_2H \longrightarrow RCOCl \xrightarrow{CH_2N_2} \underset{(2)}{R\overset{O}{\overset{\|}{C}}CH=\overset{+}{N}=\overset{-}{N}} \xrightarrow[\text{or } h\nu]{\text{heat, Ag(I)}}$$

$$\underset{(3)}{R\overset{O}{\overset{\|}{C}}\overset{..}{C}H} \longrightarrow \underset{(4)}{R\diagup\diagdown_O} \xrightarrow{\text{MeOH}} \underset{(5)}{RCH_2CO_2Me}$$

(1)

---

*Ketene chemistry is discussed more fully in Chapter 33.

259

This method, known as the Arndt-Eistert procedure,[344] is useful when the relationship between the carbonyl groups in the TM is unhelpful but becomes helpful when the chain length is shorter. We saw an example in Chapter 28 where cyanide ion was used as the chain extension reagent. Diazomethane is a more sophisticated version, needing fewer steps. The disconnection is to remove the carbene.

Analysis

$$R-CH_2-CO_2Me \implies RCO_2H + "CH_2" = CH_2N_2$$

(5a)

In Chapter 27 we analysed the synthesis of bicyclic lactone (6). In his vitamine B_{12} synthesis[305] Eschenmoser needed to lengthen the acetic acid side chain of (6) into the propionic side chain of (7). This he accomplished by the Arndt-Eistert procedure.

$$\text{(6)} \quad \xrightarrow[\substack{2.CH_2N_2 \\ 3.Ag_2O,MeOH}]{1.SOCl_2} \quad \text{(7)}$$

Unsaturated ester (8) is made by dehydration of alcohol (9). The next disconnection should be of bond (a) in (9) but this needs the unnatural synthon (10). A better strategy is to change the 1,4-relationship in (9) into a 1,3-relationship (11) by removing the CH_2 group so that natural synthon (12) can be used.

Analysis 1

(8)   (9a)   (10)

Analysis 2

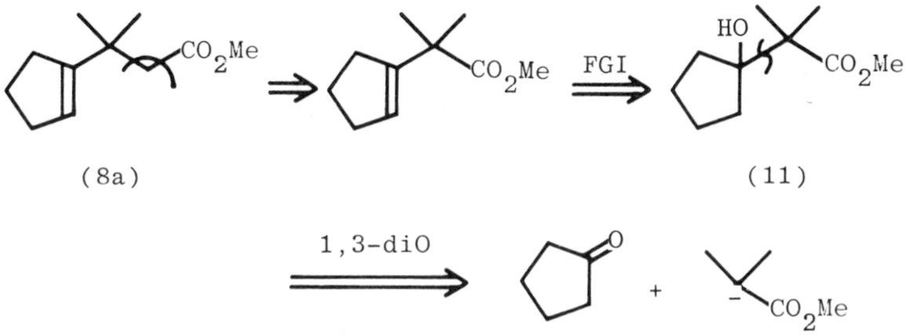

(8a)　　　　　　　　　　　　　　　　　　(11)

(12)

Control for the synthesis of β-hydroxy esters such as **(11)** is easiest by the Reformatsky method (Chapter 20). Dehydration can be carried out under the conditions for acid chloride formation (SOCl$_2$ and pyridine) thus saving a step.

Synthesis[345]

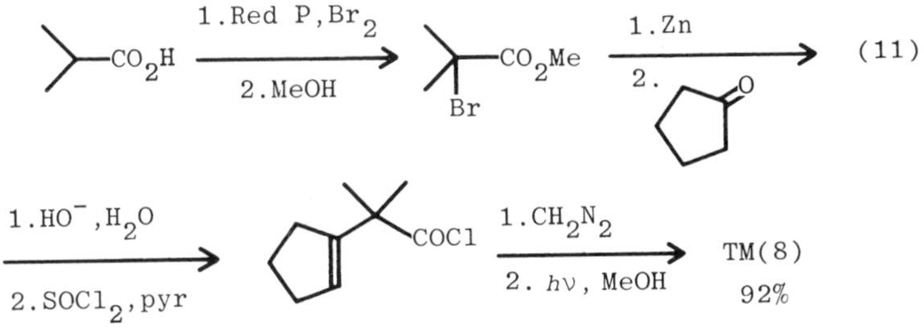

Diazoalkanes and ketones

Direct attack on ketones by diazoalkanes is a useful method of ring expansion, particularly for making seven-membered rings from easily synthesised six-membered rings. Diazoacetic esters (Chapter 30) are especially useful as the cycloheptanone[346] formed is already activated.

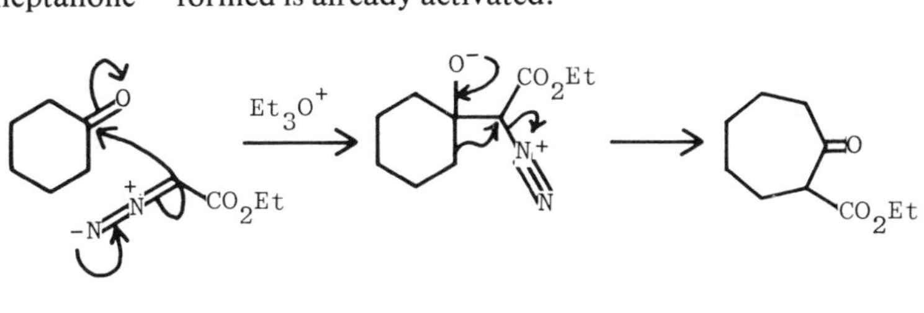

Bicyclic ketone **(13)** has an obvious first 1,3-dicarbonyl disconnection to give 1,6-dicarbonyl compound **(14)**. Reconnection to **(15)** is impossible. One solution is to use chain extension (cf. Chapter 28) but an ingenious alternative is to disconnect a carbene to reveal a 1,5-dicarbonyl compound **(16)** and a simple Michael disconnection.

Analysis

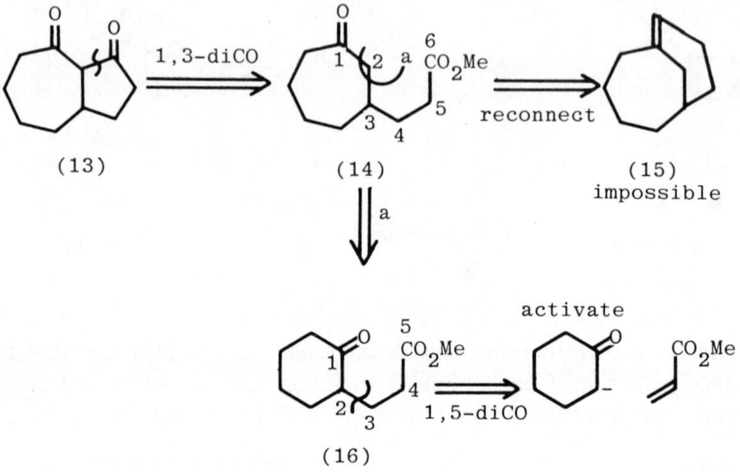

(13)   1,3-diCO   (14)   reconnect   (15) impossible

(16)   1,5-diCO   activate

Activation for the Michael reaction could be by CO_2Et group or enamine formation. Ring expansion of (16) to (14) is unambiguous as only the more substituted side chain migrates, as in the Baeyer–Villiger reaction (Chapter 27). We could simply treat (16) with diazomethane but a better method[346] is to make the reaction intramolecular by converting (16) into the diazoketone (17).

Synthesis[346]

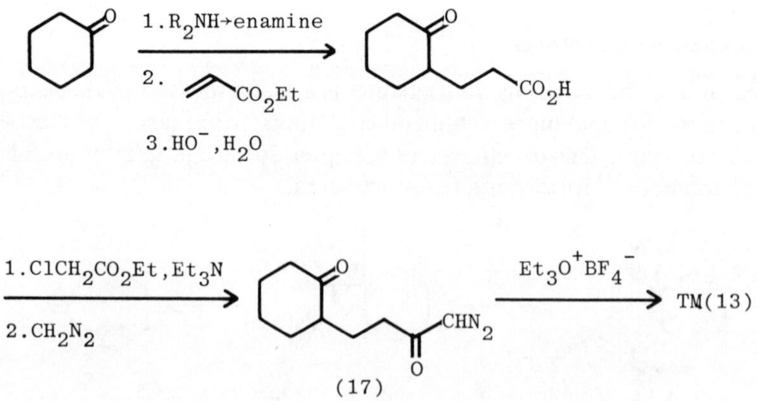

1. $R_2NH \rightarrow$ enamine

2. $\diagup\diagup CO_2Et$

3. HO^-, H_2O

1. $ClCH_2CO_2Et, Et_3N$

2. CH_2N_2

(17)

$Et_3O^+ BF_4^-$ TM(13)

The Pinacol Rearrangement

In the chapter on radical reactions (Chapter 24) we saw how to make 'pinacols', e.g. (18), by reductive dimerisation of ketones. These pinacols rearrange in acid to give *t*-alkyl ketones (19).

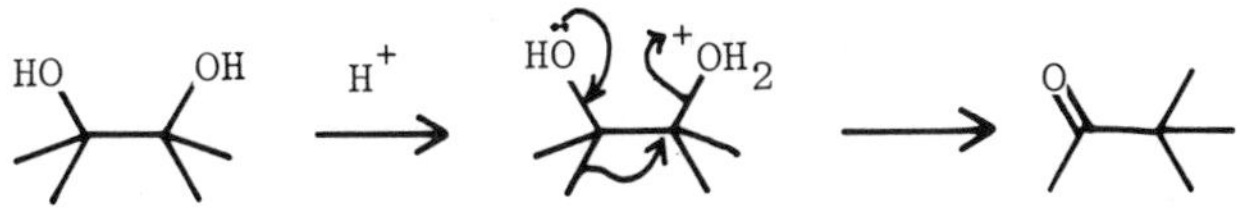

(18)　　　　　　　　　　　　　　　　　　　　(19)

Though restricted by the need for symmetry, this is a useful approach to *t*-alkyl ketones which are otherwise difficult to make.[347] The crowded alkenes[348] **(20)** must be made from alcohols **(21)** and hence from ketone **(22)** by a Grignard reaction. Ketone **(22)** has a *t*-alkyl group on one side of the carbonyl group and can be made from the symmetrical pinacol **(23)**. The easiest way to see this is to draw the rearrangement in reverse **(24)**. The alternative starting material **(25)** could not be made by pinacol reduction but could be made by hydroxylation of a partly reduced naphthalene (Chapter 36).

Analysis

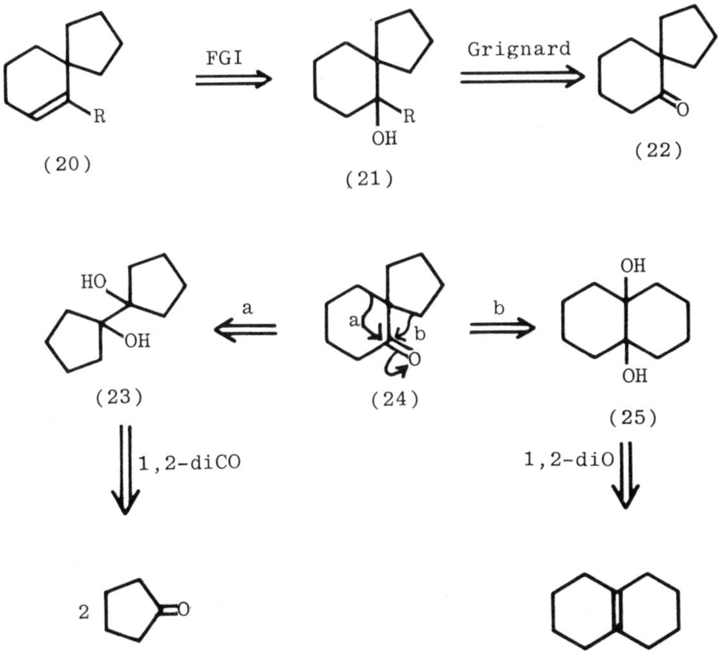

Synthesis[348]

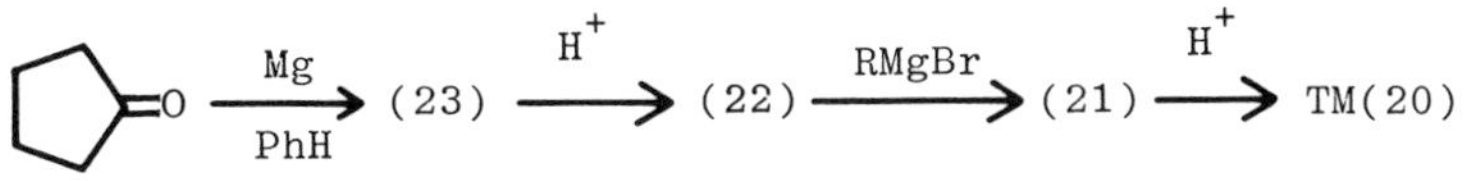

Epoxide Rearrangements

Epoxide rearrangements are closely related to the pinacol rearrangement but allow a more general synthesis of carbonyl compounds.[349] On treatment with acids or Lewis acids, even such weak ones as LiBr or $MgBr_2$, epoxides (26) open to give the more stable carbonium ion (27) which rearranges to a carbonyl compound (28). The order for migration is usually: H > Aryl > t-Alkyl > s-Alkyl > p-Alkyl.

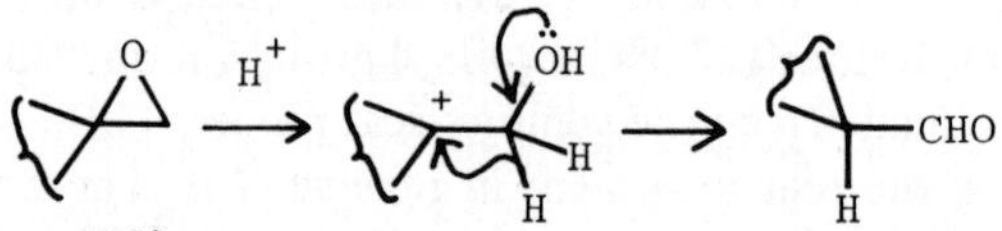

(26) (27) (28)

When the epoxide is made from a carbonyl compound, e.g. (29) (Chapter 30), the result is chain extension to the homologous aldehyde. This procedure can be carried out with sulphur ylids[350] (30) or by the Darzens reaction with decarboxylation of the intermediate acid (31).

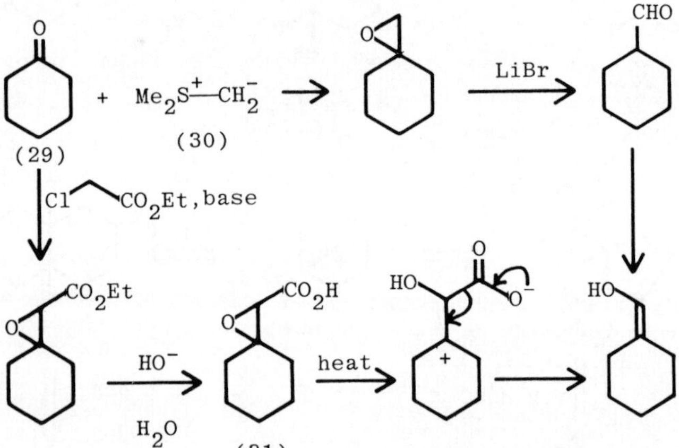

The 'pungent floral' perfumery compound (32) can be disconnected back to the aromatic aldehyde (33) by one of these methods. The Darzens route has been used successfully.

Analysis

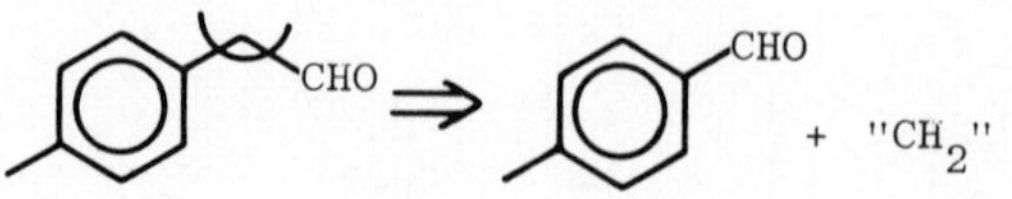

(32) (33)

Synthesis[351]

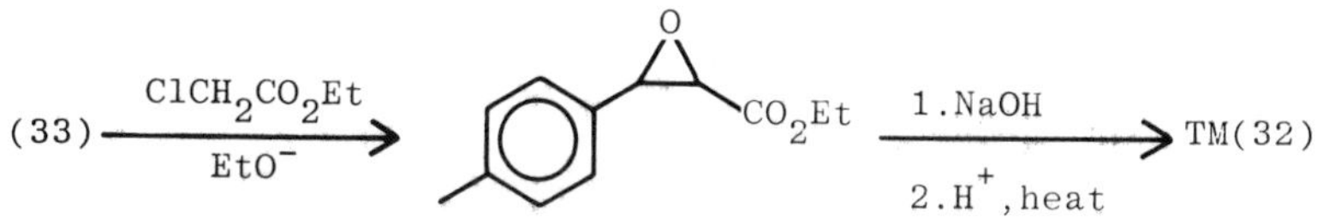

$$(33) \xrightarrow[\text{EtO}^-]{\text{ClCH}_2\text{CO}_2\text{Et}} \text{[epoxy ester intermediate]} \xrightarrow[\text{2.H}^+,\text{heat}]{\text{1.NaOH}} \text{TM(32)}$$

Anionic Rearrangements

The Faworskii rearrangement[352]

This rearrangement converts an α-halo ketone, e.g. (34), into an unstable cyclopropanone (35) which decomposes (36) into an ester (37). In the decomposition (36) the more stable carbanion acts as the leaving group.

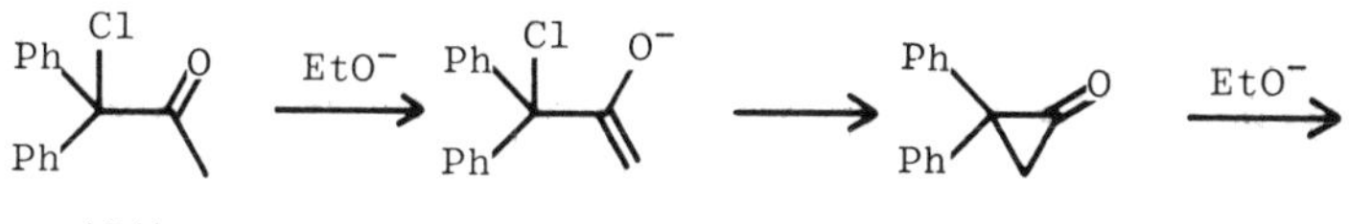

(34)   (35)

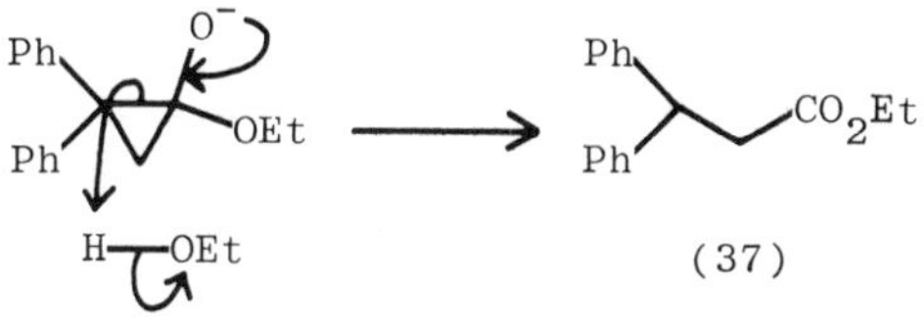

(36)   (37)

The 'disconnection' is again best seen by reversing the rearrangement. Any of the three alkyl groups on the quaternary centre • of (38a) could be moved back to the carbonyl group—moving the largest is usually the best strategy. The rest of the analysis is routine.

Analysis

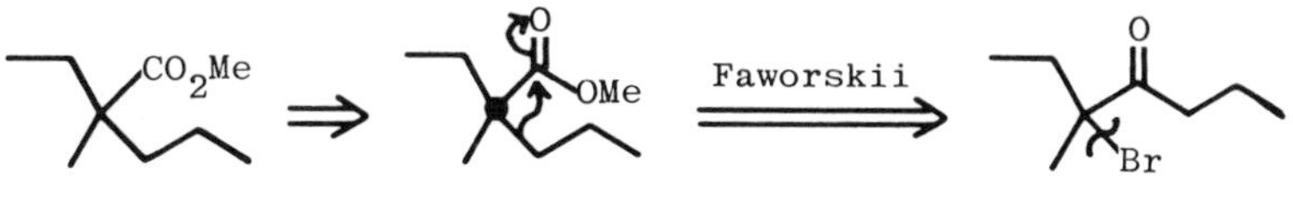

(38)   (38a)   Faworskii

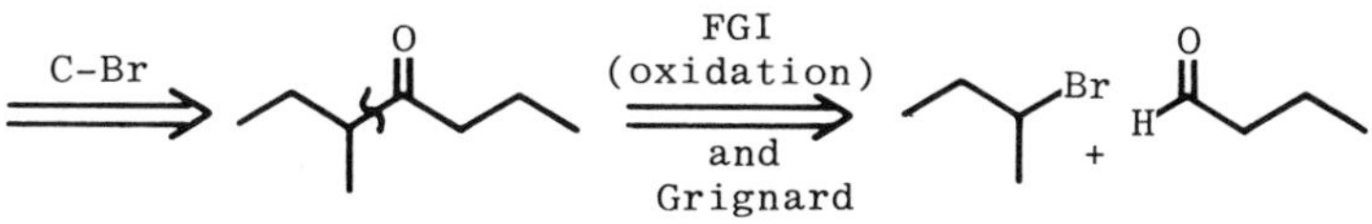

(39)

Bromination of (39) in acid solution occurs mostly on the more substituted side (Chapter 20) and the rest of the synthesis is straightforward.

Synthesis[352]

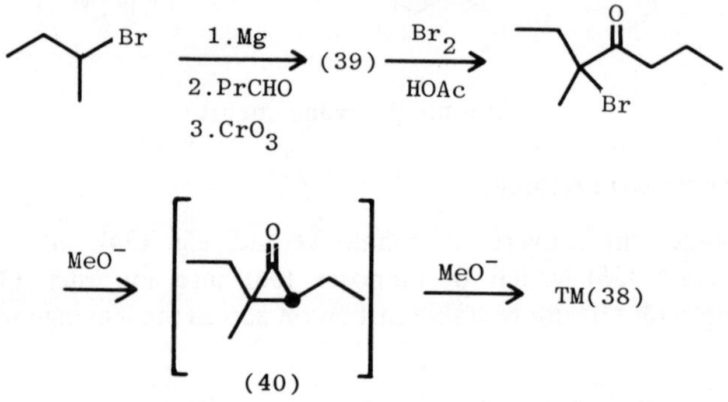

Opening of (40) with MeO⁻ is unambiguous because only the more stable carbanion, that is the one with *fewer* alkyl groups (• in 40) acts as a leaving group. This means that the Faworskii is useful for making acids with *t*-alkyl groups next to the carbonyl group—a close parallel to the pinacol rearrangement.

The spasmolytic drug component (41) is just such an acid. Symmetry suggests moving the cyclohexyl group when reversing the Faworskii so that halogenation of the symmetrical ketone (42) as well as opening of the cyclopropanone intermediate (43) will be unambiguous.

Analysis

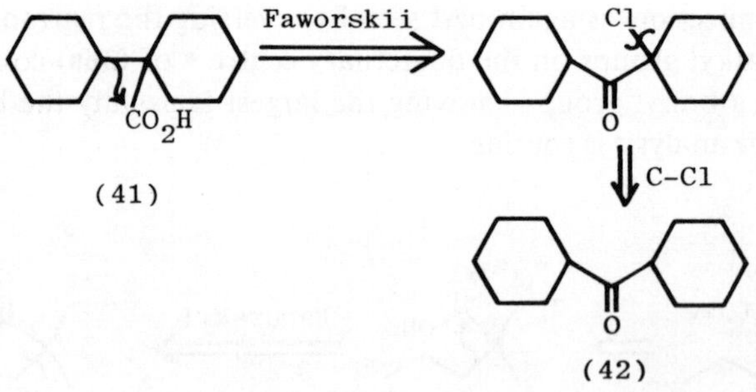

Synthesis[353]

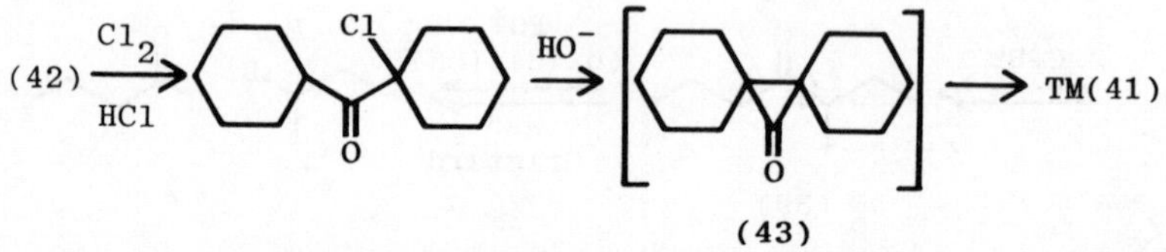

Summary

Rearrangements are difficult disconnections to see when looking at a target molecule. Two guidelines may help you:

1. compounds with a carbonyl group one atom away from a position in which it would be helpful, and
2. compounds with a *t*-alkyl group next to the carbonyl group may be synthesised by rearrangements.

Faced with one of these types of TM, you must see if you can make it by any of the rearrangements given in this chapter. Only experience will give you the confidence to apply rearrangements to unfamiliar types of molecule and we shall be building that experience in the remaining chapters.

CHAPTER 32

Four-Membered Rings: Photochemistry in Synthesis

In our analysis of the synthesis of rings of various sizes, we concluded (Chapter 29) that four-membered rings are uniquely difficult. For this reason, a special method, the photochemical 2 + 2 cycloaddition[354] is often used to make four-membered rings. Some 2 + 2 thermal cycloadditions, particularly of ketenes (Chapter 33), and some ionic reactions (page 272) are also useful.

Photochemical 2 + 2 Cycloadditions

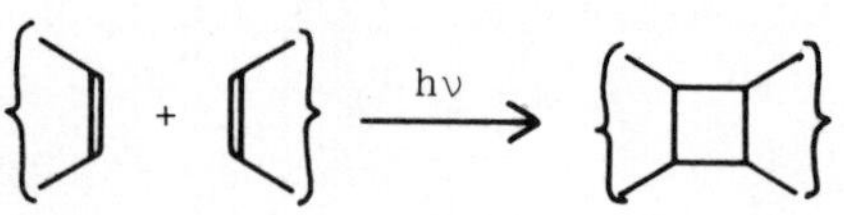

Cyclobutane formation from alkenes does not occur thermally. It is allowed by the Woodward–Hoffmann rules in the excited state and so is a photochemical reaction.[355] The reaction occurs cleanly when one alkene is conjugated, e.g. as an enone, and so absorbs u.v. light.

Disconnection of (1) by reversing the 2 + 2 cycloaddition gives ethylene and enone (2) as starting materials. Enone (2) can be made by conventional methods from three molecules of acetone.[356] In the cycloaddition,[357] (2) absorbs light and its excited state adds to ethylene to give (1).

Analysis

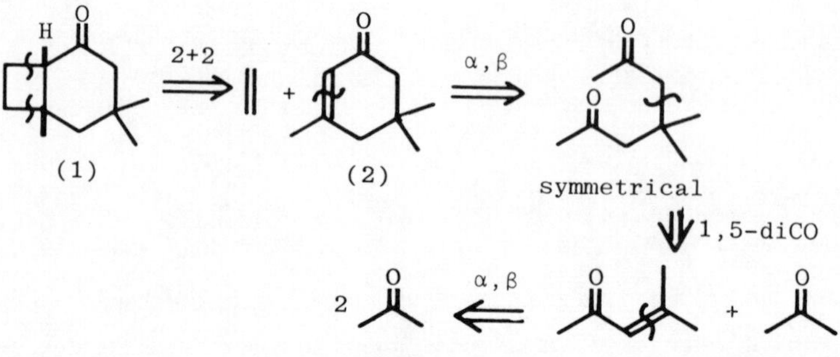

268

Synthesis[356, 357]

$$
\text{(acetone)} \xrightarrow[\text{62\%}]{\text{Al}_2\text{O}_3} \text{(2)} \xrightarrow[\text{h}\nu]{\text{CH}_2=\text{CH}_2} \text{TM(1)}
$$

The stereochemistry of the reactants is usually reproduced in the products, e.g. **(2)** → **(1)**, as in the Diels–Alder reaction (Chapter 17), though there is often little choice since a four-membered ring cannot be fused *trans* to a five-membered ring and only with difficulty to a six-membered ring. There is no 'endo rule' (cf. Chapter 17), the two components, e.g. **(3)** and **(4)**, coming together in the way that gives least steric hindrance. Thus[358] **(5)** has *cis* ring junctions A/B and B/C but a *trans* relationship between rings A and C.

$$(3) \qquad (4) \qquad\qquad (5)$$

Most cyclobutanes offer a choice of 2 + 2 disconnections and the choice can often be made by considering the availability or ease of synthesis of the starting materials. Hence the alternative disconnections (a) and (b) on **(6)** give two enals **(7)** and **(9)**, each in turn disconnected at the α,β bond to **(8)** and **(10)**. Ketone **(10)** is available—its synthesis was discussed in Chapter 1—so we prefer that route.

Analysis

$$(6a) \qquad\qquad (7) \qquad\qquad (8)$$

$$(6b) \qquad\qquad (9) \qquad\qquad (10)$$

Synthesis[359]

$$
(10) \xrightarrow[\text{base}]{\text{excess MeCHO}} (9) \xrightarrow{\text{h}\nu} \text{TM(6)}
$$

The intramolecular nature of this cycloaddition helps to make it a good reaction, an argument in favour of disconnections like (6a) and (6b). On the other hand, disconnections like (5) ⇒ (3) + (4) allow us to find simple starting materials more quickly and these two contrary considerations require a balanced judgement.

Disconnection of (12), an intermediate in Brown's synthesis[360] of bourbonene (11), could give (14) and maleic ester (13) or the single starting material (15). There is no obvious way to make (14), but further disconnections on (15) quickly take us back to simple starting materials, so that again in this case the intramolecular route is preferred.

Analysis

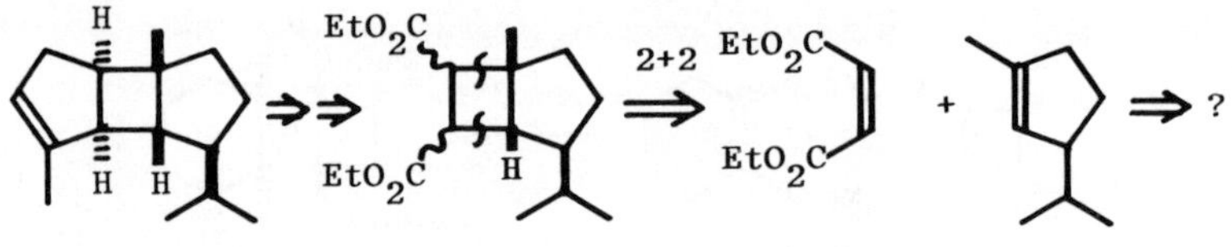

(11)   (12a)   (13)   (14)

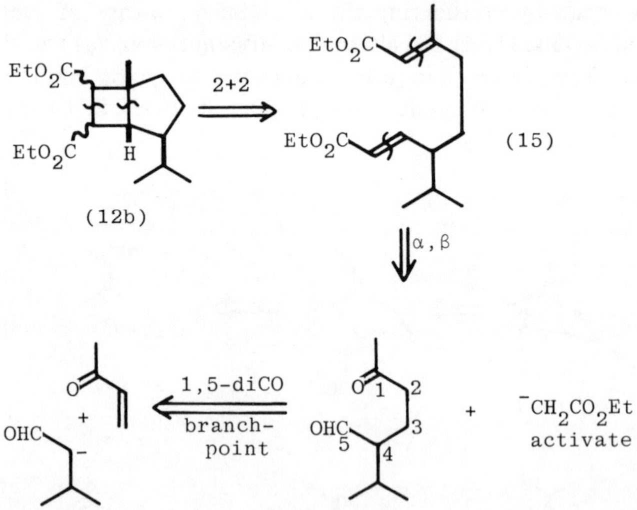

(12b)   (15)

(16)

The Michael reaction can be controlled by making the enamine of the aldehyde and the Wittig method is suitable for the conversion of (16) to (15). In the 2 + 2 cycloaddition, the isopropyl group will prefer to be *trans* to the new four-membered ring in (12) so the stereochemistry will be correct.

Synthesis[360]

OHC $\diagdown$ ... 1.R_2NH,H^+ / 2. (propenone) $\xrightarrow{}$ (16) 79% $\xrightarrow{(EtO)_2P(O)\diagup CO_2Et \text{, base}}$ (15) 50% $\xrightarrow{h\nu}$ TM(12) 53%

3. $(CO_2H)_2$

Regioselectivity of photochemical 2 + 2 cycloadditions

An explanation is beyond the scope of this book,* but the rule of thumb is to *reverse* what would be expected from an ionic reaction. Alkenes **(17)** and **(18)** each have one electrophilic end, marked (+). In an ionic reaction these would avoid each other, but in the excited state the polarity is reversed, so that when **(18)** absorbs the light, it reverses its polarity to **(19)** and the orientation of the product[361] is **(20)**. This could have been written straight away by joining the two electrophilic sites according to the rule of thumb.

(17) — $(+)$ (18) $\xrightarrow{h\nu}$ $(-)$ (19) $\xrightarrow{(17)}$ (20)

Disconnection of **(21)** suggests **(22)** and **(23)** as starting materials. The orientation is correct as the two electrophilic atoms (+) should join together. The stereochemistry is also correct (page 269).

Analysis

(21) $\Longrightarrow$ (22) $(+)$ CO_2Me + (23) $(+)$, via 2+2

Synthesis[362]

$$(22) + (23) \xrightarrow{h\nu} (21)$$

*See Fleming, *Orbitals*, p.219.

Intramolecular reactions do not necessarily follow this rule because it may be impossible for the starting material to contort itself into the required orientation. Thus **(24)** should have a preference for cyclisation to **(25)**; instead the sterically more relaxed **(26)** is the major product.[363]

(25)

(24)

hν

(26)

Four-Membered Rings by Ionic Reactions

Cyclobutene **(3)** was used on page 269 in a photochemical cycloaddition. It is made[364] by ionic reactions from readily available adipic acid **(27)** (see Chapter 27). This is something of an exception and cyclisation by carbonyl condensations is not normally a recommended route to four-membered rings (cf. Chapter 29).

(27)

1. $SOCl_2$

2. Br_2

3. MeOH

NaH
DMF

(3)
70%

One strategy for four-membered ring synthesis would be ring expansion or contraction. Contraction of five- to four-membered rings is rarely successful because of the strain that would be introduced,[365] but three-membered rings are already strained, and are easy to make (Chapter 30), so ring expansion makes good sense. Cyclobutanones such as **(28)** have the substitution pattern of rearrangement products (Chapter 31) and might be made from diol **(29)** or epoxide **(30)**.

Analysis 1

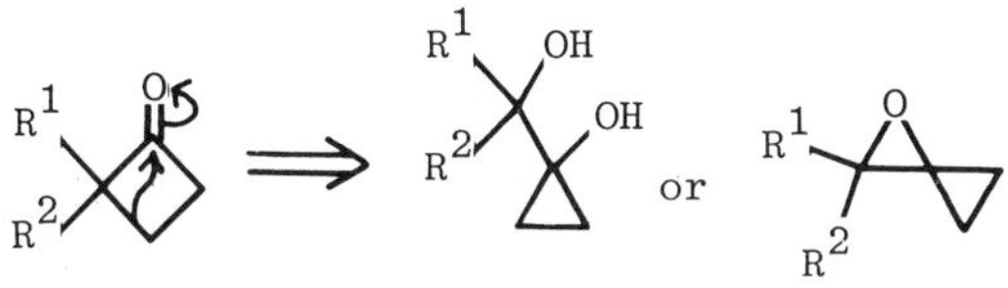

(28)   (29)   (30)

Diol **(29)** is unsymmetrical and not very promising (see Chapter 31) but epoxide **(30)** can be disconnected to ketone **(31)** and sulphur ylid **(32)** (cf. Chapter 30). This approach has been thoroughly investigated by Trost[366] and developed into an important synthetic method.

Analysis 2

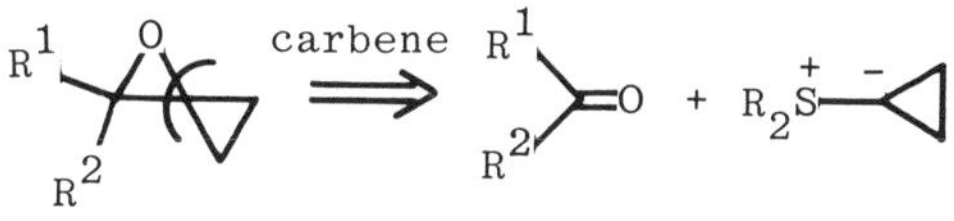

(30a)   (31)   (32)

Synthesis[366]

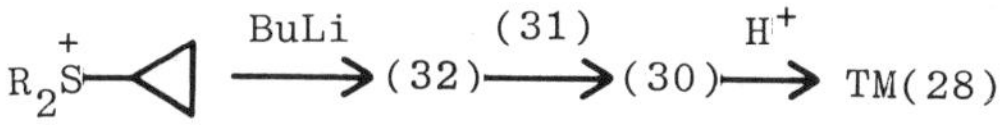

$$R_2\overset{+}{S}\!\!-\!\!\triangleleft \xrightarrow{\text{BuLi}} (32) \xrightarrow{(31)} (30) \xrightarrow{H^+} \text{TM}(28)$$

Cyclobutanone **(33)** would then be made from epoxide **(34)** and hence from aldehyde **(35)**, an obvious Diels–Alder product.

Analysis

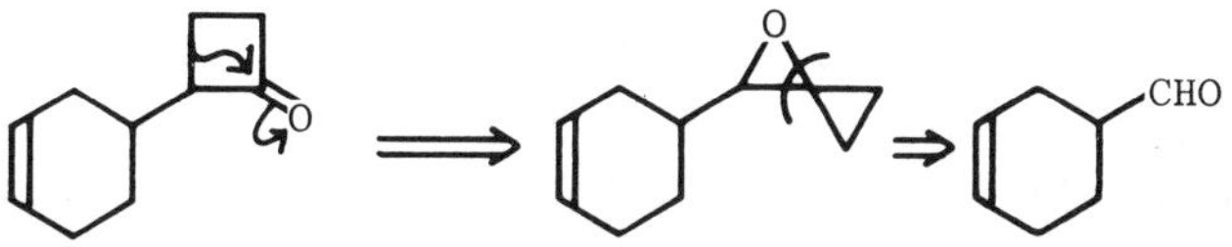

(33)   (34)   (35)

Synthesis[366]

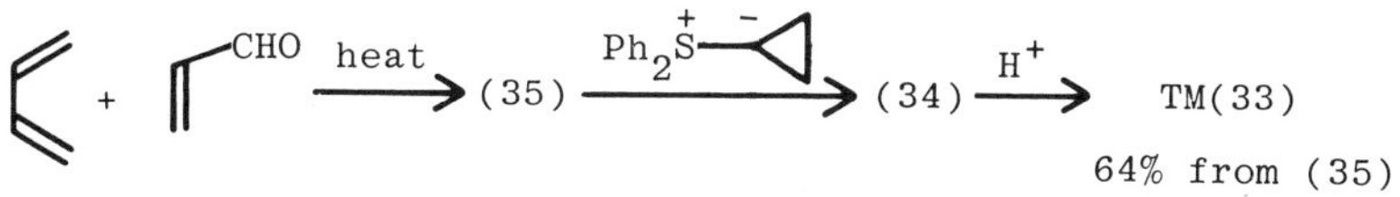

64% from (35)

Other examples based on simpler rearrangements are known.[367, 368]

CHAPTER 33

Strategy XV: Use of Ketenes in Synthesis

Summary of Ketene Chemistry

We have met ketenes (1) already as intermediates in the Arndt–Eistert procedure (Chapter 31) for chain elongation. They are highly electrophilic at the curious sp carbon atom (* in 1). Nucleophiles give acyl derivatives (3) *via* enolate (2). Ketenes are unstable and cannot be stored. In the absence of nucleophiles they dimerise: ketene itself gives ester (4) (which is available commercially) and disubstituted ketenes give cyclobutadiones (5). Monosubstituted ketenes may give either type of product.

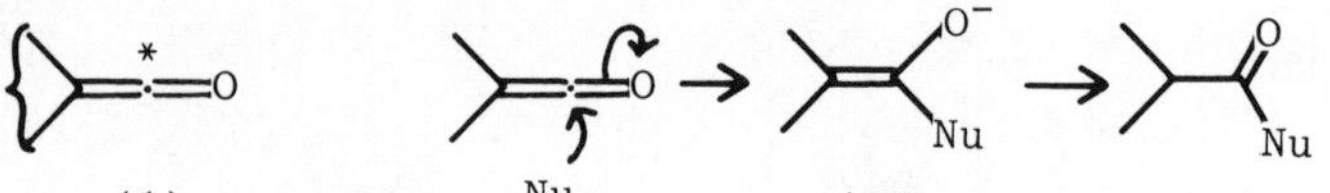

$$CH_2=C=O \longrightarrow \quad (4)$$

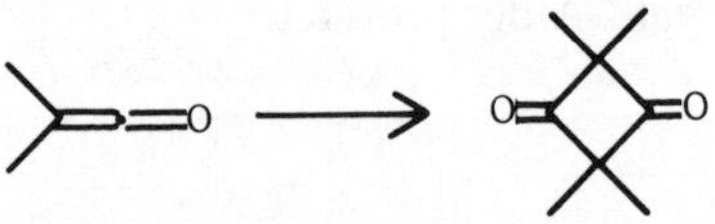

Ketenes are usually prepared *in situ* by elimination from acid chlorides with a tertiary amine. Thus (6) gives dimethyl ketene. If no other reagent is present, dimer (5) is formed. If a nucleophile is present, product (3) is formed, whilst thermal 2 + 2 cycloadditions take place with alkenes to give cyclobutanones (7).*

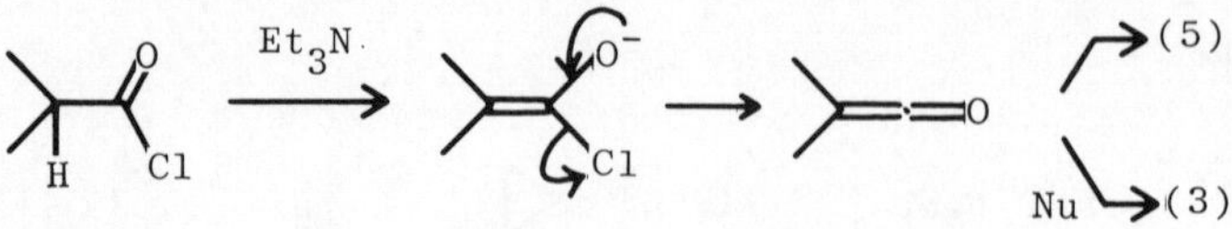

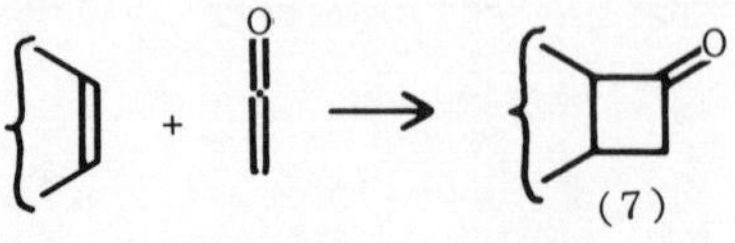

*See Fleming, *Orbitals*, p.143 for an explanation.

274

2 + 2 Thermal Cycloadditions

The disconnection for these thermal cycloadditions is **(8)**, the same as the disconnection for the photochemical cycloadditions in Chapter 32. An important difference is that these thermal reactions show ionic type regio-selectivity—in the synthesis of **(8)** the nucleophilic atom (–) in **(9)** combines with the electrophilic atom (+) of dichloroketene **(10)**.*

Analysis

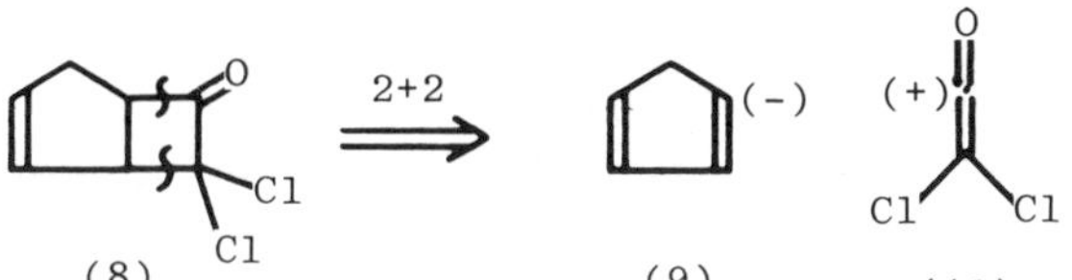

Synthesis[369]

$$Cl_2CH.COCl \xrightarrow{Et_3N} (10) \xrightarrow{(9)} (8)$$

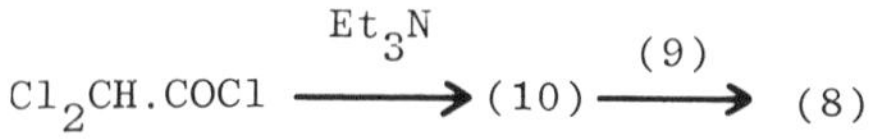

Cyclopropyl aldehyde **(11)** has a tertiary alkyl group next to the carbonyl group and might be made by contraction of a four-membered ring (Chapter 31) such as **(12)**, where X is a leaving group. If we put X = Cl, we could make **(12)** by reduction of an α-chloro cyclobutanone **(13)**, a ketene cycloadduct.

Analysis

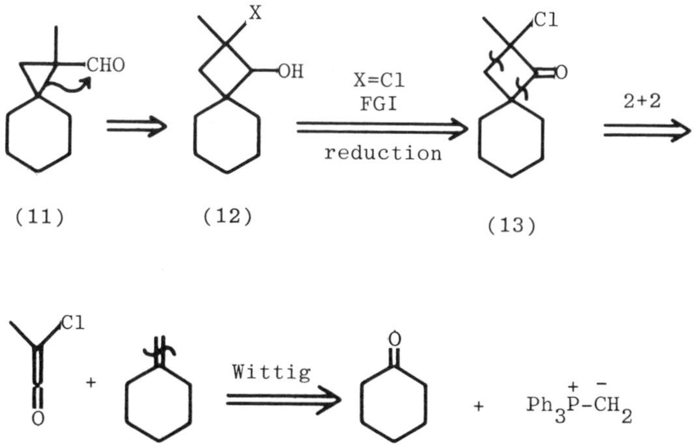

---

*These signs (+) and (–) indicate the natural polarity of the molecule and not that it carries a charge.

The synthesis of **(14)** by the Wittig reaction was discussed in Chapter 15. The ketene can be generated from the acid chloride and the orientation of the cycloaddition is correct.

Synthesis[370]

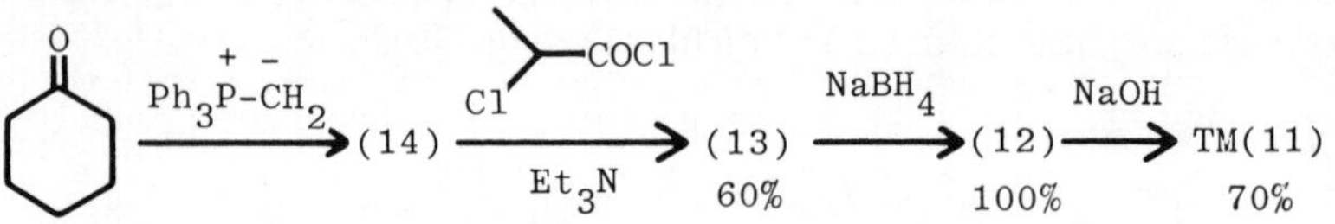

Lactones **(15)** can be made by Baeyer–Villiger rearrangement (Chapter 27) of cyclobutanones. The more highly substituted group migrates with retention of configuration.

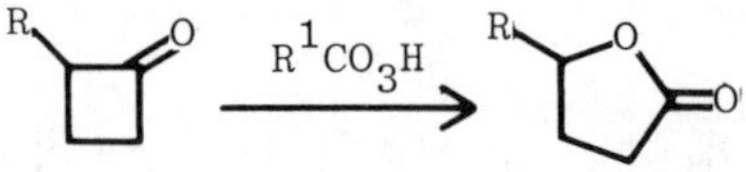

$$(15)$$

Lactone **(16)** is an important intermediate in the synthesis of prostaglandins. Reversing the Baeyer–Villiger rearrangement gives cyclobutanone **(17)** which is the adduct from ketene and cyclopentadiene. In practice, dichloroketene is much easier to make and handle than ketene itself so compound **(8)** is used as an intermediate, the chlorine atoms being removed with zinc.[371]

Analysis

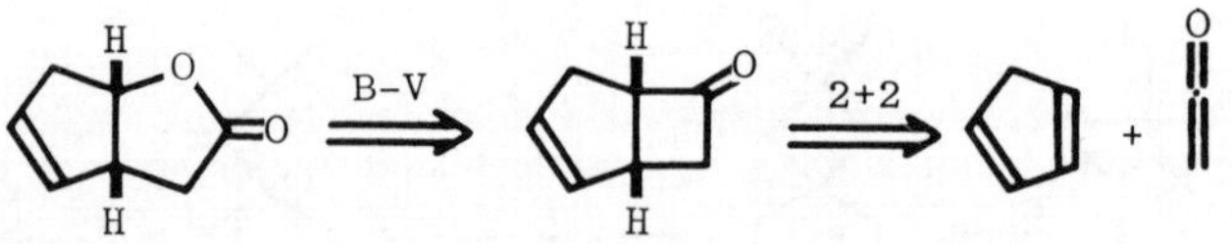

Synthesis[371]

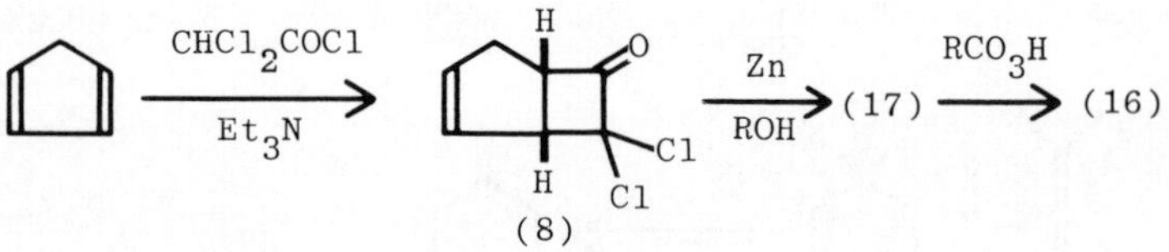

Ketene Dimers

TM**(18)** is clearly a ketene dimer and the disconnection is simply to separate the two molecules **(19)**. The *para* orientation of acid and methyl groups in **(20)** suggests a Diels–Alder reaction after FGA.

Analysis

$$(18) \quad \xrightarrow{2+2} \quad (19) \quad \Longrightarrow$$

$$(20) \quad \xrightarrow{FGA} \quad (21) \quad \xrightarrow{D-A} \quad + \quad$$

Synthesis[372]

$$+ \quad \xrightarrow{} (21) \xrightarrow{H_2, Pd} (20) \xrightarrow[2.Et_3N]{1.SOCl_2} TM(18)$$

Ketene dimer itself **(4)**, being an enol ester, reacts with nucleophiles to give acetoacetyl derivatives **(22)**. The dimer **(4)** is therefore a reagent for synthon **(23)**.

$$(4) \quad \rightarrow \quad (22) \quad \rightarrow \quad (23)$$

Heterocyclic compound **(24)**, an intermediate in a cytochalasan synthesis[373] can be made from **(25)** by a 1,3-dicarbonyl route. Amide **(25)** is an acetoacetyl derivative of the naturally occurring amino acid phenylalanine **(26)** so this is a good route to follow.

Analysis

$$(24) \quad \xrightarrow{1,3-diCO} \quad (25) \quad \Longrightarrow \quad + \quad (23)$$

Synthesis[373]

$$(26) \quad \xrightarrow[2.(4)]{1.EtOH, H^+} (25) \xrightarrow[MeOH]{MeO^-} TM(24)$$

CHAPTER 34

Five-Membered Rings

Unlike three-, four-, or six-membered rings, five-membered rings are usually made by standard carbonyl chemistry. This is partly because work on cyclopentannelation (see later in this chapter) is only just beginning, but also because five-membered rings are the easiest to make by conventional chemistry since the five-membered ring has marked kinetic and thermodynamic advantages over open chains (Chapter 29). This chapter contains a selection of conventional approaches and the next contains the special methods. In some respects, this chapter is revision.

From 1,4-Dicarbonyl Compounds

Cyclopentenones (1) disconnect to 1,4-dicarbonyl compounds (2) and any of the methods used in Chapter 25 are potential routes to cyclopentenones (see Chapters 25, 26, and 28 for examples). In the synthesis of any cyclopentenone worth discussing, the question of control (Chapter 20) is usually dominant in choosing the method.

Cyclopentenone (3) has been used[374] in photochemical syntheses of four-membered rings. Disconnection gives keto aldehyde (4) and the branch point disconnection (a) gives aldehyde (5) and synthon (6) for which we might consider α-bromacetone (Chapter 25), allylic bromide (7) (Chapter 26), or propargyl bromide (8) (Chapter 26), using in each case the enamine of aldehyde (5) to ensure control. One successful synthesis is given.

Analysis

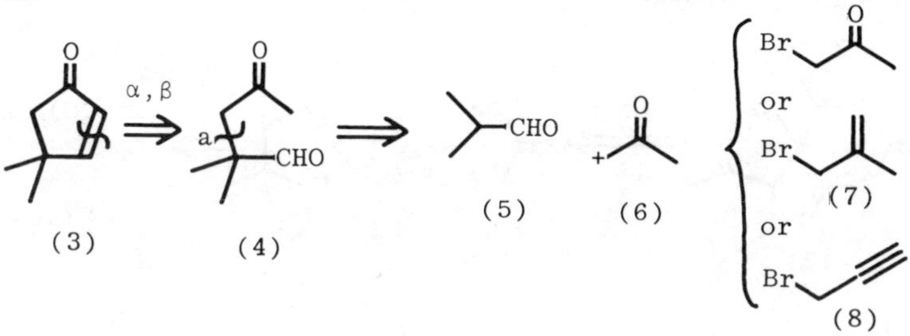

(3)   (4)   (5)   (6)   (7)   (8)

Synthesis[374, 375]

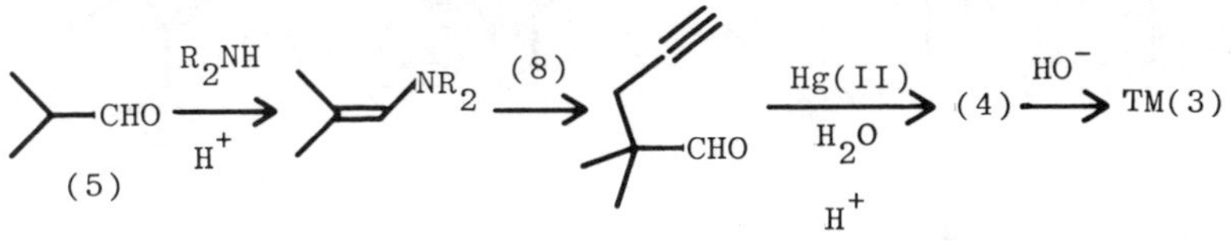

This approach to cyclopentenones, which adds a five-membered ring to a carbonyl compound, is reminiscent of Robinson annelation and is often called cyclopentannelation (see Chapter 26). Research is active in this area at the moment[376] and many new cyclopentenone syntheses as well as new reagents for synthon (6) appear each year. TM(3) has been made in five ways[374, 375] and TM(9) in at least seventeen (1980 figures).[376, 377]

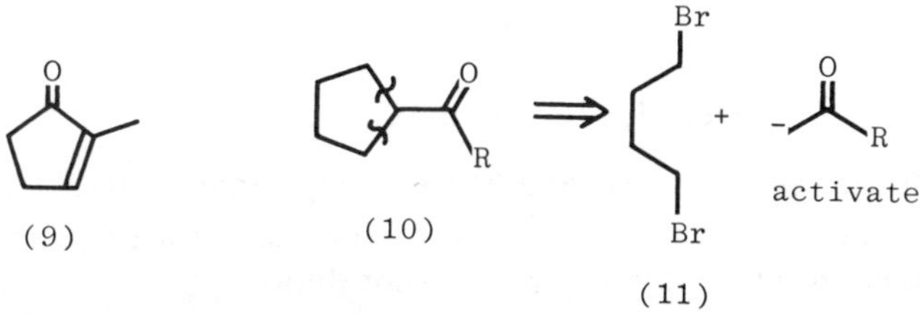

(9)   (10)   (11)

Another route to five-membered rings from 1,4-dicarbonyl compounds is summarised in the disconnection of (10) where the alkylating agent is derived from a dicarbonyl compound which may be available (see Table 25.1) or may have to be synthesised.

The barbiturate (12) can be made this way.[378] Recognising thiourea in (12), we disconnect first to diester (13). The activating group is already present in this intermediate (CO$_2$Et in 13) but the alkylating agent (14) must be synthesised from diketone (15) by FGI. Disconnection of (15) is guided by symmetry.

Spirothiobarbital: *Analysis*

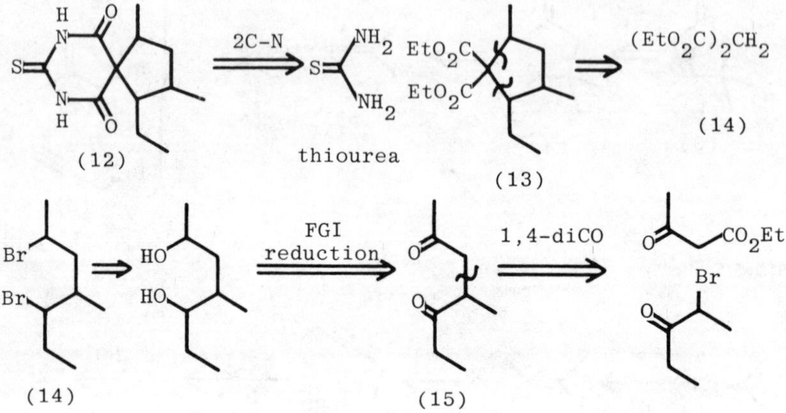

(12) thiourea (13) (14)

(14) (15)

Synthesis[378]

$$\xrightarrow[H_2O]{H^+} (15) \xrightarrow[2.HBr]{1.NaBH_4} (14) \xrightarrow[base]{CH_2(CO_2Et)_2} (13) \xrightarrow[base]{(NH_2)_2CS} TM(12)$$

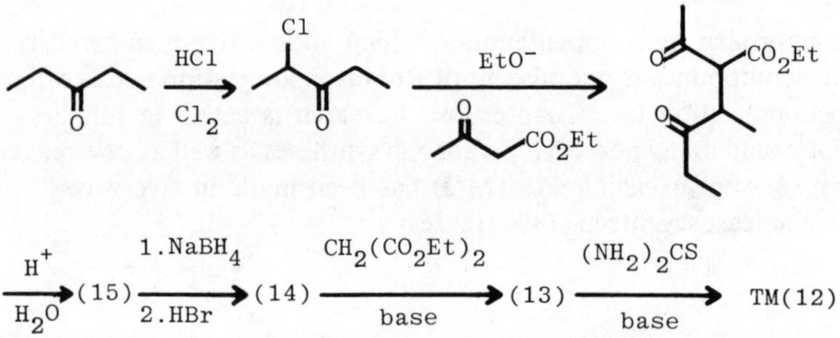

From 1,6-Dicarbonyl Compounds

The synthesis of keto ester **(17)** and hence cyclopentanone itself from adipic acid **(16)** was introduced in Chapter 19. It **(17)** can be used to make other cyclopentanones by alkylation before decarboxylation.

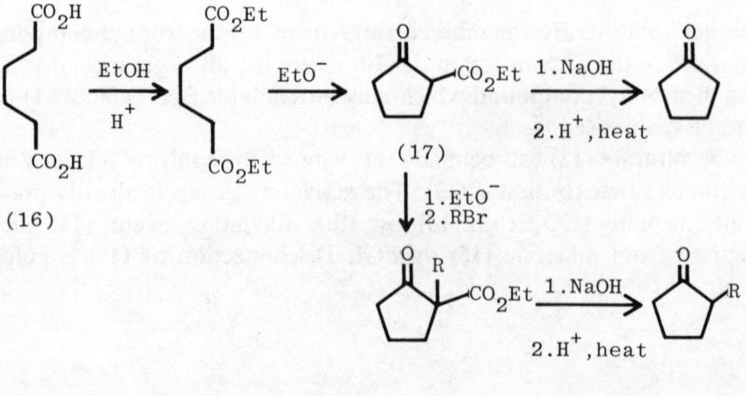

Enones such as **(18)** also disconnect to 1,6-dicarbonyl starting materials. Any method of making such compounds is also a potential synthesis of five-membered rings. We have already used this method to convert Diels–Alder adducts to cyclopentanones (Chapter 28).

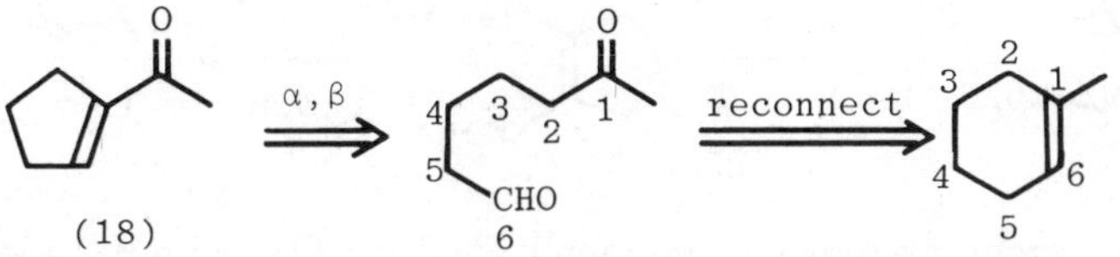

Both target molecules **(19)** and **(21)** disconnect to the same 1,6-dicarbonyl compound **(20)** and reconnection gives the naturally occurring limonene **(22)** whose synthesis we discussed in Chapter 17.

Analysis

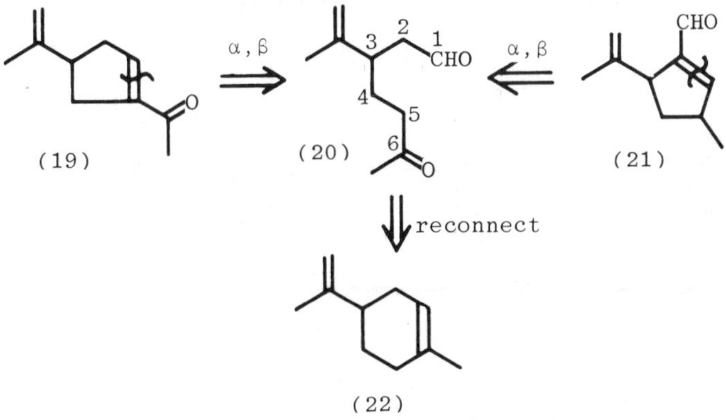

There are two problems of chemoselectivity in this synthesis. How do we cleave one double bond in **(22)** without cleaving the other, and how do we control the cyclisation of **(20)**? Epoxidation of **(22)** selectively attacks the more substituted double bond to give **(23)** which can be opened to **(20)** in two steps.[379] The cyclisation of **(20)** can be controlled by conditions: strong base gives **(19)** by thermodynamic control and weak base enolises only the aldehyde (kinetic control) to give **(21)**.

Synthesis[379]

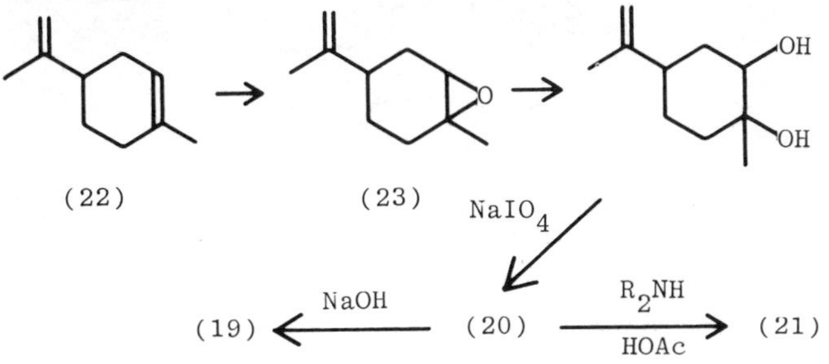

From 1,5-Dicarbonyl Compounds

The silicon modification of the acyloin reaction (see Chapter 24) is useful for making five-membered rings from 1,5-diesters.

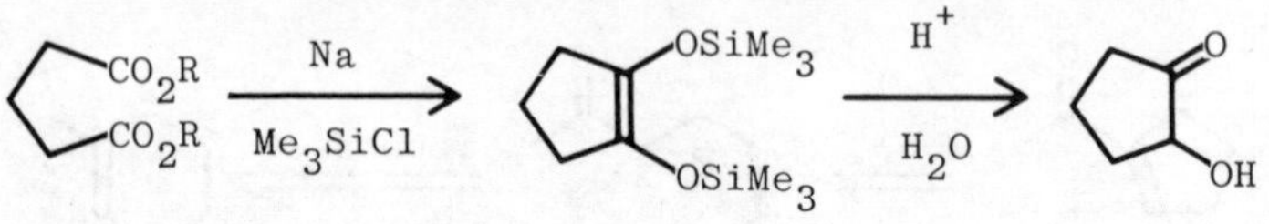

The important perfumery compound corylone (24) with a 'spicy-coffee-caramel' smell has the look of an acyloin product (25). The extra double bond is a nuisance as compounds such as (28) will not undergo the acyloin reaction. Masking this double bond with an amine (26) (cf. Chapter 21), which can be removed at the end of the synthesis, allows for an acyloin disconnection to (27). In the event, removal of the amine from (29) is very easy.

Analysis

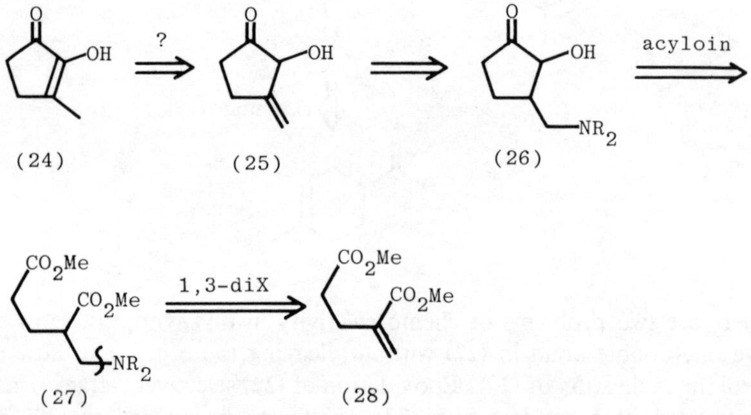

(24)   (25)   (26)

(27)   (28)

Synthesis[380]

(28) $\xrightarrow[100\%]{Me_2NH}$ (27) $\xrightarrow[\substack{toluene \\ Me_3SiCl}]{Na}$ $\xrightarrow[column]{SiO_2}$ TM(24) 70%

(29)   78%

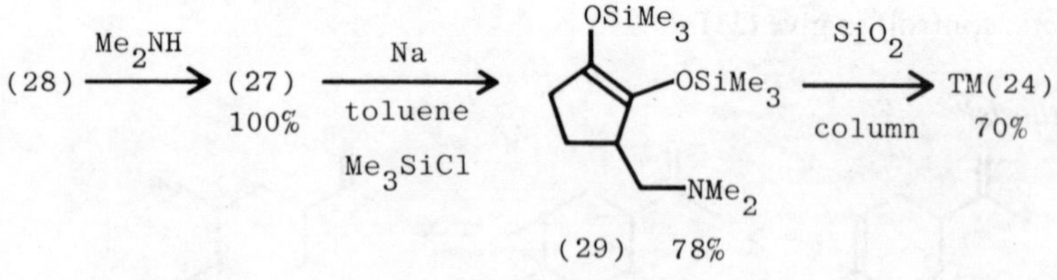

CHAPTER 35

Strategy XVI: Pericyclic Rearrangements in Synthesis. Special Methods for Five-Membered Rings

The only pericyclic reactions we have used so far have been cycloadditions: the Diels–Alder reaction (Chapter 17) and photochemical (Chapter 32) or thermal (Chapter 33) 2 + 2 cycloadditions. Electrocyclic and sigmatropic* reactions are also useful in synthesis and as each is the basis of a method of five-membered ring synthesis, they are conveniently grouped into one chapter here.

Electrocyclic Reactions

An electrocyclic reaction is the formation of a new σ bond (2) across the ends of a conjugated π system (1) or the reverse.

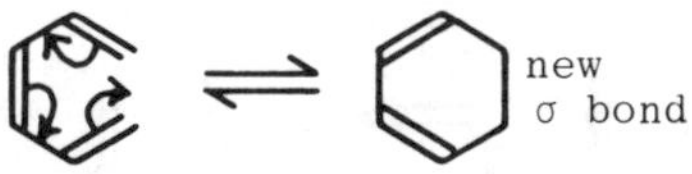

(1)   (2)

new σ bond

The dienone to cyclopentenone cyclisation †

Pentadienyl cations (3) cyclise thermally to cyclopentenyl cations (4) in a conrotatory fashion. The most important example of this reaction[381] is the formation of cyclopentenones (8) from dienones (5) *via* cations (6) and (7).

---

*The theory of these reactions is fully described in Fleming, *Orbitals*, p.98.
† Sometimes called the Nazarov reaction.

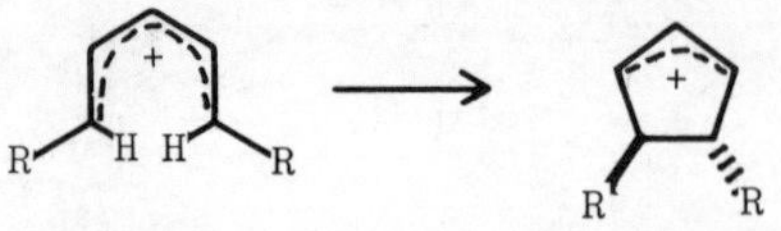

(3)　　　　　　　　(4)

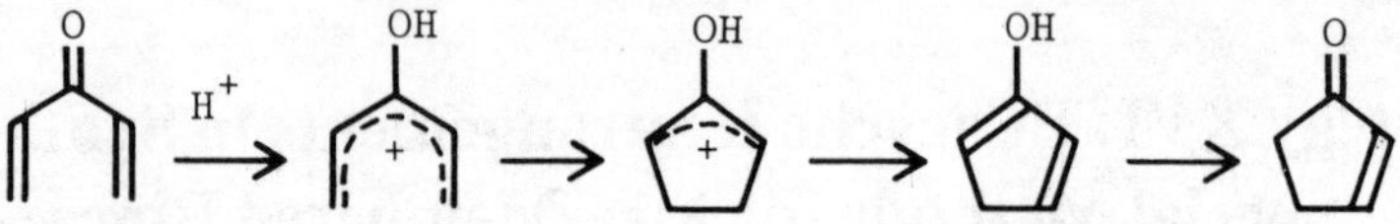

(5)　　　(6)　　　(7)　　　(8)

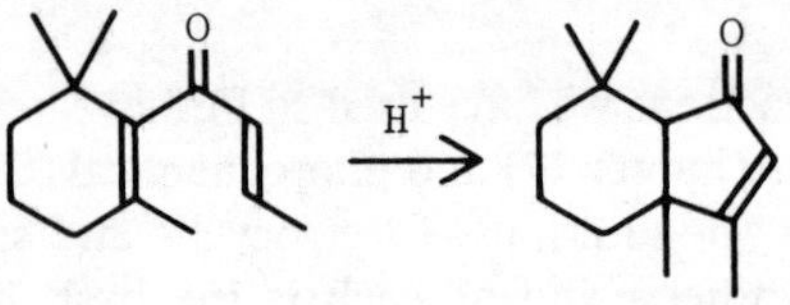

(9)　　　　　　　　(10)

So, for example, the natural product α-damascone **(9)** cyclises to **(10)** in acid solution.[382] The disconnection is of the bond (a in **11**) *opposite* the carbonyl group in the five-membered ring and the synthesis is carried out by treating the dienone **(12)** with acid or Lewis acid.[383]

Analysis

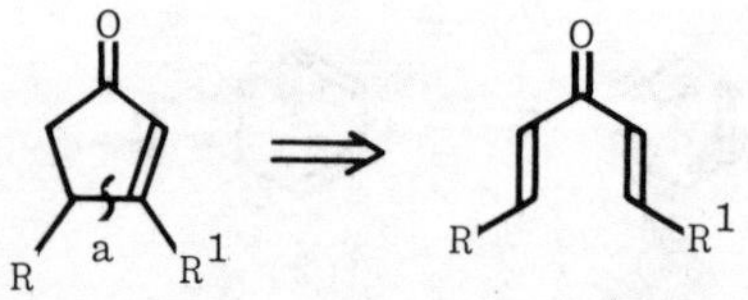

(11)　　　　　　　　(12)

Synthesis

$$(12) \xrightarrow{\text{H}^+} (11)$$

Dieneones such as **(12)** are not particularly easy to synthesise, but any method which might form them under acidic conditions usually gives cyclopentenones instead. The aromatic compound **(13)** was needed for the synthesis of steroid analogues.[384] Disconnecting the bond opposite the carbonyl group gives 'dienone' **(14)**, a Friedel–Crafts product from ether **(15)** and acid chloride **(16)**.

Analysis

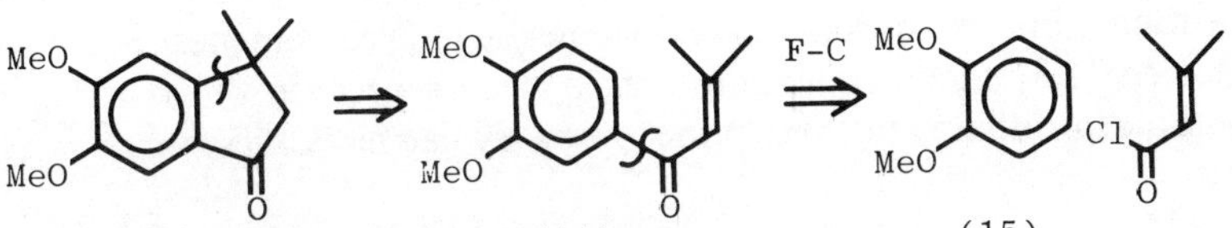

(13)   (14)   (15)   (16)

The synthesis was easier than expected: treatment of **(15)** with the free acid **(17)** and polyphosphoric acid (PPA) gave **(13)** in one step, no doubt *via* **(14)**.

Synthesis[385]

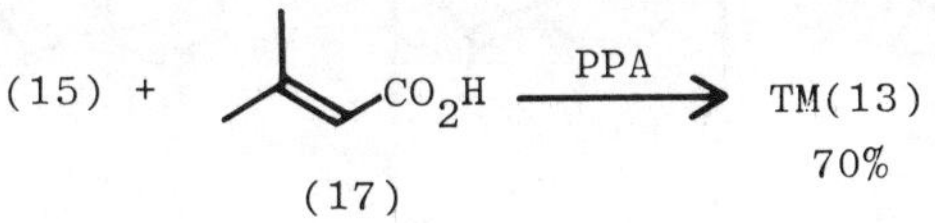

(15) + (17) $\xrightarrow{\text{PPA}}$ TM(13) 70%

The aromatic ring is not essential: the same disconnection on bicyclic ketone **(18)** gives eventually cyclohexene and unsaturated acid **(19)**. The one step synthesis was again successful.[386]

Analysis

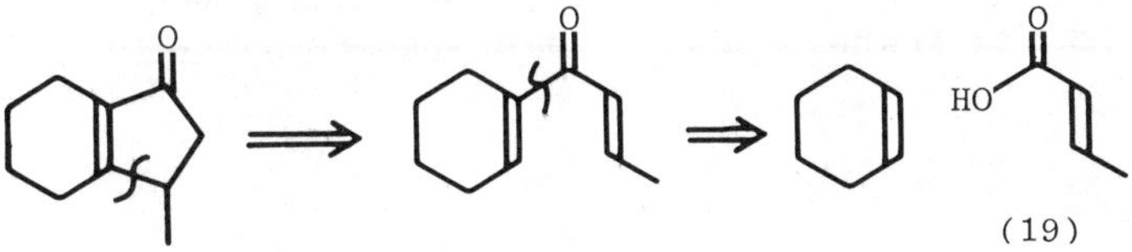

(18)   (19)

Synthesis[386]

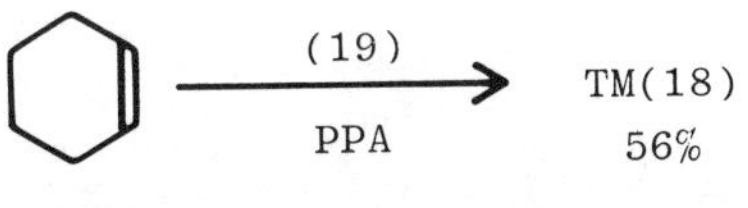

$\xrightarrow[\text{PPA}]{(19)}$ TM(18) 56%

Sigmatropic Rearrangements

A sigmatropic rearrangement is a unimolecular reaction in which a σ bond moves from one position in the molecule **(20)** to another **(21)**. Thus **(20)** → **(21)** is a [3,3] sigmatropic reaction. The numbers (see **20**) give the position of the new σ bond relative to the old.

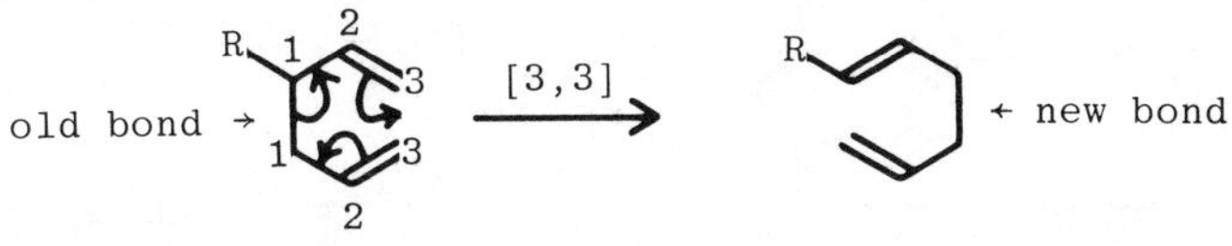

old bond →   [3,3]   ← new bond

(20)   (21)

The vinyl cyclopropane to cyclopentene rearrangement

On heating, vinyl cyclopropanes **(22)** isomerise to cyclopentenes **(23)**. This is a disallowed [1,3] sigmatropic shift which occurs because the reaction is so favourable in enthalpy terms that the symmetry barrier is overcome.[387]

(22) (23)

(24) (25) (26)

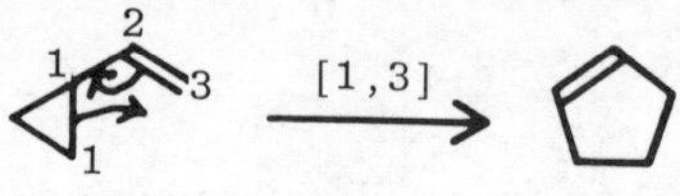

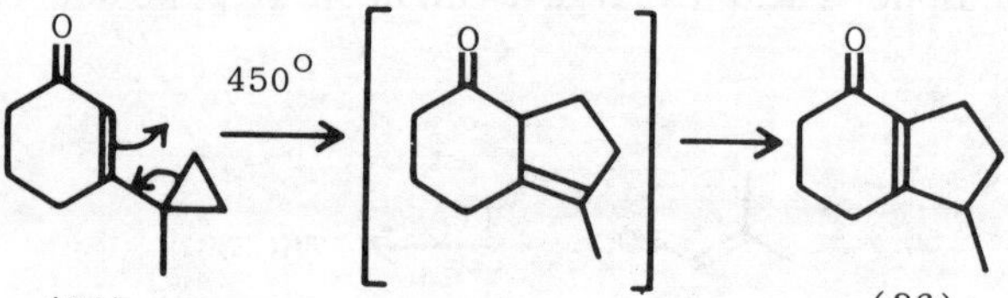

Thus **(24)** gives **(26)** on heating to 450 °C, no doubt *via* [1,3] shift to **(25)**. The product **(26)** has been used in a synthesis of the natural product zizaene.[388] The easiest way to see the disconnection is, as usual with rearrangements, to reverse the reaction. Cyclopentenone **(27)**, needed[389] for a photochemical synthesis of **(28)**, could come from two different vinyl cyclopropanes **(29)** and **(30)**. There is no obvious disconnection of **(29)** but α,β-disconnection of **(30)** reveals aldehyde **(31)** which could be made by at least two routes.

(27) (28)

Analysis

(27a) (29)

(27b) (30) (31)

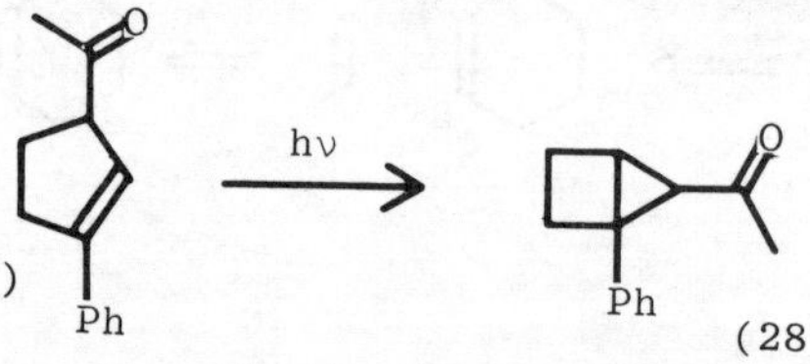

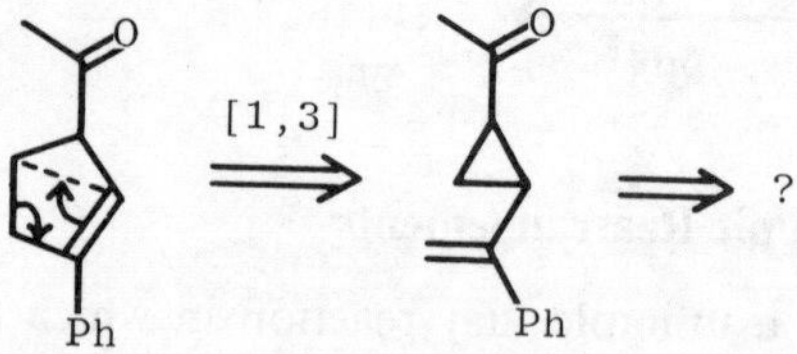

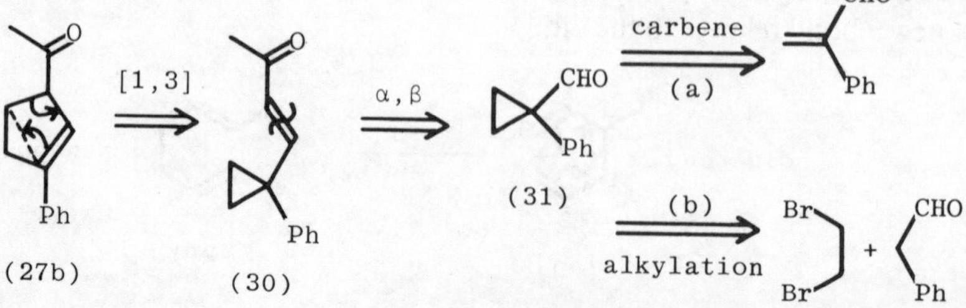

Aldehyde (31) has been made by the alkylation route,[390] using the more stable cyanide (32) and reducing with $LiAlH_4$ (we should use i-Bu_2AlH, DIBAL, nowadays). The Wittig method was used[389] to control the condensation at the acid oxidation level, making it rather a long synthesis. It is possible that the shorter route we have suggested was tried but failed.

Synthesis[389, 390]

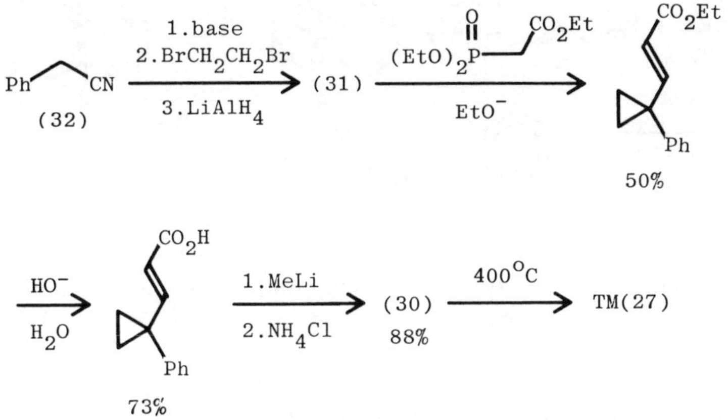

The rearrangement also works well with heterocycles: compound (33), wanted[391] for an alkaloid synthesis, disconnects to imine (34) which can be made from a benzylic cyanide (35). The synthesis is uneventful.

Analysis

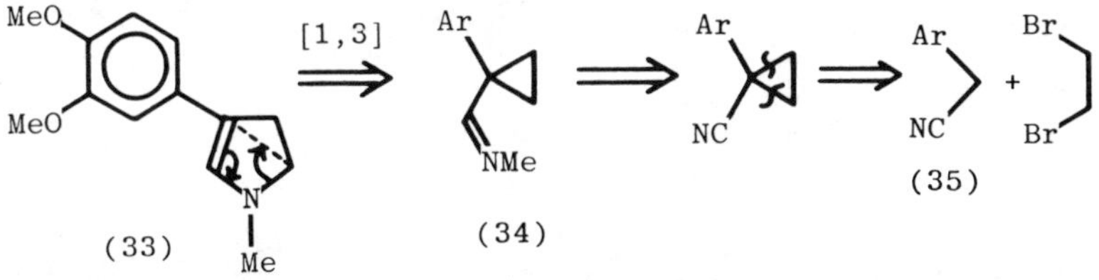

Synthesis[391]

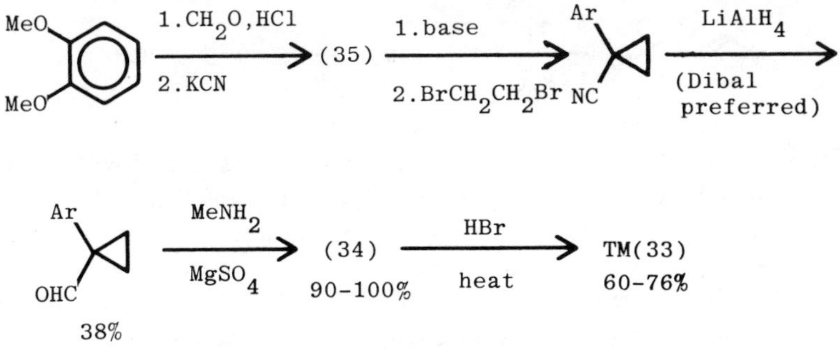

[3,3] Sigmatropic shifts: the Claisen, Cope, and Carroll rearrangements

The original [3,3] sigmatropic shift (see page 285) was the Claisen rearrangement,[392] an excellent way to introduce allyl groups onto a benzene ring. An aryl allyl ether (36), made in the usual way, rearranges on heating to the *o*-allyl phenol (37).

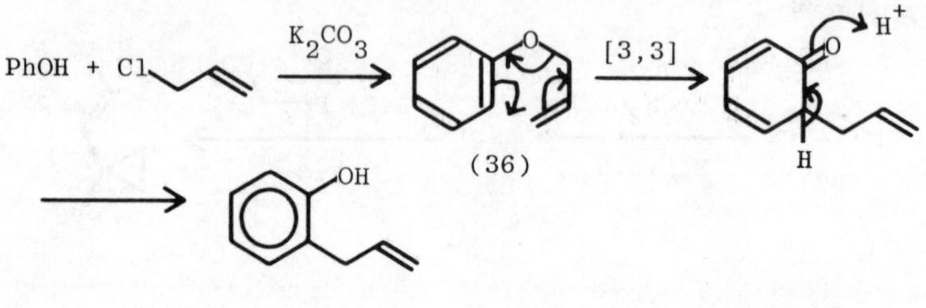

The disconnection is to replace the allyl group on oxygen, remembering to reverse it if it is unsymmetrical—note the position of R in (38) and (39).

Analysis

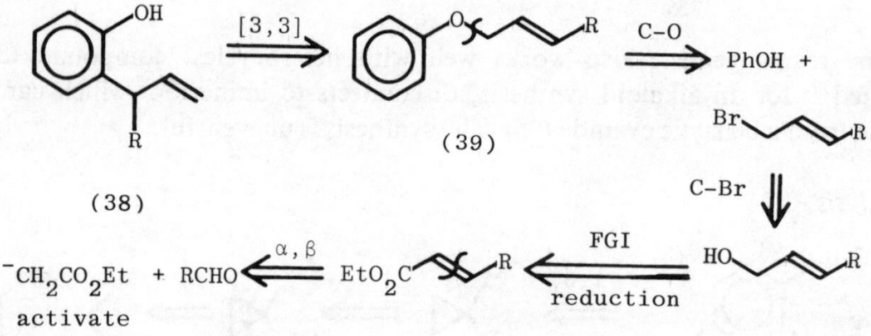

Synthesis[392]

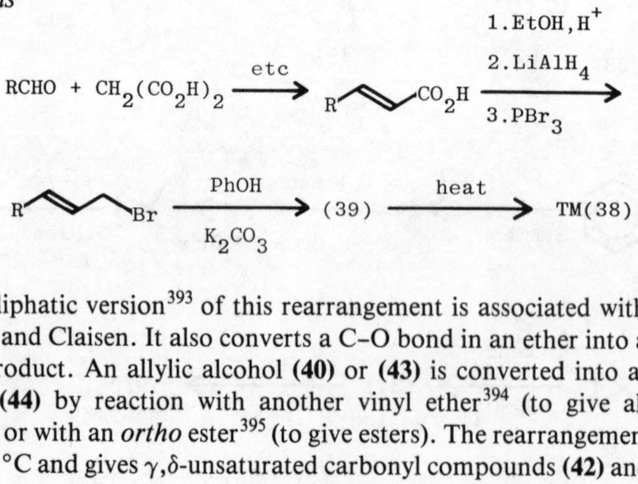

The aliphatic version[393] of this rearrangement is associated with the names of Cope and Claisen. It also converts a C–O bond in an ether into a C–C bond in the product. An allylic alcohol (40) or (43) is converted into a vinyl ether (41) or (44) by reaction with another vinyl ether[394] (to give aldehydes or ketones) or with an *ortho* ester[395] (to give esters). The rearrangement is rapid at 100–150 °C and gives γ,δ-unsaturated carbonyl compounds (42) and (45).

Synthesis 1 [394]

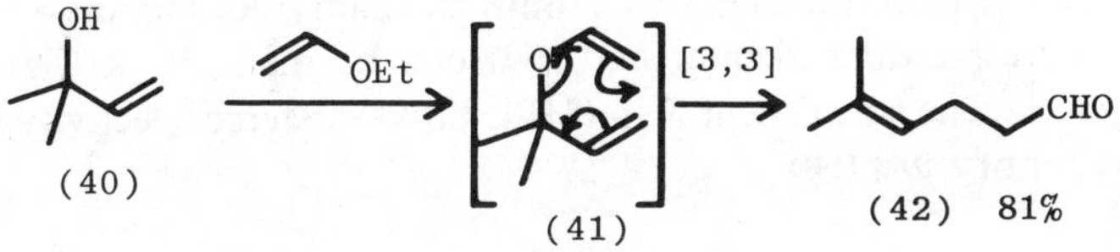

(40) → (41) [3,3] → (42) 81%

Synthesis 2 [395]

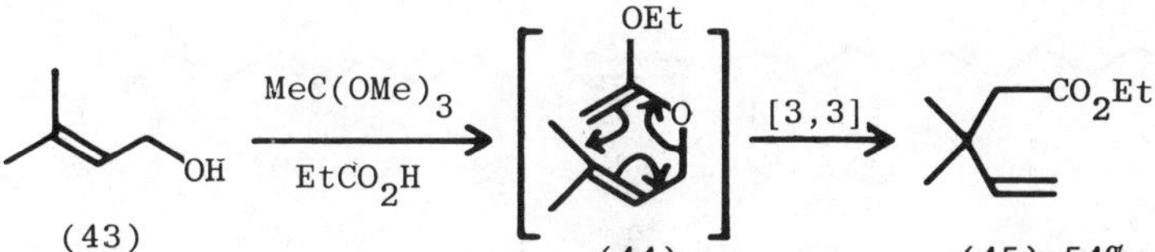

(43) → (44) [3,3] → (45) 54%

The disconnections for these reactions are not so obvious. Careful inspection of the two examples should reveal that the allyl group has been added to the enolic position of an aldehyde or ester, i.e. both (42) and (45) are γ,δ-unsaturated carbonyl compounds. These reactions are methods for regiospecific allylation of enolates, and the disconnection is therefore to separate the allyl group from the enolate, remembering to invert the allyl group.

Analysis

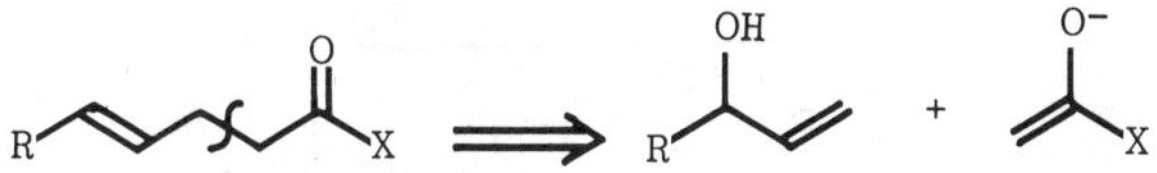

The Claisen–Cope is also highly stereoselective. [392] The transition states (46) for reactions (41) and (44) are known to be like chair cyclohexanes (47). A substituent at the vital position (• in 47) prefers to be equatorial and the new double bond in (48) must be *trans*. The H atoms marked on (47) help to make this plain.

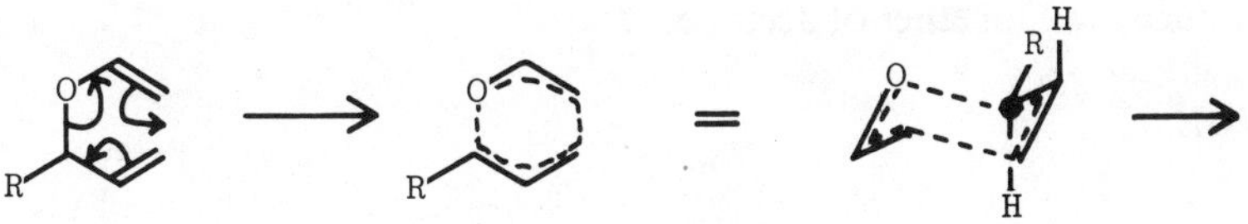

(46) (47)

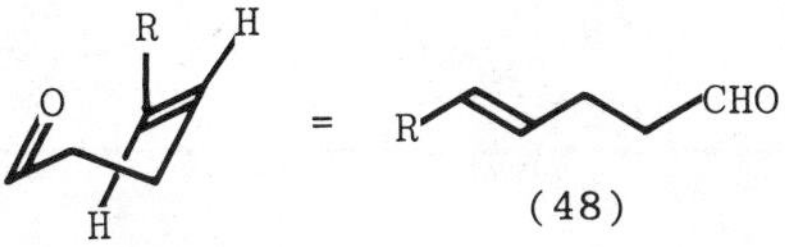

(48)

Evans needed halide **(49)** for his synthesis[396] of perhydrohistrionicotoxin (a powerful poison from the skin of Colombian toads). Reduction of ester **(50)** gives the corresponding alcohol and **(50)** can be made by a Claisen–Cope rearrangement. The substituent R in **(51)** is butyl so stereoselectivity should be high in favour of *trans* **(50)**.

Analysis

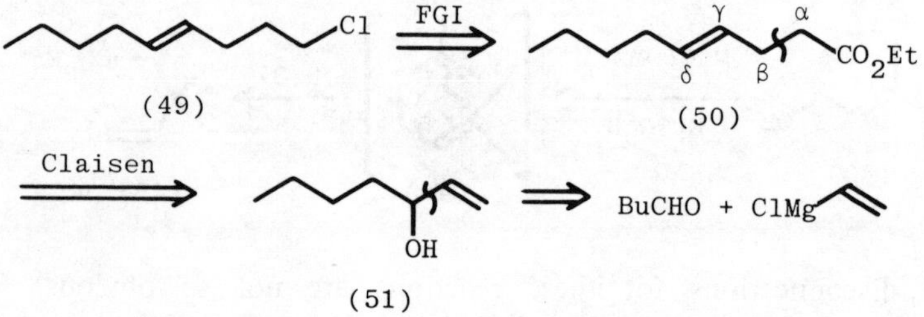

Synthesis[396]

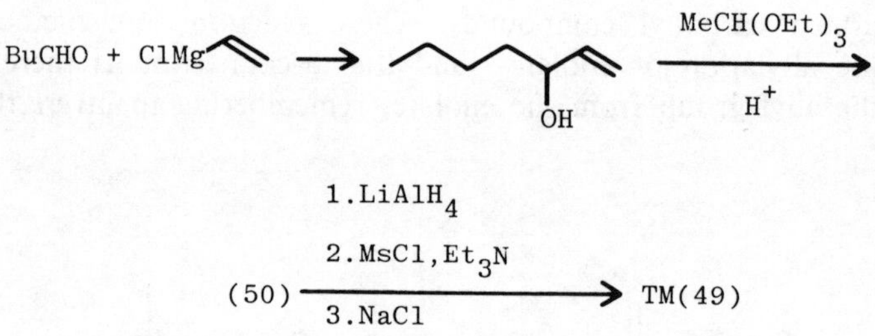

If you think these rearrangements sound like academic reactions and too refined for the rough and tumble of industry, you're in for a surprise. The Carroll rearrangement, a [3,3] sigmatropic shift, is an important industrial process[2] used to make vitamin A and some perfumery and flavouring compounds at BASF in Germany.

Ketone **(52)**, mentioned in Chapter 1, is another γ,δ-unsaturated compound, and Claisen–Cope disconnection suggests using allyl alcohol **(53)** and a reagent to produce the enol ether of acetone.

Analysis

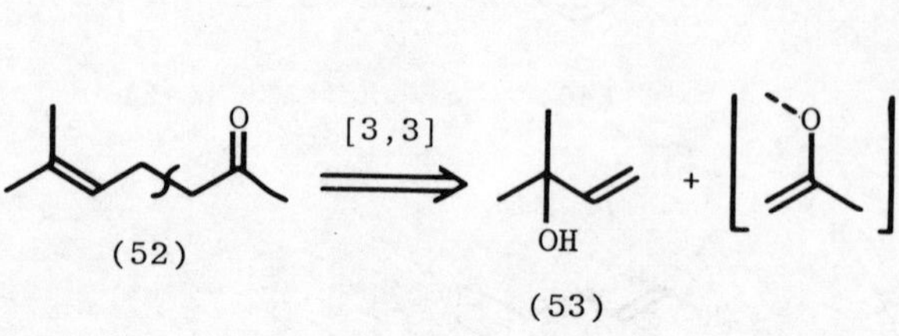

The Carroll reaction uses an acetoacetate ester **(54)**, made by ester exchange or with diketene (Chapter 33), to give enol **(55)** which can do the [3,3] sigmatropic shift and give keto acid **(56)** which decarboxylates under the reaction conditions. The synthesis of **(53)** is standard acetylene chemistry (Chapter 16).

Synthesis²

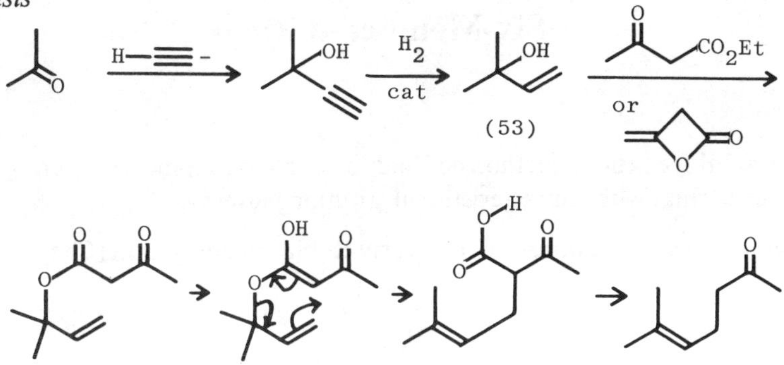

Geranyl acetone **(57)**, a natural product also manufactured at BASF, can be made the same way as it too is a γ,δ-unsaturated ketone. Disconnection with allylic inversion gives **(58)** which can be made from **(52)**.

Analysis

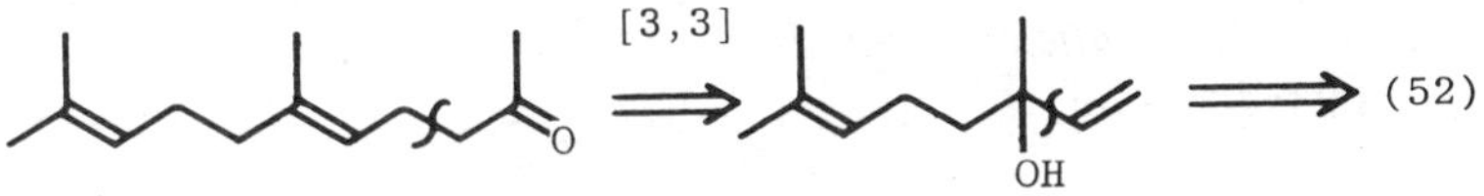

The geometry of the new double bond will be as required since the larger group prefers to be equatorial in the Carroll rearrangement transition state.

Synthesis²

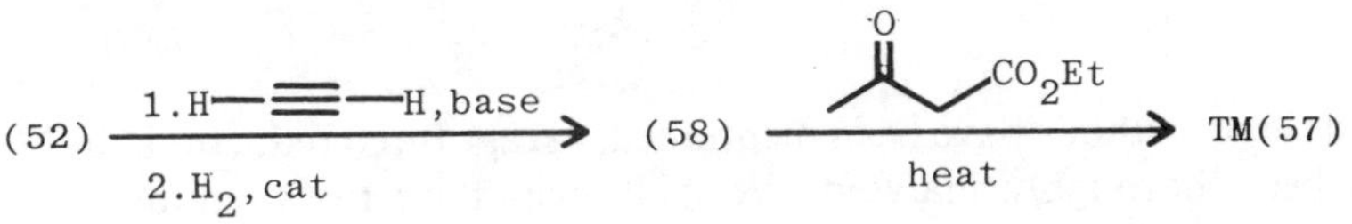

CHAPTER 36

Six-Membered Rings

There are three general methods of making aliphatic six-membered rings, each producing rings with characteristic substitution patterns:

1. *carbonyl condensations*—Particularly Robinson annelation (Chapter 21);

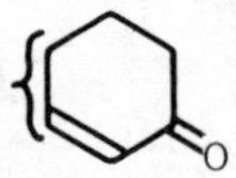

2. *Diels–Alder reactions* (Chapter 17);

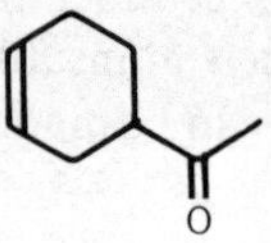

3. *reduction of aromatic compounds*
   (a) total reduction,

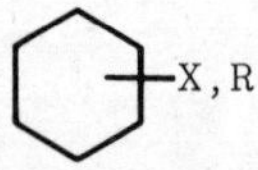

   (b) partial reduction, particularly Birch reduction.

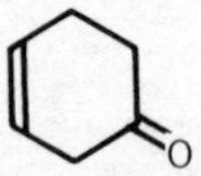

All these methods have been mentioned, except Birch reduction, and 1 and 2 have been thoroughly analysed. We shall revise these two therefore and spend more time on 3.

Carbonyl Condensations: Robinson Annelation

The Robinson annelation was introduced in Chapter 21 and is the most important of all the carbonyl condensations leading to six-membered rings.

292

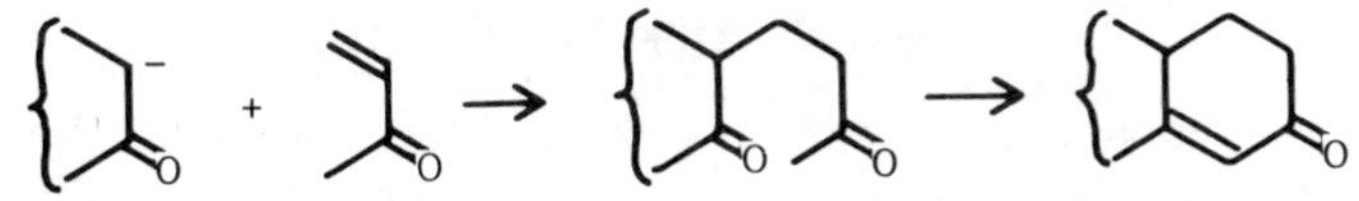

When Marx[397] wished to investigate whether alkyl groups or CO_2Et migrated better in rearrangements, he wanted compounds (1) where the R group could be varied easily. One of the double bonds could be put in by oxidation of (2)—an ideal Robinson annelation product. Disconnection reveals 1,3-dicarbonyl compound (3) as the starting material and this can be made from simple esters (4) and ethyl formate.

Analysis

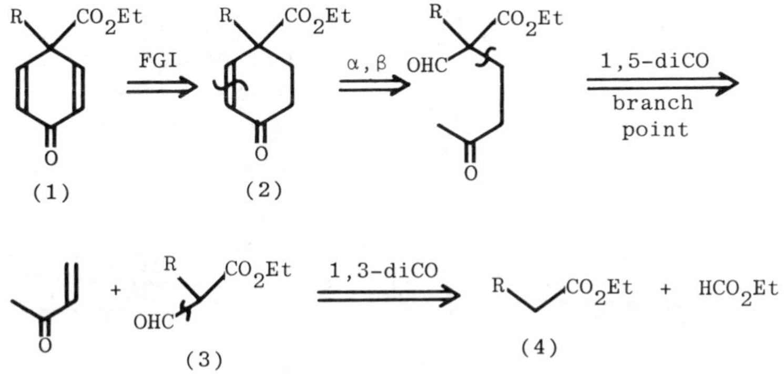

There is no ambiguity in the first step of the synthesis as only (4) can enolise and HCO_2Et is more electrophilic. The Robinson annelation was carried out in two stages, and the final oxidation used a quinone 'DDQ' (5), a good reagent for dehydrogenation.

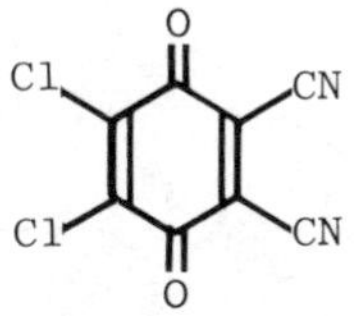

Synthesis[397]

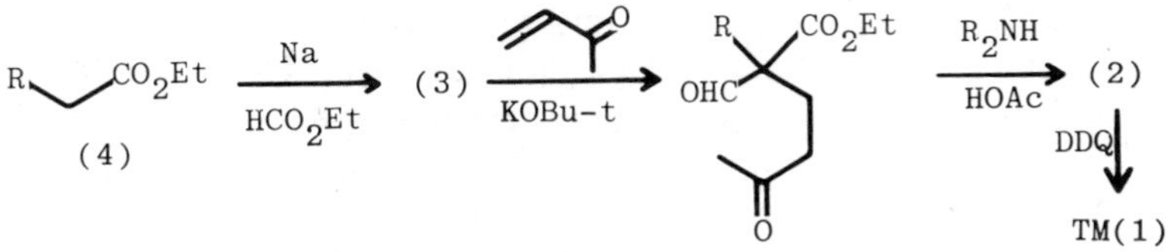

The Diels–Alder Reaction

This reaction was treated at length in Chapter 17 with analyses of its stereo-
and regioselectivity. When Büchi wished to make (6) he was naturally drawn to
the Diels–Alder reaction.[398]

Analysis 1

(6)   (7)

Unfortunately, Diels–Alder reactions give *ortho* and *para* products (see
Chapter 17) and not *meta* products like (6). This particular reaction gives[399]
(7). Alternatively, ketone (6) is a γ,δ-unsaturated carbonyl compound and so
might be made by a Claisen–Cope rearrangement (Chapter 35). The disconnection is difficult to follow but the numbers on (8) and (9) should help.

Analysis 2

(8)   (9)

Structure (9) is very close to that of the Diels–Alder dimer* (10) of methyl
vinyl ketone (Chapter 21) and a Wittig reaction connects them.

Analysis 3

(9a)   (10)

The dimerisation to give (10) is usually a nuisance but it does make (10)
readily available. The Claisen–Cope rearrangement goes from (9) to (6)
because a carbonyl group is formed at the expense of a less stable C=C double
bond.

*See Fleming, *Orbitals* p.141 for an explanation of the regioselectivity of this dimerisation.

Synthesis[398]

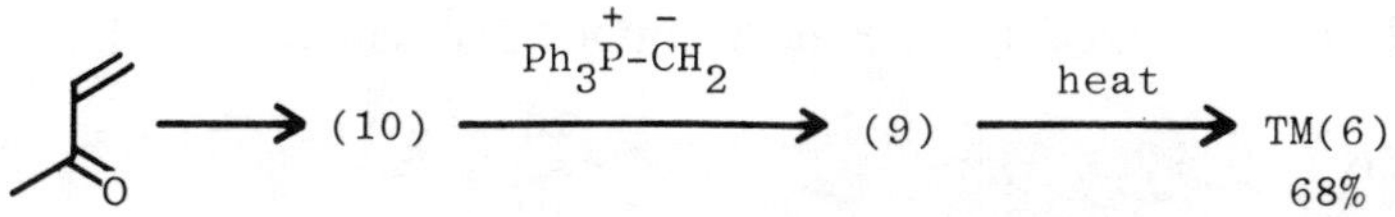

Cyclopentadiene **(11)** is also readily available so that the Diels–Alder approach is particularly suitable for compounds with the skeleton **(12)**—a common skeleton in many natural products.

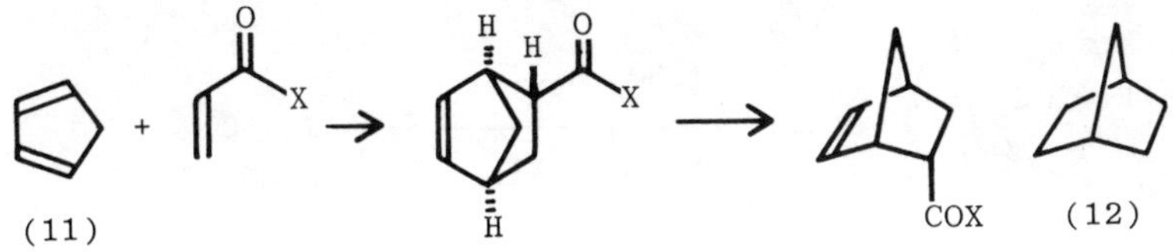

Danishefsky's pentalenolactone synthesis[400] used **(13)** as an intermediate. Immediate Diels–Alder disconnection is no good as it produces the unlikely starting materials **(14)** and **(15)**. If we first remove the acetal to give diol **(16)**, this could be made from alkene **(17)**. Diels–Alder disconnection is still impossible—**(18)** is too strained—but a minor FGI allows a good Diels–Alder reaction between **(19)** and cyclopentadiene.

Analysis

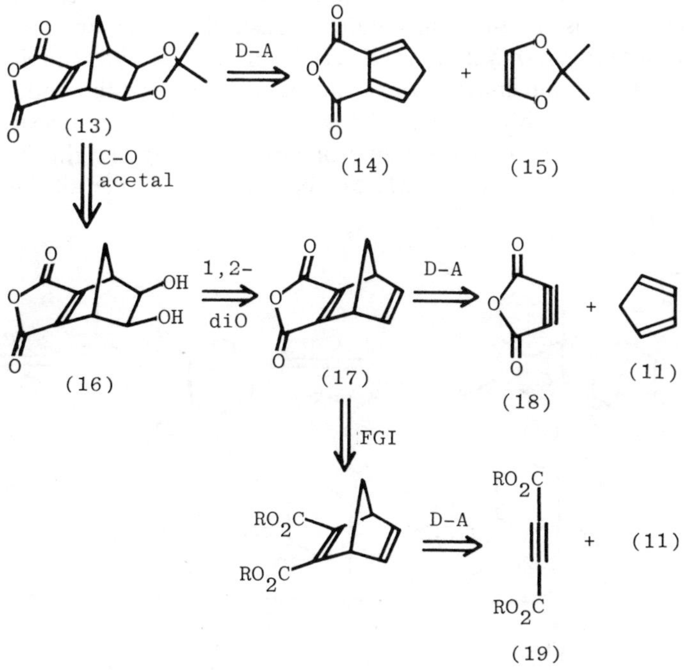

296

The hydroxylation should favour the less hindered *exo* side and Danishefsky hydroxylated before forming the anhydride. This last step proved difficult because of strain and a special reagent (EtOC≡CH) had to be used.

Synthesis[400]

EtO$_2$C—≡—CO$_2$Et

(19,R=Et)

$\xrightarrow{(11)}$

1.OsO$_4$

2.Me$_2$CO,H$^+$

1.HO$^-$,H$_2$O

2.EtO—≡

TM(13)

Reduction of Aromatic Compounds

The virtue of this approach* is that we can use the aromatic ring to set up virtually any array of substituents before reduction. Partial (Birch) reduction allows latent functionality to be exploited.

Total reduction

Total reduction of a benzene to a cyclohexane ring requires pressure and active catalysts and is more easily done industrially than in the lab. The antispasmodic drug Dicyclomine (20) can be made this way[401] and this synthesis illustrates that six-membered rings can be made by methods other than the three presented in this chapter. Disconnection of the ester reveals acid (21) with one ring that could have been aromatic (22). Note that this is FG*A* logic (Chapters 24 and 28). Acid (22) can be made by simple alkylation, using cyanide (23) (cf. Chapter 35).

Analysis

(20) $\xrightarrow[\text{ester}]{C-O}$ (21) $\xrightarrow{FGA}$

$\xrightarrow{FGI}$ Ph CN (23) $\Rightarrow$ Ph CN + Br Br

*Examples appear in Chapter 35.

In practice, better yields result from assembling the complete molecule **(25)** before reduction. The alcohol **(24)** is an amine–epoxide adduct (Chapter 6). Esters of **(21)** can also be made by Faworskii rearrangement (Chapter 31).

Synthesis[401]

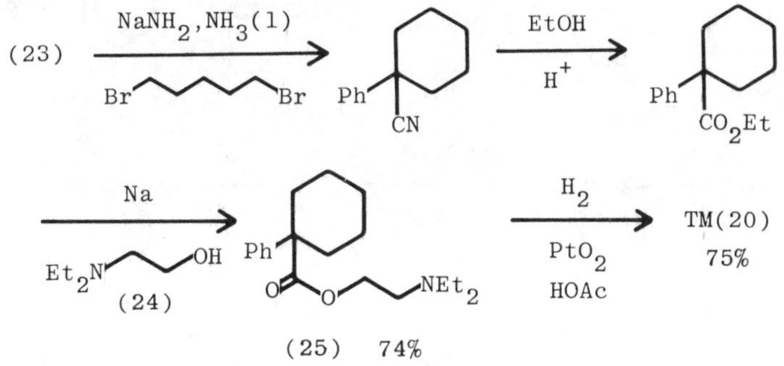

Two substituents which are remote in a saturated compound may be sensibly related in an aromatic compound. The ketone **(26)** has been used in studies on conformational analysis — we used it as a starting material in Chapter 27. We should obviously like to disconnect bond (a): this is hardly possible in **(26)**, but trivial in the aromatic **(27)**.

Analysis

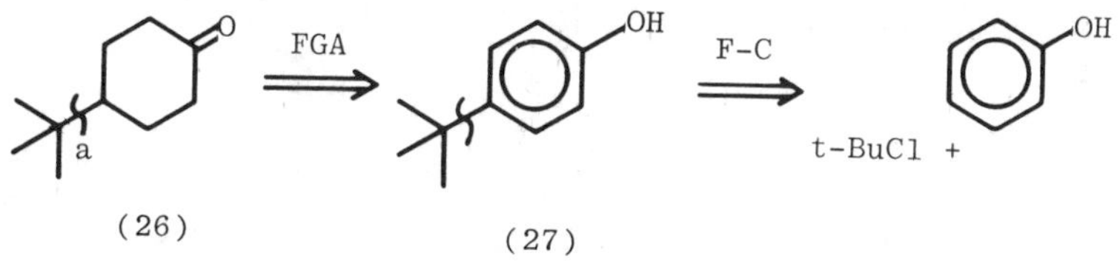

Synthesis[402]

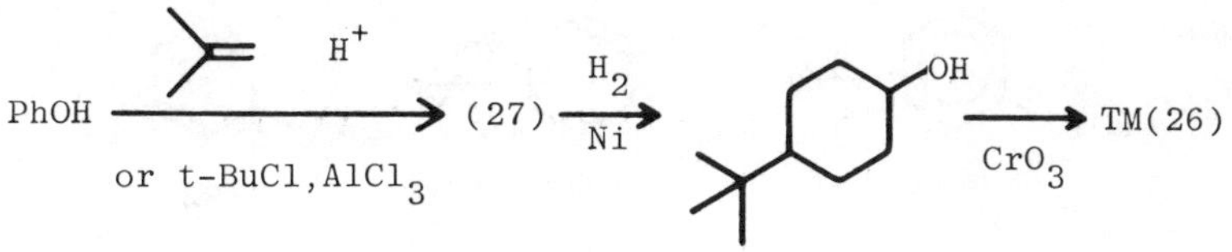

Two remote FGs may similarly be brought into revealing relationship in an aromatic compound. Amine **(28)** has no obvious disconnections but the synthesis of the aromatic amine **(29)** is a trivial exercise in substitutions (cf. Chapter 3).

298

Analysis

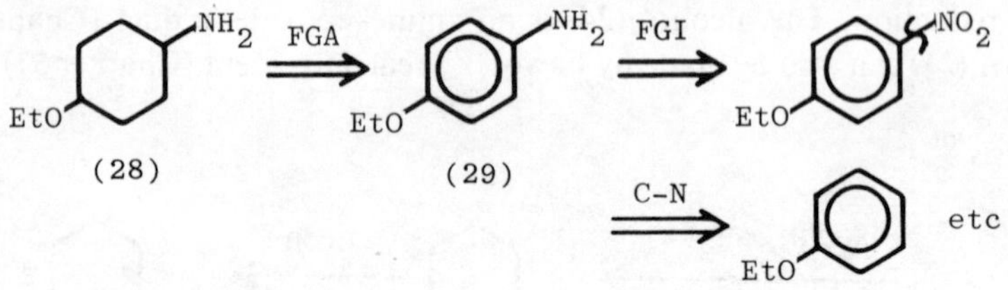

(28)   (29)

Synthesis[403]

$$PhOH \xrightarrow[EtBr]{base} PhOEt \xrightarrow[2.H_2,cat]{1.HNO_3,H^+} (29) \xrightarrow{H_2} TM(28)$$

Birch reduction

Birch reduction[404] is a partial reduction of aromatic compounds by electron transfer from dissolving metals—usually Na in liquid ammonia or Li in ethylamine—in the presence of a weak proton donor—usually an alcohol. The reaction behaves as if dianion (30) were an intermediate, giving non-conjugated dienes (31). Electron-donating substituents repel the anions (32) to give products like (33), whilst electron-withdrawing substituents attract the anions (34), to give products like (35).

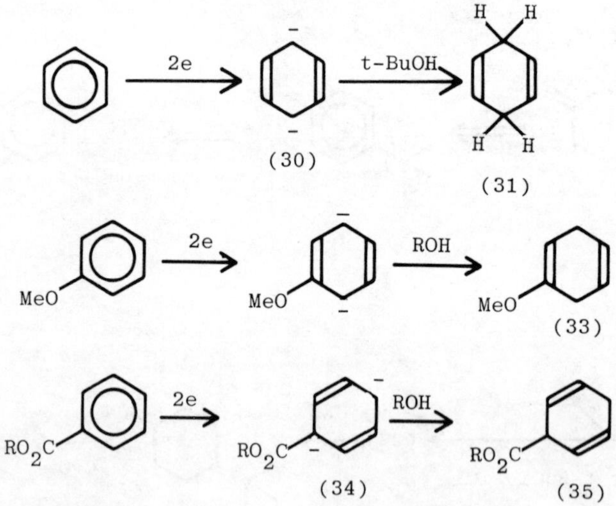

Epoxide (36) can be made from (37) by chemoselective epoxidation of the more substituted and therefore more nucleophilic double bond. Non-conjugated diene (37) is an obvious Birch reduction product as both electron-donating substituents are on the double bond and therefore away from the anions.

Analysis

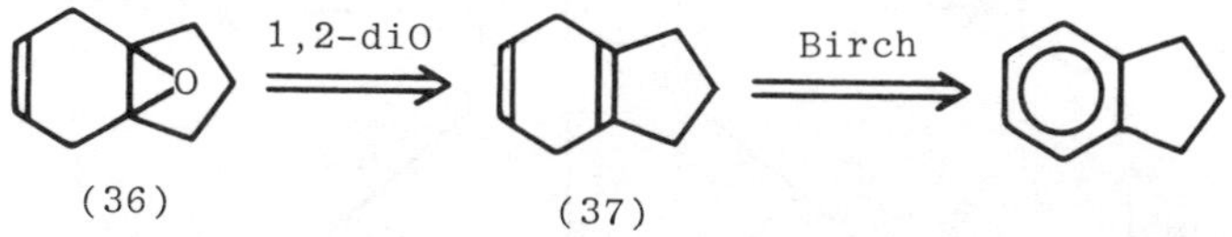

(36)　　　　　(37)

Synthesis[405]

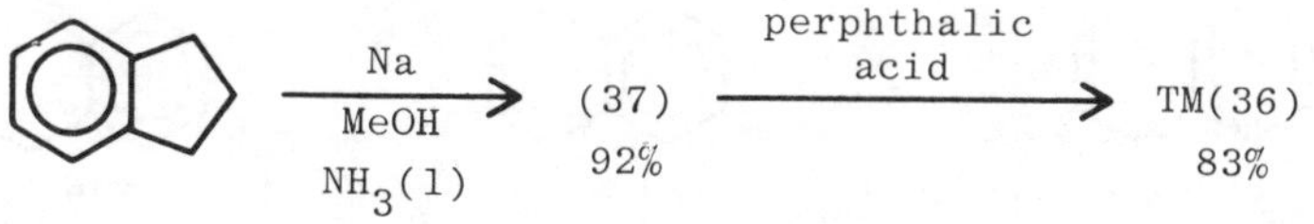

Alkoxybenzenes give vinyl ethers **(33)** by Birch reduction. These compounds are latent ketones as acid hydrolysis gives a ketone. Epoxide **(38)** can first be disconnected, by removal of the epoxide and the acetal, to ketone **(39)**. The position of the double bond, out of conjugation with the ketone but at the alkyl substituted position, is the clue for a Birch reduction. The rest is simple aromatic chemistry.

Analysis

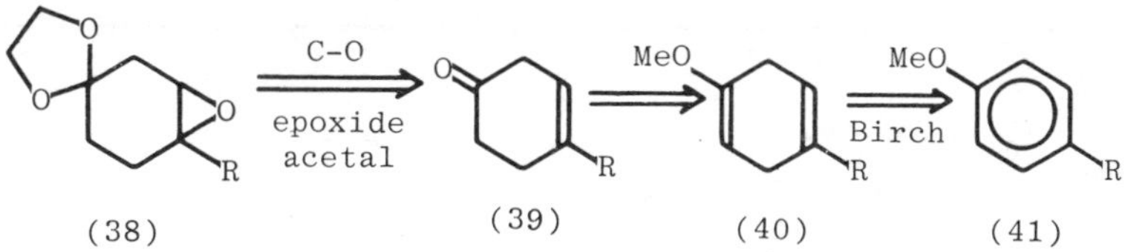

(38)　　　　　(39)　　　　(40)　　　　(41)

Synthesis[406]

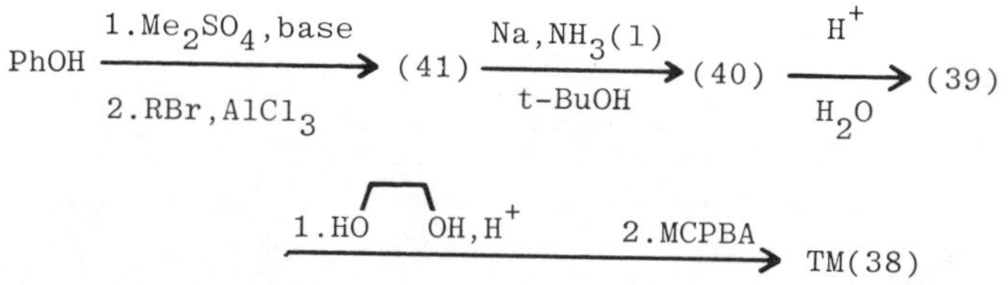

Note that **(40)** is the only Birch reduction product of **(41)** in which *both* electron-donating substituents remain on the double bonds. TM**(38)** was needed for terpene synthesis.

There are other reagents for partial reduction of aromatic systems: the reduction of naphthalene can be controlled to give any of the five products[407] **(42)–(46)**. There is no point in trying to learn these conditions: but you should be aware that these compounds are available.

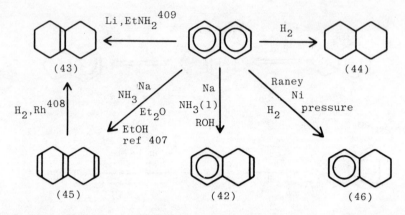

Enone **(47)** disconnects to 1,6-diketone **(48)**. Reconnection in the usual way (Chapter 27) suggests **(43)** as starting material.

Analysis

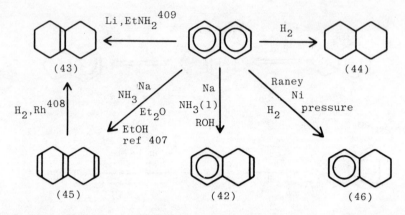

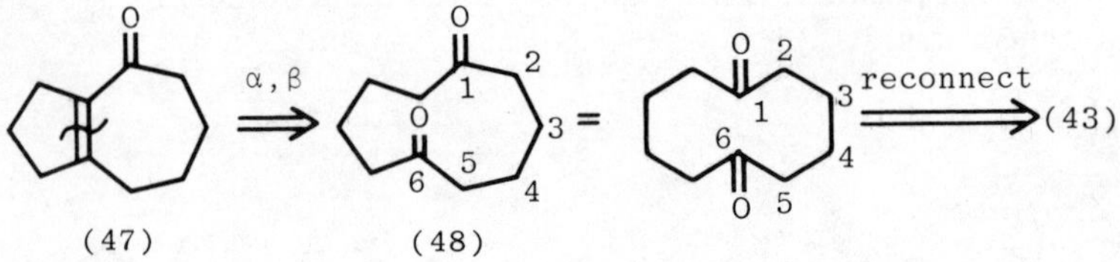

Synthesis[410]

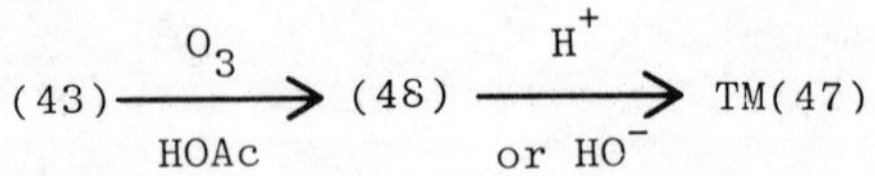

$$(43) \xrightarrow[\text{HOAc}]{O_3} (48) \xrightarrow[\text{or HO}^-]{H^+} TM(47)$$

CHAPTER 37

General Strategy C: Strategy of Ring Synthesis

This chapter collects ideas from the last eight chapters on ring synthesis and puts them in the context of our general approach to strategy. No grand new principles are needed: we shall use the same guidelines already established in Chapters 11 and 28 with one or two special guidelines for ring synthesis.

Cyclisation to Control Selectivity

Cyclisations are easy. In Chapters 7 and 20 we saw that control is unnecessary in many cyclisations as intramolecular reactions usually take precedence over intermolecular. If, therefore, a difficult step has to be incorporated into a synthesis, it is good strategy to make it a cyclisation.

Corey needed[411] ketone (2) as an intermediate in his synthesis of the marine allomone* (1). Bond (a) will be easy to make as it is *para* to the MeO group, but bond (b) will be difficult as it is *meta* to MeO. If we make the formation of bond (b) a cyclisation, the problem will be solved. We must therefore disconnect (b) first.

Analysis 1

(1)   (2)

Disconnection at the branch point (• in 3a) could be by a Michael process, removing either the aromatic ring to give (4) or the *i*-Pr group to give (5).

---

*An allomone is released by one species and used by another, e.g. a predator.

Analysis 2

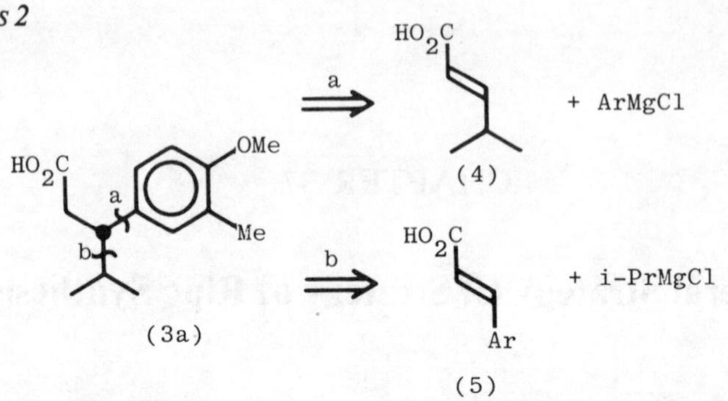

(3a)

(4) + ArMgCl

(5) + i-PrMgCl

Both unsaturated acids are easy to make by carbonyl condensations and all the starting materials are available so either route is good. We shall continue with (5): the usual α,β-disconnection gives aldehyde (6) which can be made by a one-carbon Friedel–Crafts (see Table 2.1) reaction.

Analysis 3

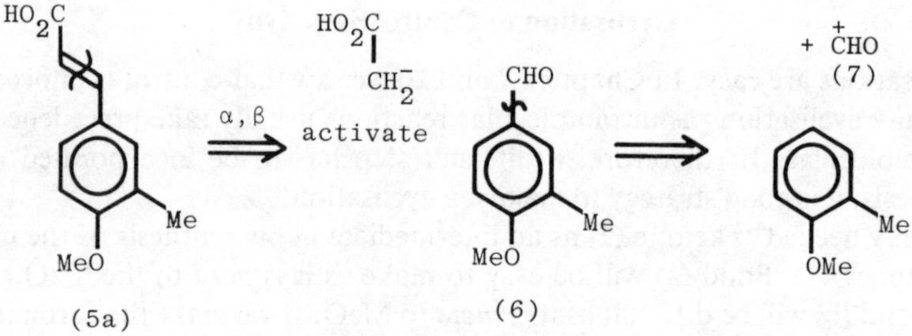

(5a)   (6)   (7)

Corey chose Me_2NCHO and $POCl_3$ (see Table 2.1) as the reagents for synthon (7), and the Knoevenagel type of control (Chapter 20) for the condensation of (6) to give (5). Copper catalysed Grignard addition to the methyl ester of (5) gave the right regioselectivity (Chapter 14), and the final cyclisation gave the target molecule as predicted.

Synthesis[411]

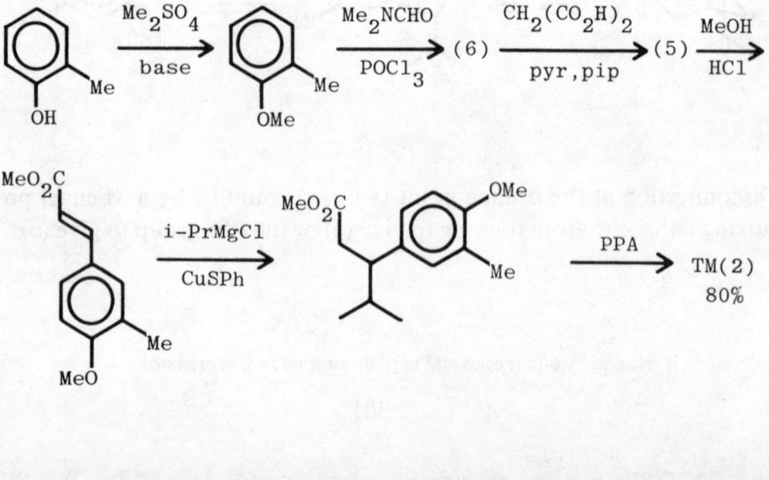

Just occasionally a disconnection is less attractive because the TM is cyclic. The normally excellent α,β-disconnection may be weak when the double bond is common to two rings (though see Chapter 36) as in TM(18) in Chapter 35. Wittig and Grignard reactions are also less good in cyclisation reactions.

Small Rings

It is often good strategy to disconnect a small (three- or four-membered) ring at an early stage in an analysis or at least consider how the small ring might be made before doing any other disconnections. The special methods needed for small rings often dominate strategy.

Ketone (8) has a three-membered ring and another, protected, ketone group. Disconnection of the three-membered ring is guided by the availability of diazoketones (9) (Chapter 31). Intermediate (10) is clearly a Birch reduction product (Chapter 37).

Analysis

The order of events in the synthesis must ensure the protection of the first ketone before the second is introduced. In practice (11) can be converted directly into (10).

Synthesis[412]

Developing Reagents for a Given Synthon

If a disconnection is demanded by strategy, it may produce a synthon for which no reagent exists. The chemist must then invent a new reagent to meet the need or abandon the strategy. Ketone (12) must obviously be made by a photochemical 2 + 2 cycloaddition (Chapter 32) from (13) or (14) as the four-membered ring dominates the strategy.

Analysis 1

The orientation of **(13)** is correct for the cycloaddition but that of **(14)** is incorrect (Chapter 32) though this may not matter in an intramolecular reaction. Nevertheless, **(13)** is the safer bet. The next disconnection ought to be **(13a)** at the ring–chain junction to give synthons **(15)** and **(16)**. The natural polarity of **(16)** is positive, so we write **(15)** negative.

Analysis 2

(13a)   (15)   (16)

A Grignard reagent will be all right for **(15)**, but **(16)** is more difficult. We cannot have **(17)** as a triple bond is impossible in a six-membered ring. The most reasonable alternative is to put a leaving group (X) at the site **(18)**.

Analysis 3

(16)   (17)   (16)   (18)

A search through the literature reveals that enol ether **(19)** is available from diketone **(20)**, available in turn from an aromatic compound **(21)**.

Analysis 4

(18) =   (19)   (20)   (21)

This was new chemistry to us, but not to the workers who carried it out,[413] and it succeeded admirably. TM(12) was used as a model for the biosynthesis of a terpene.

Synthesis[413]

$$(21) \xrightarrow[\text{2.CrO}_3]{\text{1.H}_2,\text{cat}} (20) \xrightarrow[\text{H}^+]{\text{EtOH}} (19) \xrightarrow[\text{Mg,Et}_2\text{O}]{} (13) \xrightarrow{h\nu} \text{TM(12)} \atop 92\%$$

A more challenging example is the isomeric ketone **(22)** again available by 2 + 2 cycloaddition from two possible starting materials.

Analysis 1

(22)   (23)   (24)

This time both **(23)** and **(24)** have the wrong natural polarity (Chapter 32), but again this may not matter as both reactions are intramolecular. There are more ways of making six-membered rings (Chapter 36) than of making five-membered rings (Chapter 34) so we shall continue with **(23)**, though **(24)** could no doubt also be made. The bond to disconnect in **(23)** is obvious **(23a)** but neither the polarity of the synthons nor the nature of the reagents is clear.

Analysis 2

(23a)   (25)   (26)

The halide used for synthon **(15)** in the last problem can provide a nucleophile (Grignard) or an electrophile (the halide) for synthon **(25)**, so we must choose the polarity of the disconnection from **(26)**. The natural polarity of the charged atom is negative so the first plan should be to find a nucleophile reagent for **(26)**. We have not met any such reagent, nor was one known in the literature, so one had to be invented.

An activating group could be put at the nucleophilic site **(27)** but that leaves no proton, so the double bond must be moved **(28)**. The anion **(28)** is starting to look like a Birch reduction product (Chapter 36) and the conversion of the ketone into an enol ether **(29)** completes the design. Reagent **(29)** would be derived from a salicylate **(30)** and the regiochemistry of the reduction is correct (Chapter 36).

(27)   (28)   (29)   (30)

Experiments[414] showed that the free acid (31) could be used in this way: Birch reduction quenched with halide (32) gave (33) which was hydrolysed and decarboxylated in acid solution to give (23). A new method was added to the literature of organic synthesis and was later used by Mander in his synthesis of gibberellic acid.[415]

Synthesis[414]

Alternative Strategies

I have emphasised that general guidelines on strategy are less important than a perceptive study of the particular target molecule in question. Two examples follow where three- and five-membered rings have to be disconnected after six-membered rings.

Ketone (34) looks at first like a simple product from carbene addition to (35). Unfortunately, (35) is a tautomer of naphthol (36) and cannot be made.

Analysis 1

The alternative strategy is to leave the three-membered ring alone and to use a Friedel–Crafts disconnection. Intermediate (37) would be easy to make if the α-CH₂ group were not there (38) as it would then be a diazoester addition to styrene, a route we discussed in Chapter 30. We can use our chain lengthening procedure (Chapter 31) to go from (38) to (37).

Analysis 2

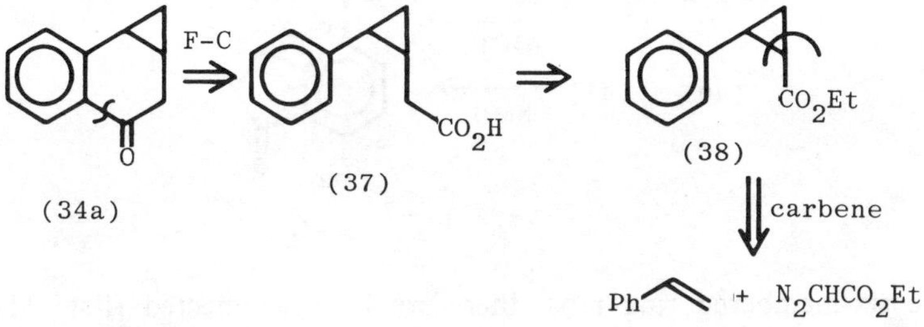

(34a)   (37)   (38)

$Ph\diagup\hspace{-0.5em}=$ + N_2CHCO_2Et

Synthesis[416] (first stages[340] from Chapter 30).

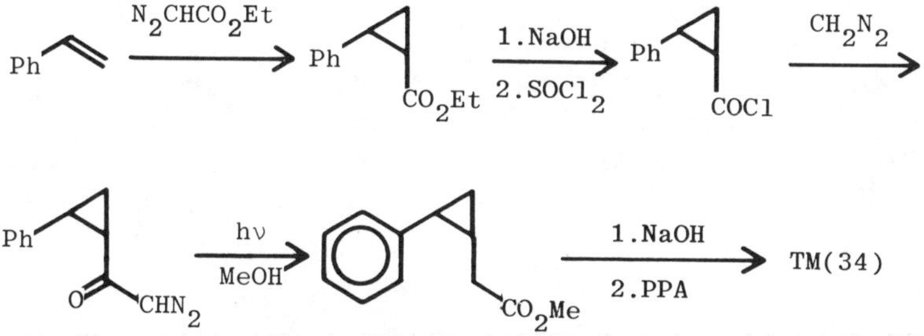

Raphael needed diketone **(39)** for his synthesis[417] of strigol — a compound which stimulates germination of the parasitic witchweed. By our previous strategy we should prefer to keep the six-membered ring and disconnect the five — a Friedel–Crafts approach *via* phenol **(40)** looks promising (cf. Chapter 35).

Analysis 1

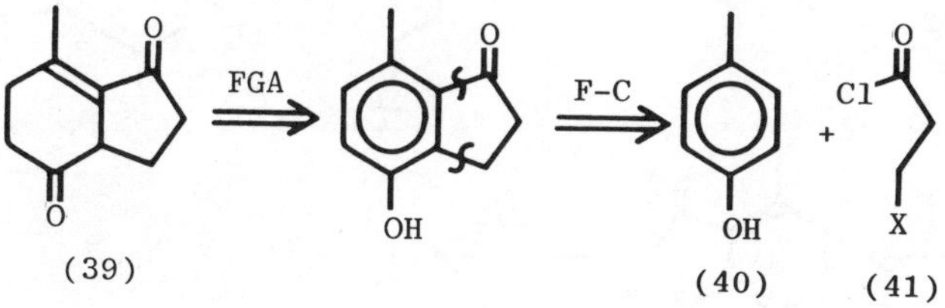

(39)   (40)   (41)

Unfortunately, the orientation of carbonyl and hydroxyl groups is incorrect: the hydroxyl dominates and directs the carbonyl *ortho* to it. Reaction[284] of **(40)** with **(41)** gives the isomeric product **(42)**: we faced a similar problem in Chapter 24.

$$(40) + (41) \xrightarrow[\text{NaCl}]{\text{AlCl}_3} (42)$$

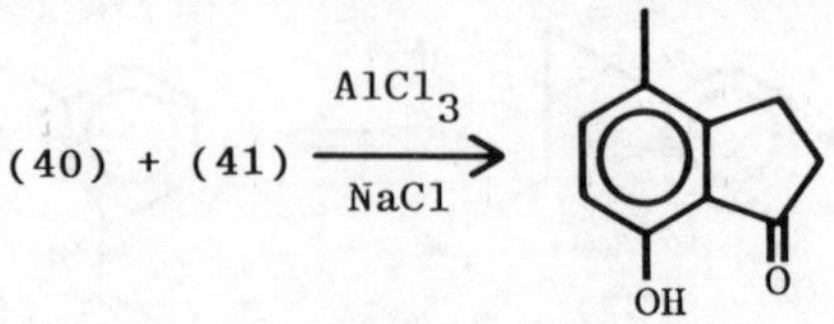

The six-membered ring must therefore be disconnected first. The α,β-disconnection gives triketone **(43)** with all 1,4-relationships, and the best strategy here is to use an acyl anion equivalent for the central carbonyl group and disconnect by two Michael reactions.

Analysis 2

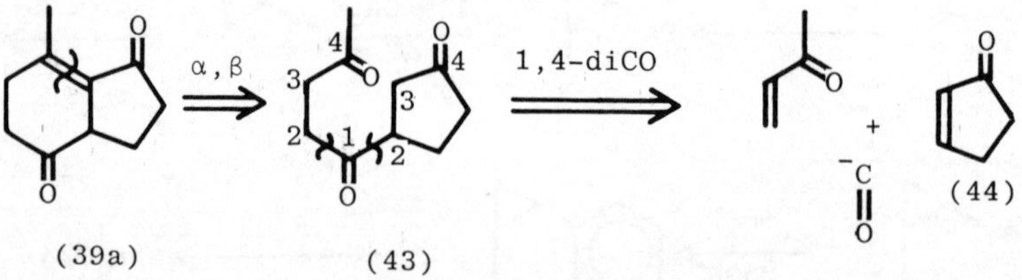

The best acyl anion equivalent for the purpose is nitromethane (Chapter 25) whose anion added cleanly first to methyl vinyl ketone and then to cyclopentenone **(44)**. No control is needed as the nitro group is as anion-stabilising as two carbonyl groups. It is removed as usual (Chapter 22) by TiCl$_3$. The cyclisation is unambiguous as no other stable ring can be formed.

Synthesis[417]

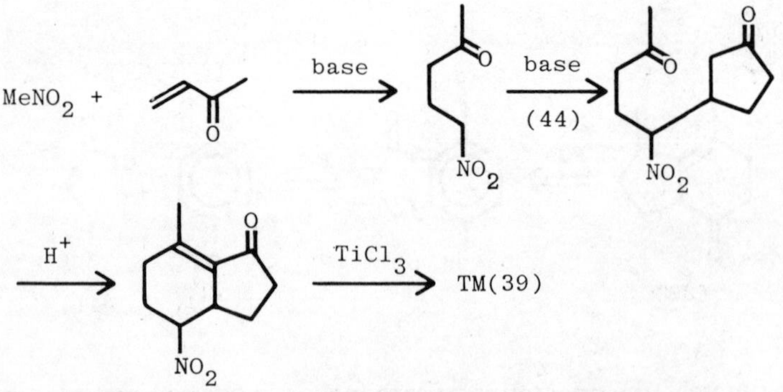

Polycyclic Compounds[418]

These should be no deterrent to logical analysis—the principles remain the same with the additional aim of reducing the number of rings as quickly as possible. This is an extension of the principle of greatest simplification (Chapter 11) and usually means disconnecting near the middle of the molecule to separate it into fragments containing only one ring.

Steroid analogue (45) has four rings. One is aromatic, so we can ignore that. The obvious first disconnection breaks ring B—the middle of the other three —so that is good strategy.

Analysis 1

(45)　　　　　　　　(46)

Disconnection of bond (a) in (46) would separate the molecule into two simple starting materials and fortunately (46) is a 1,5-diketone with a reverse Michael disconnection (Chapter 21) just where we want it.

Analysis 2

(46a)　　　　(47)　　　　(48)

The usual α,β-disconnection on (47) is not very productive as it suggests a 1,7-dicarbonyl precursor (49), but the strategically preferable ring–chain disconnection is good providing we have a reagent for acyl anion (50). We have already met the synthesis of (47) using acetylide ion as the acyl anion equivalent (Chapter 16).

310

Analysis 3

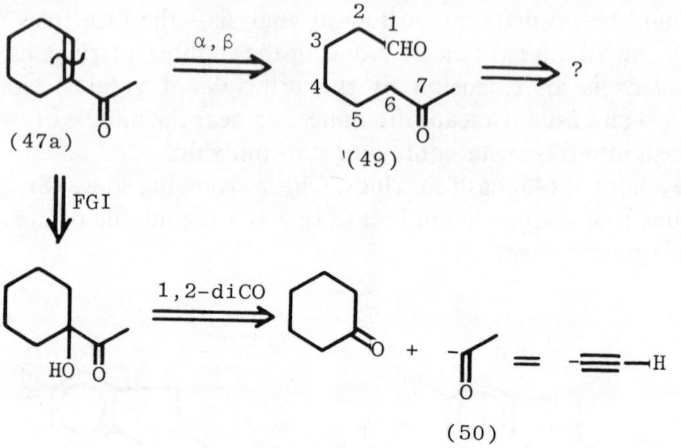

(47a)

FGI

(49)

(50)

Ketone **(48)** has the right orientation for a Friedel–Crafts reaction (in contrast to the similar one in the last problem) and the method using unsaturated acids, introduced in Chapter 35, is ideal.

Analysis 4

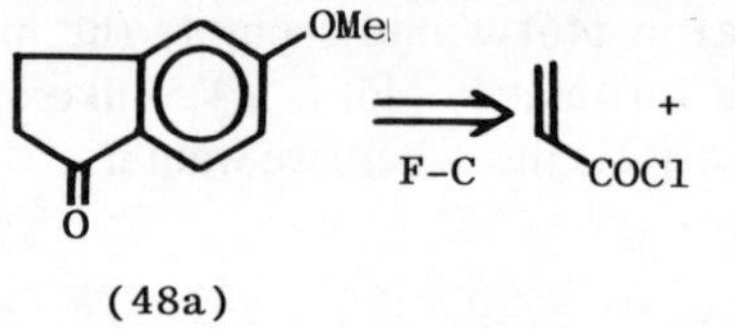

(48a)

The synthesis is more straightforward than the analysis: the dehydration of the alcohol and hydration of the triple bond in **(51)** occur in one step, the cyclisations occur spontaneously, but the final yield is poor.

Synthesis[419]

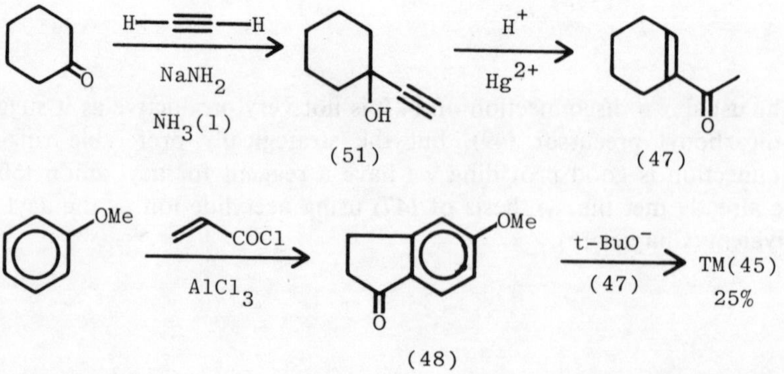

(51)

(47)

(48)

TM(45)
25%

A logical extension of this strategy, particularly important for bridged polycyclic[420] compounds, is the 'common atom' approach. Atoms common to two or more rings (the common atoms) are marked: disconnections of bonds to these atoms must inevitably reduce the number of rings.

Bridged ketone (52) was used in a synthesis of juvabione (see Chapter 38). The common atoms are marked • and disconnection of any bond to one of these atoms will give a starting material with only one ring.

(52)

The disconnections with most chemical sense are those between the functional groups (a and b) with the synthons chosen to give the carbonyl group its natural polarity.

Analysis 1

(53)   (52a)   (54)

Symmetry favours route (b) since the allyl anion in (54) is symmetrical and the alkene (55) should cyclise to (52). Loss of H^+ from cation (56) would be unambiguous as a double bond cannot be formed at a bridgehead.

Analysis 2

(54) ⟹   (55)   (56)

Intermediate (55) would be a Diels–Alder adduct if one CH_2 is removed by strategic chain shortening (Chapter 31). The orientation of the Diels–Alder is *para* and therefore correct (Chapter 17).

Analysis 3

(55a)

The synthesis was carried out using the cyanide method of chain extension (Chapter 28) rather than the Arndt–Eistert procedure (Chapter 31). The cyclisation of (55) to the target molecule was achieved by heating the acid with trifluoroacetic anhydride.

Synthesis[420, 421]

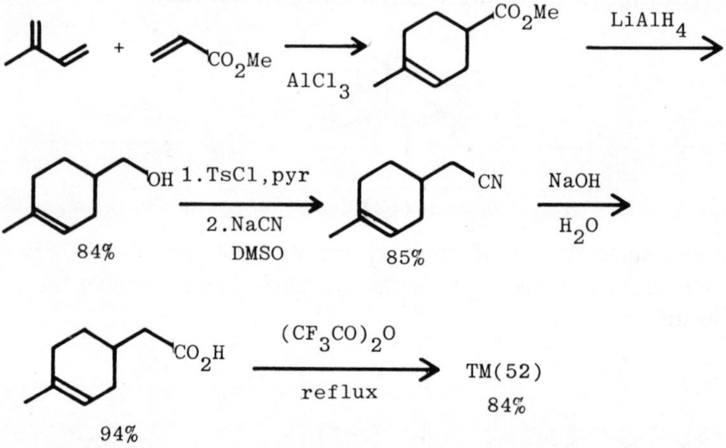

A spectacular example of the success of this strategy was the synthesis of bullvalene (57), a compound able to undergo a very large number of [3,3] sigmatropic rearrangements.* Three of the four common atoms (• in 57) are in the three-membered ring: disconnection of one common *bond* (a bond between two common atoms) will give a bicyclic system (58), but disconnection of *two* common bonds would give a single ring (59).

Analysis 1

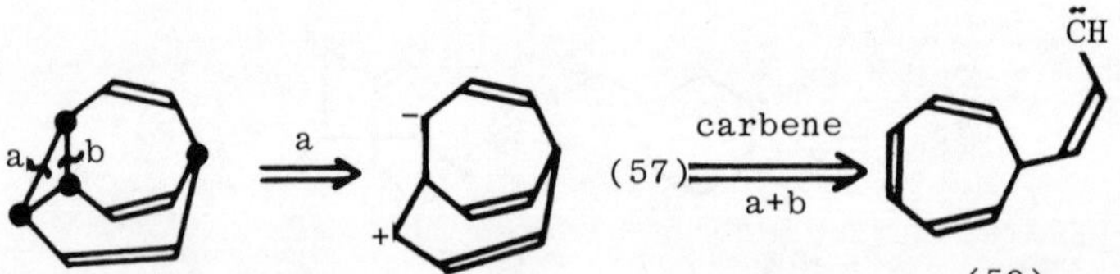

There is no chemistry corresponding to such simple disconnections, but preliminary FGI to ketones (60) or (61) looks promising as we can make α-keto carbenes (Chapter 30) from diazomethane. Since either (60) or (61) will do, a preliminary carbene disconnection (Chapter 31) to (62) allows an unambiguous common bond disconnection to (63).

*See Chapter 35 for an explanation of this term.

Analysis 2

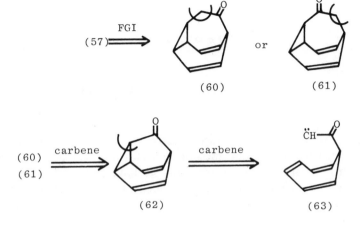

Diazoketone **(64)** will provide carbene **(63)** and this can be made in the usual way (Chapter 31) from acid **(65)**. Yet another carbene disconnection reveals that **(65)** can be made from benzene and a diazoester (Chapter 31).

Analysis 3

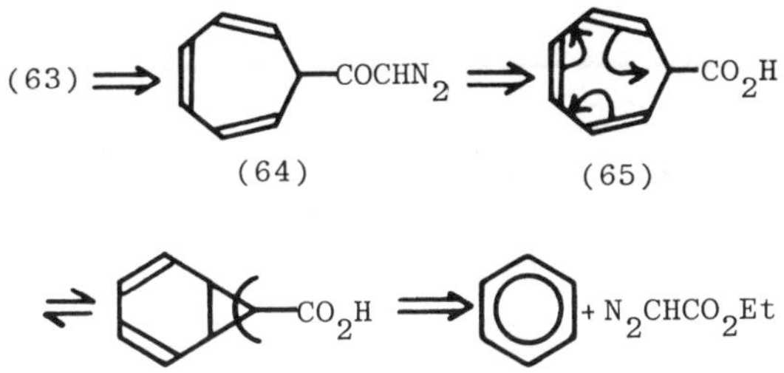

This synthesis is a sequence of carbene reactions apart from a few FGIs. In the ring expansion of ketone **(62)** it turns out that the three-membered ring migrates best, so that **(60)** is indeed formed.

Synthesis[422]

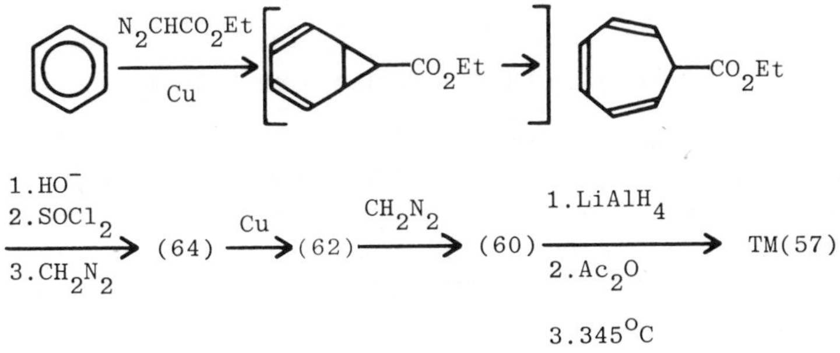

CHAPTER 38

Strategy XVII: Stereoselectivity B

In Chapter 12 we met some basic ideas on stereochemical control in synthesis. Now we reopen this important question and put into perspective the many stereospecific and stereoselective reactions we have met in the intervening chapters. Tables 38.1 and 38.2 give brief summaries. With so many methods at our disposal, a very considerable degree of stereochemical control is possible and stereochemistry is often a dominant factor in strategy.

Table 38.1   Stereospecific reactions

	Chapter
Pericyclic reactions	
Diels–Alder	17
2 + 2 Photocycloadditions	32
2 + 2 Ketene cycloadditions	33
Dienone → cyclopentenone (conrotatory)	35
Related reactions: further transformations on Diels–Alder adducts not affecting the chiral centres, e.g. oxidative cleavage	27 28
Rearrangements	
Retention of configuration in the migrating group:	
Baeyer–Villiger	27, 33
Arndt–Eistert etc.	31
Pinacol etc.	31

A supplement to Table 12.1.

Analysis of TMs with many Chiral Centres

At the start of the analysis when you have done no more than recognise the FGs and note special features or easy disconnections, note also the number of chiral centres and their position. The Prelog–Djerassi lactone (1) is an important intermediate in the synthesis of macrolide antibiotics.[423] It has a lactone ring, a carboxylic acid, and four chiral centres—three adjacent (C2–C4) and one (C6) separate.

(1)

Table 38.2   Control of sp^2 geometry

	Chapter
Making double bonds with known geometry	
Wittig reaction	15
Use of acetylenes	16
Enones: *trans* double bond	18, 19
favoured in condensations	
Claisen–Cope rearrangements	35
Conversion of sp^2 geometry to sp^3 chirality	
Diels–Alder reaction	17
stereospecific	17
stereoselective (endo adduct)	17
Synthesis of 1,2-difunctionalised compounds	23, 30
Carbene insertion	30
Photocycloaddition	32
Ketene cycloaddition	33

The adjacent chiral centres can be set up correctly by using one to set up the next and so on. For example, we have seen how an epoxide may be used to relate two chiral centres (Chapter 12), stereoselective reduction of an adjacent ketone or hydrogenation of a double bond might introduce a third, and so on. The isolated chiral centre in (1) probably cannot be set up in this way so we must start with it already in the molecule. The most appealing strategy is to have centre 6 and *one* of 2, 3, or 4 correctly set up in the starting material and put in the two remaining adjacent centres by stereoselective reactions. This strategy matches the obvious 1,3-diX (reverse Michael) disconnection (1a) as that leaves two chiral centres (4 and 6) in the starting material.

Analysis 1

(1a)   (2)

The strategy is now to synthesise (2) from some compound with the correct stereochemistry at C4 and C6 and to control the stereoselectivity of the cyclisation by adjusting the conditions. Whether this is possible can only be discovered by experiment, but one centre at least ought to be easy (C3 in 3) as the substituents round the six-membered ring are all equatorial.

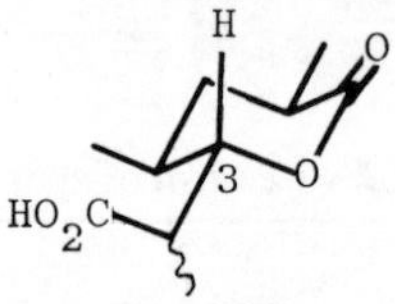

(3)

α,β-Disconnection of (2a) gives aldehyde acid (4). Available meso-diacid (6) has the right stereochemistry and its anhydride (5) is ideal for chemoselectivity (Chapter 5) between the two carbonyl groups.

Analysis 2

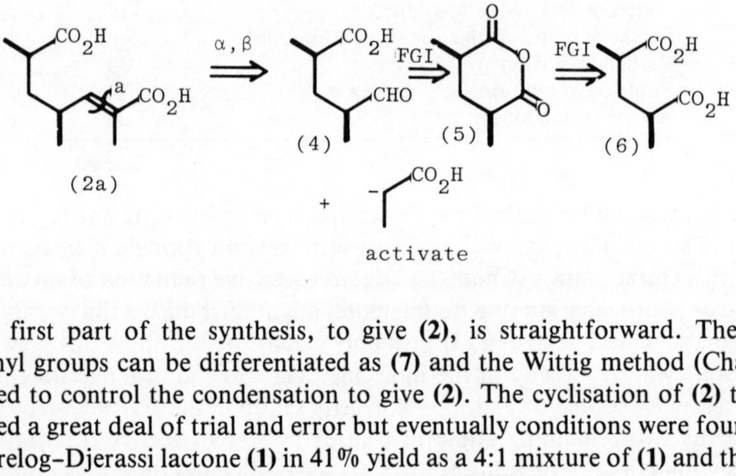

The first part of the synthesis, to give (2), is straightforward. The two carbonyl groups can be differentiated as (7) and the Wittig method (Chapter 15) used to control the condensation to give (2). The cyclisation of (2) to (1) required a great deal of trial and error but eventually conditions were found to give Prelog–Djerassi lactone (1) in 41% yield as a 4:1 mixture of (1) and the C2 epimer (see compound 3) separable by chromatography.

Synthesis[423]

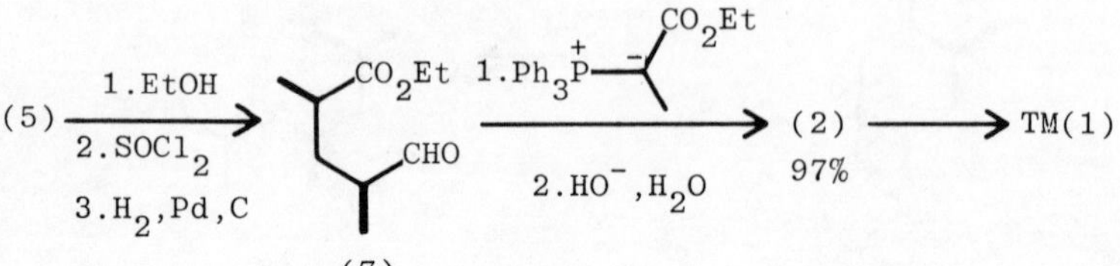

The Diels–Alder reaction offers a different approach to correct relationships of distant chiral centres. Intermediate **(8)** was needed[424] for an alkaloid synthesis. It has three chiral centres—two adjacent (C3, C4) and one remote (C8). The 1,6-dialdehyde suggests a reconnection to **(9)** and this is a Diels–Alder adduct **(10)** (after removal of the acetal) from cyclohexenone **(11)**.*

Analysis

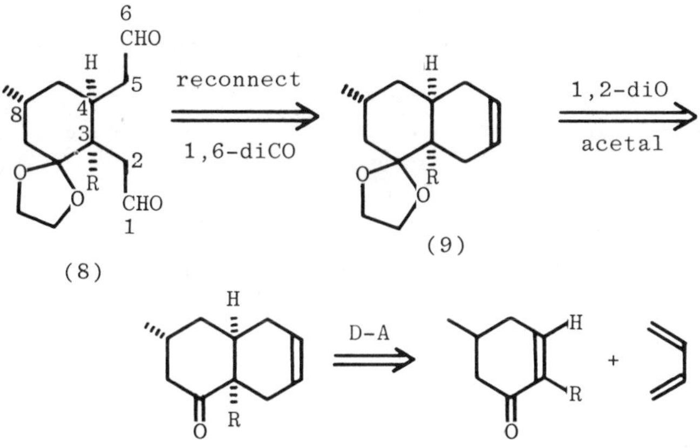

(8) (9) (10) (11)

All the stereochemical control is provided by the Diels–Alder reaction. Butadiene will add to **(11)** on the opposite side to the methyl group and give **(10)**.

Synthesis[424]

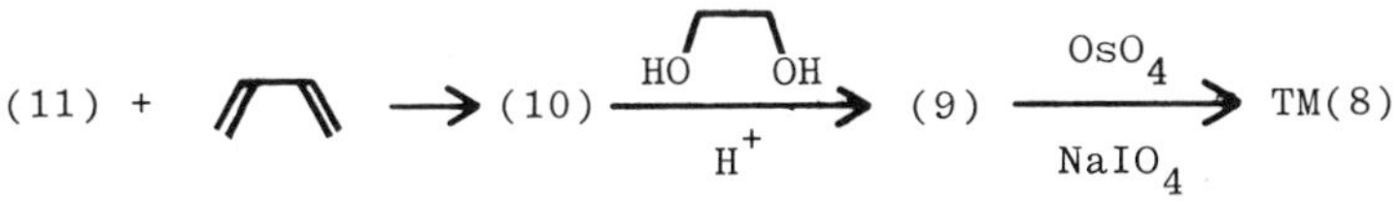

Good control is available when working with bowl shaped molecules such as *cis*-decalins **(12)**, e.g. **(9)** and **(10)**, as approach of reagents must be from outside the 'bowl' **(13)**, i.e. from the *same* side as the ring junction hydrogens.

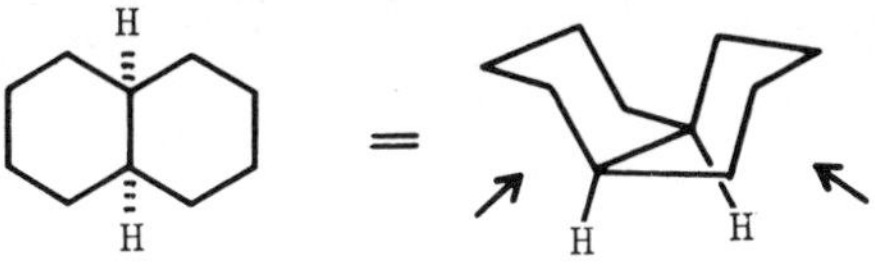

(12) (13)

*See Chapter 37 for the synthesis of molecules of this type.

Even the Robinson annelation product (14), made in Chapter 21, is reduced selectively to (15), though it is only a shallow bowl. Note the chemoselectivity: the conjugated ketone is not reduced.

(14) (15)

Heathcock's synthesis of the strange polycyclic terpene copaene (16) illustrates this approach well. Copaene was made from ketone (17) by a series of relatively trivial reactions: it is the synthesis of the skeleton (17) which is interesting. The central four-membered ring contains all the common atoms (• in 17) but disconnection of any two bonds, e.g. by a 2 + 2 approach to give (18), gives a difficult ten-membered ring.

Analysis 1

(16) (17) (18)

The carbonyl group in (17) suggests an alternative disconnection corresponding to the intramolecular alkylation of enolate (19), and this gives a *cis*-decalin (20) as an intermediate.

Analysis 2

(17a) (19) (20)

Cis-decalone (20) has four adjacent chiral centres. One (C6) needs no control as it is to be oxidised to a ketone in the transformation to (16). The other three must be as shown in (20) if the cyclisation of (19) to (17) is to occur because the S_N2 displacement goes with inversion. Disconnection (17a) also appeals because the two adjacent FGs at C5 and C6 in **(20)** can come from a double bond (Chapter 23), suggesting the next disconnection to **(21)**.

Analysis 3

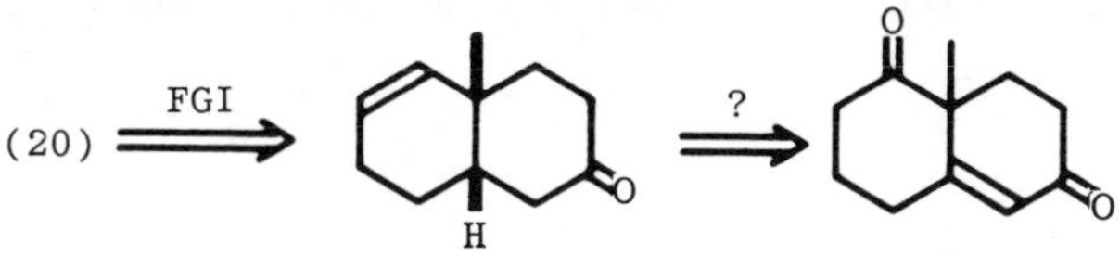

(21) (14)

The structure of **(21)** is reminiscent of Robinson annelation product **(14)**, a readily available starting material (Chapter 21). Reduction of one ketone (see page 318) gives **(15)**, converted at once to a tosylate **(22)** for later elimination. Hydrogenation requires the double bond of **(22)** to lie flat on the catalyst surface (Chapter 12) and this is possible only from the outside of the 'bowl' (see **22a**) giving *cis*-decalone **(23)**.

Synthesis[425] *1*

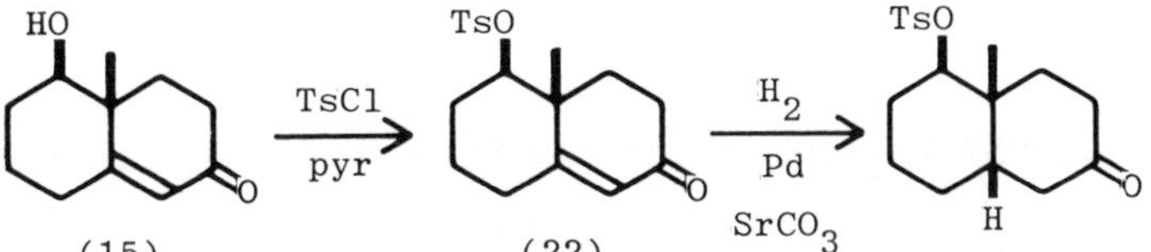

(15) (22) (23)

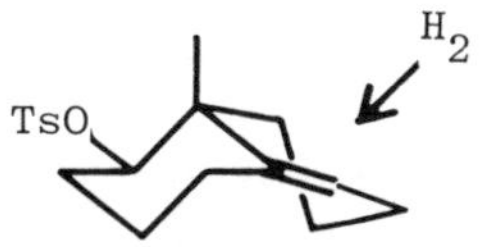

(22a)

The three essential chiral centres in **(20)** are now correctly set up. The extra oxygen FG must now be introduced. Elimination gives **(21)**, and epoxidation of the double bond after protection of the ketone takes place from the outside of the 'bowl' to give **(24)**.

Synthesis[425] *2*

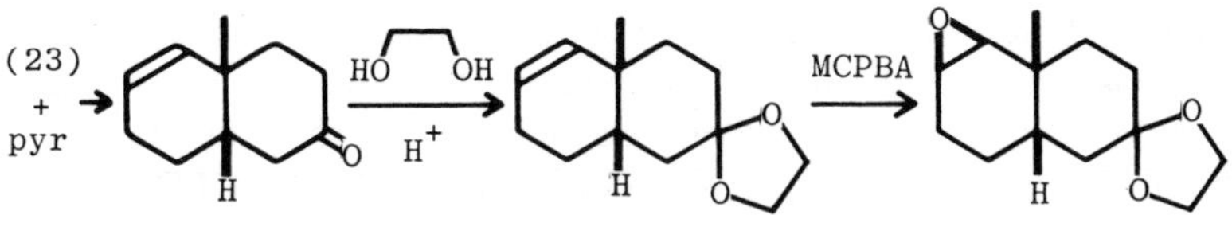

(21) (24)

320

Introduction of the protected hydroxyl group OR is regioselective at C6 (*trans*-diaxial opening at the less hindered centre—see Chapter 12), giving a compound (25) with all the correct FGs and stereochemistry. Tosylation of the free hydroxyl group provides the leaving group and liberation of the ketone allows cyclisation to (27). Compound (27) was converted into copaene (16).[425]

Synthesis[425] *3*

(24) $\xrightarrow[\text{PhCH}_2\text{OH}]{\text{PhCH}_2\text{O}^-}$ (25, R=CH$_2$Ph) $\xrightarrow[\text{2.H}^+,\text{H}_2\text{O}]{\text{1.TsCl,pyr}}$

(26) $\xrightarrow[\text{DMSO}]{\text{base}}$ (27)

Synthesis to Establish Stereochemistry of Natural Products

Natural products are often isolated in amounts too small to permit the determination of their stereochemistry: if we can get the gross structure correct then we are more than pleased. The stereochemistry is then found by stereo-controlled synthesis of the various possible isomers and comparison of authentic synthetic samples with the natural product. The importance of precise and predictable stereochemistry is obvious here.

Büchi[426] carried out this daunting task on the flavouring compound aromadendrene (28); he synthesised different isomers, all with known stereochemistry, and (28) proved to be identical with the natural product. We shall discuss only this isomer.

(28)

Aromadendrene (28) has only one FG—an exomethylene double bond. The skeleton has three fused rings: three-, seven-, and five-membered, and five chiral centres, all adjacent, so we can probably use one to set up the rest. Only one disconnection—the Wittig—is available without FGA, so the ketone (29) must be an intermediate.

Analysis 1

(29)

We know very little about controlling the stereochemistry in five- or seven-membered rings, but a great deal about controlling the stereochemistry in six-membered rings. The best strategy for ketone **(29)** is to make it by stereospecific rearrangement **(29a)** (see Chapter 31) from a decalin **(30)** and set up all five chiral centres correctly in the decalin. The stereochemistry of the hydroxyl group at the ring junction in **(30)** is unimportant.

Analysis 2

(29a)

(30, X=leaving group)

As in the copaene synthesis **(20)**, the two adjacent FGs can be made from a double bond, i.e. from intermediate **(31)**. Hydroxylation of bowl-shaped **(31)** ought to occur from the outside, giving the correct stereochemistry of **(30)**. The cyclohexene in **(31)** could be made by a Diels–Alder reaction if there were a carbonyl group instead of the methyl group, as in **(32)**, and FGA here is worthwhile as it leads to such a great simplification.

Analysis 3

(30)   (31)   (32)   (33)

Worthwhile, that is, if the regio- and stereoselectivity of the Diels–Alder reaction are correct. Ths two alkyl groups attached to the diene in **(33)** direct different ways (Chapter 17) but experiments had shown that the three-membered ring at the end of the diene is more powerful than the CH_2 near the middle. The stereoselectivity is more complicated. The dienophile should add to the underside of (33) to keep away from the methyl groups on the three-membered ring. This pushes the ring junction hydrogen up. *Endo* selectivity should then put the COR group down (see 34).

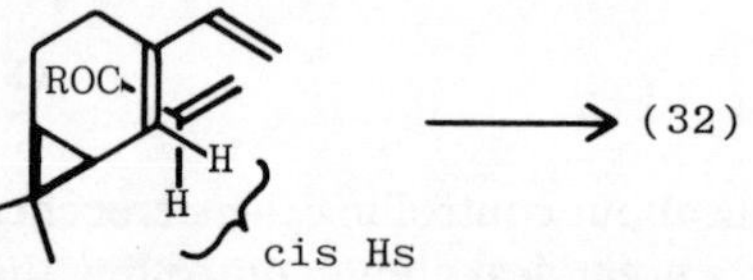

(34)

Further Wittig analysis of (33) gives aldehyde (35). Optically active (–)-perillaldehyde (36) is available (it is a natural product) so disconnection (35a) becomes very attractive. Formation of three-membered rings by displacement at a tertiary centre is known (Chapter 30).

Analysis 4

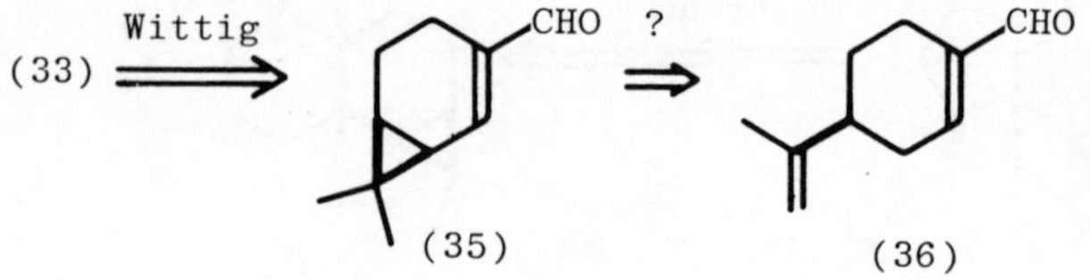

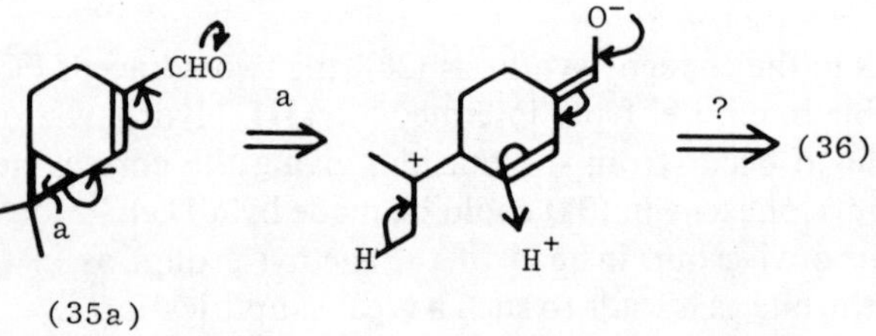

In the event,[426] treatment of (36) with HBr and then strong base gave (35) with two centres established. The diene (33), duly synthesised by a Wittig reaction, gave an 85:15 ratio of *endo* (37) to *exo* Diels–Alder adducts with acrolein ($CH_2=CH.CHO$). All the addition was from underneath.

Synthesis 1

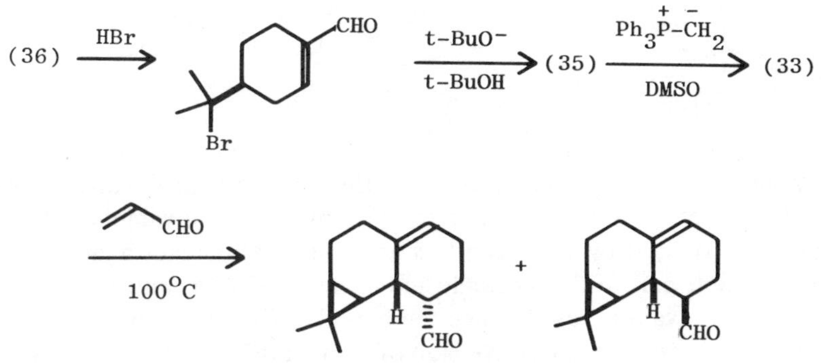

(37)endo,85: :15,exo

The aldehyde was reduced to a methyl group in stages and hydroxylation put in the remaining chiral centres as predicted. Chemoselective tosylation of the less hindered alcohol provided the leaving group in (38) which rearranged as planned to give (29) and hence optically active aromadendrene.

Synthesis 2[426]

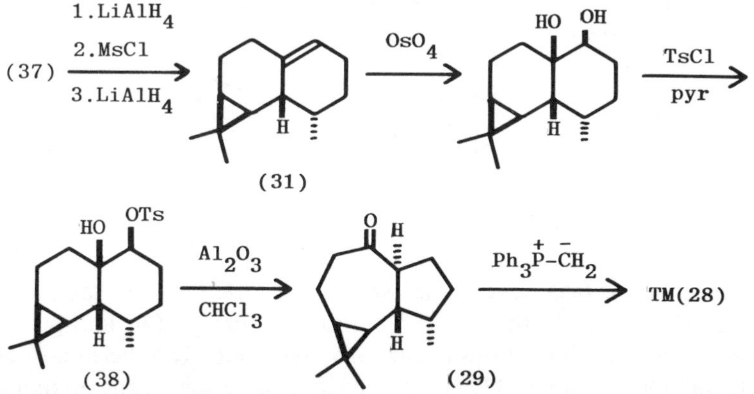

So far all our examples have been cyclic compounds as stereochemistry is much easier to control in rings than in chains. One strategy for open chain compounds is to use a rigid structure — an alkene, aromatic compound, or ring — to set up the stereochemistry correctly and then to break down the rigid structure to reveal the open chain.

Cage molecules are particularly suitable for this strategy as very often only one stereochemistry is possible if the molecule is to exist. Camphor (39) has two chiral centres, but only one diastereoisomer can exist. Fragmentation *via* (40) must give *cis* diacid (41).

(39) → SeO₂ → (40) → H₂O₂ / HO⁻ → (41)

Camphoric acid **(41)** is still cyclic, but the method applies equally well to open chain compounds. The two chiral centres in juvabione **(42)**, the compound produced by some conifers as a chemical defence against moths, are on a flexible part of the molecule and cannot easily be controlled. Monti's synthesis[427] of juvabione used a cage compound to fix the stereochemistry.

Minor reactions allow the conversion of alcohol **(43)** into juvabione without affecting the two chiral centres. Alcohol **(43)** could be made by reduction of **(44)** which can be reconnected to **(45)**. The reaction **(45 → 44)** is a fragmentation, shown by the arrows on **(45a)**.

Juvabione: *Analysis*

(42)   (43)   (44)

reconnect ⟹ (45, Y=leaving group)   (45a)   → (44)

Rewriting **(45)** as **(45a)** reveals the two chiral centres (•) for juvabione. Note also that the stereochemistry of Y must be correct for anti-periplanar elimination, that is with Y *trans* to the carbonyl group. The skeleton of **(45)** is available as **(46)** (Chapter 37) but methylation gave the wrong stereoisomer **(48)**.

Synthesis 1 [427]

(46)   → base →   (47)   →   (48)   84%

This suggests that enolate **(47)** is attacked by electrophiles from the less hindered side and so formation of the enolate **(49)** and its capture by a proton gave the correct stereoisomer **(50)**.

Synthesis 2

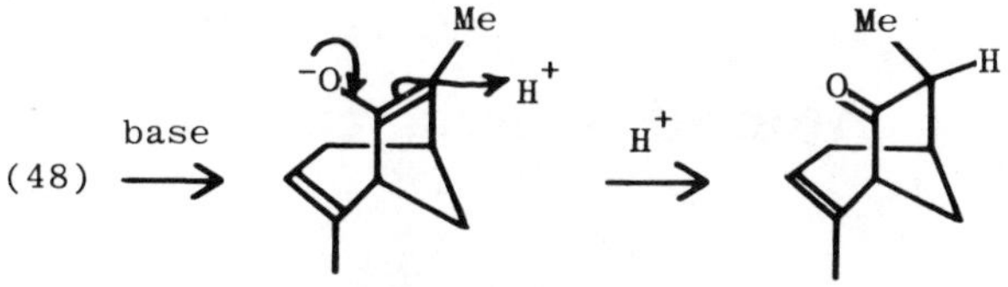

Addition of HCl to **(50)** occurs on the less hindered underside of the double bond to give **(51)** which duly fragmented to **(44)** with nucleophiles. A short cut was to use LiAlH$_4$ for the fragmentation so that the aldehyde **(44,** X = H) is formed and reduced to **(43)** in the same step.

Synthesis 3

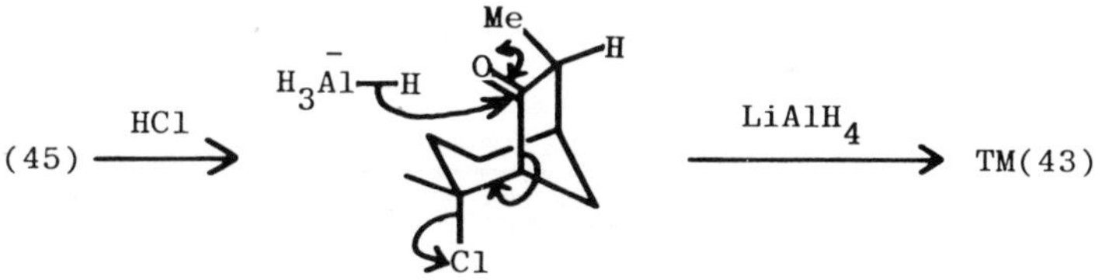

CHAPTER 39

Aromatic Heterocycles

Substituted furans (1), isoquinolines (2), and other aromatic heterocycles have wide applications as pharmaceuticals and in perfumery, agricultural, and colour chemistry. One important role is to help convey a biologically active substitutent into the living cell.

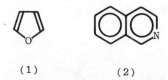

(1)　　　　　(2)

We use the same disconnections as we used for saturated heterocycles in Chapter 29, but our task is easier because the thermodynamic stability of the aromatic rings allows short cuts. This chapters introduces few new principles and the synthesis of the carbon frameworks can be useful revision.

It is particularly important to recognise the oxidation levels of the carbon atoms bonded to the heteroatom. We already know how to disconnect at three oxidation levels (3), (4), and (5) (using nitrogen as an example). In the aromatic series, compounds such as (6), (7), and (8) all have carbon atoms (•) at the oxidation level of a carboxylic acid.

Alcohol

(3)

Aldehyde and ketone

(4a)　　　　or　　　　(4b)

326

Carboxylic acid

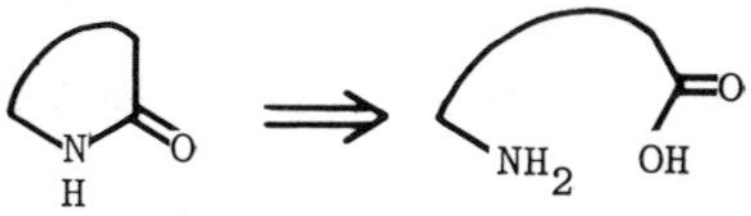

(5)

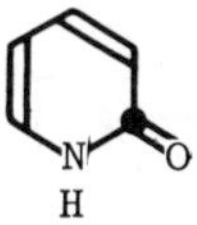

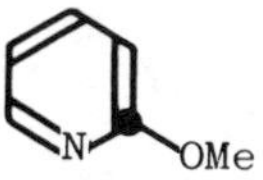

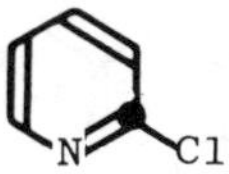

(6)   (7)   (8)

It is impossible to discuss all the aromatic heterocycles so I shall select examples from some of the most important types.

Five-Membered Rings: Furan and Pyrrole

The simplest disconnection of furans **(1)** or pyrroles **(9)** is the removal of the heteroatom leaving a 1,4-dicarbonyl compound **(10)** to be made by the methods of Chapter 25.

Analysis

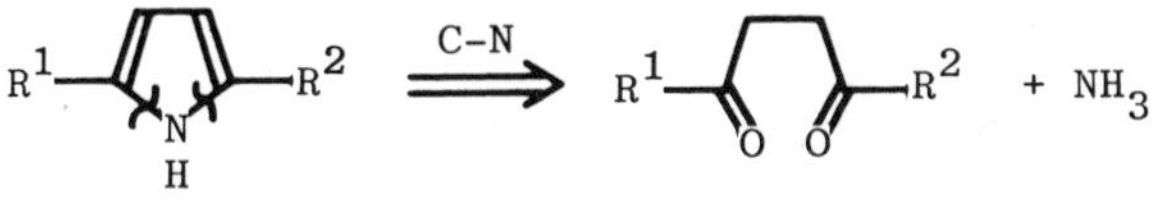

(9)   (10)

Bicyclic furan **(11)** comes from diketone **(12)** and the ring–chain disconnection gives simple starting materials. The enamine method of control (Chapter 20) gives good results[428] and the cyclisation needs no added oxygen nucleophiles as there are two oxygen atoms available in **(12)**.

Analysis

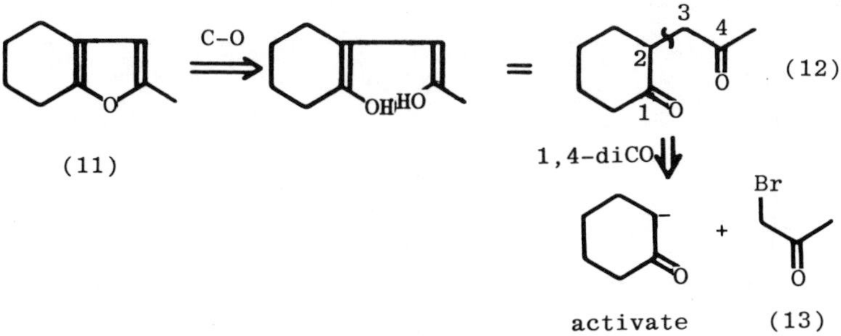

(11)   1,4-diCO

activate   (13)

Synthesis[428]

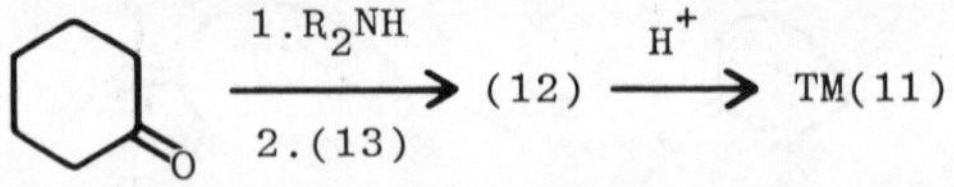

Cyclisation to furans occurs readily in acid: base must be avoided as it gives cyclopentenones, e.g. **(14)**, instead by carbonyl condensations (Chapter 34).

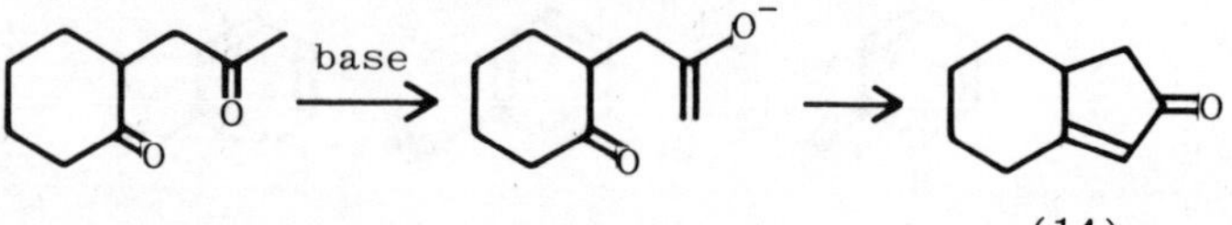

(12)   (14)

Pyrroles **(10)** can be made the same way, the cyclisation being carried out with ammonia, but an alternative strategy is particularly valuable for carbonyl-substituted pyrroles. Pyrrole esters such as **(15)** are needed for the synthesis of porphyrins (as in haemoglobin), chlorins (as in chlorophyll), and corrins (vitamin B_{12}). Ester **(15)** has the haem side chain and can be converted by hydrolysis and decarboxylation into a pyrrole **(16)** with a reactive free position (H in **16**).

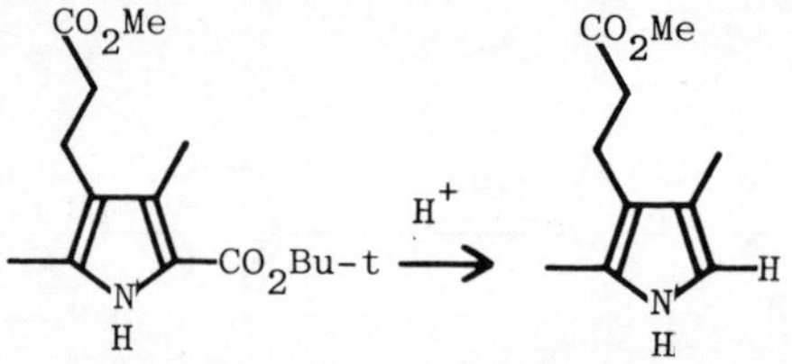

(15)   (16)

Disconnection of the C–N bond on the other side from the ester group gives **(17)** or **(18)**. It does not matter where the double bond is to start with since it will migrate when required.

Analysis 1

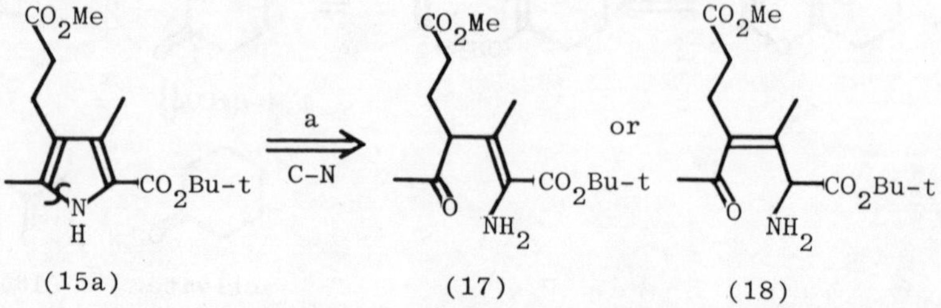

(15a)   (17)   (18)

α,β-Disconnection of enone (18) gives a simple 1,5-dicarbonyl compound (19) and an amino ketone (20) which can be made by reduction of oxime (21) and hence from acetoacetate by nitrosation (Chapter 23).

Analysis 2

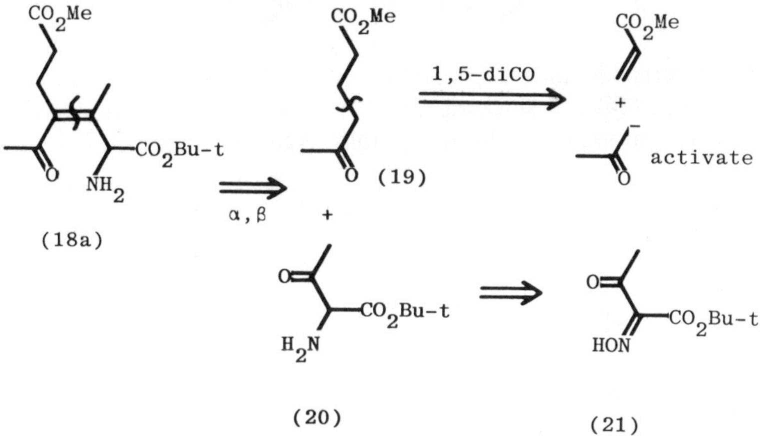

Such a long analysis suggests a daunting synthesis, but the stability of aromatic compounds means that reaction of (20) and (19) gives (15) in one step. All we have to do is to construct (22) by standard CO_2Et activated methods and combine it with (21) under reducing conditions. Reduction, condensation, decarboxylation all occur together. This is the Hantszch pyrrole synthesis.[429]

Synthesis

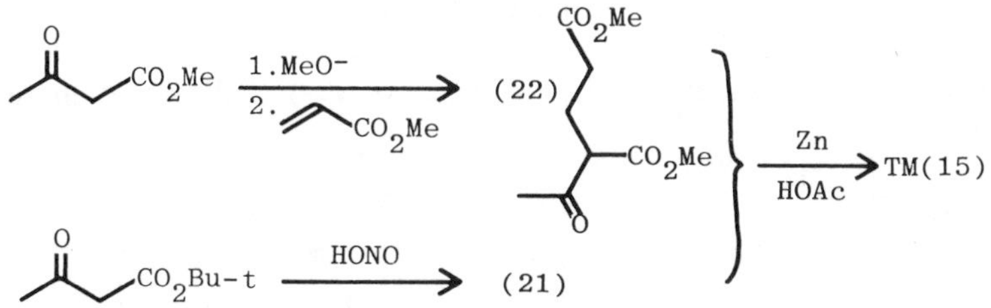

Electrophilic substitution

Electrophilic substitution is less useful on pyrroles and furans than on benzenes (see Chapters 2 and 3) as the less stable five-membered rings are destroyed by strong acids. Nitrations and Friedel–Crafts acylations can still be carried out under mild conditions; the α-positions are attacked first, or, if these are blocked, the β-positions are attacked almost equally as well.

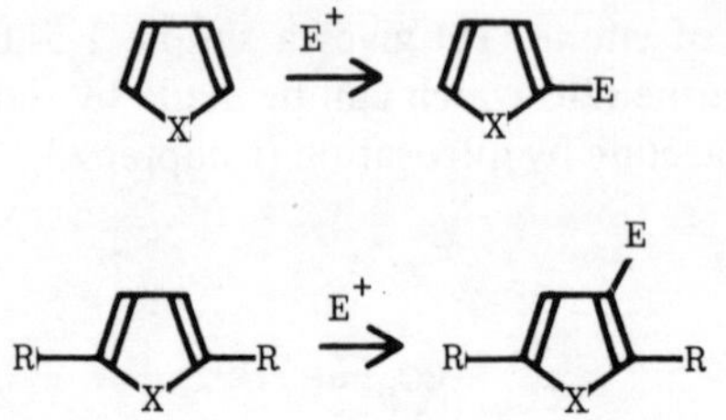

The bacteriocide nitrofurazone **(23)** is an imide made from available semicarbazide **(24)** and aldehyde **(25)**. The nitro group is in an α-position so C–N disconnection to very cheap furfuraldehyde **(26)** is good.

Analysis

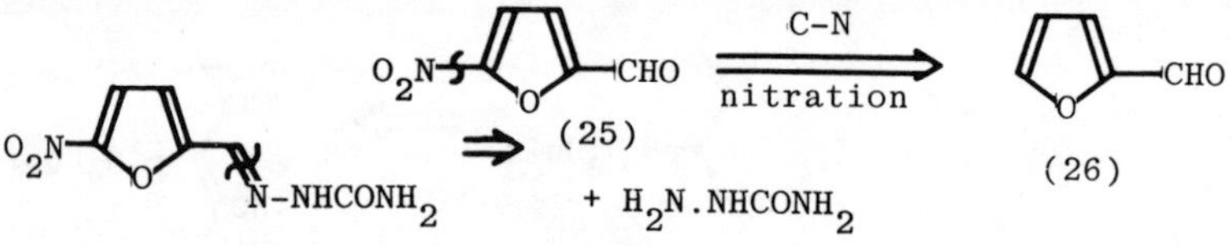

Nitration of these heterocycles with the salt NO_2^+ AcO^- formed *in situ* from nitric acid and acetic anhydride avoids decomposition of the furan ring.

Synthesis[430]

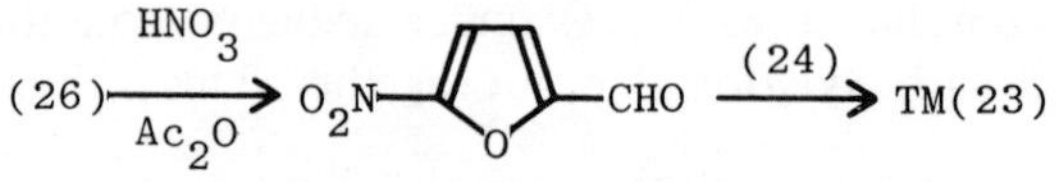

The coronary vasodilator Benziodarone **(27)** is an iodination product of phenol **(28)**. Friedel–Crafts disconnection is sensible as the α-position on the benzofuran **(29)** is blocked. The synthesis of **(29)** is discussed in the next section.

Analysis

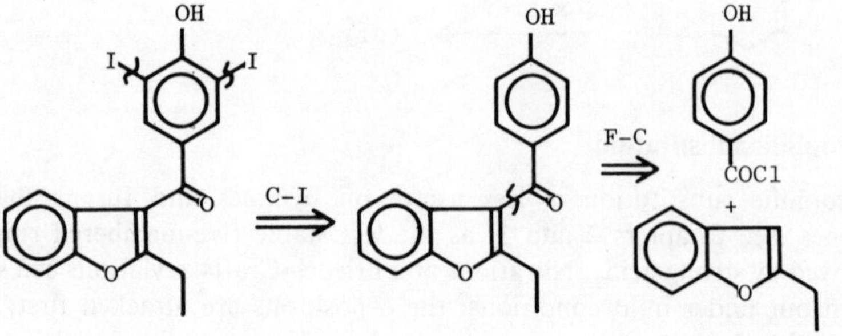

Protection of the hydroxyl group will be needed during the Friedel–Crafts reaction and a methyl ether is easiest as acid **(30)** is available. The benzofuran **(29)** is reactive enough for a Friedel–Crafts catalyst to be unnecessary. Methyl ether protecting groups are usually removed with HBr or HI (see Table 9.1) but these strong acids would destroy the furan. Pyridinium chloride proves satisfactory in practice.

Synthesis[431]

Benzofurans and indoles

The benzo derivatives benzofuran **(31)** and indole **(32)** cannot be analysed in quite such a simple fashion as the parent molecules.

Disconnection of the benzofuran **(29)** by the usual methods gives a phenolic ketone **(33)**. Further Friedel–Crafts disconnection, e.g. to **(34)**, is unpromising as the carbonyl group in **(33)** is in a most unhelpful position. If it were one nearer the ring, **(33)** could be made by a Friedel–Crafts reaction: one atom further away and we could use FGA (Chapter 24).

Analysis 1

FGA provides a successful strategy. If we add a carbonyl group to **(29)**, we get enone **(35)**, with a helpful α,β-disconnection to **(36)** and thence salicyl-aldehyde **(37)** and chloroacetone as the available starting materials.

Analysis 2

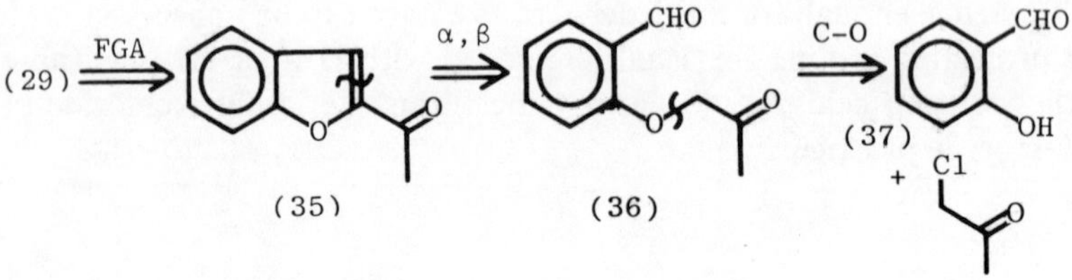

$$(29) \xrightarrow{\text{FGA}} (35) \xrightarrow{\alpha,\beta} (36) \xrightarrow{\text{C-O}} (37) + Cl$$

The synthesis of **(35)** is straightforward. Removal of the carbonyl group cannot be done under acidic conditions or the furan will decompose: the Wolf–Kishner method (see Table 24.1) works very well.

Synthesis[431]

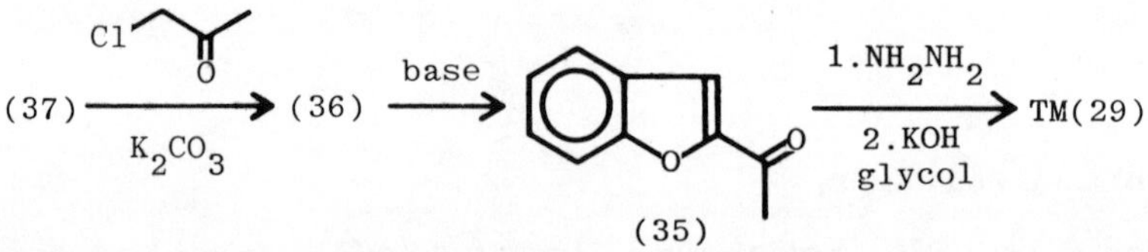

$$(37) \xrightarrow[K_2CO_3]{Cl} (36) \xrightarrow{\text{base}} (35) \xrightarrow[\substack{2.KOH \\ glycol}]{1.NH_2NH_2} TM(29)$$

Indoles are a special case as there is one pre-eminent method: the Fischer indole synthesis. The phenylhydrazone **(38)** of a ketone is treated with acid (or Lewis acid—ZnCl$_2$ is often used). Tautomerisation to **(39)** is followed by a [3,3] sigmatropic shift (Chapter 35) to exchange the weak N–N bond for a strong C–C bond. The rest is downhill to the stable aromatic indole **(40)**.

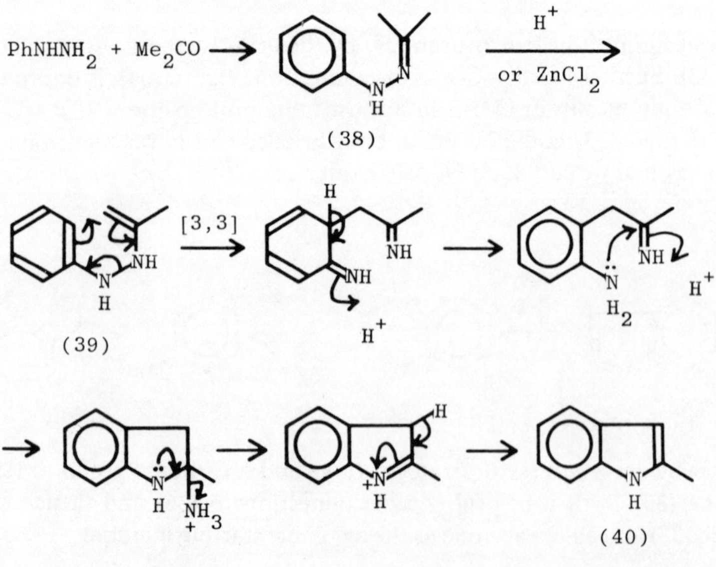

$$PhNHNH_2 + Me_2CO \longrightarrow (38) \xrightarrow[\text{or } ZnCl_2]{H^+}$$

$$(39) \xrightarrow{[3,3]} \longrightarrow$$

$$\longrightarrow \longrightarrow \longrightarrow (40)$$

Working backwards through the Fischer indole synthesis, the 'obvious' disconnection **(41)** reveals the position of the carbonyl group, and with the disconnection of the C–C bond **(42)** gives the starting materials. These were the disconnections we explored and rejected for **(29)**, but which now lead to an excellent indole synthesis.

Analysis

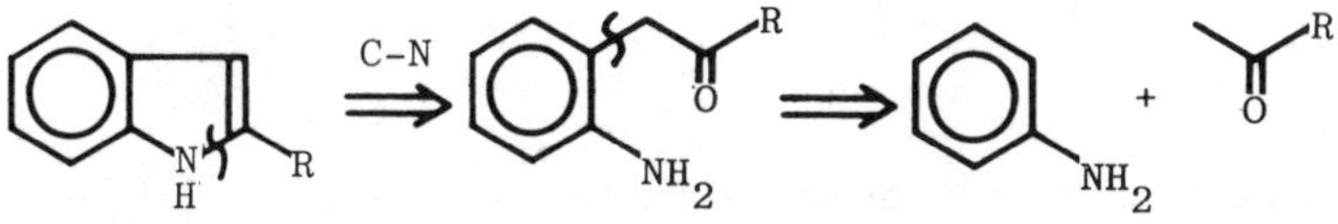

(41)   (42)   (-NHNH$_2$)

The antidepressant Iprindole **(43)** is obviously made from an indole **(44)** by alkylation with **(46)**. Fischer indole disconnection reveals available cyclic ketone **(45)** as the starting material. Disconnection of **(46)** is standard C–X work from Chapter 6.

Analysis

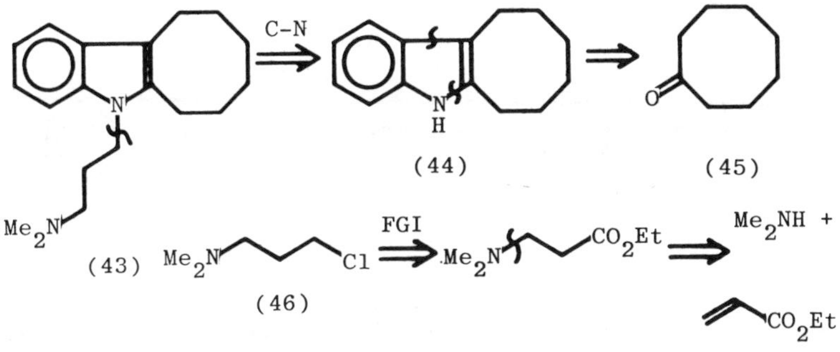

(44)   (45)

(43)   (46)

Zinc chloride was used for the Fischer indole synthesis and the sodium salt of **(44)** was used in the alkylation step.

Synthesis[432]

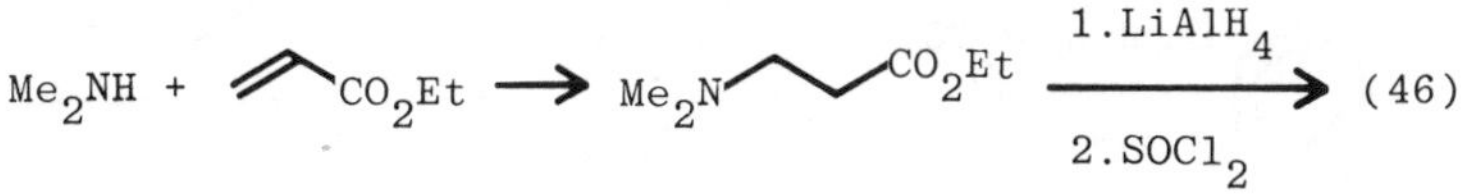

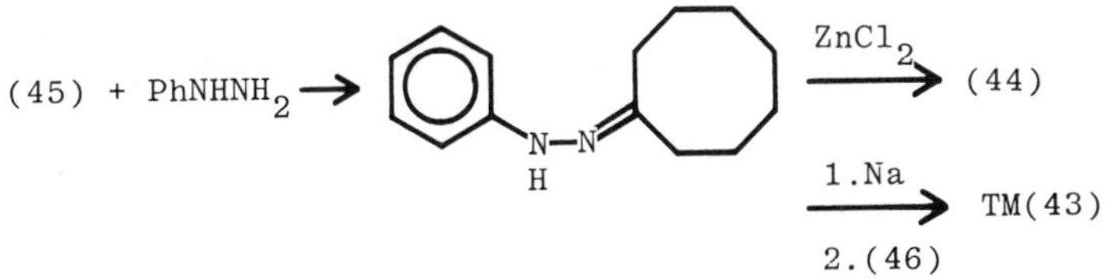

Me$_2$NH + $\longrightarrow$CO$_2$Et $\rightarrow$ Me$_2$N$\longrightarrow$CO$_2$Et $\xrightarrow[\text{2.SOCl}_2]{\text{1.LiAlH}_4}$ (46)

(45) + PhNHNH$_2$ $\rightarrow$ $\xrightarrow{\text{ZnCl}_2}$ (44)

$\xrightarrow[\text{2.(46)}]{\text{1.Na}}$ TM(43)

If the ketone is unsymmetrical, the more stable enamine—that is the more substituted—is formed. Indoles of type (47) can be made by this route.

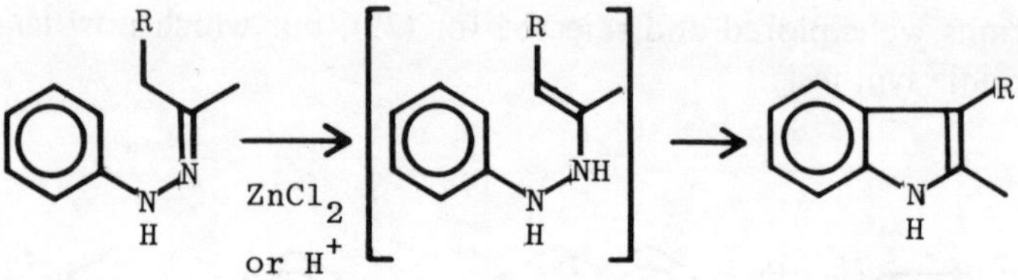

(47)

Fischer indole disconnection of (48) requires keto acid (49) and a substituted hydrazine (50). Keto acid (49) is a standard 1,4-dicarbonyl problem (Chapter 25). Phenylhydrazines such as (50) are made by reduction of diazonium cations derived from aromatic amines (51).

Analysis

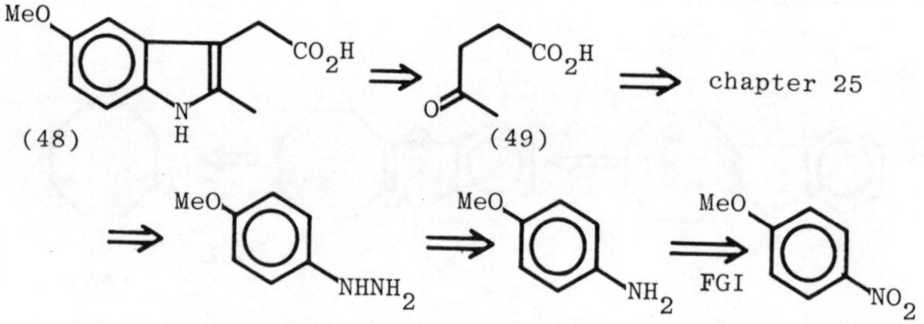

(48)   (49)   chapter 25

(50)   (51)

This type of indole, with a 5-MeO group, is the basis of many pharmaceuticals—acid (48) is used to make Indomethacin, a non-steroidal anti-inflammatory agent.[433] The synthesis is straightforward if the acid is protected as an ester, HCl being used to catalyse the Fischer step.

Synthesis[434]

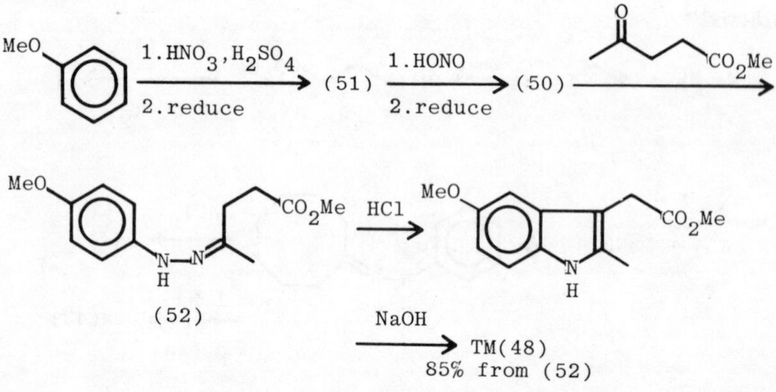

(51)

(50)

(52)   NaOH
$\longrightarrow$ TM(48)
85% from (52)

Six-Membered Rings

Pyridines

Pyridine itself and a variety of substituted pyridines are available. Other pyridines have to be synthesised and the usual C–N disconnections, this time to 1,5-dicarbonyls (53) lead to good syntheses. It is often easier to leave out the remaining double bond (54) as dihydropyridines are easily oxidised to pyridines.

(53)   (54)

Pyridine diester (55) comes from (56) by this logic. Further 1,5-disconnection (Chapter 21) of symmetrical (56) eventually reveals two acetoacetones and formaldehyde as starting materials.

Analysis

(55)   (56)

The synthesis is easier than expected as heating together a mixture of acetoacetate ester, formaldehyde, and ammonium acetate gives dihydropyridine (57). Oxidation with a quinone such as DDQ (Chapter 36) gives the pyridine.

Synthesis[435]

(57)

With unsymmetrical double bonds, one position of the remaining double bond in (53) may be more helpful than the other. Disconnection of (58) could give either (59) or (60): only (60) has an α,β-disconnection to available reagents. Again the synthesis is very simple.

Analysis

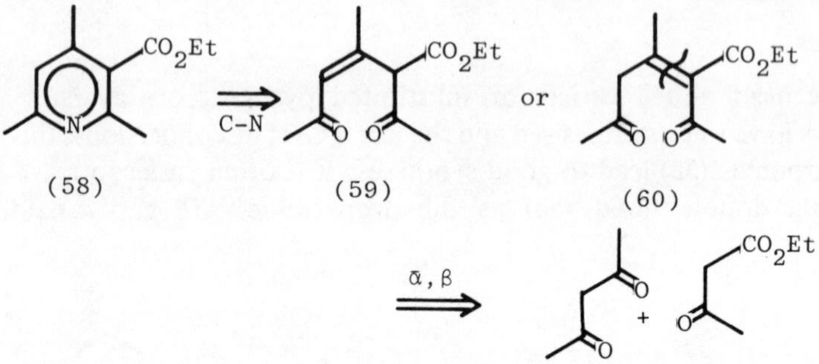

(58) (59)   or   (60)

Synthesis[436]

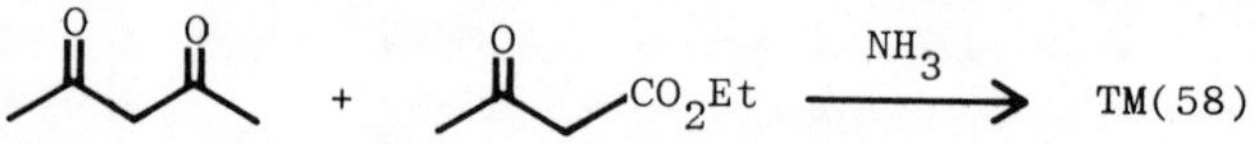

Quinolines

Here again a special method predominates—the Skraup synthesis—which proves to be quite logical in spite of its reputation as a 'witches brew' reaction. Quinoline (61), a perfumery compound with a 'civet-honey' smell,[437] might be disconnected to give an amine and a synthon (63).

Analysis

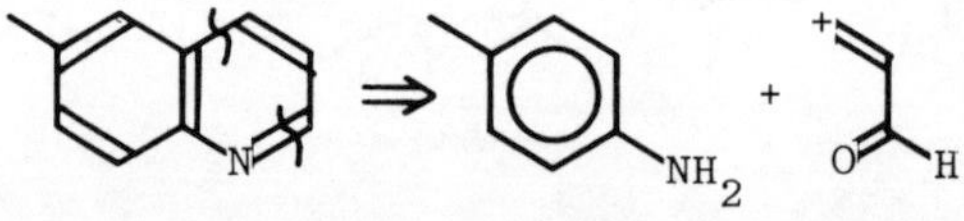

(61)   (62)   (63)

The simplest answer for (63) is to use acrolein, $CH_2{=}CH.CHO$, and oxidise. This is the basis of the Skraup synthesis. He made (61) in 1871, generating acrolein from glycerol, though nowadays it is readily available. His oxidising agent—*p*-nitrotoluene (64)—was chosen to regenerate the other reagent (62), an early form of recycling.

Synthesis[438]

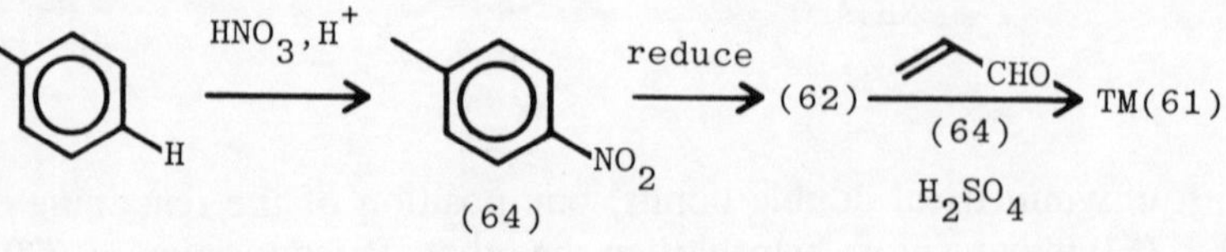

(64)

Isoquinolines

Isoquinolines and their relatives, e.g. **(65)**, occur widely in nature and are easier to make than quinolines. An obvious disconnection on **(65)** removes one carbon atom from the ring at the aldehyde oxidation level.

Analysis 1

MeO

MeO

NH

R

(65)

⟹

MeO

MeO

NH₂

CHO

R

(66)

The reaction is an internal Mannich reaction (cf. Chapter 20) and goes under mild conditions. The amine **(66)** is an ether of the physiologically active dopamine and is made *via* cyanide **(67)**.

Analysis 2

(66) $\xrightarrow[\text{FGI}]{\text{reduction}}$ Ar⌒CN ⟹ Ar⌒Cl ⟹ ArH + CH₂O + HCl

(67)   chloromethylation

The first step is chloromethylation (Chapter 2) at the most reactive position (less hindered). Cyanide displacement and reduction give **(66)** which cyclises with the aldehyde to give **(65)** in acid solution.

Synthesis[439]

MeO

MeO

$\xrightarrow[\text{2.KCN}]{\text{1.CH}_2\text{O,HCl}}$
3.reduce

MeO

MeO

NH₂

(66)

$\xrightarrow[\text{H}^+]{\text{RCHO}}$ TM(65)

An alternative is to make amide **(68)** from **(66)** and to cyclise it with POCl₃. The product can be reduced to **(65)** or oxidised to the isoquinoline **(69)**.

Synthesis[440]

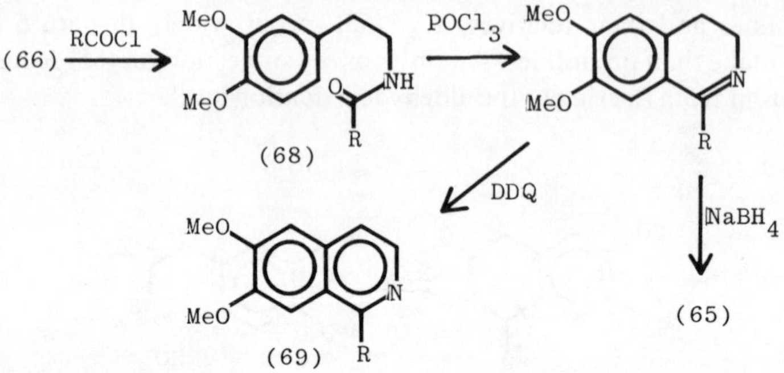

Coumarins and chromones

Coumarins **(70)** and chromones **(71)** with other benzo derivatives number some important commercial compounds among their ranks. Coumarin itself **(70)** is the new-mown hay smell and the rat poison Warfarin **(72)** is a substituted coumarin. No special methods are needed for these easily made compounds as syntheses of Warfarin should demonstrate.

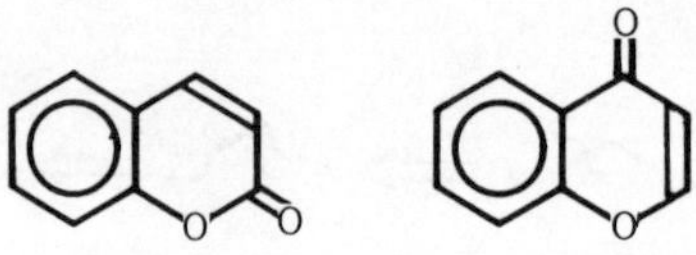

(70) (71)

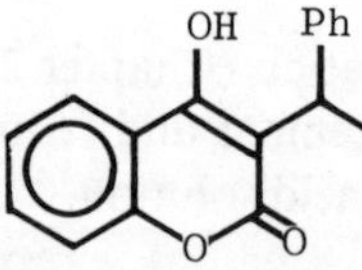

(72)

The 1,5-dicarbonyl disconnection (a), seen more clearly on the keto tautomer **(73)** of Warfarin, is strategically excellent as it separates two branch points and gives simple starting materials, coumarin **(74)** and the enone **(75)** (Chapter 20).

Warfarin: *Analysis*

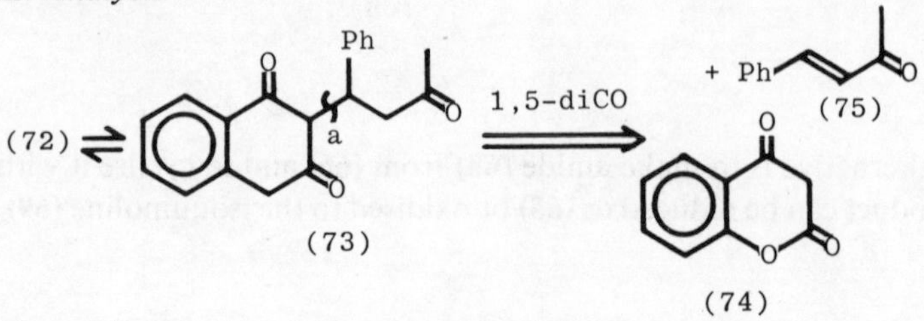

Disconnection of **(74)** by C–O and 1,3-dicarbonyl methods suggests salicylate ester **(76)** and an acetic acid derivative as starting materials. The synthesis gives good yields if acetic anhydride is used — ester formation making the C–C formation intramolecular.

Analysis 1

(74a)　　　　　　　　　　　　　　　(76)

Synthesis 1[441]

$$(76) \xrightarrow{Ac_2O} \quad \xrightarrow{MeO^-} \quad TM(74)$$

An alternative analysis **(74b)** removes malonic acid from phenol — the logic being that ester formation will make the Friedel–Crafts part of the reaction intramolecular. In the event, this is a one step procedure[442] with acid catalysis. The Warfarin synthesis uses weak base catalysis.[443]

Analysis 2

(74b)

Warfarin: *Synthesis*

$$PhOH \xrightarrow[POCl_3]{CH_2(CO_2H)_2} (74) \xrightarrow[pyridine]{(75)} TM(72)$$

Aromatic Heterocycles with Two Heteroatoms

There is a wide range of such compounds from five-membered rings such as oxazole **(77)** to seven-membered rings such as diazepine **(78)**. We shall treat all three six-membered ring compounds with two nitrogen atoms and one five-membered ring compound, again with two nitrogen atoms, as these classes contain many important compounds.

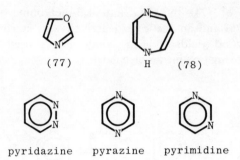

(77)   H   (78)

pyridazine   pyrazine   pyrimidine

Pyridazines

These are six-membered rings with two adjacent nitrogen atoms like the antihypotensive Hydrallazine **(79)**. Tautomer **(80)** shows the first disconnection clearly and removal of a recognisable fragment (hydrazine, cf. Chapter 29) leaves aldehyde acid **(82)**. Again the oxidation level of each carbon atom is important.

Hydrallazine: *Analysis*

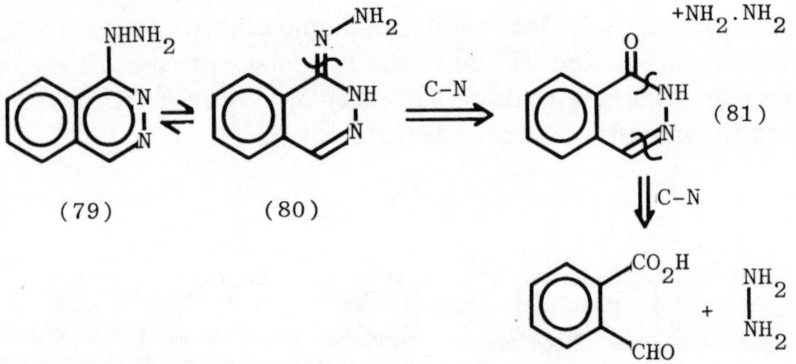

(79)   (80)   (81)

(82)

Partial reduction of phthalic anhydride **(83)** is a reliable method of ensuring chemoselectivity (cf. Chapter 5). As often with aldehydes, it is easier to reduce to the alcohol and reoxidise. Amide **(81)** doesn't react with hydrazine, but the chloride **(84)** does.

Synthesis[444]

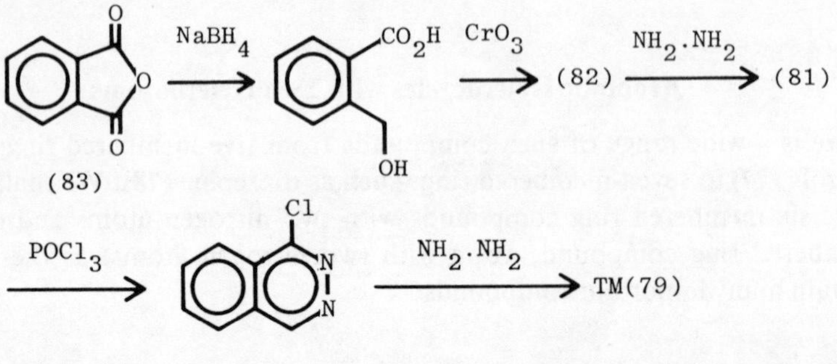

(83)   (82) ⟶ (81)

POCl₃ ⟶   NH₂.NH₂ ⟶ TM(79)

Pyrazines

The same principles apply to the synthesis of pyrazines such as **(85)**, the odour principle of green peppers. The enol ether must be made from amide **(86)** and we can then disconnect into available (Chapter 23) glyoxal **(88)** and compound **(87)** which is the amide of the naturally occurring amino acid leucine.

Analysis

$$(85) \quad (86) \quad (87) \quad + \quad (88)$$

The cyclisation is of course a good reaction and *O*-methylation of **(86)** is achieved with diazomethane.

Synthesis[445]

$$(87) + (88) \longrightarrow (86) \xrightarrow{\text{CH}_2\text{N}_2} \text{TM}(85)$$

Many other pyrazines are odour principles of foods as diverse as coffee and roast beef. They are amazingly powerful—the bench where **(85)** was synthesised still smelt of green peppers years afterwards.[446] Humans can detect one part in 10^{12} of water.

Pyrimidines

We looked at the synthesis of a pyrimidine base from the nucleic acids in Chapter 29. Another example of a pyrimidine is Aphox **(89)**, the ICI insecticide which kills aphids (greenfly) without harming their natural predators, the ladybirds. Disconnection of the ester side chain gives acid chloride **(90)** (cf. Chapter 5) and **(91)**, more easily disconnected as tautomer **(92)** into keto ester **(93)** and guanidine **(94)**.

Aphox: *Analysis*

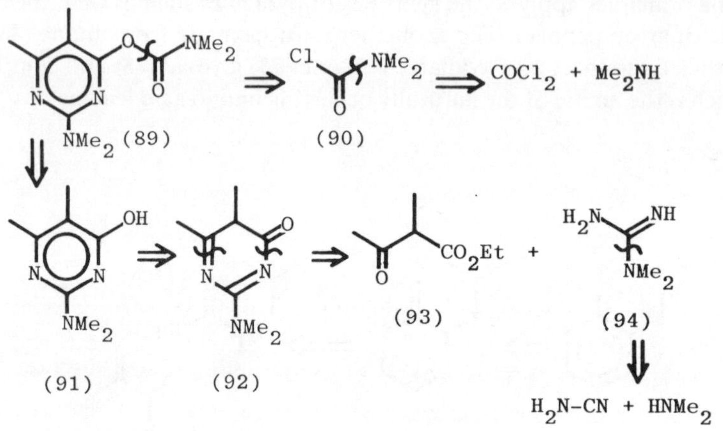

The guanidine **(94)** is made by addition of Me_2NH to available cyanamide $NH_2.CN$, and keto ester **(93)** is an acetoacetate product. Cyclisation goes well and acylation with **(90)** completes the synthesis.

Synthesis[447]

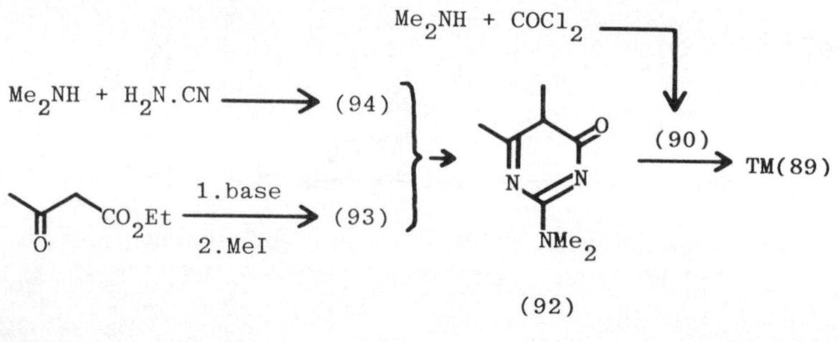

Imidazoles

These five-membered rings with two nitrogen atoms **(95)** are present in many natural products and many synthetic biologically active compounds. The obvious disconnection gives a 1,2-dicarbonyl compound, an acid, and two molecules of ammonia.

Analysis

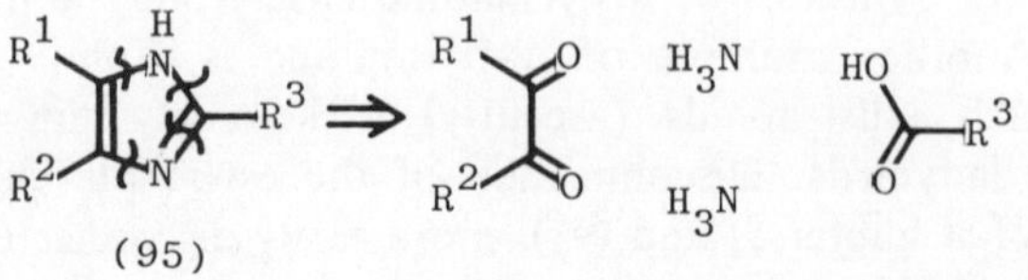

Amazingly, this method works, though an aldehyde usually replaces the carboxylic acid for greater reactivity. Oxidation to the aromatic imidazole occurs easily. Imidazole (96) is used in the synthesis of pentostatin, a compound which enhances the effects of anti-cancer drugs. An 'enone' disconnection—the methyl group in (97) will easily form an anion stabilised by the nitro group—gives nitroimidazole (97), made by nitration of (98). Disconnection as of (95) gives simple starting materials.

Analysis

Keto aldehyde (99) is best made by the α-functionalisation approach[448] (Chapter 23), and imidazole formation occurs with $Zn(OH)_2$ as catalyst.[449] Nitration at the right position occurs[450] with nitric acid and a final condensation[451] with benzaldehyde gives (96).

Synthesis[448-451]

An alternative imidazole synthesis supplies both nitrogen atoms as formamide $HCONH_2$, and uses an α-halo ketone (100) for the rest. This is the Bredereck reaction[452] another testimony to the ease of heterocyclic synthesis.

The peptic ulcer drug Tagamet (101) was made by Smith, Kline, and French using this route.[453] Tagamet is another guanidine and disconnection in the way we used for (94) gives amine (102) and some as yet unknown electrophilic reagent akin to $R_2N.CN$.

Tagamet: *Analysis 1*

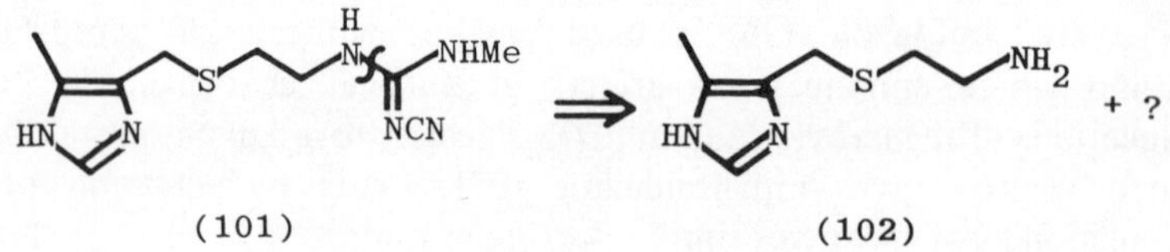

(101) ⟹ (102) + ?

Disconnection of the thiol amine **(103)** leaves the simple imidazole **(104)** where the hydroxyl group will provide the leaving group for displacement by sulphur. Disconnection of the imidazole at this stage would require a difficult α-hydroxy ketone so a preliminary FGI to ester **(105)** makes this step much simpler.

Analysis 2

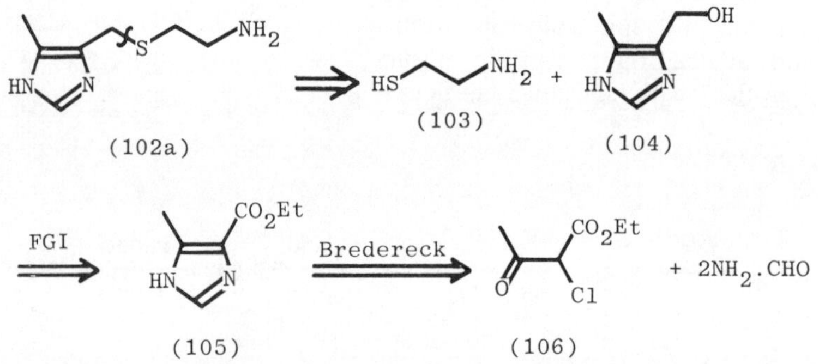

(102a) ⟹ HS⌒NH₂ (103) + (104)

FGI ⟹ (105) Bredereck⟹ (106) + 2NH₂.CHO

The chloroketone **(106)** can be made directly from acetoacetate and the rest of the synthesis is routine until the last step. Reagent **(107)** was developed for this task after many experiments.

Synthesis[453]

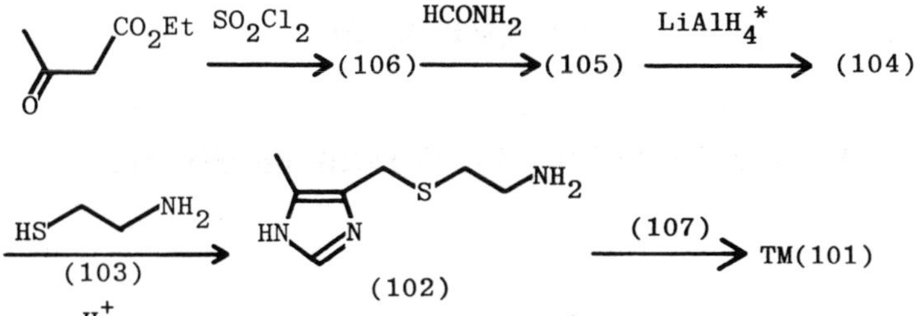

$$\xrightarrow{\text{SO}_2\text{Cl}_2} (106) \xrightarrow{\text{HCONH}_2} (105) \xrightarrow{\text{LiAlH}_4^*} (104)$$

$$\xrightarrow[\text{H}^+]{(103)}$$

(102)

$$\xrightarrow{(107)} \text{TM}(101)$$

*Na,NH$_3$(l) in production

$$\text{NH}_2\text{CN} \xrightarrow[\text{2.MeI}]{\text{1.CS}_2} \underset{\text{MeS}}{\overset{\text{S}}{\diagup}}\text{-NHCN} \xrightarrow{\text{MeNH}_2} \underset{\text{MeS}}{\overset{\text{MeNH}}{\diagup}}\text{N.CN}$$

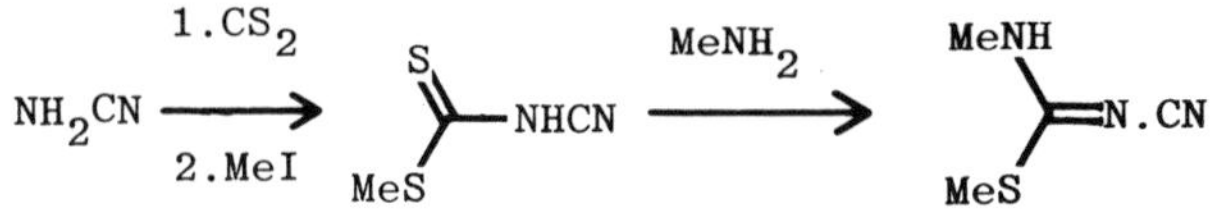

(107)

CHAPTER 40

General Strategy D: Advanced Strategy

A number of useful guidelines on strategy is collected in this final chapter and applied to examples from a wide range of the types of molecules we have been discussing.

Convergence

A synthesis of ten steps, each having 90% yield, gives an overall yield of 34% if the synthesis follows a linear pattern (i). A single branch in the plan (ii) increases the yield to 53%, while a more branched plan (iii) gives 66%. These branched strategies are called *convergent* and are preferred to linear multistep synthesis on principle (though other factors may outweigh this principle).

$$A \to B \to C \to D \to E \to F \to G \to H \to I \to J \to TM \qquad (i)$$

$$\left. \begin{array}{l} A \to B \to C \to D \to E \\ F \to G \to H \to I \to J \end{array} \right\} \ \to K \to TM \qquad (ii)$$

$$\left. \begin{array}{l} \left. \begin{array}{l} A \to B \to C \\ D \to E \to F \end{array} \right\} \ \to M \\[4pt] \left. \begin{array}{l} G \to H \to I \\ K \to L \end{array} \right\} \ \to N \end{array} \right\} \to TM \qquad (iii)$$

Some of the guidelines established in earlier general strategy chapters (12, 28, 37) amount to convergence: disconnections in the middle of a molecule or a at a branch point are likely to lead to convergent syntheses. Examples of convergent syntheses appear in Chapters 5, 6, 10, 21, 23, and 39.

The synthesis of compounds with three rings (1) can be tackled in a convergent manner if rings A and C are built first and joined afterwards. The linear approach is to build A first, then B, and then C. If a convergent synthesis is wanted, ring B should be disconnected first.

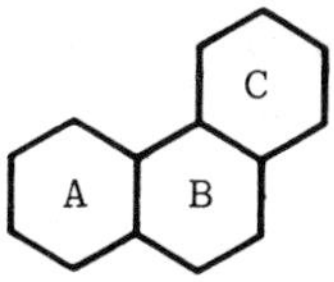

(1)

Ferruginol (2) is a simple example. Disconnection of the central ring by Friedel–Crafts reaction requires a carbonium ion which can be made by protonation of olefin (3).

Ferruginol: *Analysis 1*

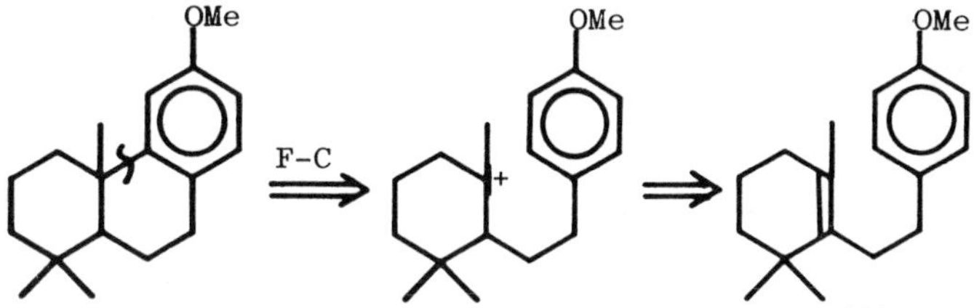

We shall want to disconnect at (a) in (3a), so making (3) from alcohol (4) is sensible. Disconnection at (a) now gives ketone (5) and synthon (6) which could be a Grignard reagent but FGA provides an alternative in acetylene (7). This is the point of convergence as both (5) and (7) must be made.

Analysis 2

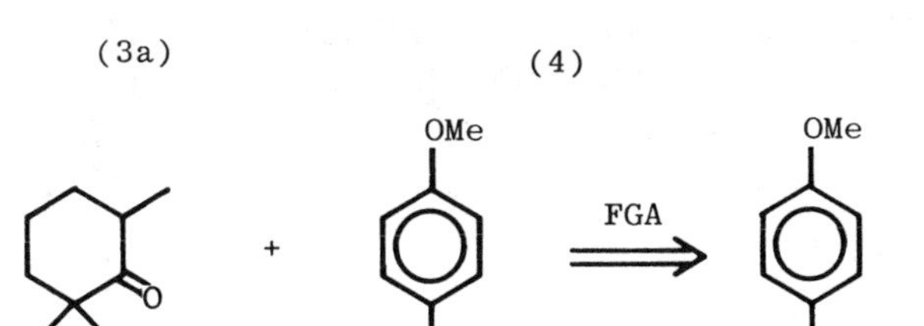

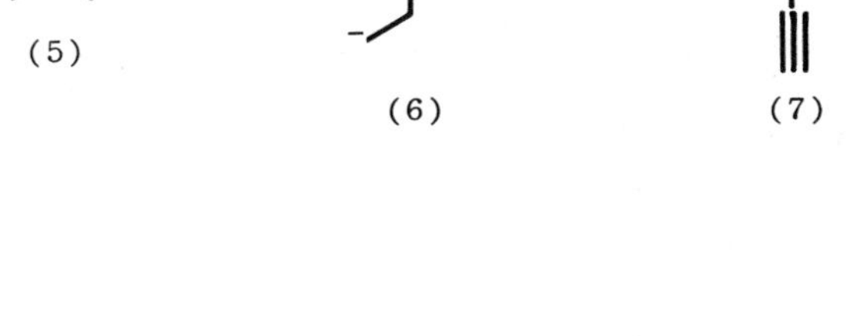

Alkylation of cyclohexanone under vigorous conditions gives **(5)**, while **(7)** can be made from ketone **(9)** by elimination on the gem-dihalide **(8)**.

Analysis 3

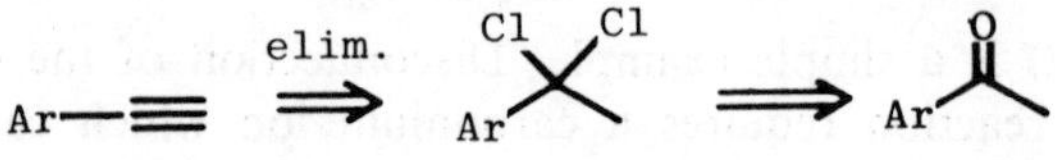

The synthesis is now complete[454] with a single branch. The closure of ring B can be carried out directly on alcohol **(4)**, no doubt *via* **(3)**.

Synthesis[454]

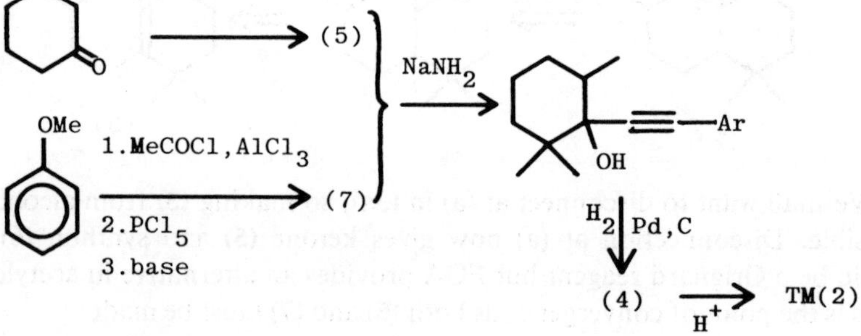

α-Bisabolene **(10)** is an example of another group of molecules where convergence is a help in analysis. It has three double bonds, essentially unrelated. Disconnection of the central bond (a) (Wittig) giving two roughly equally sized fragments is more likely to lead to a convergent synthesis than disconnection of either of the other double bonds.

α-Bisabolene: *Analysis*

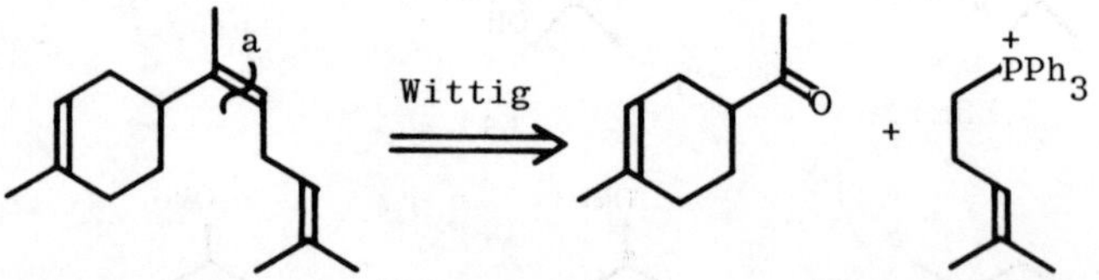

Ketone **(11)** is an obvious Diels–Alder adduct whilst phosphonium salt **(12)** needs halide **(13)**, easily made by chain extension (Chapter 28) from halide **(14)** discussed in Chapter 1.

Analysis 2

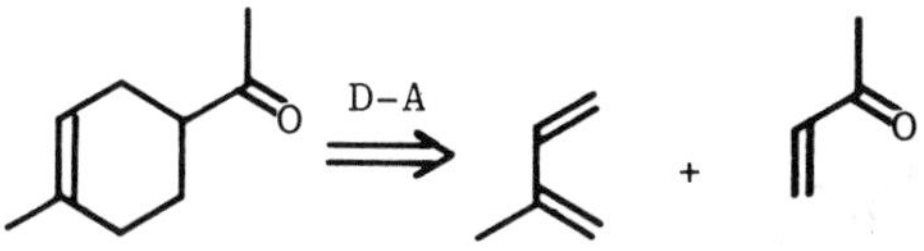

(11a)

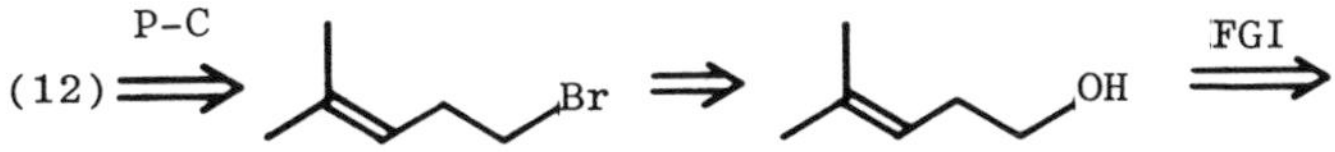

(12) (13)

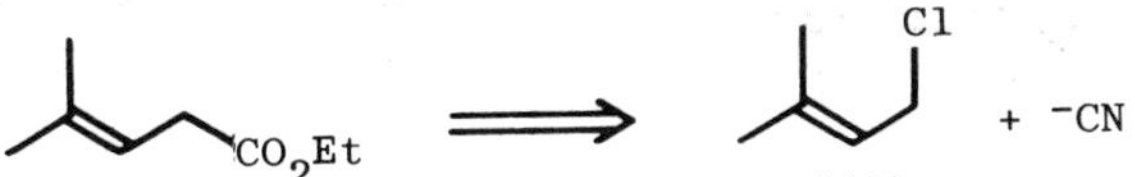

(14)

The orientation of (11) is correctly *para* for a Diels–Alder reaction (Chapter 17) and the Wittig reaction gives mostly *Z*-alkene (Chapter 15) as required. Curiously, both branches of this synthesis start with isoprene.

Synthesis[455]

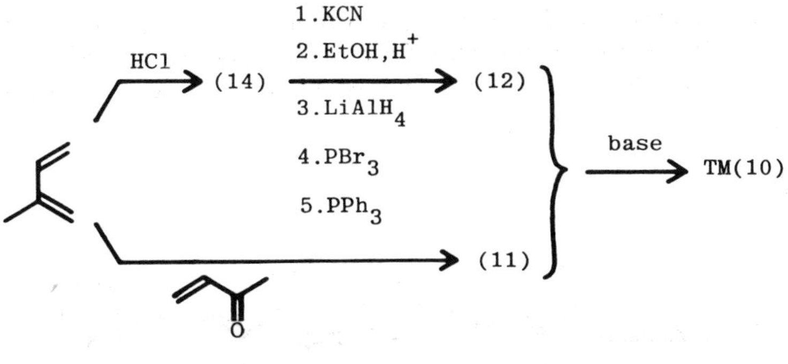

Molecules like bisabolene with several alternative first disconnections are typical cases where it is helpful to consider convergence. The fungal metabolite trisporic acid (15) is a more testing example. The two central double bonds (a) and (b) offer possible Wittig disconnections. The less central (b) leads to an easily synthesised halide (17) (Chapter 25) and a difficult ketone (16), but the more central (a) leads to two nearly equal fragments (18) and (19) and to a more convergent synthesis.

Trisporic acid: *Analysis 1*

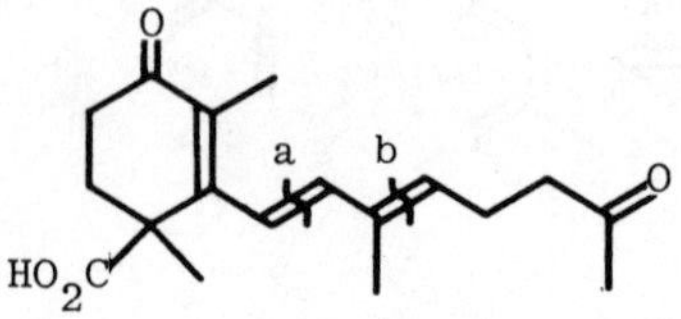

(15)

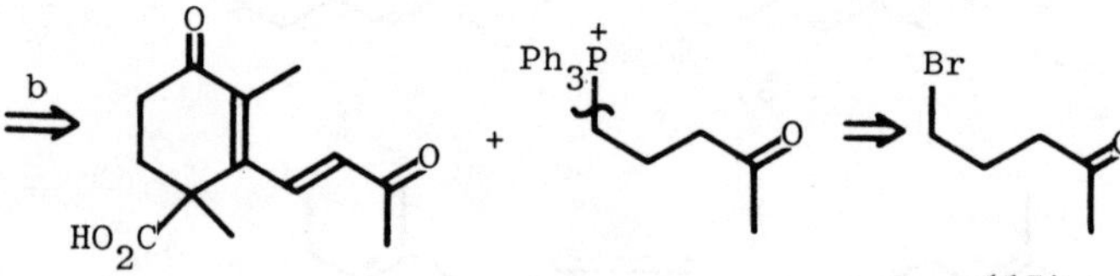

(16)   (17)

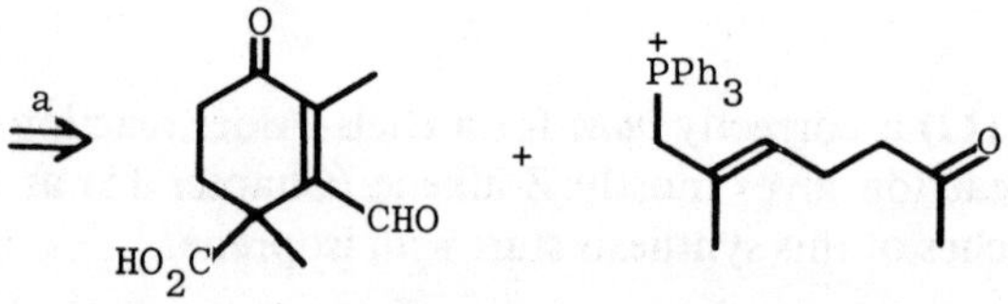

(18)   (19)

Further analysis of (18) and (19) incorporates many points from earlier chapters. The phosphonium salt (19) must be protected and will be derived from alcohol (20) *via* a halide. This is the same skeleton as ketone (21) (Chapter 1).

Analysis 2

(19)   (20)

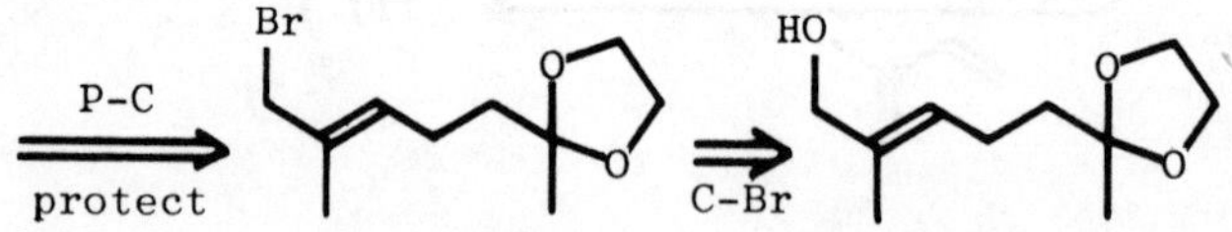

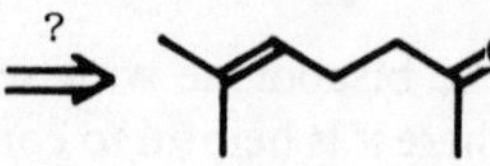

(21)

In practice,[456] allylic oxidation of (22) with SeO_2 (Chapter 24) gives aldehyde (23) and hence the phosphonium salt (19) by simple steps.

Phosphonium salt **(19)**: *Synthesis 1*[456]

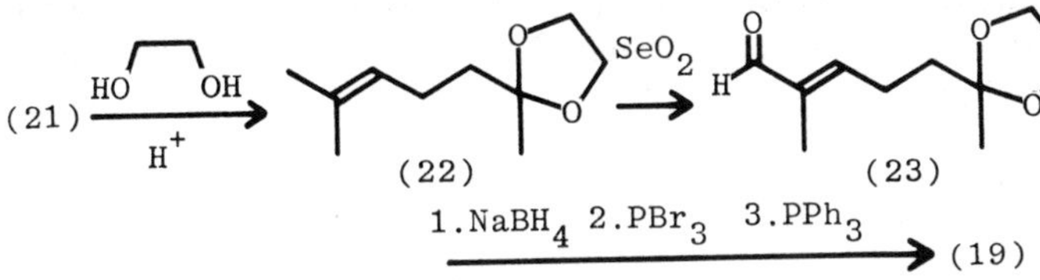

The other branch of this convergent analysis starts at the cyclohexenone **(18)**. This looks like a Robinson annelation product (Chapter 21) and requires available ethyl vinyl ketone and tricarbonyl compound **(24)** as starting materials.

Analysis 3

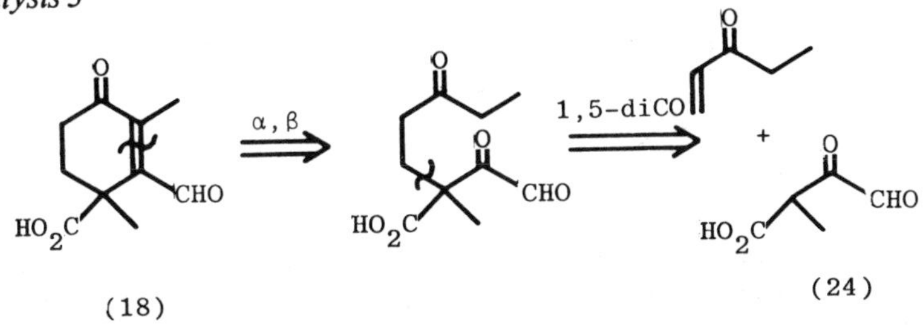

Further disconnection of **(24)** by 1,3-dicarbonyl methods (Chapter 19) gives aldehyde ester **(25)** as starting material. The aldehyde group will certainly need protection here and probably also during the Robinson annelation so available dichloracetic acid is used.

Analysis 4

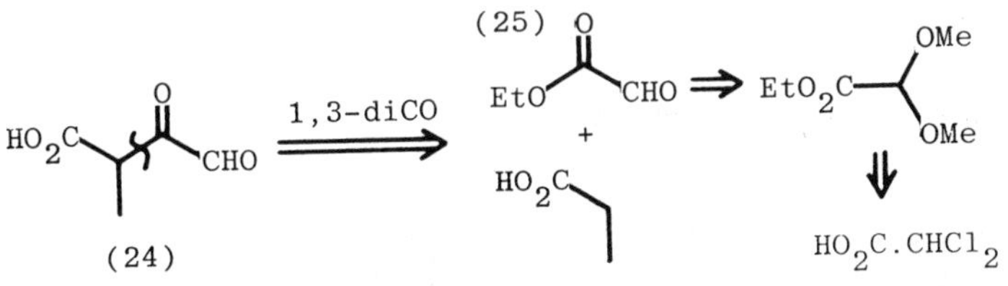

The acid will also need protection so we start by making **(26)**, the protected version of **(24)**, by the planned route.

Synthesis 2[457]

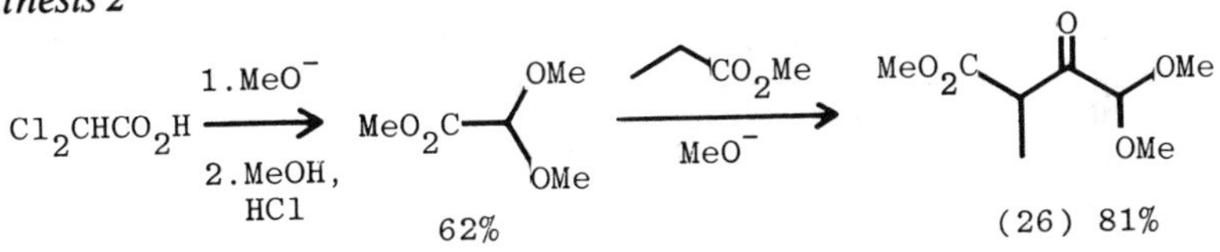

The Robinson annelation goes in one step in base and acid hydrolysis releases aldehyde (27) which gives trisporic acid after the convergent Wittig reaction and hydrolysis. The double bond is *E* as the ylid from (19) is conjugated (Chapter 15).

Synthesis 3[456]

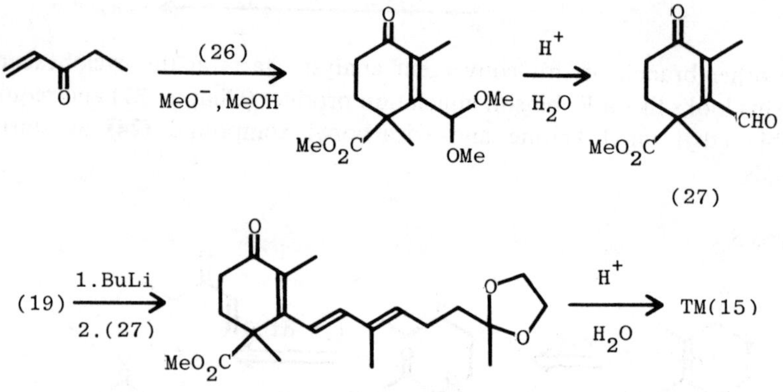

One area where convergent analysis has been particularly useful is in peptide synthesis.[458] The disconnections are all of amide bonds to give the individual amino acids so that linear or convergent planning is simple. Details are beyond the scope of this book.

A word of warning is appropriate here as we have spent some time on convergence. Convergent syntheses are better than linear ones only if all other things are equal. There is no magic about convergence and a bad step can be just as disastrous here as elsewhere. One synthesis of the bark beetle pheromone multistriatin (28) is convergent[459] but the yield in the convergent step is terrible (5%). The linear synthesis in Chapter 12 is much better.

Multistriatin: *Convergent Synthesis*[459]

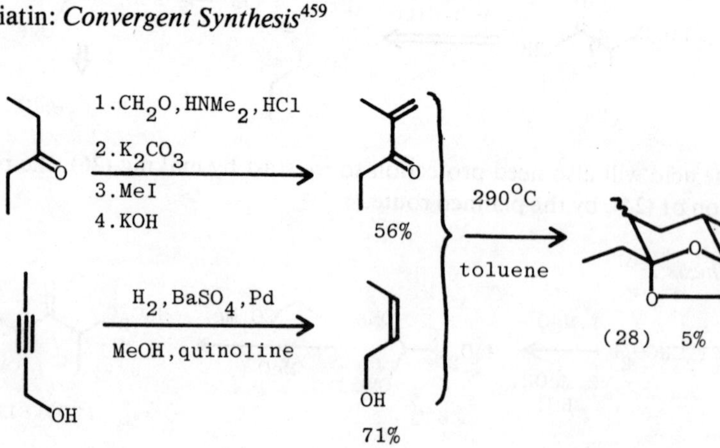

Key Reaction Strategy

The purely analytical approach we have used throughout this book can be supplemented by 'forwards and backwards' thinking from a key reaction. If a reaction is good enough—the Diels–Alder reaction is an obvious example—a synthesis may come from thinking forwards from a possible Diels–Alder product as well as backwards from the TM.

The anion of **(29)** is being developed as a new reagent in synthesis.* There was no good synthesis of this simple compound, nor of the even simpler compound **(30)** from which it is obviously made. Thinking of the Diels–Alder reaction as a key reaction led to the idea of adduct **(31)** as starting material which might be converted into **(30)** *via* condensation and reverse Diels–Alder reaction.

Analysis

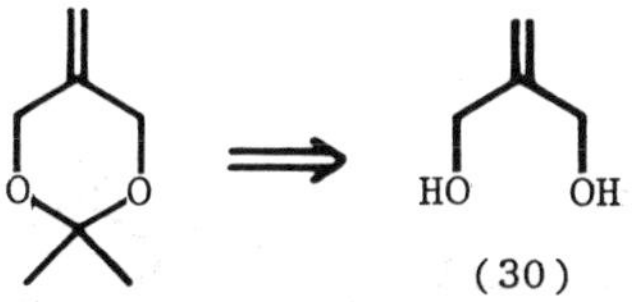

(29) (30)

Possible Synthesis

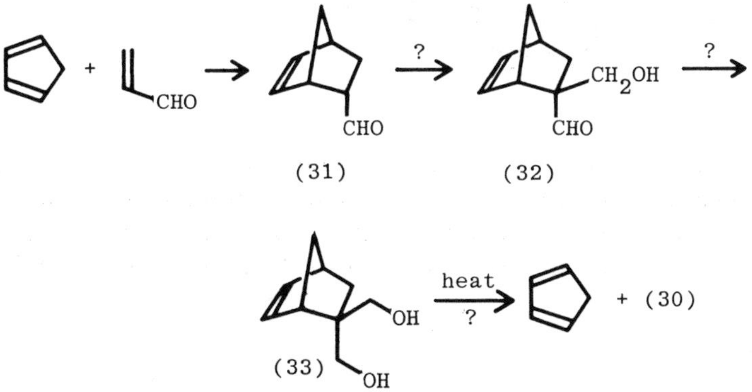

(31) (32) (33)

Condensation of **(31)** to **(32)** requires formaldehyde and base. Reduction to **(33)** occurs under these conditions by the Cannizarro reaction so the synthesis is shorter than seemed likely. The reverse Diels–Alder reaction to **(30)** goes at high temperature.

*It behaves as the synthon:

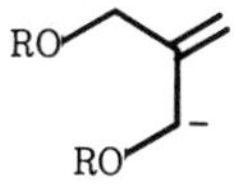

$$(31) \xrightarrow[\text{base}]{CH_2O} (33) \xrightarrow{520^\circ C} (30) \xrightarrow[\text{TsOH}]{} TM(29)$$

(with the reagent drawn above the last arrow bearing two OMe groups)

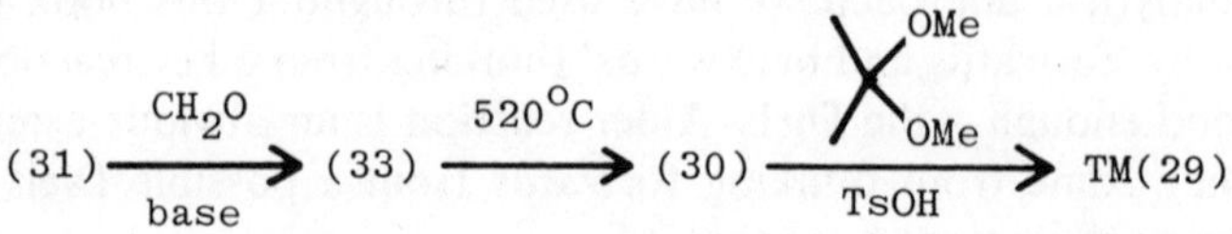

The alkaloid lycorane **(34)** provides a more challenging example. It has three adjacent chiral centres, one next to nitrogen. We shall want to disconnect C–N bonds but it it bad strategy to disconnect the chiral centres from nitrogen as we should lose control over the stereochemistry. As **(34)** is an isoquinoline we can use the Mannich disconnection (Chapter 39) giving **(35)** and then a simple C–N disconnection to intermediate **(36)**.

Lycorane: *Analysis 1*

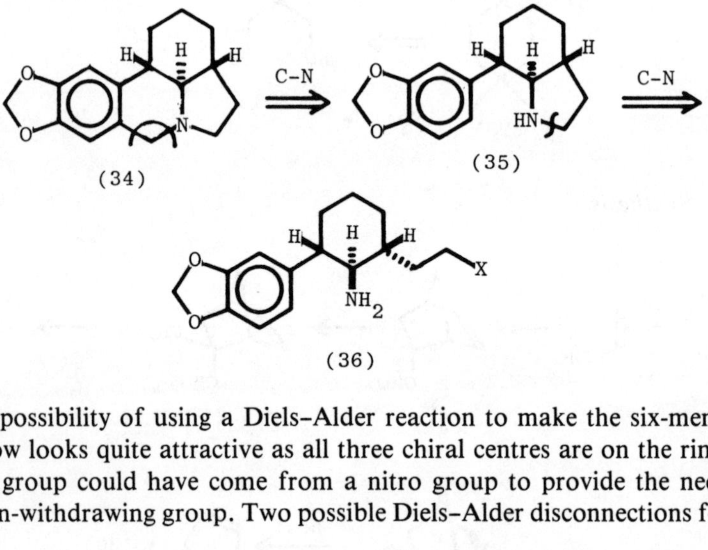

(34) $\xrightarrow{C-N}$ (35) $\xrightarrow{C-N}$

(36)

The possibility of using a Diels–Alder reaction to make the six-membered ring now looks quite attractive as all three chiral centres are on the ring. The amino group could have come from a nitro group to provide the necessary electron-withdrawing group. Two possible Diels–Alder disconnections follow.

Analysis 2

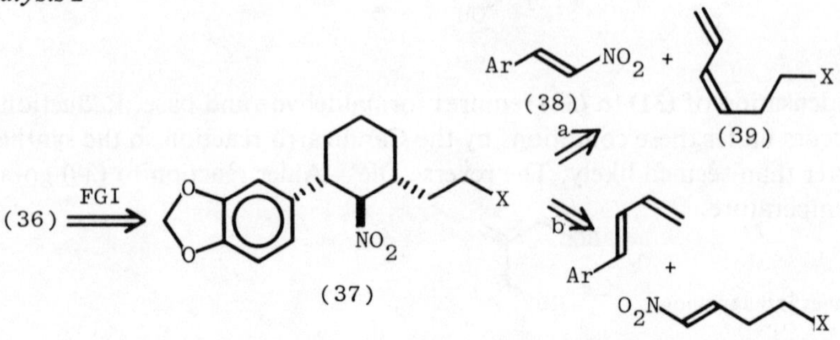

(36) $\xrightarrow{FGI}$ (37)

$Ar \diagdown NO_2$ + (38) $\diagdown\diagdown X$ (39)

a.   b.

$Ar \diagdown$ + $O_2N \diagdown\diagdown X$

The regioselectivity is correct (*ortho* NO_2 and diene substituent) in each case and the stereochemistry (Chapter 15) of the nitro olefin should be *trans*— which is good as we can make only *trans* nitro olefins (Chapter 22). Aromatic nitro olefins are easy to make so strategy (a) can be taken further.

Though the stereochemistry of (38) is correct, we shall need *cis* diene (39) to get the right endo adduct (Chapter 17). This is an unattractive idea and we can avoid it by preliminary disconnection of the side chain of (37) *via* FGI to carbonyl and FGA to abolish the offending centre (see Chapter 24).

Analysis 3

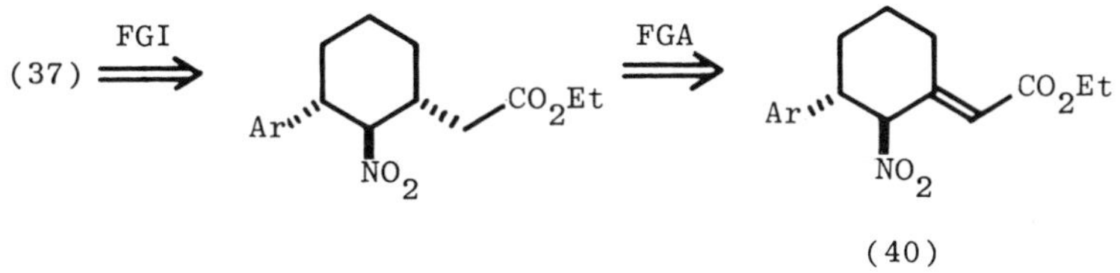

(40)

α,β-Disconnection of (40a) and then FGI suggests a butadiene with a heteroatom in the 1-position as starting material, so available 1-acetoxy-butadiene can be used.

Analysis 4

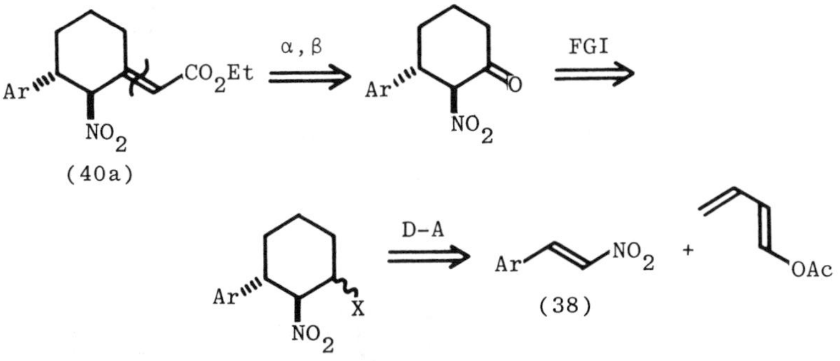

The diene is *trans*, so it will give one wrong chiral centre, but this was part of the plan and it is to become a carbonyl group. We made the starting material in Chapter 2.

Synthesis[461] *1*

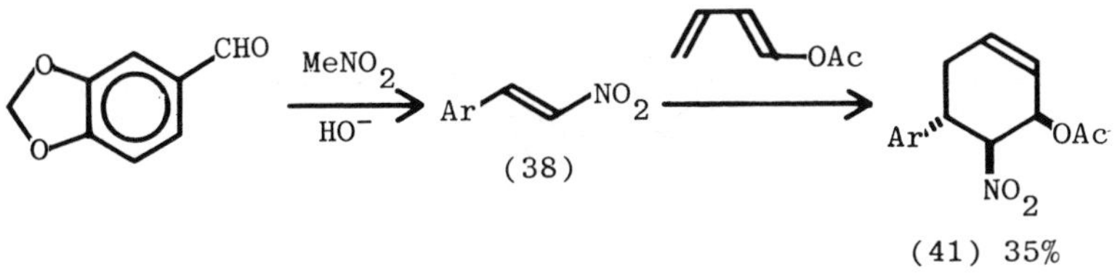

The nitro group must be reduced at once to avoid 'enolisation' at the chiral centre and the double bond is reduced at the same time. The acetyl group transfers spontaneously to nitrogen and is allowed to remain there for protection during the condensations. Oxidation gives the necessary ketone (42) and abolishes the 'wrong' chiral centre.

Synthesis 2

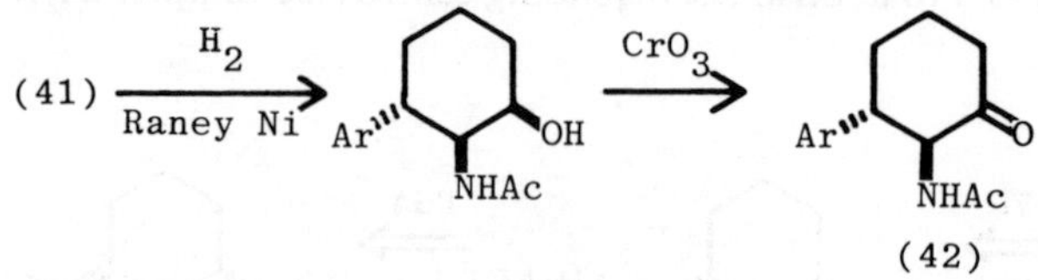

The condensation is controlled by the Wittig method and hydrogenation adds hydrogen on the opposite side from the large aryl group to give all-equatorial (43).

Synthesis 3

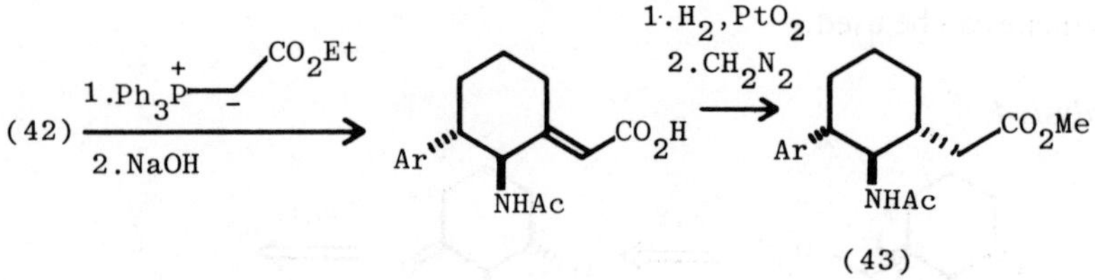

Reduction of the ester, to provide the leaving group for cyclisation, also reduces the acetyl group to an ethyl group. Cyclisation occurs according to plan and the ethyl group is eliminated at the end.

Synthesis 4

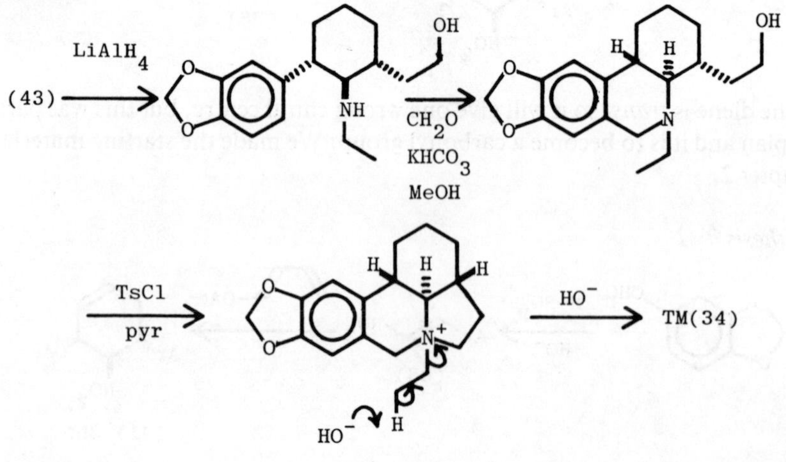

Strategy of Available Starting Materials

We have already used this strategy. It was used specifically for aromatic compounds (Chapter 3), 1,2- and 1,4-difunctionalised compounds (Chapters 23 and 25) and formed one of our earliest guidelines to good disconnections (Chapter 11). In this chapter we look at some less obviously available starting materials and TMs made from them.

Epichlorhydrin (44) is useful when 1,2,3-trifunctionalised compounds are needed. It reacts with nucleophiles at the less hindered end of the epoxide to give (45) and treatment of this with base forms a new epoxide which can react with a second nucleophile to give (46).

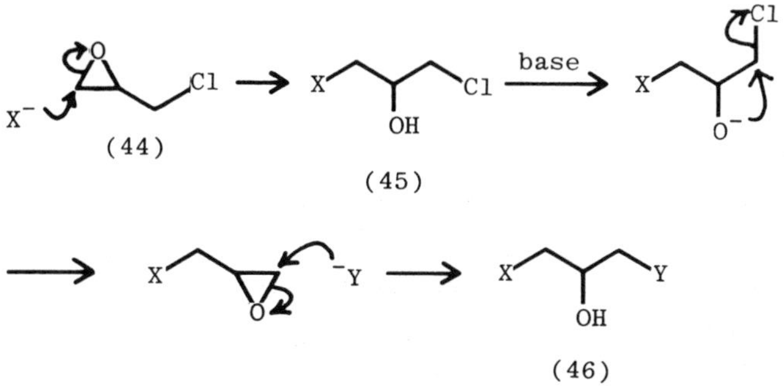

The antidepressant Vivalan[462] (47) contains a 1,2,3-trifunctionalised fragment (•) revealed by C–O and C–N disconnections.

Vivalan: *Analysis*

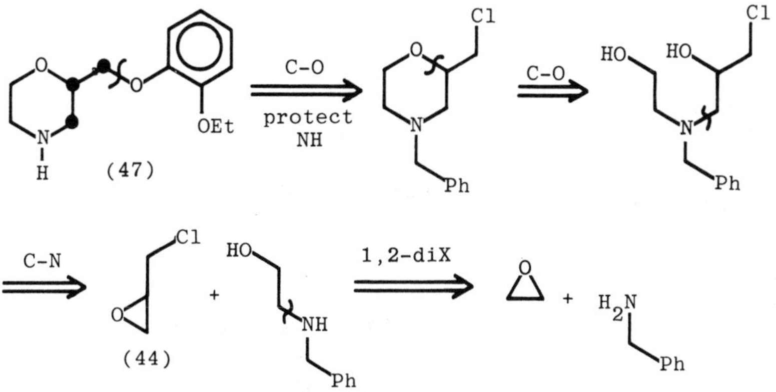

Monoalkylation of available catechol (48) can be accomplished in fair yield in basic solution (cf. Chapter 5) and the synthesis is convergent.

358

Synthesis[462]

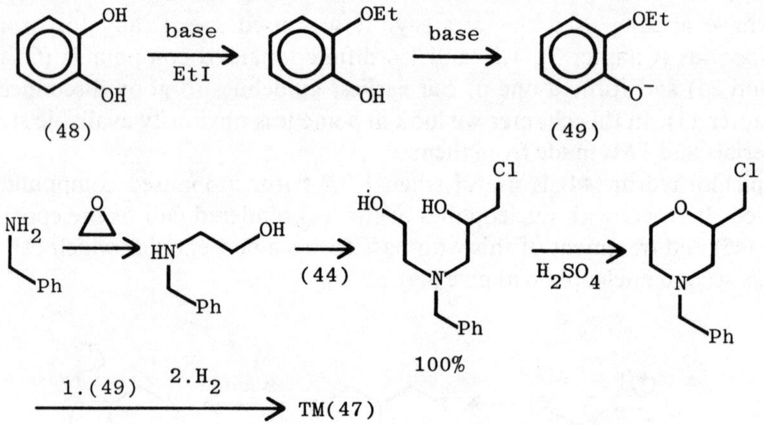

Table 40.1 Molecules whose synthesis was designed to start from a readily available starting material

Target molecule	Starting material	Reference
Corey lactone and hence prostaglandins 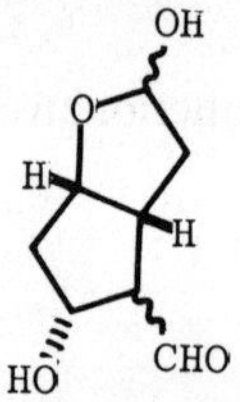	Cyclopentadiene (dimer) 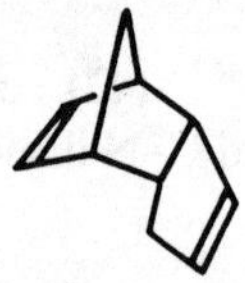	463
Cis-jasmone	(See Chapter 33)	464
Prostaglandins *via*: 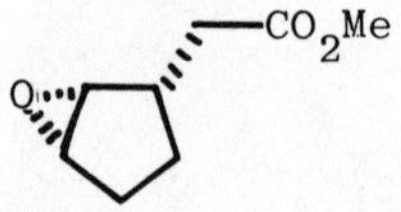		465
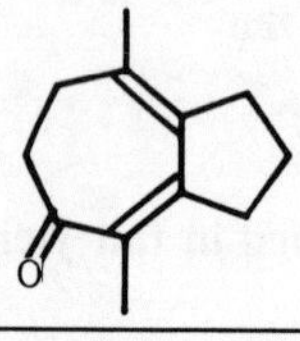	Wieland–Miescher ketone 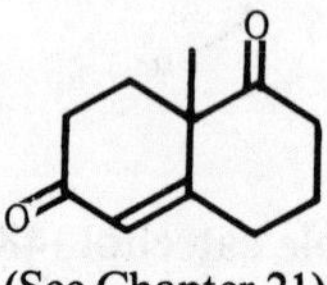 (See Chapter 21)	466

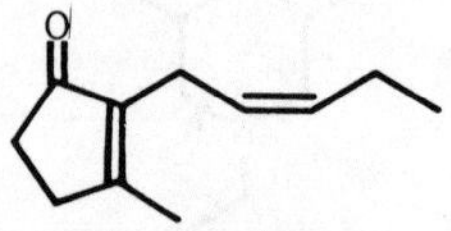

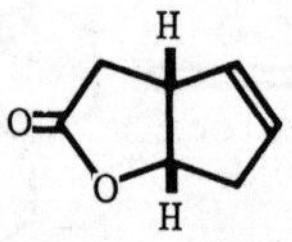

Products made from epichlorhydrin are easy to see. In other cases syntheses have been designed by 'forwards and backwards' thinking (cf. page 353) to use readily available compounds. Tables 40.1 and 40.2 give some examples to show both the type of TMs and the molecules regarded as being 'readily available'. Most of the starting materials have themselves been discussed in this book.

Table 40.2 Some readily available starting materials

Starting material	Made by	Applications
Corylone (cyclotene)	Chapter 34	467

Butyne-diol and *cis*-butene-diol	Chapter 16	Chapter 12

	Chapter 19	Chapter 12

Epichlorhydrin		Chapter 40

Hagemann's ester		Progesterone[469]
		Strigol[470]
		Refs[471]

	Ref. 468	
Cyclo-octa-1,5-diene		472

One starting material, furfural (**50**), is so cheap (Chapter 25) that it is worth going to some lengths to use it. It can be reduced to alcohol (**51**) which rearranges to dihydropyran (**52**) used to put on the THP protecting group

(Chapter 9). Hydration of (52) gives hemiacetal (53) which can be used for (54) in base-catalysed reactions. All these compounds are therefore readily available. Other surprising sequences are given in Table 40.3.

Table 40.3 Starting materials available by simple routes from other cheap compounds

Cyclopentadiene routes[473]

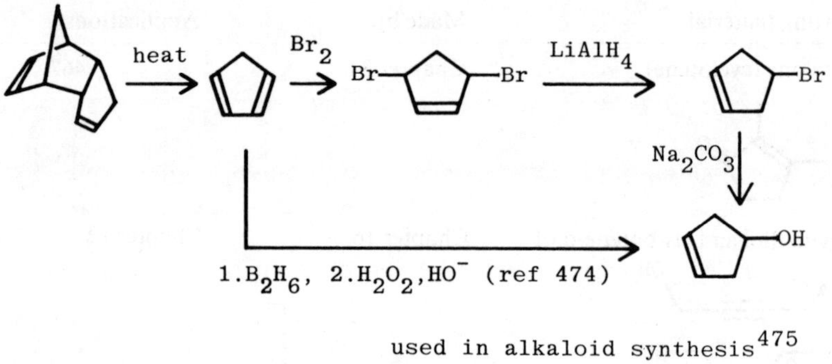

$$\text{(dimer)} \xrightarrow{\text{heat}} \text{(cyclopentadiene)} \xrightarrow{\text{Br}_2} \text{Br}-\text{(cyclopentene)}-\text{Br} \xrightarrow{\text{LiAlH}_4} \text{(cyclopentene)}-\text{Br}$$

$$\xrightarrow[\text{1.B}_2\text{H}_6, \ 2.\text{H}_2\text{O}_2,\text{HO}^- \ (\text{ref } 474)]{}$$

$$\xrightarrow{\text{Na}_2\text{CO}_3} \text{(cyclopentenol)}-\text{OH}$$

used in alkaloid synthesis[475]

Citric acid routes[476]

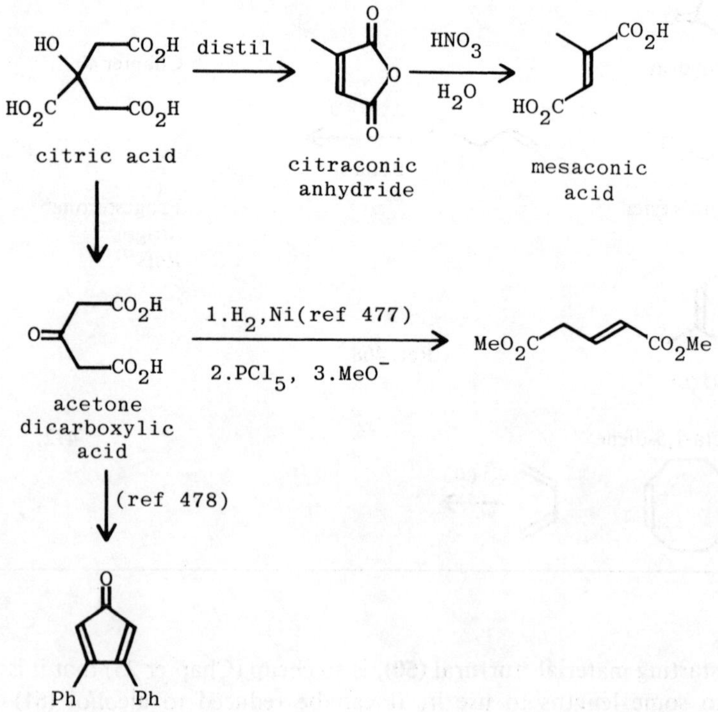

citric acid $\xrightarrow{\text{distil}}$ citraconic anhydride $\xrightarrow[\text{H}_2\text{O}]{\text{HNO}_3}$ mesaconic acid

acetone dicarboxylic acid $\xrightarrow[\text{2.PCl}_5, \ 3.\text{MeO}^-]{\text{1.H}_2,\text{Ni(ref 477)}}$ $\text{MeO}_2\text{C}-\text{CO}_2\text{Me}$

(ref 478)

Synthesis[479]

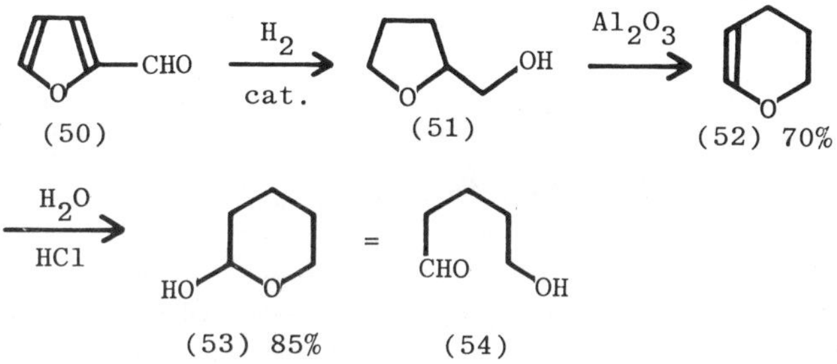

CHO
H_2
cat.
OH
Al_2O_3
(50)
(51)
(52) 70%
H_2O
HCl
HO
=
CHO
OH
(53) 85%
(54)

When Baldwin[480] wished to study cyclisations by Michael reactions—studies which led him to formulate the Baldwin rules—he needed hydroxy ester **(55)**. This is clearly just a condensation reaction away from **(54)**.

Analysis

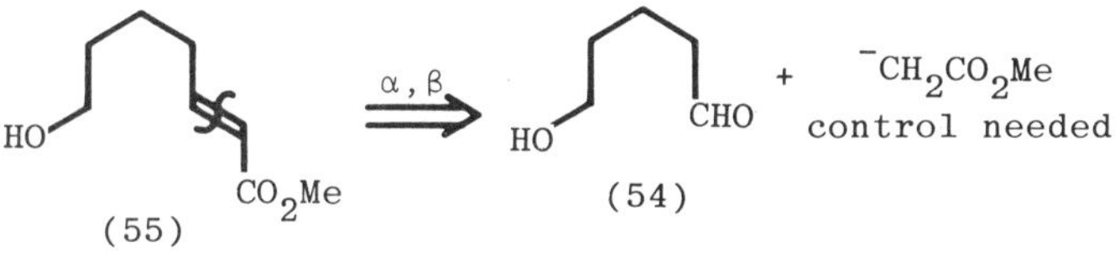

HO
CO_2Me
(55)
α, β
HO
CHO
(54)
+
$^-CH_2CO_2Me$
control needed

The condensation has been carried out both by the Wittig[481] and by the malonate[482] methods.

Synthesis

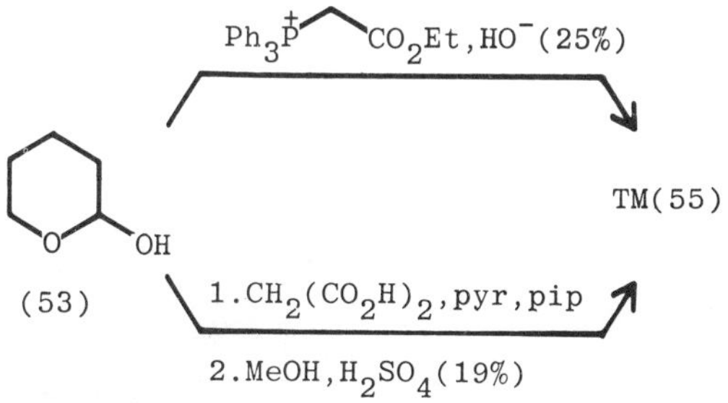

$Ph_3\overset{+}{P}$ CO_2Et, HO^- (25%)
TM(55)
OH
(53)
1. $CH_2(CO_2H)_2$, pyr, pip
2. MeOH, H_2SO_4 (19%)

An Industrial Example

Strategic arguments apply equally to industrial and laboratory syntheses. The BASF company manufactures *α* **(56)** and *β* **(57)** sinensals, compounds responsible for the odour of orange oil. Analysis of the syntheses combines aspects of key reactions, linear and convergent strategies, and available starting materials.

(56)

(57)

α-Sinensal (56) contains three separate sets of functional groups: an enal (C1–C3), a double bond (C6–C7), and a conjugated diene (C9–C12). Key reactions could be the Wittig for any of the double bonds (though the stereochemistry of C6–C7 would be a problem) and the Claisen–Cope rearrangement for C6–C7 if C10 is a carbonyl group at any stage (Chapter 35).

A starting material available at BASF is ketone (58), made for their vitamin A synthesis[2] by the oxidation of acetone (an example of α-functionalisation, see Chapter 23) which could supply C1 and C2. Disconnection at C2–C3 can therefore be left to the end of the analysis.

(58)

The convergent approach (page 362) requires disconnection (a) in the middle of the molecule. A Wittig reaction between ketone (59) and phosphonium salt (60) should be the answer.

Convergent Analysis 1

(56a)

(59)

+PPh$_3$ (60)

The aldehyde group in (59) must be protected, as it will be if it is made from (58). The protected version (61) of (59) can be made from acetoacetate and allylic halide (62) which can come from (58) (cf. Chapters 16 and 35) using acetylene for the vinyl anion synthon.

Convergent Analysis 2

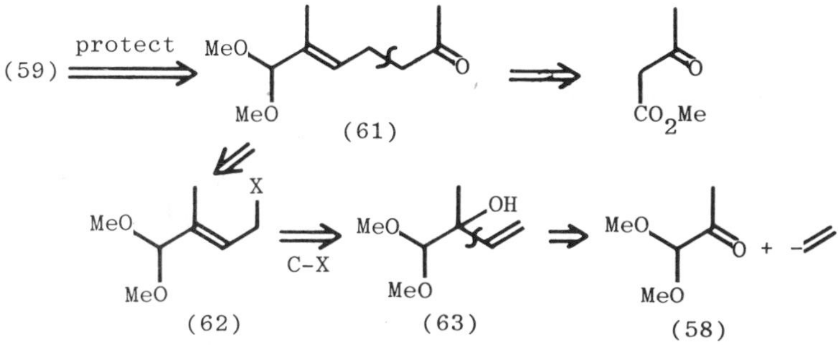

This part of the synthesis is straightforward, the only change from the plan being the loss of the original protecting group during the allylic rearrangement and its necessary replacement.

Ketone **(64)**: *Synthesis*[483]

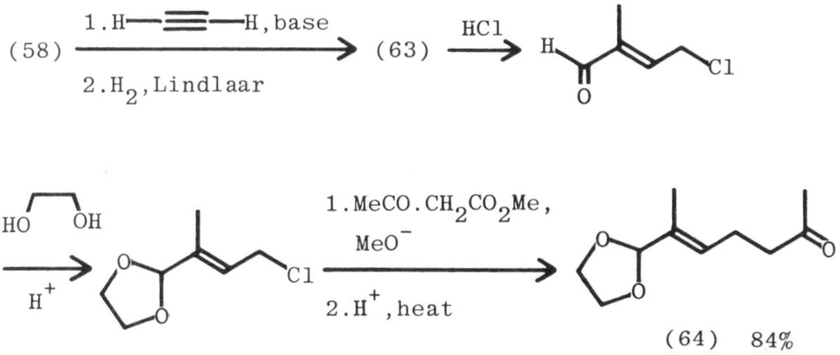

You may like to compare this synthesis with the allylic functionalisation route to **(59)** given on page 350.

We must now return to the other convergent branch, the phosphonium salt **(60)**. It must be made from halide **(65)** and a branch point disconnection can be arranged by FGI to alcohol **(66)**. Chloroketone **(67)** is discussed in Chapter 25.

Convergent Analysis 3

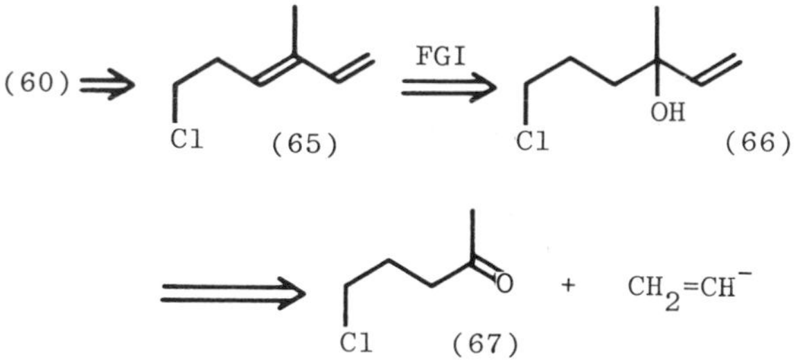

This time vinyl Grignard was used for the vinyl anion and dehydration of **(66)** gave a mixture of **(68)** and **(69)**. These were not separated as a mixture of α and β sinensals is acceptable as the product.

Phosphonium salt **(60)**: *Synthesis*[2]

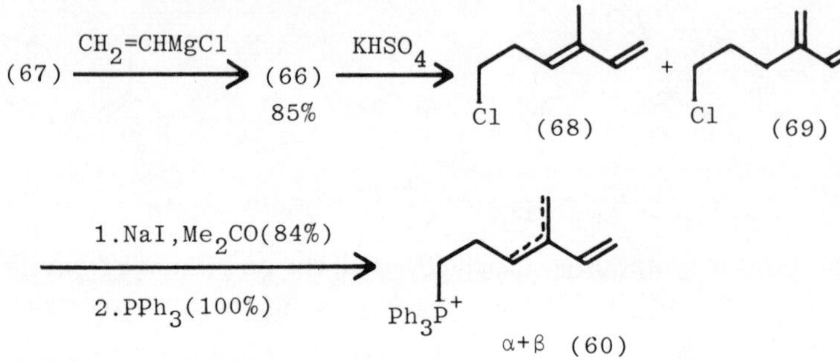

α and β *Sinensals*: *Synthesis*[2]

(60) —— 1.BuLi; 2.(64); 3.H^+,H_2O ——→ (56) + (57) 65%

A linear analysis of the sinensals starts at the right hand end with a rehydration and a Grignard disconnection to ketone **(70)**. This is γ,δ-unsaturated and so can be made by a Carroll rearrangement (Chapter 35) (remembering to invert the allylic alcohol). This allylic alcohol can again be made by vinyl Grignard addition to a ketone **(64)** already made for the convergent approach.

Linear Analysis

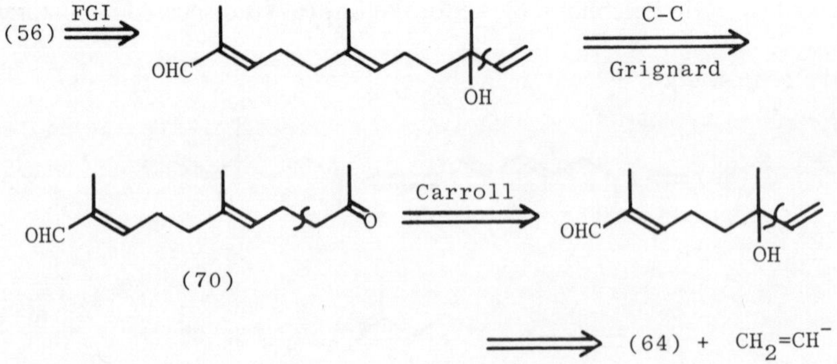

Industrially, diketone (Chapter 33) is used in the Carroll rearrangement step and the dehydration again gives a mixture of α and β sinensals.

Synthesis[2]

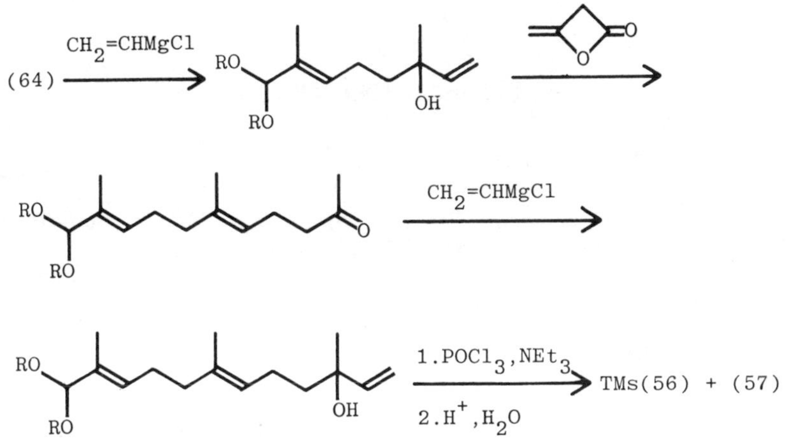

$$(64) \xrightarrow{\text{CH}_2=\text{CHMgCl}} \quad \xrightarrow{} $$

These two syntheses look very similar in length and indeed many of the reactions are the same. BASF give no indication which they prefer and it is appropriate to end with two alternative syntheses for the same compound. Synthesis is creative, and new syntheses are carried out daily. You too should now be able to devise new syntheses yourself.

General References

Ap Simon: J. Ap Simon, ed., *The Total Synthesis of Natural Products*, Wiley-Interscience, New York, 3 volumes, 1973–1977.

Carruthers: W. Carruthers, *Some Modern Methods of Organic Synthesis*, Second Edition, Cambridge University Press, 1978.

Drug Synthesis: D. Lednicer and L. A. Mitscher, *The Organic Chemistry of Drug Synthesis*, Wiley, New York, 1977.

Fleming, Orbitals: I. Fleming, *Frontier Orbitals and Organic Chemical Reactions*, Wiley, London, 1976.

Fleming, Synthesis: I. Fleming, *Selected Organic Syntheses*, Wiley, London, 1973.

Houben-Weyl: E. Müller, ed., *Methoden der Organischen Chemie*, Fourth Edition, Thieme, Stuttgart, Many Volumes, 1952–1981.

House: H. O. House, *Modern Synthetic Reactions*, Second Edition, Benjamin, Menlo Park, 1972.

Perfumes: T. F. West, H. J. Strausz, and D. H. R. Barton, *Synthetic Perfumes*, Arnold, London, 1949.

Pesticide Manual: C. A. Worthing ed., *The Pesticide Manual*, Sixth Edition, British Crop Protection Council, Croydon, 1979.

Pesticides: R. J. W. Cremlyn, *Pesticides: Preparation and Mode of Action*, Wiley, Chichester, 1978.

Vogel: B. S. Furniss, A. J. Hannaford, V. Rogers, P. W. G. Smith, and A. R. Tatchell, *Vogel's Textbook of Practical Organic Chemistry*, Fourth Edition, Longman, London, 1978.

References

1. S. Warren, *Designing Organic Syntheses*, Wiley, Chichester, 1978.
2. H. Pommer and A. Nürrenbach, *Pure Appl. Chem.*, 1975, **43**, 527; *Angew. Chem. Int. Ed. Engl.*, 1977, **16**, 423.
3. A. F. Thomas in *Ap Simon*, Vol.2, pp.4–7.
4. G. T. Pearce, W. E. Gore, and R. M. Silverstein, *J. Org. Chem.*, 1976, **41**, 2797.
5. *Drug Synthesis*, pp.9–10; H. Salkowski, *Ber.*, 1895, **28**, 1917.
6. P. H. Gore in *Friedel–Crafts and Related Reactions*, ed. G. A. Olah, Vol.III, part 1, Interscience, New York, 1964, p.180.
7. W. Weinrich, *Ind. Eng. Chem.*, 1943, **35**, 264; S. H. Patinkin and B. S. Friedman in ref.6, Vol.II, part 1, p.81.
8. E. L. Martin, *Org. React.*, 1942, **1**, 155.
9. R. C. Fuson and C. H. McKeever, *Org. React.*, 1942, **1**, 63.
10. *Perfumes*, p.141; cf. J. R. Holum, *J. Org. Chem.*, 1961, **26**, 4814.
11. Ref.6, Vol.II.
12. Ref.6, Vol.III.
13. H. S. Booth, H. M. Elsey, and P. E. Burchfield, *J. Am. Chem. Soc.*, 1935, **57**, 2066.
14. J. R. Johnson and L. T. Sandborn, *Org. Synth. Coll.*, 1932, **1**, 111; H. E. Ungnade and E. F. Orwoll, *Ibid.*, 1955, **3**, 130; *Vogel*, p.660.
15. H. T. Clarke and R. R. Read, *Org. Synth. Coll.*, 1932, **1**, 514.
16. *Pesticides*, p.154; *Pesticide Manual*, p.537.
17. J. M. Tedder, A. Nechvatal, and A. H. Jubb, *Basic Organic Chemistry*, Part 5, Industrial Products, Wiley, London, 1975, p.574.
18. Ref.17, pp.463–467.
19. *Perfumes*, p.162; G. Baddeley, G. Holt, and W. Pickles, *J. Chem. Soc.*, 1952, 4162.
20. Ref.17, p.564.
21. M. S. Carpenter, W. M. Easter, and T. F. Wood, *J. Org. Chem.*, 1951, **16**, 586.
22. D. J. Byron, G. W. Gray, A. Ibbotson, and B. M. Worrall, *J. Chem. Soc.*, 1963, 2253; D. J. Byron, G. W. Gray, and R. C. Wilson, *J. Chem. Soc.* (C), 1966, 840.
23. H. A. Scarborough and W. A. Waters, *J. Chem. Soc.*, 1926, 559.
24. *Drug Synthesis*, pp.9–10; *U.S. Pat.*, 1954, 2,689,248; *Chem. Abstr.*, 1956, **50**, 2671b.
25. J. L. Simonsen and M. G. Rau, *J. Chem. Soc.*, 1917, 220; R. O. Clinton, U. J. Salvador, S. C. Laskowski, and M. Wilson, *J. Am. Chem. Soc.*, 1952, **74**, 592.
26. D. T. Collin, D. Hartley, D. Jack, L. H. C. Lunts, J. C. Press, A. C. Ritchie, and P. Toon, *J. Med. Chem.*, 1970, **13**, 674.
27. J. H. Burckhalter, F. H. Tendick, E. M. Jones, P. A. Jones, W. F. Holcomb, and A. L. Rawlins, *J. Am. Chem. Soc.*, 1948, **70**, 1363.

28. *Pesticides*, p.118; *Pesticide Manual*, p.208.
29. A. H. Blatt, *Org. React.*, 1942, **1**, 342.
30. H. Wynberg, *Chem. Rev.*, 1960, **60**, 169.
31. *Pesticides*, p.198.
32. *Perfumes*, p.260.
33. *Pesticides*, p.152; *Pesticide Manual*, p.446; W. Schäfer, L. Eue, and P. Wegler, *Ger. Pat.*, 1958, 1,039,779; *Chem. Abstr.*, 1960, **54**, 20060i.
34. A. G. Davies, J. Kenyon, and L. W. F. Salamé, *J. Chem. Soc.*, 1957, 3148.
35. M. H. Benn and M. G. Ettlinger, *Chem. Commun.*, 1965, 445.
36. J. H. Boyer and J. Hamer, *J. Am. Chem. Soc.*, 1955, **77**, 951.
37. O. M. Halse, *J. Prakt. Chem.*, 1914, **(2)89**, 451.
38. W. T. Olson, H. F. Hipsher, C. M. Buess, I. A. Goodman, I. Hart, J. H. Lamneck, and L. C. Gibbons, *J. Am. Chem. Soc.*, 1947, **69**, 2451; *Perfumes*, p.219; *Vogel*, p.755.
39. *Perfumes*, p.226.
40. J. E. Baldwin, J. de Bernardis, and J. E. Patrick, *Tetrahedron Lett.*, 1970, 353.
41. J. E. Cranham, D. J. Higgons, and H. A. Stevenson, *Chem. Ind. (London)*, 1953, 1206; cf. H. A. Stevenson, R. F. Brookes, D. J. Higgons, and J. E. Cranham, *Brit. Pat.*, 1955, 738,170; *Chem. Abstr.*, 1956, **50**, 10334b.
42. *Drug Synthesis*, p.14; S. M. McElvain and T. P. Carney, *J. Am. Chem. Soc.*, 1946, **68**, 2592.
43. *Drug Synthesis*, p.111; S. Veibel, *Ber.*, 1930, **63**, 1582, 2074; L. Spiegler, *U.S. Pat.*, 1960, 2,947,781; *Chem. Abstr.*, 1961, **55**, 7353f; M. Freifelder, *J. Org. Chem.*, 1962, **27**, 1092; G. Wilbert and J. de Angelis, *U.S. Pat.*, 1958, 2,998,450; *Chem. Abstr.*, 1962, **56**, 2381e.
44. M. Bergmann and L. Zervas, *Ber.*, 1932, **65**, 1192.
45. *Drug Synthesis*, p.44; O. H. Hubner and P. V. Petersen, *U.S. Pat.*, 1958, 2,830,088; *Chem. Abstr.*, 1958, **52**, 14690i.
46. H. H. Hodgson and E. R. Ward, *J. Chem. Soc.*, 1949, 1316.
47. Dr. A. Topham, Personal Communication.
48. L. I. Smith and J. A. Sprung, *J. Am. Chem. Soc.*, 1943, **65**, 1276.
49. P. Ruggli and A. Maeder, *Helv. Chim. Acta*, 1942, **25**, 936.
50. T. Tojima, H. Takeshiba, and T. Kinoto, *Bull. Chem. Soc. Jpn.*, 1979, **52**, 2441.
51. *Perfumes*, p.315.
52. F. Näf, R. Decorzant, and W. Thommen, *Helv. Chim. Acta*, 1975, **58**, 1808; M. Rosenberger, D. Andrews, F. DiMaria, A. J. Duggan, and G. Saucy, *Ibid.*, 1972, **55**, 249; C. Feugas, *Bull. Soc. Chim. Fr.*, 1963, 2568.
53. H. J. Dauben, H. R. Ringold, R. H. Wade, D. L. Pearson, and A. G. Anderson, *Org. Synth. Coll.*, 1963, **4**, 221.
54. B. Tchoubar, *Bull. Soc. Chim. Fr.*, 1949, 160, 164, 169.
55. *Drug Synthesis*, pp.95–96; D. F. Reinhold, R. A. Firestone, W. A. Gaines, J. M. Chemerda, and M. Sletzinger, *J. Org. Chem.*, 1968, **33**, 1209.
56. E. J. Corey, J.-L. Gras, and P. Ulrich, *Tetrahedron Lett.*, 1976, 809.
57. R. O. Clinton, U. J. Salvador, S. C. Laskowski, and M. Wilson, *J. Am. Chem. Soc.*, 1952, **74**, 592.
58. J. L. Everett, J. J. Roberts, and W. C. J. Ross, *J. Chem. Soc.*, 1953, 2386.
59. J. L. C. Kachinsky and R. G. Salomon, *Tetrahedron Lett.*, 1977, 3235.
60. *Pesticides*, p.142; *Pesticide Manual*, p.145; *Vogel*, p.754.
61. J. B. Hendrickson and C. Kandall, *Tetrahedron Lett.*, 1970, 343; W. L. Judefind, and E. E. Reid, *J. Am. Chem. Soc.*, 1920, **42**, 1043.
62. M. Freifelder, *J. Am. Chem. Soc.*, 1960, **82**, 2386.
63. G. Büchi and H. Wüest, *J. Org. Chem.*, 1969, **34**, 1122.
64. T. Nakajima, S. Masuda, S. Nakashima, T. Kondo, Y. Nakamoto, and S. Suga, *Bull. Chem. Soc. Jpn.*, 1979, **52**, 2377.

369

65. R. Adams and C. R. Noller, *Org. Synth. Coll.*, 1932, **1**, 109; W. D. Langley, *Ibid.*, 127.
66. P. A. Levene, *Org. Synth. Coll.*, 1943, **2**, 88.
67. O. Widman and E. Wahlberg, *Ber.*, 1911, **44**, 2065.
68. E. M. Schultz and S. Mickey, *Org. Synth. Coll.*, 1955, **3**, 343.
69. C. S. Marvel, *Org. Synth. Coll.*, 1955, **3**, 523, 848.
70. G. F. MacKenzie and E. K. Morris, *U.S. Pat.*, 1958, 2,848,491; *Chem. Abstr.*, 1959, **53**, 1151b.
71. E. K. Harvill, R. M. Herbst, and E. G. Schreiner, *J. Org. Chem.*, 1952, **17**, 1597.
72. *Fleming, Orbitals*, p.34.
73. *Vogel, pp.684-687*.
74. M. Bockmühl and G. Ehrhart, *Liebigs Ann. Chem.*, 1948, **561**, 52.
75. N. H. Cromwell, in *Heterocyclic Compounds*, ed., R. C. Elderfield, Vol. 6, 1957, Wiley, New York, pp.502-517.
76. A. Rieche and E. Schmitz, *Chem. Ber.*, 1956, **89**, 1254.
77. D. T. Collin, D. Hartley, D. Jack, L. H. C. Lunts, J. C. Press, A. C. Ritchie, and P. Toon, *J. Med. Chem.*, 1970, **13**, 674.
78. R. Schröter and F. Möller, *Houben-Weyl*, XI/1, pp.341-731.
79. K. A. Schellenberg, *J. Org. Chem.*, 1963, **28**, 3259.
80. B. Wojcik and H. Adkins, *J. Am. Chem. Soc.*, 1934, **56**, 2419.
81. W. H. Hartung, *J. Am. Chem. Soc.*, 1928, **50**, 3370; M. Rabinowitz, in *The Chemistry of the Cyano Group*, ed., Z. Rappoport, Interscience, London, 1970, pp.319-331.
82. R. F. Nystrom, *J. Am. Chem. Soc.*, 1955, **77**, 2544.
83. *Drug Synthesis*, p.70; L. Beregi, P. Hugon, J. C. LeDouarec, and H. Schmitt, *French Pat.*, 1963, M1658. *Chem. Abstr.*, 1963, **59**, 3831f.
84. T. Sheradsky, in *The Chemistry of the Azido Group*, ed., S. Patai, Interscience, London, 1971, pp.333-342.
85. C. A. VanderWerf, R. Y. Heisler, and W. E. McEwen, *J. Am. Chem. Soc.*, 1954, **76**, 1231.
86. W. A. Noyes and P. K. Porter, *Org. Synth. Coll.*, 1932, **1**, 457; P. L. Salzberg and J. V. Supniewski, *Ibid.*, 119.
87. J. F. W. McOmie, *Protective Groups in Organic Chemistry*, Plenum Press, London, 1973.
88. A. Amann, H. Koenig, P. C. Thieme, and H. Giertz, *Ger. Offen.*, 1974, 2,310,140; *Chem. Abstr.*, 1975, **82**, 31115; H. Koenig, P. C. Thieme, and H. Giertz, *Ger. Offen.*, 1974, 2,310,141; *Chem. Abstr.*, 1975, **82**, 31118.
89. T. Ishida and K. Wada, *J. Chem. Soc., Chem. Commun.*, 1977, 337.
90. T. Ishida and K. Wada, *J. Chem. Soc. Perkin Trans. 1*, 1979, 323.
91. E. J. Corey, R. L. Danheiser, S. Chandrasekaran, P. Siret, G. E. Keck, and J.-L. Gras, *J. Am. Chem. Soc.*, 1978, **100**, 8031.
92. E. J. Corey, J.-L. Gras, and P. Ulrich, *Tetrahedron Lett.*, 1976, 809.
93. R. H. Mazur, J. M. Schlatter, and A. H. Goldkamp, *J. Am. Chem. Soc.*, 1969, **91**, 2684.
94. J. M. Davey, A. H. Laird, and J. S. Morley, *J. Chem. Soc. (C)*, 1966, 555.
95. R. A. Boissonas, St. Guttmann, P.-A. Jaquenoud, and J.-P. Waller, *Helv. Chim. Acta*, 1956, **39**, 1421.
96. S. Searles, H. R. Hays, and E. F. Lutz, *J. Org. Chem.*, 1962, **27**, 2828. B. M. Trost, W. L. Schinski, and I. B. Mantz, *J. Am. Chem. Soc.*, 1969, **91**, 4320.
97. J. K. Kochi, *J. Am. Chem. Soc.*, 1963, **85**, 1958.
98. M. P. Mertes, P. E. Hanna, and A. A. Ramsey, *J. Med. Chem.*, 1970, **13**, 125.
99. *Perfumes*, p.66; V. Grignard, *Compt. Rend.*, 1900, **130**, 1322; A. Klages, *Ber.*, 1904, **37**, 1721.

370

100. *Drug Synthesis*, p.45.
101. D. W. Adamson, *Brit. Pat.*, 1949, 624,118. *Chem. Abstr.*, 1950, **44**, 662h.
102. B. Lythgoe and I. Waterhouse, *J. Chem. Soc. Perkin Trans. 1*, 1979, 2429.
103. E. J. Corey and J. W. Suggs, *Tetrahedron Lett.*, 1975, 2647.
104. E. J. Corey and G. Schmidt, *Tetrahedron Lett.*, 1979, 399.
105. M. S. Kharasch and O. Reinmuth, *Grignard Reactions of Non-Metallic Substances*, Prentice-Hall, New York, 1954, pp.913–960 and 767–845.
106. K. Friedrich and K. Wallenfels, in *The Chemistry of the Cyano Group*, ed., Z. Rappoport, Interscience, London, 1970, pp.67–110. F. C. Schaeffer, *Ibid.*, pp.256–262.
107. R. Adams and A. F. Thal, *Org. Synth. Coll.*, 1932, **1**, 107, 270. J. V. Supniewski and P. L. Salzberg, *Ibid.*, 46; E. Rietz, *Ibid.*, 1955, **3**, 851.
108. L. A. Bigelow, *Org. Synth. Coll.*, 1932, **1**, 136; E. A. Coulson, *J. Chem. Soc.*, 1934, 1406.
109. R. K. Smalley and H. Suschitzky, *J. Chem. Soc.*, 1964, 755.
110. *Perfumes*, p.70; P. A. Levene and A. Walti, *J. Biol. Chem.*, 1931, **94**, 367.
111. R. C. Huston and A. H. Agett, *J. Org. Chem.*, 1941, **6**, 123; T. G. Clarke, N. A. Hampson, J. B. Lee, J. R. Morley, and B. Scanlon, *Tetrahedron Lett.*, 1968, 5685.
112. J. F. Bunnett and S. Sridharan, *J. Org. Chem.*, 1979, **44**, 1458.
113. *Drug Synthesis*, p.44; H. Arnold, N. Brock, E. Kuhas, and D. Lorenz, *Arzneimittel-Forsch.*, 1954, **4**, 189; *Chem. Abstr.*, 1954, **48**, 7795e.
114. G. L. Dorough, H. B. Glass, T. L. Gresham, G. B. Malone, and E. E. Reid, *J. Am. Chem. Soc.*, 1941, **63**, 3100.
115. R. M. Einterz, J. W. Ponder, and R. S. Lenox, *J. Chem. Educ.*, 1977, **54**, 382.
116. G. M. Badger, H. C. Carrington, and J. A. Hendry, *Brit. Pat.*, 1946, 576,962; *Chem. Abstr.*, 1948, **42**, 3782g.
117. Ref.105, p.597; V. Grignard, *Ann. Chim. (Paris)*, 1901 (7), **24**, 475–476.
118. S. F. Acree, *Am. Chem. J.*, 1905, **33**, 180; *J. Chem. Soc.*, 1905, Ai, 216; Ref.105, p.609.
119. S. Siegel, W. M. Boyer, and R. R. Jay, *J. Am. Chem. Soc.*, 1951, **73**, 3237.
120. P. Schorigin, W. Issaguljanz, A. Gussewa, V. Ossipowa, and C. Poljakowa, *Ber.*, 1931, **64**, 2584.
121. P. J. Pearce, D. H. Richards, and N. F. Scilly, *J. Chem. Soc., Chem. Commun.*, 1970, 1160.
122. G. Büchi and H. Wüest, *J. Org. Chem.*, 1969, **34**, 1122.
123. For the Friedel–Crafts step see p.000 and for bromine addition see p.000.
124. G. Ohloff and W. Giersch, *Helv. Chim. Acta*, 1980, **63**, 76.
125. D. J. Cram and F. A. A. Elhafez, *J. Am. Chem. Soc.*, 1952, **74**, 5851.
126. U. Ravid and R. M. Silverstein, *Tetrahedron Lett.*, 1977, 423.
127. D. H. G. Crout, Personal Communication; cf. M. Schröder, *Chem. Rev.*, 1980, **80**, 187.
128. W. J. Elliott and J. Fried, *J. Org. Chem.*, 1976, **41**, 2475.
129. E. L. Eliel and R. S. Ro, *J. Am. Chem. Soc.*, 1957, **79**, 5995.
130. J. Klein, E. Dunkelblum, E. L. Eliel, and Y. Senda, *Tetrahedron Lett.* 1968, 6127; see also *House*, pp.54–70.
131. G. Kinast and L.-F. Tietze, *Chem. Ber.*, 1976, **109**, 3626.
132. R. L. Augustine, *J. Org. Chem.*, 1958, **23**, 1853; S. Nishimura and M. Shimahara, *Chem. Ind. (London)*, 1966, 1796.
133. cf. B. Carnmalm, T. De Paulis, E. Jacupovic, L. Johansson, U. H. Lindberg, B. Ulff, N. E. Stjernstrom, A. L. Renyi, S. B. Ross, and S. O. Ogren, *Acta Pharm. Suec.*, 1975, **12**, 149; *Chem. Abstr.*, 1975, **83**, 157704.
134. R. Rossi and P. A. Salvadori, *Synthesis*, 1979, 209.
135. Ref.105, pp.767–845.

371

136. M. J. Jorgenson, *Org. React.*, 1970, **18**, 1.
137. *House*, pp.492–595.
138. E. B. Vliet, C. S. Marvel, and C. M. Hsueh, *Org. Synth. Coll.*, 1943, **2**, 416.
139. C. S. Marvel, *Org. Synth. Coll.*, 1955, **3**, 495.
140. J. Weichet, L. Novak, J. Stribrny, and L. Blaha, *Czech. Pat.*, 1964, 112,243; *Chem. Abstr.*, 1965, **62**, 13049e.
141. B. Glatz, G. Helmchen, H. Muxfeldt, H. Porcher, R. Prewo, J. Senn, J. J. Stezowski, R. J. Stojda, and D. R. White, *J. Am. Chem. Soc.*, 1979, **101**, 2171.
142. M. Suzuki, T. Suzuki, T. Kawagishi, and R. Noyori, *Tetrahedron Lett.*, 1980, **21**, 1247.
143. J. Munch-Petersen, *Org. Synth. Coll.*, 1973, **5**, 762.
144. P. Pfeiffer and H. L. de Waal, *Liebigs Ann. Chem.*, 1935, **520**, 185.
145. T. Kametani and H. Nemoto, *Tetrahedron Lett.*,. 1979, 3309.
146. Method of J. Cason, *Chem. Rev.*, 1947, **40**, 15; D. A. Shirley, *Org. React.*, 1954, **8**, 28.
147. Method of L. Clarke, *J. Am. Chem. Soc.*, 1911, **33**, 529; W. B. Renfrew, *Ibid.*, 1944, **66**, 144; C. S. Marvel and F. D. Hager, *Org. Synth. Coll.*, 1932, **1**, 248; J. R. Johnson and F. D. Hager, *Ibid.*, 351.
148. E. Urion, *Compt. Rend.*, 1932, **194**, 2311; *Chem. Abstr.*, 1932, **26**, 5079.
149. *Fleming, Orbitals*, p.70.
150. *House*, pp.1–34.
151. W. G. Brown, *Org. React.*, 1951, **6**, 469; N. G. Gaylord, *Reduction with Complex Metal Hydrides*, Interscience, New York, 1956, p.446.
152. J. Auerbach and S. M. Weinreb, *J. Org. Chem.*, 1975, **40**, 3311.
153. G. H. Posner, *Org. React.*, 1972, **19**, 1; 1975, **22**, 253; H. O. House, *Acc. Chem. Res.*, 1976, **9**, 59; J. F. Normant, *Synthesis*, 1972, 63.
154. C. P. Casey and R. A. Boggs, *Tetrahedron Lett.*, 1971, 2455.
155. H. E. Zimmerman, T. P. Gannett, and G. E. Keck, *J. Org. Chem.*, 1979, **44**, 1982.
156. R. P. Thummel, *J. Am. Chem. Soc.*, 1976, **98**, 628.
157. *House*, pp.682–709.
158. G. Wittig and U. Schöllkopf, *Org. Synth. Coll.*, 1973, **5**, 751.
159. C. F. Hauser, T. W. Brooks, M. L. Miles, M. A. Raymond, and G. B. Butler, *J. Org. Chem.*, 1963, **28**, 372.
160. W. S. Wadsworth and W. D. Emmons, *Org. Synth. Coll.*, 1973, **5**, 547.
161. C. A. Henrick, *Tetrahedron*, 1977, **33**, 1845; B. A. Bierl, M. Beroza, and C. W. Collier, *Science*, 1970, **170**, 87.
162. R. N. McDonald and T. W. Campbell, *J. Org. Chem.*, 1959, **24**, 1969.
163. G. H. Whitfield, *Brit. Pat.*, 1955, 735,118; *Chem. Abstr.*, 1956, **50**, 8721f.
164. H. Pasedach, *Ger. Offen.*, 1972, 2,047,446; *Chem. Abstr.*, 1972, **77**, 4876.
165. J. W. Copenhaver and M. H. Bigelow, *Acetylene and Carbon Monoxide Chemistry*, Reinhold, New York, 1949, pp.130–142.
166. A. W. Johnson, *J. Chem. Soc.*, 1946, 1014.
167. A. A. Kraevskii, I. K. Sarycheva, and N. A. Preobrazhenskii, *Zh. Obsch. Khim.*, 1963, **33**, 1831; *Chem. Abstr.*, 1964, **61**, 14518f.
168. R. A. Baker and D. A. Evans, *Annu. Rep. Prog. Chem., Sect. B*, 1975, 347; Aliphatic and Related Natural Product Chemistry, *Chem. Soc. Sp. Per. Rep.1*, 1979, pp.102–127.
169. V. F. Kucherov and E. P. Serebryakov, *Izv. Akad. Nauk. SSSR, Ser. Khim.*, 1960, 1057; *Chem. Abstr.*, 1961, **55**, 475h.
170. R. J. Thomas, K. N. Campbell, and G. F. Hennion, *J. Am. Chem. Soc.*, 1938, **60**, 718.
171. *Perfumes*, p.164; J. V. Nef, *Liebigs Ann. Chem.*, 1899, **308**, 277.

372

172. M. F. Ansell, W. J. Hickinbottom, and A. A. Hyatt, *J. Chem. Soc.*, 1955, 1592.
173. S. Swaminathan and K. V. Narayanan, *Chem. Rev.*, 1971, **71**, 429; J. H. Saunders, *Org. Synth. Coll.*, 1955, **3**, 22; G. W. Stacy and R. A. Mikulec, *Ibid.*, 1963, **4**, 13.
174. M. C. Kloetzel, *Org. React.*, 1948, **4**, 1; H. L. Holmes, *Ibid.*, 60; L. W. Butz, *Ibid.*, 1949, **5**, 136; J. Sauer, *Angew. Chem., Int. Ed. Engl.*, 1966, **5**, 211.
175. O. Diels and K. Alder, *Liebigs Ann. Chem.*, 1929, **470**, 62.
176. *Carruthers*, pp.202–211; J. G. Martin and R. K. Hill, *Chem. Rev.*, 1961, **61**, 537.
177. F. V. Brutcher and D. D. Rosenfeld, *J. Org. Chem.*, 1964, **29**, 3154.
178. N. Green, M. Beroza, and S. A. Hall, *Adv. Pest Control Res.*, 1960, **3**, 129; *Pesticides*, p.196.
179. M. Y. Kim and S. M. Weinreb, *Tetrahedron Lett.*, 1979, 579.
180. E. F. Lutz and G. M. Bailey, *J. Am. Chem. Soc.*, 1964, **86**, 3899; T. Inukai and T. Kojima, *J. Org. Chem.*, 1967, **32**, 869, 872; 1971, **36**, 924; *Carruthers*, pp.198–202, 211–215.
181. G. Satzinger, *Liebigs Ann. Chem.*, 1969, **728**, 64.
182. N. L. Wendler and H. L. Slates, *J. Am. Chem. Soc.*, 1958, **80**, 3937.
183. *House*, pp.734–786.
184. *Perfumes*, p.164.
185. C. Beyer and L. Claisen, *Ber.*, 1887, **20**, 2178.
186. J. K. H. Inglis and K. C. Roberts, *Org. Synth. Coll.*, 1932, **1**, 235.
187. *Pesticide Manual*, p.428; N. J. A. Gutteridge, *Chem. Soc. Rev.*, 1972, **1**, 381.
188. L. B. Kilgore, J. H. Ford, and W. C. Wolfe, *Ind. Eng. Chem.*, 1942, **34**, 494.
189. P. S. Pinkney, *Org. Synth. Coll.*, 1943, **2**, 116; W. Dieckmann, *Ber.*, 1894, **27**, 102.
190. S. M. McElvain and K. Rorig, *J. Am. Chem. Soc.*, 1948, **70**, 1820; J. P. Schaefer and J. J. Bloomfield, *Org. React.*, 1967, **15**, 1; cf. p.000.
191. *House*, pp.629–733.
192. A. T. Nielsen and W. J. Houlihan, *Org. React.*, 1968, **16**, 115.
193. J. B. Conant and N. Tuttle, *Org. Synth. Coll.*, 1932, **1**, 199; H. Adkins and H. I. Cramer, *J. Am. Chem. Soc.*, 1930, **52**, 4349; See A. I. Meyers, *Heterocycles in Organic Synthesis*, Wiley, New York, 1974, for applications of the reagent.
194. W. G. Dauben, M. S. Kellogg, J. I. Seeman, and W. A. Spitzer, *J. Am. Chem. Soc.*, 1970, **92**, 1786.
195. N. B. Lorette, *J. Org. Chem.*, 1957, **22**, 346.
196. O. E. Curtis, J. M. Sandri, R. E. Crocker, and H. Hart, *Org. Synth. Coll.*, 1963, **4**, 278.
197. *Perfumes*, p.142.
198. *Drug Synthesis*, pp.220–221; K. W. Wheeler, M. G. Van Campen, and R. S. Shelton, *J. Org. Chem.*, 1960, **25**, 1021.
199. Ref.192, p.86.
200. Ref.192, p.117.
201. J. B. Conant and A. H. Blatt, *J. Am. Chem. Soc.*, 1929, **51**, 1227.
202. Ref.192, pp.112–113.
203. A. E. Abbott, G. A. R. Kon, and R. D. Satchell, *J. Chem. Soc.*, 1928, 2514.
204. Ref.192, p.125.
205. H. Paul and I. Wendel, *Chem. Ber.*, 1957, **90**, 1342.
206. cf. H. R. Snyder, L. A. Brooks, and S. H. Shapiro, *Org. Synth. Coll.*, 1943, **2**, 531; A. P. Krapcho, J. Diamanti, C. Cayen, and R. Bingham, *Ibid.*, 1973, **5**, 198.
207. cf. P. A. Levene and G. M. Meyer, *Org. Synth. Coll.*, 1943, **2**, 288; G. R. Zellars, and R. Levine, *J. Org. Chem.*, 1948, **13**, 160.
208. Ref.192, pp.272–330.
209. A. Eschenmoser and C. E. Wintner, *Science*, 1977, **196**, 1418.
210. B. Puetzer, C. H. Nield, and R. H. Barry, *J. Am. Chem. Soc.*, 1945, **67**, 833.

373

211. M. Tramontini, *Synthesis*, 1973, 703; *House*, pp.654–660.
212. F. F. Blicke, *Org. React.*, 1942, **1**, 303.
213. A. P. Beracierta and D. A. Whiting, *J. Chem. Soc., Perkin Trans. 1*, 1978, 1257.
214. G. Jones, *Org. React.*, 1967, **15**, 204; *House*, pp.632–653.
215. J. W. Opie, J. Seifter, W. F. Bruce, and G . Mueller, *U.S. Pat.*, 1951, 2,538,322; *Chem. Abstr.*, 1951, **45**, 6657c.
216. C. A. Kingsbury and G. Max, *J. Org. Chem.*, 1978, **43**, 3131.
217. K. Alder and H. F. Rickert, *Ber.*, 1939, **72**, 1983.
218. T. Reffstrup and P. M. Boll, *Acta Chem. Scand., Ser.B*, 1977, **31**, 727.
219. R. L. Shriner, *Org. React.*, 1942, **1**, 1; M. W. Rathke, *Ibid.*, 1974, **22**, 423; *House*, pp.671–677.
220. S. N. Reformatsky, *J. Prakt. Chem.*, 1896, **54**, 469, 477; M. S. Newman and A. Kutner, *J. Am. Chem. Soc.*, 1951, **73**, 4199.
221. J. W. Cornforth and R. H. Cornforth, in *Natural Substances from Mevalonic Acid*, ed. T. W. Goodwin, Biochem. Soc. Symposium 29, Academic Press, London, 1970, pp.5–17.
222. E. C. DuFeu, F. J. McQuillin, and R. Robinson, *J. Chem. Soc.*, 1937, 53.
223. J. W. Cornforth, R. H. Cornforth, G. Popják, and I. Y. Gore, *Biochem. J.*, 1958, **69**, 146.
224. G. Stork, A. Brizzolara, H. Landesman, J. Szmuszkovicz, and R. Terrell, *J. Am. Chem. Soc.*, 1963, **85**, 207.
225. L. Birkofer, S. M. Kim, and H. D. Engels, *Chem. Ber.*, 1962, **95**, 1495.
226. S.-R. Kuhlmey, H. Adolph, K. Rieth, and G. Opitz, *Liebigs Ann. Chem.*, 1979, 617; L. Nilsson, *Acta Chem. Scand., Ser.B*, 1979, **33**, 203.
227. Ref.192, p.211.
228. C. R. Hauser, F. W. Swamer, and J. T. Adams, *Org. React.*, 1954, **8**, 165.
229. G. Jones, *Org. React.*, 1967, **15**, 274–325.
230. E. T. Borrows and B. A. Hems, *J. Chem. Soc.*, 1945, 577.
231. P. J. Kocienski, J. M. Ansell, and R. W. Ostrow, *J. Org. Chem.*, 1976, **41**, 3625.
232. *Pesticides*, p.196; C. F. H. Allen and J. Van Allen, *Org. Synth. Coll.*, 1955, **3**, 783.
233. *House*, pp.595–623; E. D. Bergmann, D. Ginsburg, and R. Pappo, *Org. React.*, 1959, **10**, 179.
234. P. D. Bartlett and G. F. Woods, *J. Am. Chem. Soc.*, 1940, **62**, 2933.
235. *House*, pp.596–597.
236. R. V. Stevens and A. W. M. Lee, *J. Am. Chem. Soc.*, 1979, **101**, 7032.
237. J. P. Bays, M. V. Encinas, R. D. Small, and J. C. Sciano, *J. Am. Chem. Soc.*, 1980, **102**, 727.
238. R. L. Frank and R. C. Pierle, *J. Am. Chem. Soc.*, 1951, **73**, 724.
239. M. E. Jung, *Tetrahedron*, 1976, **32**, 3; *House*, pp.606–611.
240. R. Connor and D. B. Andrews, *J. Am. Chem. Soc.*, 1934, **56**, 2713.
241. S. Ramachandran and M. S. Newman, *Org. Synth. Coll.*, 1973, **5**, 486; T. A. Spencer, H. S. Neel, D. C. Ward, and K. L. Williamson, *J. Org. Chem.*, 1966, **31**, 434.
242. J. K. Roy, *Science and Culture* (India), 1953, **19**, 156; *Chem. Abstr.*, 1954, **48**, 13660g.
243. R. L. Shriner and H. R. Todd, *Org. Synth. Coll.*, 1943, **2**, 200.
244. J. E. McMurry, *Acc. Chem. Res.*, 1974, **7**, 281; J. E. McMurry and J. Melton, *J. Org. Chem.*, 1973, **38**, 4367.
245. *Drug Synthesis*, p.73; G. B. Bachman, H. B. Hass, and G. O. Platau, *J. Am. Chem. Soc.*, 1954, **76**, 3972.
246. J. R. Butterick and A. M. Unrau, *J. Chem. Soc., Chem. Commun.*, 1974, 307.
247. R. Gabler, H. Müller, G. E. Ashby, E. R. Agouri, H.-R. Meyer, and G. Kabas, *Chimia*, 1967, **21**, 65.

374

248. G. Poidevin, P. Foy, and T. Rull, *Bull. Soc. Chim. Fr.*, 1979, II—196.
249. G. I. Poos, J. Kleis, R. R. Wittekind, and J. D. Rosenau, *J. Org. Chem.*, 1961, **26**, 4898; J. Thesing, G. Seitz, R. Hotovy, and S. Sommer, *Ger. Pat.*, 1961, 1,110,159; *Chem. Abstr.*, 1962, **56**, 2352h.
250. J. H. Clark, J. M. Miller, and K.-H. So, *J. Chem. Soc., Perkin Trans. 1*, 1978, 941; E. Bellasio, B. Cavalleri, T. La Noce, and E. Testa, *Farmaco Ed. Sci.*, 1976, **31**, 471; *Chem. Abstr.*, 1976, **85**, 177173; See also T.-L. Ho, *J. Chem. Soc., Chem. Commun.*, 1980, 1149.
251. O. W. Lever, *Tetrahedron*, 1976, **32**, 1943.
252. G. F. Hennion and F. P. Kupiecki, *J. Org. Chem.*, 1953, **18**, 1601.
253. G. L. Lange, D. J. Wallace, and S. So, *J. Org. Chem.*, 1979, **44**, 3066.
254. E. Wenkert, N. F. Golob, and R. A. J. Smith, *J. Org. Chem.*, 1973, **38**, 4068; E. Wenkert, D. A. Berges, and N. F. Golob, *J. Am. Chem. Soc.*, 1978, **100**, 1263.
255. *Drug Synthesis*, pp.219-220; J. Mills, *U.S. Pat.*, 1957, 2,812,363; *Chem. Abstr.*, 1961, **55**, 5427d.
256. W. S. Ide and J. S. Buck, *Org. React.*, 1948, **4**, 269.
257. J. B. Lambert, H. W. Mark, A. G. Holcomb, and E. S. Magyar, *Acc. Chem. Res.*, 1979, **12**, 321; J. B. Lambert, H. W. Mark, and E. S. Magyar, *J. Am. Chem. Soc.*, 1977, **99**, 3059.
258. *Drug Synthesis*, p.46; L. Stein and E. Lindner, *U.S. Pat.*, 1958, 2,827,460; *Chem. Abstr.*, 1959, **53**, 415f.
259. *House*, pp.407-411.
260. O. Truster, *Org. React.*, 1953, **7**, 327.
261. *Drug Synthesis*, pp.64-65; C. H. Boeringer Sohn, *Belg. Pat.*, 1961, 611,502; *Chem. Abstr.*, 1962, **57**, 13678i.
262. B. B. Snider, D. M. Roush, and T. A. Killinger, *J. Am. Chem. Soc.*, 1979, **101**, 6023.
263. Ref.105, p.688.
264. A. B. Smith and P. J. Jerris, *Synth. Commun.*, 1978, **8**, 421; S. W. Baldwin and M. T. Crimmins, *J. Am. Chem. Soc.*, 1980, **102**, 1198; *Tetrahedron Lett.*, 1978, 4197.
265. P. Margaretha, *Tetrahedron Lett.*, 1971, 4891.
266. *House*, pp.478-491.
267. P. N. Confalone, E. D. Lollar, G. Pizzolato, and M. R. Uskoković, *J. Am. Chem. Soc.*, 1978, **100**, 6291; P. N. Confalone, G. Pizzolato, D. L. Confalone, and M. R. Uskoković, *Ibid.*, 1980, **102**, 1954.
268. M. Verhage, D. A. Hoogwater, J. Reedijk, and H. van Bekkum, *Tetrahedron Lett.*, 1979, 1267.
269. H. Moureu, P. Chovin, and R. Sabourin, *Bull. Soc. Chim. Fr.*, 1955, 1255.
270. W. G. Dauben, M. Lorber, and D. S. Fullerton, *J. Org. Chem.*, 1969, **34**, 3587.
271. Ref.17, pp.197-245.
272. D. Holland and D. J. Milner, *Chem. Ind. (London)*, 1979, 707.
273. E. C. Dodds and R. Robinson, *Proc. R. Soc. London, Ser.B*, 1939, **127**, 148.
274. J. J. Bloomfield, D. C. Owsley, and J. M. Nelke, *Org. React.*, 1976, **23**, 259.
275. K. Rühlmann, *Synthesis*, 1971, 236; J. J. Bloomfield, D. C. Owsley, C. Ainsworth, and R. E. Robertson, *J. Org. Chem.*, 1975, **40**, 393.
276. P. J. Jerris, P. M. Wovkulich, and A. B. Smith, *Tetrahedron Lett.*, 1979, 4517.
277. P. Ruggli and P. Zeller, *Helv. Chim. Acta*, 1945, **28**, 741; I. Hagedorn, U. Eholzer, and A. Lüttringhaus, *Ber.*, 1960, **93**, 1584.
278. A. C. Cope, D. S. Smith, and R. J. Cotter, *Org. Synth. Coll.*, 1963, **4**, 377; T. Mukaiyama, H. Nambu, and T. Kumamoto, *J. Org. Chem.*, 1964, **29**, 2243.
279. A. Krebs, *Tetrahedron Lett.*, 1968, 4511.

280. G. Edgar, G. Calingaert, and R. E. Marker, *J. Am. Chem. Soc.*, 1929, **51**, 1483; G. W. Moersch and F. C. Whitmore, *Ibid.*, 1949, **71**, 819.
281. *House*, pp.163-167.
282. *House*, pp.228-239.
283. *House*, pp.15-16.
284. Ref.6, p.961; B. R. Davis and I. R. N. McCormick, *J. Chem. Soc., Perkin Trans.* 1, 1979, 3001.
285. E. C. Horning and D. B. Reisner, *J. Am. Chem. Soc.*, 1949, **71**, 1036; E. L. Martin, *Org. React.*, 1942, **1**, 155.
286. C. Mannich and K. W. Merz, *Arch. Pharm. (Weinheim, Ger.)*, 1927, **265**, 15, 104; *Chem. Abstr.*, 1927, **21**, 1449, 1803.
287. Ref.214, p.282; J. T. Plati, W. H. Strain, and S. L. Warren, *J. Am. Chem. Soc.*, 1943, **65**, 1273; F. S. Kipping and A. E. Hunter, *J. Chem. Soc.*, 1901, **79**, 602.
288. H. Fritz and E. Stock, *Tetrahedron*, 1970, **26**, 5821.
289. R. P. Linstead and E. M. Meade, *J. Chem. Soc.*, 1934, 935.
290. J. Jernow, W. Tautz, P. Rosen, and J. F. Blount, *J. Org. Chem.*, 1979, **44**, 4210.
291. M. S. Newman and C. A. VanderWerf, *J. Am. Chem. Soc.*, 1945, **67**, 233.
292. L. Johnson, *U.S. Pat.*, 1948, 2,443,827; *Chem. Abstr.*, 1949, **43**, 678; T. E. Bellas, R. G. Brownlee, and R. M. Silverstein, *Tetrahedron*, 1969, **25**, 5149; G. W. Cannon, R. C. Ellis, and J. R. Leal, *Org. Synth. Coll.*, 1963, **4**, 597.
293. *Drug Synthesis*, p.226; C. A. Miller and L. M. Long, *J. Am. Chem. Soc.*, 1951, **73**, 4895.
294. O. Moldenhauer, W. Irion, D. Mastaglio, R. Pfluger, and H. Döser, *Liebigs Ann. Chem.*, 1953, **583**, 50.
295. J. P. Vigneron and V. Bloy, *Tetrahedron Lett.*, 1980, 1735; J. P. Vigneron and J. M. Blanchard, *Ibid.*, 1739; J. P. Vigneron, R. Méric, and M. Dhaenens, *Ibid.*, 2057.
296. D. Lednicer in *Advances in Organic Chemistry*, ed. E. C. Taylor, Wiley-Interscience, New York, 1972, Vol.8, pp.180-245; *House*, pp.275-278, 353-363.
297. M. T. Edgar, G. R. Pettit, and T. H. Smith, *J. Org. Chem.*, 1978, **43**, 4115.
298. W. G. Taylor, *J. Org. Chem.*, 1979, **44**, 1020.
299. R. Mitschka and J. M. Cook, *J. Am. Chem. Soc.*, 1978, **100**, 3973.
300. K. Y. Geetha, K. Rajagopalan, and S. Swiminathan, *Tetrahedron*, 1978, **34**, 2201.
301. N. R. Easton, J. H. Gardner, and J. R. Stevens, *J. Am. Chem. Soc.*, 1947, **69**, 2941.
302. H. O. House, C.-C. Yau, and D. Vanderveer, *J. Org. Chem.*, 1979, **44**, 3031.
303. H. O. House and M. J. Umen, *J. Org. Chem.*, 1973, **38**, 1000.
304. F. Plavac and C. H. Heathcock, *Tetrahedron Lett.*, 1979, 2115.
305. A. Eschenmoser and C. E. Winter, *Science*, 1977, **196**, 1410.
306. *House*, pp.321-329.
307. J. Vasilevskis, J. A. Gualtieri, S. D. Hutchings, R. C. West, J. W. Scott, D. R. Parrish, F. T. Bizzarro, and G. F. Field, *J. Am. Chem. Soc.*, 1978, **100**, 7423.
308. G. Magnusson, *Tetrahedron*, 1978, **34**, 1385.
309. W. E. Bachmann and W. S. Struve, *J. Am. Chem. Soc.*, 1941, **63**, 2590; W. Carruthers and A. Orridge, *J. Chem. Soc., Perkin Trans. 1*, 1977, 2411; H. Gerlach and W. Müller, *Helv. Chim. Acta*, 1972, **55**, 2277.
310. R. A. Pratt and R. A. Raphael, *Unpublished Observations*.
311. A. S. Dreiding and A. J. Tomasewski, *J. Am. Chem. Soc.*, 1954, **76**, 540.
312. H. Nakai, Y. Arai, N. Hamanaka, and M. Hayashi, *Tetrahedron Lett.*, 1979, 805.
313. E. J. Corey and J. G. Smith, *J. Am. Chem. Soc.*, 1979, **101**, 1038.
314. D. A. Prins, *Helv. Chim. Acta*, 1957, **40**, 1621.
315. H. Stetter, I. Krüger-Hansen, and M. Rizk, *Chem. Ber.*, 1961, **94**, 2702.

376

316. W. H. Perkin, *J. Chem. Soc.*, 1885, 801; 1886, 806; 1887, 1; E. Haworth and W. H. Perkin, *Ibid.*, 1894, 591; *House*, pp.541–544.
317. R. A. Barnes and W. M. Budde, *J. Am. Chem. Soc.*, 1946, **68**, 2339.
318. N. H. Cromwell and B. Phillips, *Chem. Rev.*, 1979, **79**, 331.
319. R. M. Rodebaugh and N. H. Cromwell, *J. Heterocycl. Chem.*, 1968, **5**, 309; 1969, **6**, 439; B. Wladislaw, *J. Org. Chem.*, 1961, **26**, 711.
320. F. Nerdel and H. Kressin, *Liebigs Ann. Chem.*, 1967, **707**, 1.
321. R. H. Everhardus, R. Gräfing, and L. Brandsma, *Recl. Trav. Chim. Pays-Bas*, 1976, **95**, 153; B. A. Trofimov, S. V. Amosova, G. K. Musorin, and M. G. Voronkov, *Zh. Org. Khim.*, 1978, **14**, 667; *Chem. Abstr.*, 1978, **88**, 190507.
322. R. B. Woodward and R. H. Eastman, *J. Am. Chem. Soc.*, 1946, **68**, 2229.
323. R. Huisgen, *Angew. Chem., Int. Ed. Engl.*, 1963, **2**, 563, 633.
324. *Fleming, Orbitals*, pp.148–161.
325. J. Thesing and W. Sirrenberg, *Chem. Ber.*, 1959, **92**, 1748; J. J. Tufariello and J. P. Tette, *J. Chem. Soc., Chem. Commun.*, 1971, 469; J. J. Tufariello and G. E. Lee, *J. Am. Chem. Soc.*, 1980, **102**, 373.
326. C. A. Grob and H. R. Kiefer, *Helv. Chim. Acta*, 1965, **48**, 799.
327. G. Sosnovsky and M. Konieczny, *Synthesis*, 1976, 735.
328. *Drug Synthesis*, pp.256–260, 298–310.
329. B. Elpern, W. Wetterau, P. Carabateas, and L. Grumbach, *J. Am. Chem. Soc.*, 1958, **80**, 4916.
330. R. J. DePasquale, *J. Org. Chem.*, 1977, **42**, 2185.
331. *Drug Synthesis*, p.363; Ref.17, p.414; L. H. Sternbach and E. Reeder, *J. Org. Chem.*, 1961, **26**, 4936.
332. J. M. Conia, *Angew. Chem., Int. Ed. Engl.*, 1968, **7**, 570.
333. M. Elliott, A. W. Farnham, N. F. Janes, P. H. Needham, D. A. Pulman, and J. H. Stevenson, *Nature (London)*, 1973, **246**, 169.
334. P. D. Klemmensen, H. Kolind-Andersen, H. B. Madsen, and A. Svendsen, *J. Org. Chem.*, 1979, **44**, 416.
335. *House*, pp.666–671.
336. H. Achenbach and J. Witzke, *Tetrahedron Lett.*, 1979, 1579.
337. *House*, pp.709–733.
338. S. D. Burke and P. A. Grieco, *Org. React.*, 1979, **26**, 261.
339. J. F. Ruppert and J. D. White, *J. Chem. Soc., Chem. Commun.*, 1976, 976.
340. A. Burger and W. L. Yost, *J. Am. Chem. Soc.*, 1948, **70**, 2198.
341. H. E. Simmons, T. L. Cairns, and S. A. Vladuchick, *Org. React.*, 1973, **20**, 1.
342. S. A. Monti and T. W. McAninch, *Tetrahedron Lett.*, 1974, 3239.
343. E. Piers and E. H. Ruediger, *J. Chem. Soc., Chem. Commun.*, 1979, 166.
344. W. E. Bachman and W. S. Struve, *Org. React.*, 1942, **1**, 38.
345. A. B. Smith, *J. Chem. Soc., Chem. Commun.*, 1974, 695.
346. W. L. Mock and M. E. Hartman, *J. Am. Chem. Soc.*, 1970, **92**, 5767.
347. A. P. Krapcho, *Synthesis*, 1976, 425.
348. E. J. Corey, J. F. Arnett, and G. N. Widiger, *J. Am. Chem. Soc.*, 1975, **97**, 430.
349. D. Dieterich, *Houben-Weyl*, VII/2a/1, pp.927–1048.
350. B. M. Trost and L. S. Melvin, *Sulfur Ylides*, Academic Press, New York, 1975.
351. *Perfumes*, p.124; A. Knorr, E. Laage, and A. Weissenborn, *Ger. Pat.*, 1930, 591,452; *Chem. Abstr.*, 1934, **28**, 2367.
352. A. S. Kende, *Org. React.*, 1960, **11**, 261.
353. J. G. Aston, J. T. Clarke, K. A. Burgess, and R. B. Greenburg, *J. Am. Chem. Soc.*, 1942, **64**, 300.
354. P. G. Sammes, *Quart. Rev.*, 1970, **24**, 37; J. Kossanyi, *Pure Appl. Chem.*, 1979, **51**, 181; A. B. Holmes in General and Synthetic Methods, *Chem. Soc., Spec. Per. Reports*, ed. G. Pattenden, 1980, p.329.
355. *Fleming, Orbitals*, p.86 and 208.

356. Ref.192, p.112.
357. D. C. Owsley and J. J. Bloomfield, *J. Chem. Soc.*, 1971 (C), 3445.
358. G. L. Lange, M.-A. Huggins, and E. Neidert, *Tetrahedron Lett.*, 1976, 4409.
359. R. C. Cookson, J. Hudec, S. A. Knight, and B. Whitear, *Tetrahedron Lett.*, 1962, 79.
360. M. Brown, *J. Org. Chem.*, 1968, **33**, 162.
361. E. J. Corey, R. B. Mitra, and H. Uda, *J. Am. Chem. Soc.*, 1963, **85**, 362; 1964, **86**, 485.
362. P. A. Wender and J. C. Lechleiter, *J. Am. Chem. Soc.*, 1978, **100**, 4321.
363. M. Fétizon, S. Lazare, C. Pascard, and T. Prange, *J. Chem. Soc., Perkin Trans. 1*, 1979, 1407.
364. R. N. McDonald and R. R. Reitz, *J. Org. Chem.*, 1972, **37**, 2418.
365. D. Redmore and C. D. Gutsche in *Adv. in Alicyclic Chem.*, 1971, **3**, 10.
366. B. M. Trost, M. J. Bogdanowicz, and J. Kern, *J. Am. Chem. Soc.*, 1975, **97**, 2218; B. M. Trost, M. Preckel, and L. M. Leichter, *Ibid.*, 2224.
367. C. D. Gusche and D. Redmore, *Carbocyclic Ring Expansion Reactions*, Academic Press, New York, 1968, pp.9, 63, 83, 131.
368. J. M. Conia and M. J. Robson, *Angew. Chem. Int. Ed. Engl.*, 1975, **14**, 473.
369. R. W. Holder, *J. Chem. Educ.*, 1976, **53**, 81.
370. W. T. Brady and A. D. Patel, *J. Org. Chem.*, 1973, **38**, 4106.
371. E. J. Corey, Z. Arnold, and J. Hutton, *Tetrahedron Lett.*, 1970, 307; M. J. Dimsdale, R. F. Newton, D. K. Rainey, C. F. Webb, T. V. Lee, and S. M. Roberts, *J. Chem. Soc., Chem. Commun.*, 1977, 716.
372. M. Farina and G. DiSilvestro, *Tetrahedron Lett.*, 1975, 183.
373. T. Schmidlin and C. Tamm, *Helv. Chim. Acta*, 1980, **63**, 121.
374. S. Wolff, W. L. Schreiber, A. B. Smith, and W. C. Agosta, *J. Am. Chem. Soc.*, 1972, **94**, 7797.
375. P. D. Magnus and M. S. Nobbs, *Synth. Commun.*, 1980, **10**, 273.
376. R. L. Danheiser, D. J. Carini, and A. Basak, *J. Am. Chem. Soc.*, 1981, **103**, 1604; P. Prempree, T. Siwapinyoyos, C. Thebtaranonth, and Y. Thebtaranonth, *Tetrahedron Lett.*, 1980, **21**, 1169; R. A. Ellison, *Synthesis*, 1973, 397; M. E. Jung, *Tetrahedron*, 1976, **32**, 3.
377. R. L. Funk and K. P. C. Vollhardt, *Synthesis*, 1980, 118.
378. *Drug Synthesis*, p.275; W. J. Doran and E. M. VanHeyningen, *U.S. Pat.*, 1951, 2,561,689.
379. J. Meinwald and T. H. Jones, *J. Am. Chem. Soc.*, 1978, **100**, 1883.
380. R. C. Cookson and S. A. Smith, *J. Chem. Soc., Perkin Trans. 1*, 1979, 2447.
381. T. S. Sorensen and A. Rauk, in *Pericyclic Reactions*, eds. A. P. Marchand and R. E. Lehr, Academic Press, New York, 1977, Vol.2, pp.21–33.
382. G. Ohloff, K. H. Schulte-Elte, and E. Demole, *Helv. Chim. Acta*, 1971, **54**, 2913.
383. J. M. Allen, K. M. Johnston, J. F. Jones, and R. G. Shotter, *Tetrahedron*, 1977, **33**, 2083.
384. F.-H. Marquardt, *Helv. Chim. Acta*, 1965, **48**, 1476.
385. T. R. Kasturi and S. Parvathi, *J. Chem. Soc., Perkin Trans. 1*, 1980, 448.
386. S. Dev, *J. Indian Chem. Soc.*, 1957, **34**, 169.
387. E. Piers, C. K. Lau, and I. Nagakura, *Tetrahedron Lett.*, 1976, 3233; J. E. Baldwin in ref.381, p.273 and ref.367, pp.163–170.
388. E. Piers and J. Banville, *J. Chem. Soc., Chem. Commun.*, 1979, 1138.
389. H.-U. Gonzenbach, I.-M. Tegmo-Larsson, J.-P. Grosclaude, and K. Schaffner, *Helv. Chim. Acta*, 1977, **60**, 1091.
390. H. Küntzel, H. Wolf, and K. Schaffner, *Helv. Chim. Acta*, 1971, **54**, 868.
391. C. P. Forbes, G. L. Wenteler, and A. Wiechers, *Tetrahedron*, 1978, **34**, 487; *Ap Simon*, Vol.3, p.447 ff, 527 ff.
392. D. S. Tarbell, *Org. React.*, 1944, **2**, 1.

378

393. G. B. Bennett, *Synthesis*, 1977, 589; S. J. Rhoads and N. R. Raulins, *Org. React.*, 1975, **22**, 1.
394. R. Marbet and G. Saucy, *Helv. Chim. Acta*, 1967, **50**, 2095; A. W. Burgstahler and I. C. Nordin, *J. Am. Chem. Soc.*, 1961, **83**, 198.
395. Y. Nakada, R. Endo, S. Muramatsu, J. Ide, and Y. Yura, *Bull. Chem. Soc. Jpn.*, 1979, **52**, 1511.
396. D. A. Evans and E. W. Thomas, *Tetrahedron Lett.*, 1979, 411.
397. J. N. Marx, J. C. Argyle, and L. R. Norman, *J. Am. Chem. Soc.*, 1974, **96**, 2121.
398. G. Büchi and J. E. Powell, *J. Am. Chem. Soc.*, 1970, **92**, 3126.
399. O. P. Vig, K. L. Matta, A. Lal, and I. Raj, *J. Indian Chem. Soc.*, 1964, **41**, 142; *Chem. Abstr.*, 1964, **61**, 1895e.
400. S. Danishefsky, M. Hirama, K. Gombatz, T. Harayama, E. Berman, and P. Schuda, *J. Am. Chem. Soc.*, 1978, **100**, 6536.
401. C. H. Tilford, M. G. Van Campen, and R. S. Shelton, *J. Am. Chem. Soc.*, 1947, **69**, 2902.
402. V. I. Isagulyants and P. P. Bagryantseva, *Neftyanoe Khoz.*, 1938, 36; *Chem. Abstr.*, 1939, **33**, 8183; S. H. Patinkin and B. S. Friedman in ref.6, Vol.II/I, pp.199–201; E. L. Eliel, R. J. L. Martin, and D. Nasipuri, *Org. Synth. Coll.*, 1973, **5**, 175; S. Winstein and N. J. Holness, *J. Am. Chem. Soc.*, 1955, **77**, 5562.
403. R. Schröter, *Houben-Weyl*, XI/I, p.688; *Vogel*, p.755; R. W. West, *J. Chem. Soc.*, 1925, **127**, 494.
404. A. J. Birch and G. Subba Rao, *Adv. Org. Chem., Methods and Results*, 1972, **8**, 1; *House*, pp.190–205.
405. E. Giovanni and H. Wegmüller, *Helv. Chim. Acta*, 1958, **41**, 933.
406. W. C. Still, A. J. Lewis, and D. Goldsmith, *Tetrahedron Lett.*, 1971, 1421.
407. *House*, pp.7, 192–202; A. J. Birch, A. R. Murray, and H. Smith, *J. Chem. Soc.*, 1951, 1945; W. Hückel and H. Schlee, *Chem. Ber.*, 1955, **88**, 346.
408. J. J. Sims, V. K. Honwad, and L. H. Selman, *Tetrahedron Lett.*, 1969, 87.
409. E. M. Kaiser and R. A. Benkeser, *Org. Synth.*, 1970, **50**, 88.
410. W. Hückel, R. Danneel, A. Schwartz, and A. Gercke, *Liebigs Ann. Chem.*, 1929, **474**, 121; W. Hückel, A. Gercke, and A. Gross, *Ber.*, 1933, **66**, 563.
411. E. J. Corey, M. Behforouz, and M. Ishiguro, *J. Am. Chem. Soc.*, 1979, **101**, 1608.
412. G. Stork, D. F. Taber, and M. Marx, *Tetrahedron Lett.*, 1978, 2445; see footnote 3.
413. J.-M. Conia and P. Beslin, *Bull Soc. Chim. France*, 1969, 483; R. L. Cargill, J. R. Dalton, S. O'Connor, and D. G. Michels, *Tetrahedron Lett.*, 1978, 4465.
414. D. F. Taber, *J. Org. Chem.*, 1976, **41**, 2649.
415. J. M. Hook and L. N. Mander, *J. Org. Chem.*, 1980, **45**, 1722; J. M. Hook, L. N. Mander, and R. Urech, *J. Am. Chem. Soc.*, 1980, **102**, 6628.
416. M. J. Perkins, N. B. Peynircioglu, and B. V. Smith, *J. Chem. Soc., Perkin Trans.* 2, 1978, 1025.
417. G. A. Macalpine, R. A. Raphael, A. Shaw, A. W. Taylor, and H.-J. Wild, *J. Chem. Soc., Chem. Commun.*, 1974, 834.
418. E. J. Corey and W. L. Jorgensen, *J. Am. Chem. Soc.*, 1976, **98**, 189, 203; E. J. Corey, H. W. Orf, and D. A. Pensak, *Ibid.*, 210.
419. A. Chatterjee, R. C. Chatterjee, and B. K. Bhattacharyya, *J. Indian Chem. Soc.*, 1957, **34**, 855.
420. E. J. Corey, W. J. Howe, H. W. Orf, D. A. Pensak, and G. Petersson, *J. Am. Chem. Soc.*, 1975, **97**, 6116.
421. T. Inukai and T. Kojima, *J. Org. Chem.*, 1966, **31**, 1121; S. A. Monti and G. L. White, *Ibid.*, 1975, **40**, 215; S. D. Larsen and S. A. Monti, *J. Am. Chem. Soc.*, 1977, **99**, 8015.

422. W. von E. Doering, B. M. Ferrier, E. T. Fossel, J. H. Hartenstein, M. Jones, G. Klumpp, R. M. Rubin, and M. Saunders, *Tetrahedron*, 1967, **23**, 3943.
423. P. A. Bartlett and J. L. Adams, *J. Am. Chem. Soc.*, 1980, **102**, 337.
424. T. Harayama, M. Takatani, and Y. Inubushi, *Tetrahedron Lett.*, 1979, 4307.
425. C. H. Heathcock, R. A. Badger, and J. W. Patterson, *J. Am. Chem. Soc.*, 1967, **89**, 4133.
426. G. Büchi, W. Hofheinz, and J. W. Paukstelis, *J. Am. Chem. Soc.*, 1966, **88**, 4113; 1969, **91**, 6473.
427. S. D. Larsen and S. A. Monti, *Synth. Commun.*, 1979, **9**, 143.
428. H. E. Baumgarten, P. L. Creger, and C. E. Villars, *J. Am. Chem. Soc.*, 1958, **80**, 6609.
429. A. W. Johnson, E. Markham, R. Price, and K. B. Shaw, *J. Chem. Soc.*, 1958, 4254.
430. *Drug Synthesis*, p.228.
431. *Drug Synthesis*, p.313; N. P. Buu Hoi and C. Beaudet, *U.S. Pat.*, 1961, 3,012,042; *Chem. Abstr.*, 1962, **57**, 11168c.
432. L. M. Rice, E. Hertz, and M. E. Freed, *J. Med. Chem.*, 1964, **7**, 313.
433. T. Y. Shen, R. L. Ellis, T. B. Windholz, A. R. Matzuk, A. Rosegay, S. Lucas, B. E. Witzel, C. H. Stammer, A. N. Wilson, F. W. Holly, J. D. Willet, L. H. Sarett, W. J. Holtz, E. A. Rislay, G. W. Nuss, and M. E. Freed, *J. Am. Chem. Soc.*, 1963, **85**, 488.
434. E. Shaw, *J. Am. Chem. Soc.*, 1955, **77**, 4319.
435. F. Brody and P. R. Ruby in *Pyridine and Its Derivatives*, ed. A. Weissberger, Interscience, New York, 1960, Part I, pp.500–533.
436. Ref.435, pp.355–434.
437. *Perfumes*, p.306.
438. Z. H. Skraup, *Monatsh. Chem.*, 1881, **2**, 139; (details on p.158).
439. W. M. Whaley and T. R. Govindachari, *Org. React.*, 1951, **6**, 151.
440. W. M. Whaley and T. R. Govindachari, *Org. React.*, 1951, **6**, 74.
441. *Pesticide Manual*, p.545; Ref.17, p.400.
442. E. Ziegler, H. Junek, and G. Wiltgrube, *Monatsh. Chem.*, 1956, **87**, 386.
443. M. Ikawa, M. A. Stahmann, and K. P. Link, *J. Am. Chem. Soc.*, 1944, **66**, 902; N. J. A. Gutteridge, *Chem. Soc. Rev.*, 1972, **1**, 381.
444. J. Druey and B. H. Ringier, *Helv. Chim. Acta*, 1951, **34**, 195.
445. R. G. Buttery, R. M. Seifert, R. E. Lundin, D. G. Guadagni, and L. C. Ling, *Chem. Ind. (London)*, 1969, 490.
446. *Gustation and Olfaction*, eds. G. Ohloff and A. F. Thomas, Academic Press, London, 1971, p.173.
447. F. L. C. Baranyovits and R. Ghosh, *Chem. Ind. (London)*, 1969, 1018; *Pesticide Manual*, p.432.
448. H. L. Riley, J. F. Morley, and N. A. C. Friend, *J. Chem. Soc.*, 1932, 1875.
449. J. M. Gulland and T. F. Macrae, *J. Chem. Soc.*, 1933, 662.
450. R. G. Fargher and F. L. Pyman, *J. Chem. Soc.*, 1919, 217.
451. D. C. Baker and S. R. Putt, *J. Am. Chem. Soc.*, 1979, **101**, 6127.
452. H. Bredereck and G. Theilig, *Chem. Ber.*, 1953, **86**, 88; M. R. Grimmett, *Adv. Heterocycl. Chem.*, 1970, **12**, 113.
453. G. J. Durant, J. C. Emmett, C. R. Ganellin, P. D. Miles, M. E. Parsons, H. D. Brain, and G. R. White, *J. Med. Chem.*, 1977, **20**, 901.
454. D. L. Snitman, R. J. Himmelsbach, and D. S. Watt, *J. Org. Chem.*, 1978, **43**, 4758.
455. F. Delay and G. Ohloff, *Helv. Chim. Acta*, 1979, **62**, 369.
456. J. A. Secrist, C. J. Hickey, and R. E. Norris, *J. Org. Chem.*, 1977, **42**, 525.
457. E. E. Royals and A. G. Robinson, *J. Am. Chem. Soc.*, 1956, **78**, 4162.

380

458. Fleming, *Synthesis*, p.98.
459. W. E. Gore, G. T. Pearce, and R. M. Silverstein, *J. Org. Chem.*, 1975, **40**, 1705.
460. A. P. Kozikowski and K. Isobe, *Tetrahedron Lett.*, 1979, 833.
461. R. K. Hill, J. A. Joule, and L. J. Loeffler, *J. Am. Chem. Soc.*, 1962, **84**, 4951.
462. F. Loftus, *Synth. Commun.*, 1980, **10**, 59.
463. D. Brewster, M. Myers, J. Ormerod, P. Otter, A. C. B. Smith, M. E. Spinner, and S. Turner, *J. Chem. Soc., Perkin Trans. 1*, 1973, 2796.
464. P. A. Grieco, *J. Org. Chem.*, 1972, **37**, 2363.
465. C. B. Chapleo, M. A. W. Finch, T. V. Lee, and S. M. Roberts, *J. Chem. Soc., Chem. Commun.*, 1979, 676; M. A. W. Finch, T. V. Lee, and S. M. Roberts, *Ibid.*, p.677.
466. P. Geetha, K. Narasimhan, and S. Swaminathan, *Tetrahedron Lett.*, 1979, 565.
467. G. Traverso, G. P. Polloni, G. de Giuli, A. Barco, and A. Invernizzi Gamba, *Gazz. Chim. Ital.*, 1971, **101**, 225; A. Fischli, M. Klaus, H. Mayer, P. Schönholzer, and H. Rüegg, *Helv. Chim. Acta*, 1975, **58**, 564; R. K. Boeckman, D. M. Blum, and S. D. Arthur, *J. Am. Chem. Soc.*, 1979, **101**, 5060.
468. L. I. Smith and G. F. Rouault, *J. Am. Chem. Soc.*, 1943, **65**, 631.
469. R. L. Markezich, W. E. Willy, B. E. McCarry, and W. S. Johnson, *J. Am. Chem. Soc.*, 1973, **95**, 4414; 4416; D. R. Morton, M. B. Gravestock, R. J. Parry, and W. S. Johnson, *Ibid.*, 4417; D. R. Morton and W. S. Johnson, *Ibid.*, 4419.
470. J. B. Heather, R. S. D. Mittal, and C. J. Sih, *J. Am. Chem. Soc.*, 1974, **96**, 1976.
471. J. D. White and W. L. Sung, *J. Org. Chem.*, 1974, **39**, 2323; T. Kametani and H. Nemoto, *Tetrahedron Lett.*, 1979, 3309.
472. C. B. Quinn and J. R. Wiseman, *J. Am. Chem. Soc.*, 1973, **95**, 1342.
473. P. D. Bartlett and M. R. Rice, *J. Org. Chem.*, 1963, **28**, 3351.
474. S. Winstein, E. L. Allred, and J. Sonnenberg, *J. Am. Chem. Soc.*, 1959, **81**, 5833.
475. E. E. van Tamelen, L. J. Dolby, and R. G. Lawton, *Tetrahedron Lett.*, 1960, No.19, 30; E. E. van Tamelen, J. P. Yardley, and M. Miyano, *Ibid.*, 1963, 1011; E. E. van Tamelen, J. P. Yardley, M. Miyano, and W. B. Hinshaw, *J. Am. Chem. Soc.*, 1969, **91**, 7333.
476. H. H. Mottern and G. L. Keenan, *J. Am. Chem. Soc.*, 1931, **53**, 2347.
477. E. E. van Tamelen, C. Placeway, G. P. Schiemenz, and I. G. Wright, *J. Am. Chem. Soc.*, 1969, **91**, 7359.
478. D. M. White, *J. Org. Chem.*, 1974, **39**, 1951.
479. R. L. Sawyer and D. W. Andrus, *Org. Synth. Coll.*, 1955, **3**, 276; G. F. Woods, *Ibid.*, 470.
480. J. E. Baldwin, *J. Chem. Soc., Chem. Commun.*, 1976, 734.
481. L. D. Bergel'son, E. V. Dyatlovitskaya, and M. M. Shemyakin, *Izv. Akad. Nauk SSSR, Ser. Khim.*, 1963, 506; *Chem. Abstr.*, 1963, **59**, 3766d.
482. G. I. Fray, R. H. Jaeger, E. D. Morgan, R. Robinson, and A. D. B. Sloan, *Tetrahedron*, 1961, **15**, 20; J. Kennedy, N. J. McCorkindale, and R. A. Raphael, *J. Chem. Soc.*, 1961, 3813.
483. M. Baumann, W. Hoffmann, and H. Pommer, *Liebigs Ann. Chem.*, 1976, 1626.

Index

The appearance of a compound name in the index usually means that its synthesis appears in the book. Sub-divisions 'synthesis of' and 'use in synthesis' appear only when the synthesis of a type of compound is discussed and it is also used as a starting material.

A formula index for all target molecules in this book and the accompanying workbook appears in the workbook.

Note on nomenclature: No systematic nomenclature is used in this book. This does not imply a preference for trivial nomenclature but rather one for the structural formula—the universal language of organic chemistry. Trivial names are given so that readers may talk about complex molecules and look up simple ones in suppliers' catalogues.

Index of Abbreviations

The first mention of each abbreviation is given on the page where the abbreviation is explained.